Fundamentals of Process Safety Engineering

Fundamentals of Process Safety Engineering

Samarendra Kumar Biswas
Umesh Mathur
Swapan Kumar Hazra

CRC Press
Taylor & Francis Group
Boca Raton London New York

CRC Press is an imprint of the
Taylor & Francis Group, an informa business

First edition published 2021
by CRC Press
6000 Broken Sound Parkway NW, Suite 300, Boca Raton, FL 33487-2742

and by CRC Press
2 Park Square, Milton Park, Abingdon, Oxon, OX14 4RN

First edition published by CRC Press 2021

CRC Press is an imprint of Taylor & Francis Group, LLC

ISBN: 9780367620769 (hbk)
ISBN: 9780367620899 (pbk)
ISBN: 9781003107873 (ebk)

DOI: 10.1201/9781003107873

Typeset in Times
by codeMantra

Contents

Foreword

The longstanding failure of universities to include Process Safety in the undergraduate curriculum has mystified me. Graduating engineers could begin their careers utterly unaware of the tragedies of Bhopal, Seveso, Mexico City, Pasadena, Texas City, Buncefield, Toulouse, and many others, and without the barest understanding of how to prevent them. Academics indeed rose to the challenge to study and teach energy and environmental conservation, security, alternative fuels, and other societally important topics. But Process Safety Engineering (PSE), with its potential to avert widespread human and economic catastrophe, remained absent from engineering curricula worldwide, with very few exceptions.

Fortunately, this is beginning to change. As universities begin to shoulder their responsibilities to prepare engineers to design and operate safe facilities, the need for textbooks and other teaching tools is becoming more critical. This new textbook, written by Dr. S.K. Biswas, Mr. Umesh Mathur, and Mr. S.K. Hazra, will help fulfill this vital need. And there remain many engineers in the industry with a substantial gap in their Process Safety knowledge.

I have known Mr. Hazra and Dr. Biswas for more than 15 years. Out of a chance encounter in 2002 grew a fruitful collaboration to help raise the knowledge and practice of Process Safety in India and around the world. Both gentlemen came to Process Safety out of a recognition of its importance that stood well out in front of their colleagues at the time. And like many of the Process Safety champions that I have had the pleasure to know over the years, they motivated those around them to join them on their quest to educate others and enlist them to help eliminate Process Safety incidents.

This book contains a great deal of technical detail, of which all engineers should have a basic understanding. But I hope that readers using this book retain these three critical messages throughout their careers:

- You must understand the hazards of your materials and processes and the consequences that could arise if you don't manage them properly.
- You must know the risk of your process and implement all barriers necessary to ensure that it meets your corporate risk criteria.
- Having implemented the necessary barriers, you must maintain them, so they perform as designed, all the time.

This is easy to say, but not as easy to do. But your life, and the lives of your co-workers and plant neighbors, depend on your knowledge of Process Safety and your professionalism in making sure it is done faithfully throughout your career.

Scott Berger
Former Executive Director
Center for Chemical Process Safety (CCPS, AIChE)
February 2019

Preface

This book addresses the vital subject of Process Safety in the petroleum refining, petrochemical, and other similar processing plants. A good understanding of this subject is of great importance for chemical engineers in the aftermath of the world's worst industrial disaster in Bhopal, India, and numerous other catastrophic accidents in the last quarter of the previous century. These have led to the widespread adoption of the commendable concept of "Responsible Care" in the process industries. Despite this development, many significant accidents have continued to occur – most regrettably – in the chemical process industries, despite the lessons learned over the past forty years. These include the nineteen accidents that we have described in some detail in this book. Our purpose is to explain the circumstances that led to these accidents, including failures in management, training, instrumentation, and control systems. In particular, we describe many types of human errors of judgment, neglect, and undue haste that led to disaster.

We aim to provide readers a comprehensive background in PSE, as it plays a central role in achieving "manufacturing excellence" in the planning, engineering, and operation of oil, gas, and petrochemical industries, or generally, in any sophisticated process plant. This book is a practitioner's guide to the many technically diverse PSE aspects that are encountered by those active in this field. We hope that this book will also provide chemical engineering and technology students, at the undergraduate and postgraduate levels, with an understanding of hazards in process plants and, especially, how PSE should be incorporated in the design and operation of safe and efficient facilities.

The development of international guidelines and regulations related to Process Safety has been quite remarkable over the past four decades. Today, there is a vast body of technical literature, regulatory codes, and several major books related to this subject. Unfortunately, much of the technical literature is scattered in technical journals, which are not easily accessible for most engineers. Major reference books are also available that are too expensive for average practitioners to own; they are available only in university libraries or at a few major global organizations that provide specialized services related to Process Safety. Fortunately, there is an excellent multivolume reference book on the subject (by Lees), but several other standard references, in our view, do not include discussion of some essential topics. Examples include the proper thermodynamic basis for computing release rates from ruptures in pipelines or vessels; the evaluation of relief valve capacity under sonic or subsonic conditions at the valve throat; and the computation of phase equilibria (1) at high pressures or low temperatures for hydrocarbons and petroleum mixtures, or (2) for complex chemical mixtures.

The sheer volume of available information is daunting. Regrettably, with few exceptions even today, most chemical engineering curricula do not include serious instruction on Process Safety and related disciplines. Practitioners working in manufacturing facilities find themselves in a difficult position because it is not self-evident to them how and where they should start their PSE studies. Corporate safety documentation usually is scattered, and there are few attempts to organize such information understandably. Short courses of a few days' duration are also generally insufficient to learn about many important aspects.

We hope that this book will inform practicing professionals about important PSE topics and generate significant interest with educators in teaching these topics comprehensively and systematically. We would recommend considering a three- or four-hour credit course (48–60 class hours) in most chemical engineering/technology curricula (the final undergraduate year or first year of postgraduate study).

We wrote this book to enable easier access to – and comprehension of – much relevant technical information and the main issues that should be emphasized when performing Process Safety-related engineering analyses. We have provided many numerical examples to illustrate practical applications. The solutions to these were verified using Microsoft Excel® and also by other mathematical packages such as Mathcad®, Maple®, and Polymath®. In some instances that require extensive trial and error, Fortran 95 programs were written and validated using several compilers to ensure that the results were consistent. Examples of computations requiring extensive trial and error include relief valve sizing and rating for real gases, the thermodynamics of phase equilibria, and the estimation of release rates from a rupture in a pipeline. We wish to acknowledge the cooperation of the Numerical Algorithms Group, Oxford, U.K., whose NAG Fortran® compiler proved to be an excellent resource, especially for debugging complex code. The mathematical libraries provided with the NAG compiler are of outstanding quality. We also used the excellent Silverfrost FTN95, Absoft, and Compaq's Digital Visual Fortran compilers for programming various numerical examples.

The subject of PSE is of vital interest, especially to chemical and mechanical engineers in the process industries, thanks to the significant number of governmental regulations issued in the wake of the Bhopal disaster and numerous other catastrophic accidents in the world. The onset of new rules has been particularly intense in the last quarter of the previous century and even more frequent in this century's first two decades. A careful study of the causes of the numerous accidents that we have described in some detail in this book would, therefore, be valuable. These newly formulated laws and regulations have led to the development and widespread adoption of the commendable concept of "Responsible Care" in the chemical process industries.

It is impractical to cover all possible aspects and issues in a field as diverse as Process Safety. The authors have relied on their experience as practicing engineers, going back over 50 years each, selecting those items that deserve attention in a wide variety of industrially significant situations. We have outlined sources and methods but have avoided theoretical derivations, focusing instead primarily on applications. We would invite suggestions from readers on items they would like to see covered in future editions.

We hope that this book will be a useful resource for technical professionals responsible for Process Safety methods, procedures, and field implementation as part of their responsibilities. These methods are valuable when considering the design of new facilities or evaluating changes in existing systems to enhance their safety. Students should be motivated to study this material carefully as they would undoubtedly find it useful in their academic and professional pursuits. Our objective is to empower them to understand the underlying fundamentals and the assumptions and limitations inherent in various techniques and methodologies. PSE is a rapidly evolving field and calls for the exercise of considerable judgment in interpreting the results of any analysis.

The last chapter incorporates a significant discussion of how the methods and procedures required for achieving manufacturing excellence also provide significant Process Safety benefits. These methods and procedures have been adopted meticulously by those corporations that have accrued the most successful track record in achieving the most profitable operations while also running their plants uniformly in a safe and environmentally responsible manner. These initiatives are also consistent with the management requirements for complying with international quality standards for manufacturing. Accordingly, we stress the benefits of implementing corporate-wide continuous improvement processes that highlight the importance of unrelenting vigilance at all levels of the manufacturing environment. Many such initiatives rely on high-quality instrumentation and automation: process control, real-time optimization, intelligent alarm management, and safety systems designed to minimize exposure of plant personnel to hazards – while also protecting valuable equipment and other assets. In our opinion, this aspect is not covered adequately in other textbooks.

We have tried to be meticulous in proofreading the text, having spent countless hours making revisions, and fine-tuning the discussion of most of the technically elaborate topics. A large number of numerical examples have been provided throughout this book to illustrate the underlying principles. Below each equation, we define each term and its units to ensure proper usage. While we have tried to check all computations thoroughly, we are fully aware that errors could still exist. We would be most grateful to readers who bring these to our attention.

We also welcome comments and suggestions to improve the contents of this book. As with all such endeavors, we expect readers will find areas for improvement or corrections. While we have relied on many reliable sources, we accept sole responsibility for any errors or omissions and would be most grateful to readers who point them out to us.

We are immensely grateful to our colleagues and mentors, too numerous to mention individually, who have shared their wealth of knowledge and experience with us most generously. We hope that this book proves worthy of the confidence they have placed in us and the time they invested in our maturation and growth.

Finally, we wish profusely to thank Mr. Scott Berger, a noted authority in Process Safety Engineering and Management, for agreeing to write a foreword for this book. Mr. Berger also made numerous suggestions to improve the contents of this book, and we are very much in his debt.

S. K. Biswas, samarbiswas36@gmail.com
Umesh Mathur, u.mathur@yahoo.com
S.K. Hazra, skhazra@riskenergy.co.in

Acknowledgments

The authors were encouraged to write this book by many illustrious Process Safety professionals worldwide, many of whom are personal friends, some for over 20 years. Their inspiration and support have encouraged us to take up the challenging and daunting task of authoring this book on "Fundamentals of Process Safety Engineering". Our objective was to start from fundamental concepts and cover current and recommended techniques to reduce risk to enable practicing Process Safety professionals to achieve manufacturing excellence meticulously. We are most grateful to all the safety professionals who have advised us directly or have contributed to the vast technical literature in this area.

We would first like to express their gratitude to Mr. Scott Berger, former Executive Director of the Center for Chemical Process Safety (CCPS), AIChE. Mr. Berger dedicated great efforts and time to review the entire manuscript, more than once. Mr. Berger provided many valuable suggestions, and we hope we have incorporated changes that reflect his thinking adequately.

We wish to thank Mr. Laurence G. Britton of AIChE's CCPS (now retired); his review and comments on the entirety of Chapter 4 enabled us to make numerous improvements.

Professor Geoffrey Chamberlain is a globally recognized expert on explosion phenomena. Dr. Chamberlain has served as Manager, Hazard & Risk, at Shell Global Solutions. He is a Professor at Loughborough University, and we are grateful for his very detailed critical review of Chapter 9 ("Explosion"). Professor Chamberlain's comments and suggestions significantly improved our discussion of vapor cloud explosions (VCE). Extensive research has been reported in the literature on this subject following three severe process industry accidents between 2005 and 2009. Today, the modeling of VCEs is far more comprehensive, thanks to Professor Chamberlain's contributions. In this book, we have added a discussion of several semi-empirical models with examples. Also, brief discussions of commercially available CFD models were added; these are extremely detailed mathematically, and a fuller treatment is beyond the scope of this book.

Mr. Peter Waite is a Process Safety veteran and former Senior Hazard Consultant at Cremer and Warner, and Technical Director of Entec, U.K. Mr. Waite provided us valuable suggestions concerning the importance of "Human Factors" in Process Safety; these encouraged us to add Chapter 14 to this book.

We also wish to thank Mr. Asit Hazra (no relation), formerly Director, Environment & Chemical Emergencies, Govt. of Canada, for his comments and suggestions.

We wish to record our thanks to the safety professionals of various organizations who accorded us copyright permissions for relevant portions of their respective publications. In this context, we first thank Dr. Anil Gokhale of AIChE's CCPS, who was extraordinarily gracious. Mr. Stein Hauge of SINTEF Norway accorded us permission on behalf of SINTEF promptly, and we express our thanks to him. Ms. Paula Bohlander of PGS and Gexcon (TNO) Netherlands B.V., granted us numerous permissions. Mr. Ron Gerstel of GEXCON was also most helpful. Mr. Adrian Piarorazio kindly accorded us permission from Baker Engineering and Risk Consultants.

We thank the Indian Institute of Chemical Engineers (IIChE), especially President Prof. V.V. Basabrao, and Mr. Abhijit Mitra, Secretary Eastern Region. IIChE has supported and encouraged Dr. S.K. Biswas throughout the period when he conducted several Process Safety investigations on their behalf, and also S.K. Hazra, who later stepped into Dr. Biswas' shoes at IIChE.

Finally, we wish to acknowledge Engineer Mohammad Hasan for his magnificent rendering of all drawings, figures, and tables in this book using Microsoft Visio®. Mr. Hasan worked long hours on very tight schedules, and we can only say: "Thank you, Hasan. You worked tirelessly on all our sketches to convert them to professional-quality Visio drawings, and your contributions have enhanced our efforts markedly. We will always be in your debt".

NOTE TO READERS

The purpose of this book is primarily to provide practicing engineers and undergraduate- or graduate-level Chemical Engineering and Technology students with an understanding of safety hazards in process plants and how they should be managed safely and professionally. The second chapter of this book describes many tragic accidents – in Bhopal, Seveso, Mexico City, Pasadena, Texas City, Toulouse, and many others. In our discussions of these catastrophes, we emphasize the importance of acquiring a clear understanding of how accidents occur and the methods and procedures that need to be followed to prevent such occurrences. The teaching of this subject comprehensively and professionally has been handicapped by a lack of suitable textbooks and trained teachers in academia. We hope that this book will address this shortcoming and generate significant interest in students to study this important subject, even motivating some to pursue it to an advanced level. We trust that this book will encourage faculty members to appreciate the potential of the subject and to develop it for teaching, research, and consultancy work.

This book assumes no prior training in Process Safety Engineering (PSE); however, knowledge of fluid flow, heat transfer, and mass transfer operations, essential parts of standard chemical engineering or chemical technology curricula, is a prerequisite. Therefore, this book should be suitable for a three- or four-credit one-semester course in the final undergraduate or first postgraduate year of study.

Although designed primarily as a textbook for students, we are confident that engineers and other technical personnel in the industry will find this book extremely useful. Concepts and methodologies covered in this book should be useful to research and development experts in developing inherently safer processes, to process and project engineers in developing safer plant designs, instrumentation, and layouts, and to plant personnel in operating and maintaining plants in a safe, profitable, and socially responsible manner.

List of Figures

List of Tables

Acronyms and Abbreviations

ACC	American Chemical Council
ACDS	Advisory Committee on Dangerous Substances
ACIGH	American Conference of Government Industrial Hygienists
ACMH	Advisory Committee on Major Hazards
AIChE	American Institute of Chemical Engineers
AEGL	Acute Exposure Guideline Levels
AFPM	American Fuels and Petrochemical Manufacturer's Association (former NPRA)
ALARA	As Low As Reasonably Acceptable
ALARP	As Low As Reasonably Practicable
ANSI	American National Standard Institute
APELL	Awareness and Preparedness for Emergencies at Local Level
API	American Petroleum Institute
ASME	American Society of Mechanical Engineers
BI	Business Interruption
BLEVE	Boiling Liquid, Expanding Vapor Explosion
BP	Boiling Point
BPCS	Basic Process Control System
BSP	Barrier Status Panel
CAEPPR	Chemical Accidents (Emergency Planning, Preparedness, and Response) Rules, India
CAM	Congestion Assessment Method
CBM	Condition-Based Maintenance
CCF	Common Cause Failure
CCR	Central Control Room
CCPS	Center for Chemical Process Safety (AIChE)
CEI	Dow Chemical Exposure Index
CFD	Computational Fluid Dynamics
CHEM Services	Chemical Hazard & Emergency Management Services, Queensland, Australia
CM	Corrective Maintenance
CMMS	Computerized Maintenance Management System
COMAH	Control of Major Accident Hazards – Regulation in the U.K.
CPCB	Central Pollution Control Board (India)
CPQRA	Chemical Process Quantitative Risk Assessment
CSB	Chemical Safety Board (USA)
CSChE	Canadian Society of Chemical Engineering
CW	Cooling Water
DCS	Distributed Control System
DDT	Deflagration to Detonation
DIERS	Design Institute for Emergency Relief Systems (AIChE)
DISH	Directorate of Industrial Safety and Health, India

DSHA	Defined Situations of Hazard and Accident
DNV GL	DNV GL Det Norske Veritas Germanischer Lloyd
DOE	Department of Energy (USA)
EIA	Environmental Impact Assessment
EIS	Environmental Impact Statement
EBV	Emergency Block Valve
EPP	Emergency Preparedness Plan
ERPG	Emergency Response Planning Guidelines
ESD	Emergency Shutdown System
ESV	Emergency Shutdown Valve
ESRA	European Safety and Reliability Association
F&EI	Dow Fire & Explosion Index
FAR	Fatal Accident Rate
FCE	Final Control Element
FAIR	Focused Asset Integrity Review
FMEA	Failure Mode and Effects Analysis
F-N	Fatality Frequency-Cumulative Number (Curve)
FTA	Fault Tree Analysis
GHS	The Globally Harmonized System of Classification and Labeling of Chemicals
HAZAN	Hazard Analysis
HAZID	Hazard Identification
HAZOP	Hazard and Operability Analysis
HE	Hazard Evaluation
HEMP	Hazard and Effects Management
HF	Human Factor
HID	Hazardous Installations Directorate (U.K.)
HIRAC	Hazard Identification, Risk Assessment, and Control
HMI	Human–Machine Interface
HRA	Human Reliability Analysis
HSE	Health and Safety Executive (U.K.)
HSE-MS	Health, Safety and Environment Management System
HTRI	Heat Transfer Research Institute
ICC	Indian Chemical Council
IEC	International Electrotechnical Commission
IEEM	International Conference on Industrial Engineering and Engineering Management
IDLH	Immediate Danger to Life or Health
IMO	International Maritime Organization
IOGP	International Association of Oil & Gas Producers
IOMC	The Inter-Organization Programme for the Sound Management of Chemicals
IPEEE	Individual Plant Examination for External Events
IPL	Individual Protection Level
IPL	Independent Protection Layer
IPS	Instrumented Protective System

ISA	International Society of Automation
LFL	Lower Flammability Limit
LAH	Level Alarm-High
LAMPS	Local Accident Mitigation And Prevention (U.N. Model)
LEPC	Local Emergency Planning Committee (USA Model)
LI	Level Indicator
LT	Level Transmitter
LIC	Level Indicator-Controller
LNG	Liquefied Natural Gas
LPG	Liquefied Petroleum Gas
LOPA	Layer of Protection Analysis
LOTO	Lockout, Tag-Out
MAHB	Major Accident Hazards Bureau (E.U.) of JRC (Joint Research Centre)
MSIHC	Manufacture, Storage and Import of Hazardous Chemicals Rule, India
MAWP	Maximum Allowable Working Pressure
MOC	Management of Change
MHF	Major Hazard Facilities Regulation (Victoria, Australia)
MHIDAS	Major Hazard Incident Data Service (E.U.)
MI	Mechanical Integrity
MIACC	Major Industrial Accidents Council of Canada
MOEF	Ministry of Environment & Forests, India
MARG	Mutual Aid Resource Group (India Model)
NASA	National Aeronautics and Space Administration (USA)
NDMA	National Disaster Management Authority (India)
NFPA	National Fire Protection Association
NIOSH	National Institute of Occupational Safety and Health (USA)
NPRA	National Petoleum Refiners Association
NRC	Nuclear Regulatory Commission (USA)
NSC	National Safety Council (USA)
NSCI	National Safety Council of India
OSBL	Outside Battery Limits
OREDA	The Offshore and Onshore Reliability Data Project
OSHA	Occupational Safety and Health Administration (U.S.)
PADHI	Planning Advice for Developments near Hazardous Installations (U.K.)
PCB	Pollution Control Board (India)
PEIM	Process Equipment Integrity Management
P&ID	Piping And Instrumentation Diagram
PFD	Probability of Failure on Demand
PHA	Process Hazard Analysis
PHI	Potential Hazardous Installation
PI	Pressure Indicator
PL	Protection Layer
PLL	Potential Loss of Life

PM	Preventive Maintenance or Predictive Maintenance
PPRT	Plans de Prévention des Risques Technologiques (France)
PRA	Probabilistic Risk Assessment
PRV	Pressure Relief Valve
PSA	Petroleum Safety Authority, Norway
PSE	Process Safety Engineering
PSV	Pressure Safety Valve
PSM	Process Safety Management
PTSC	Partnership Towards A Safer Community (Canada Model)
QRA	Quantitative Risk Assessment
RBI	Risk-Based Inspection
RBPS	Risk-Based Process Safety
RC	Responsible Care
RNNP	Trends in Risk Level in the Petroleum Industry (Norway)
RV	Relief Valve
RSSG	Royal Society Study Group (U.K.)
SCBA	Self-Contained Breathing Apparatus
SCE	Safety-Critical Equipment
SCTA	Safety-Critical Task Analysis
SFARP	So Far As is Reasonably Practicable
SWSS	Process Safety Regulation China
SIF	Safety Instrumented Function
SIGTTO	Society of International Gas Tankers and Terminals Limited
SIL	Safety Integrity Level
SIS	Safety Instrumented System
SRS	Safety Requirement Specification
TEEL	Temporary Emergency Exposure Limit
TIMP	Technical Integrity Management Project
TOR	Tolerability of Risk
TO&O	Technical, Operational and Organizational
TRIF	Temporary Refuge Impairment Frequency
TQ	Threshold Quantity
UVCE	Unconfined Vapor Cloud Explosion
VCE	Vapor Cloud Explosion
VLE	Vapor–Liquid Equilibrium
VLLE	Vapor–Liquid–Liquid Equilibrium
XV	Remote-Activated/Controlled Valve

Authors

The senior author, Dr. S.K. Biswas, obtained his bachelor's degree in chemical engineering in 1957 from Jadavpur University, Calcutta, India. He went on to get his Ph.D. in chemical engineering at the University of Wisconsin, Madison, USA, in 1962. He then began his academic career as a professor of chemical engineering in Delhi University, India, and, in 1963, at the Indian Institute of Technology, Delhi, India. After 1966, he began working in product development research and engineering for Imperial Chemical Industries (ICI), India. This book's origins date back to the events that followed immediately after the 1984 Bhopal disaster in Madhya Pradesh, India, widely considered the worst industrial accident in history. Upon learning of the magnitude of this event, its profound and widespread impact, and the likely causes that led to it, Dr. Biswas approached the management of ICI and requested permission to devote his energies thenceforth to the subject of Process Safety Engineering (PSE) in the manufacturing industries. To their credit, ICI immediately accepted his proposal to review and revise existing methods and procedures and to develop sound guidelines for their effective implementation at their facilities. Over the next 30 years, Dr. Biswas consulted with many noted authorities in the field of industrial Process Safety. He focused on the most practicable methods and procedures to ensure not only inherently safe designs for new installations, but also the most effective field remediation techniques that should be applied, wherever feasible, in existing plants. After retiring from ICI, Dr. Biswas worked with Aegis Group in Bombay for many years. At Aegis, he was responsible for design and engineering for a major LPG storage terminal. Finally, he worked as Consultant to Nitrex Chemicals India Limited in Gujarat where he provided guidance, from design to commercial production, for a new Industrial Nitrocellulose plant.

The second author, Umesh Mathur, was an undergraduate student of Dr. S. K. Biswas at the Indian Institute of Technology (IIT), Delhi, India. He graduated with a B.Tech. degree in 1966, just before Dr. Biswas left IIT to join ICI, India. He then obtained a Post-Graduate Diploma in Petroleum Refining and Petrochemicals at the Indian Institute of Petroleum, Dehradun, India. He then worked as a Process and Project Technologist for over 4 years at the Burmah-Shell Refineries in Bombay. He obtained an M.S. in chemical engineering at the University of Tulsa, OK, USA, in 1980. Mathur is a licensed Professional Engineer in Oklahoma and is a Life Member of the American Institute of Chemical Engineers. He has been a practicing chemical engineer in the process industries for over 54 years and has implemented numerous industrial projects in process engineering, process control, troubleshooting, and process optimization. He has published several technical papers and has made presentations before the AIChE, NPRA (now AFPM), GPSA, HTRI, and other professional bodies. Since the mid-1970s, he has served as an expert chemical engineering witness before federal, state, and local governmental agencies in the USA, and at the International Court of Arbitration of the International Chamber of Commerce, Mexico City.

Umesh Mathur would like to acknowledge his profound debt of gratitude to the primary author, Dr. S.K. Biswas, and the late Professors Dr. P.K. Mukhopadhyay and Dr. Frank Rumford, his most influential undergraduate chemical engineering professors at the Indian Institute of Technology, Delhi, India, until 1966. Drs. Biswas and Mukhopadhyay taught at the very highest academic standards. At the University of Tulsa, he was fortunate to come under the influence of Professors Wilbur L. Nelson, Francis S. Manning, Richard E. Thompson, and A. Paul Buthod who provided intensive indoctrination in petroleum refining technology, mathematical modeling, process simulation, thermodynamics, and numerical methods in chemical engineering. These have remained anchors in his professional life for almost 50 years. To all these admirable teachers, he would like to express his thanks and indebtedness. Finally, he remains grateful to his colleagues and mentors, too numerous to acknowledge individually, who provided guidance, training, and support in all aspects of his life as a practicing chemical engineer.

Over the decades, the most significant advances in all aspects of the thermodynamics of phase equilibria have originated with Professor John M. Prausnitz of the University of California, Berkeley, CA, USA. Professor Prausnitz very kindly provided advice and software to Umesh Mathur (parameter estimation of activity coefficient models and VLE/LLE calculations for complex chemical mixtures and polymer systems.) The late Professor Aage Fredenslund of the Danish Technical University, Lyngby, Denmark, who. developed the original UNIFAC method with Professor Prausnitz, provided much advice and software to Mathur in the 1970s. Professor Jürgen Gmehling, formerly of the Universities of Dortmund, Germany and Oldenburg, Germany, assembled DDBSP, the most extensive and reliable pure component and mixture properties databanks in the world and, among other developments, also originated the Dortmund modification of the UNIFAC method (D-UNIFAC) that is generally acknowledged to be the best group contribution method today for phase equilibria. Professor Gmehling very kindly provided the DDBSP software and databanks to Mathur on highly favorable terms.

The third author, Swapan Kumar Hazra, is a mechanical engineer who obtained a B.Tech. degree in mechanical engineering from the Indian Institute of Technology, Kharagpur, India. He then studied for another year in the graduate school at the Indian Institute of Technology, Delhi, India. Hazra was a junior colleague of Dr. S.K. Biswas in ICI, India, for several years. Later, he was the CEO at Aegis Group (Chemical, LPG & Logistics). After Dr. Biswas became the Technical Director, Aegis Group, Hazra continued a close collaboration for another 14 years with Dr. S.K. Biswas on many engineered projects, with special emphasis on Process Safety. Hazra has chaired the Indian Chemical Council's (ICC) Safety, Health, and Environment Committee for almost ten years. He has been a member of the American Society of Mechanical Engineers (ASME). Hazra was active in numerous professional safety organizations, including the National Fire Protection Association (NFPA) in the USA and BL NCE in Canada. He was India's nominee to the Responsible Care Leadership Group of the International Council of Chemical Association (ICCA, Arlington ,US). He was also a member of Working Group on Safety of the Society of International Gas Tankers and Terminals (SIGTTO, U.K.). He has an extensive and close ongoing collaboration with many noted PSE experts worldwide, both industrial and academic. He was a member of the Government of India's Chemical Emergency Group and served as a National Disaster Management Group Expert to deal with chemical emergencies.

S.K. Hazra acknowledges the encouragement provided by the ICC, and especially Director-General Mr. H.S. Karangale, Secretary-General Mr. R.R. Gokhale, Director (Responsible Care & SHE), and Mr. V.G. Bukkawar (retired). We also thank ICC Presidents Mr. S.N. Singh, Mr. Jai Hiremath, Mr. Yogesh Kothari, Mr. Ravi Kapoor, and Mr. Rangaswamy Parthasarathy, all of whom kindly supported S.K. Hazra, during his tenure as Chairman (Safety, Health, and Environment Committee of the ICC), and in his efforts to promote the Process Safety discipline and inculcate the concept of "Responsible Care" among ICC members. The SHE Committee members contributed individually to this cause and deserve our commendation.

Both Mathur and Hazra have been in senior technological positions in the Oil, Gas, Refining, and Petrochemical Industries for over 50 years (Umesh Mathur executed projects primarily in the USA, but also in India, Canada, Latin America, Europe, and South Korea, while Swapan Hazra's projects were in India and Nigeria).

1 Hazards in the Process Industries

The term "process industries" is used in this book to mean those industries that transform raw materials into intermediates or end products by physical or chemical means. Some examples are gas and oil exploration and production, petroleum refineries, petrochemicals, plastics, fiber, heavy and fine chemicals, fertilizer units, ferrous and nonferrous metal production units, agricultural production units, etc. The process industries engage in manufacturing, storage, handling, and transportation (bulk or packaged) of chemical compounds and mixtures. Although not strictly considered as process industries, many service or utility industries such as power plants or water treatment plants that use hazardous chemicals also need to implement process safety engineering and management procedures.

The term "hazard" refers to a material or a condition that has the potential to cause harm to human life, health, property, the environment, or some combination of these. Some common examples of hazards in our homes are LPG/natural gas (cooking/heating fuel), hot water, electric power, slippery floors, and weak roofs. In any process industry, hazards are classified as "process hazard" and "occupational hazard".

Process hazards are caused by the release (or potential release) of materials or energy caused by processing activities that might lead to fire, explosion, or toxic exposure. Often, catastrophic consequences may result, such as personnel injuries or deaths, with severe environmental and financial impacts. Management failures in process safety result in process hazards, generally have a low probability of occurrence, and have high consequential impacts.

Occupational hazards generally refer to workplace incidents affecting a single individual or a small group of workers. These are caused by unsafe working practices, failure to use adequate personal protective equipment (PPE), or other human failures. Examples include falls while working at heights (no safety belt), head injury (no helmet), asphyxiation upon vessel entry (no breathing apparatus), electrocution (failure to observe lockout, tag-out procedures), hearing losses at high noise levels (no hearing protection), and so forth. Occupational hazard-related incidents generally have a high probability of occurrence and relatively low consequential impacts.

Process hazard analysis is a thorough, orderly, and systematic discipline for identifying and evaluating the dangers inherent in the processing of highly hazardous materials.

Process safety engineering (PSE) refers to a set of interrelated methods and procedures for managing hazards associated with the process industries. Its goal is to minimize, if not eliminate, the frequency and severity of incidents resulting from releases of chemicals and other energy sources (US Occupational Safety and Health Administration – OSHA, 1993).

Risk assessment is an essential discipline in process safety engineering and management and is addressed in some detail in Chapter 13. An early introduction to

DOI: 10.1201/9781003107873-1

TABLE 1.1
List of Common Process Plant Hazards

Chemical hazards	Flammable chemicals
	Explosive chemicals
	Reactive chemicals
	Toxic chemicals
Physical hazards	Physical explosion
	Electrostatic charges
	Rollover of liquids
	Boilover of liquids
Environmental hazards	Air pollution
	Water pollution
	Hazardous wastes
Other hazards	Static electricity-induced ignition
	Faulty electrical system hazards

the term is necessary owing to its extreme importance. Risk is a function of the probability of a hazard being realized multiplied by the potential consequences. Risk is expressed in terms of an expected frequency or probability. The higher the frequency, the more likely is the event. The frequency of occurrence is the inverse of the probability of occurrence. Risk may also be defined as the ratio of potential consequence to frequency.

The process safety hazards commonly encountered in process plants are listed in Table 1.1.

1.1 CHEMICAL HAZARDS

1.1.1 Flammable Chemicals

A flammable chemical is a gas or a vaporized liquid which, when ignited in the presence of oxygen, continues to react with oxygen, giving rise to a flame that emits heat and light. The formal definition of a flammable substance is one having a flash point below 93°C.

A flame is the visible portion of the volume within which the oxidation of fuels occurs in gaseous form. Thus, a flame is a phenomenon that occurs in the gaseous phase only. A flammable liquid must first vaporize; the vapors so generated can undergo combustion in a flame. Chemical decomposition or pyrolysis on solid surfaces yields flammable, volatile products that can sustain combustion.

In the combustion of solids, if pyrolysis does not yield volatile products at a rate higher than some minimum value required for sustaining the flame, the combustion proceeds at the surface of the solid, producing smoke but no flame. Such non-flaming combustion is known as smoldering.

Flammable gases or vapors burn in air only over a limited range of composition. Below the lower flammability limit (LFL), the mixture is too lean to ignite and propagate the flame. Similarly, above the upper flammability limit (UFL), the mixture is too rich to ignite and propagate the flame. The concentrations between these two limits constitute the flammability range. Further, by lowering the oxygen concentration (LOC) in air by adding an inert gas, the UFL can be brought down, as explained in Chapter 3.

For flammable liquids, the flash point (discussed in detail in Section 3.1.2) is a crucial property determining how easily ignition would occur. At the flash point temperature, the liquid's vapor pressure is sufficient to yield a vapor concentration in the air that corresponds to the LFL. Therefore, regulatory bodies use the flash point as an essential parameter for classifying the hazard category of flammable chemicals, using terms such as:

i. Flammable gases
ii. Extremely flammable liquids
iii. Very highly flammable liquids
iv. Highly flammable liquids
v. Flammable liquids.

1.1.2 Explosive Chemicals

Upon initiation by shock, impact, friction, fire, chemical reaction, or other ignition sources, an explosive material releases energy at a very high rate, causing a sudden increase in atmospheric pressure in the surroundings and, typically, a flash or a loud noise. An explosive may be a pure substance (such as nitroglycerine, trinitrotoluene (TNT), or pentaerythritol tetranitrate (PETN)) or a "preparation" (such as dynamite that contains nitroglycerine, nitrocellulose, and ammonium nitrate as essential ingredients). Such formulations are known as "condensed explosives" and are used for military purposes or as commercial blasting explosives for mining and demolition work. They are generally safer to handle than their main ingredients.

Condensed explosives fall into two categories: "high explosives" such as dynamite or TNT are known as "detonating explosives" whose decomposition reactions proceed very rapidly. On the other hand, in black powder (a mixture of potassium nitrate, sulfur, and charcoal), the decomposition reaction occurs relatively slowly, simulating rapid burning or combustion. Such explosives are known as "deflagrating explosives". The terms "detonation" and "deflagration" are explained in greater detail in Chapter 9.

The manufacture, storage, and handling of condensed explosives are governed by strict safety norms stipulated in the appropriate laws and regulations in various countries. People working in such industries are generally well trained and have a high safety awareness level compared to those in other process industries. Despite this, accidents do occur in explosives plants owing to the inherent properties and instability of explosives.

However, unintended and accidental explosions also occur in the process industries and isolated storage facilities of flammable liquids and liquefied gases. Some accidents have resulted in large numbers of casualties. Explosions in this second category are described as follows:

- Confined explosions in reaction vessels caused by uncontrolled chemical reactions

- Confined explosions in systems involving reactive materials, such as ammonium nitrate, chlorates, or peroxides
- Vapor cloud explosions
- Boiling liquid, expanding vapor (BLEVE) explosions
- Dust explosions.

These have been discussed in detail in Chapters 8 and 9.

1.1.3 Reactive Chemicals

A chemical reactivity hazard exists when an uncontrolled chemical reaction can lead to high temperatures, high pressures, explosion, or release of flammable or toxic substances. Such hazards can be classified into two main categories: reactive materials and reactive interactions.

Reactive materials are those that are inherently hazardous or become unstable when contacted with air or water. Examples of such materials are as follows:

- Unstable materials that undergo exothermic decomposition at high temperatures or in the presence of a catalyst. Such highly reactive chemicals include peroxides, ethylene oxide, sodium chlorate, etc.
- Polymerizing monomers, such as acrylic acid or styrene, that tend to self-polymerize at high temperatures or in the presence of impurities
- Pyrophoric materials, such as phosphorus or spent catalysts that ignite spontaneously when exposed to air
- Substances such as metallic sodium or potassium, calcium carbide, or oleum that react violently with water.

However, many materials not considered reactive can give rise to dangerous situations when combined with incompatible materials. Such incompatible interactions may occur during physical operations, such as mixing water with concentrated sulfuric acid, or chemical reactions (e.g., calcium carbide and water react to yield acetylene).

Incompatible situations may not be restricted to two-component systems: adding a third component may sometimes catalyze a hazardous reaction.

When incompatible materials not stored in specifically designated areas are released accidentally in chemical warehouses, serious incidents can occur. In a fire, a spill of a corrosive material from its damaged container could eat through several adjacent steel drums of incompatible hazardous materials and cause releases that, after mixing, could lead to a disastrous domino effect.

It is necessary to identify such situations well in advance, at the planning stage, and incorporate adequate safeguards as part of routine operational systems, methods, and procedures. Hazard and operability (HAZOP) and risk control studies, necessary for such situations, are addressed in Chapters 12 and 13.

1.1.4 Toxic Chemicals

Toxic chemicals are those that adversely affect our health. These chemicals enter our body in three ways: (1) inhalation, (2) absorption through the skin, and (3) ingestion.

TABLE 1.2
Example Regulatory Categories for Acute Toxicity Levels (Major Accident Hazards)

Toxicity	Oral Toxicity, LD_{50} (mg/kg)	Dermal Toxicity, LD_{50} (mg/kg)	Inhalation Toxicity, LC_{50} (mg/L)
Extremely toxic	<5	<40	<0.5
Highly toxic	>5–50	>40–200	0.5–2.0
Toxic	>50–200	>200–1000	>2–10

The toxicity of a chemical is defined based on the quantity required to cause death in 50% of test animals, usually laboratory rats. The lethal concentration (milligrams per liter) for an inhaled gas is defined as the LC_{50}, typically for an exposure of 30 minutes. For an orally administered substance, the "lethal dose" (or LD50) is defined in milligrams per kg of body mass.

The effects of exposure to toxic chemicals may be acute or chronic. Acute effects result from a single exposure to a high concentration of the chemical. In contrast, chronic effects result from repeated exposures to low concentrations, perhaps over a significant duration in a worker's lifetime. Thus, acute effects are pertinent for a significant accidental release, and chronic effects for determining the allowable threshold concentration levels in the workplace. Acute effects are considered for "major accident hazards" and chronic effects for "occupational hygiene". These topics are discussed in detail in Chapter 10.

Various governmental regulations (for the manufacture, storage, and import of hazardous chemicals) list the values for acute toxicity at which the chemicals can produce catastrophic accident hazards (see Table 1.2).

1.2 PHYSICAL HAZARDS

1.2.1 Physical Explosion

There are many everyday examples of physical explosions in the process industries that do not involve chemical reactions:

- Rupture of pressure vessels (containing compressed gases, flammable, or toxic chemicals) when subjected to excessive pressure
- Rupture of a transport container, overfilled with liquefied gas and subjected to high ambient temperatures while in transit
- "Vessel burst" caused by BLEVE in a tank containing a liquefied gas under pressure and engulfed in a fire (covered in greater detail in Chapter 8)
- Rupture of a pressure vessel that has been weakened by corrosion, erosion, or brittle failure
- Rupture of a pipeline containing a stagnant mass of liquid or liquefied gas between two closed valves (without proper pressure relief), whose temperature rises excessively.

1.2.2 Electrostatic Charges

Electrostatic (or "static") charge can be a dangerous ignition source for flammable gases and has caused many fires and explosions in process plants. Operations such as handling of low electrical conductivity liquids, pneumatic transport of gas–solid mixtures, or charging of liquids and powders into batch mixers or reaction vessels are particularly prone to static hazards. Effective earthing (i.e., grounding) provisions must be made to avoid fires and explosions in such operations. Details are covered in Chapter 4.

1.2.3 Rollover/Boilover of Liquids

"Rollover" is a hazardous phenomenon associated with liquefied gas storage, such as LNG (liquefied natural gas). LNG is generally stored in tanks near atmospheric pressure under cryogenic conditions (approximately −160°C). LNG's chief constituent is methane, while nitrogen and C_2 to C_5 hydrocarbons may also be present in small quantities. There are appreciable differences in the composition of LNG from different sources. Consequently, their densities differ. Stratification can occur if the density of a liquid cargo charged to a tank is significantly different from that of the "heel" already in the tank.

Heat leakage into the tank can cause a slow circulation in the tank (free convection). Removal of the boiloff LNG vapors to maintain pressure in the tank increases the density of the topmost layer of liquid. At the same time, continued heat ingress through the walls and the bottom of the tank reduces the density of the bottom layer. When liquid density at the bottom becomes less than that at the top, the result can be a rollover of the tank contents with the rapid evolution of vapor. This phenomenon has the potential to overpressure and damage the tank.

Lees[1] has recommended some effective measures to reduce the risk of tank rollover. These include limiting the variation in LNG composition, mixing the tank contents using the top and bottom filling points, the physical stirring of the tank contents, monitoring of parameters related to stratification, and provision of a high-capacity vent.

"Boilover" refers to a phenomenon that sometimes occurs in distillation columns or when attempting to extinguish a fire in a tank containing different grades of crude oil or products, in case the roof was blown out because of an internal fire.

For an open tank under fire, if firewater is sprayed onto the burning oil, it quickly sinks to the bottom of the tank owing to its higher density and generally does not extinguishing the fire. This phenomenon happened in a fuel tank at a Venezuelan power plant. Once the lighter components have evaporated and burned at the liquid surface, the residue becomes denser than the oil immediately underneath. This residue forms a hot layer and sinks, progressing downward much faster than the regression of the liquid surface. When this hot layer, called a "heat wave", reaches the water or water-in-oil emulsion at the bottom of the tank, the water boils violently, resulting in a sudden ejection and scattering of burning oil over a large area.

Boilover must be distinguished from another phenomenon called "slop-over", in which minor frothing occurs when water is sprayed onto the surface of burning oil.

1.3 ENVIRONMENTAL HAZARDS

Environmental hazards in the process industries include the following:

- Atmospheric discharges (air pollutants)
- Water-soluble and suspended or particulate wastes discharged into water bodies, causing pollution of surface and groundwaters
- Solid wastes that pollute both the soil and groundwater.

The management of environmental hazards is the domain of environmental engineering and has not been discussed in this book.

1.3.1 Air Pollutants

The primary air pollutants and toxic substances from the process industries are defined explicitly in governmental regulations in various countries. These include sulfur oxides (SO_x), nitrogen oxides (NO_x), carbon monoxide, ammonia, volatile organic compounds (VOCs), and particulate matter, all of which have adverse effects on human health and the environment. Stacks and vents are significant sources of SO_x, NO_x, carbon monoxide, and dust. These pollutants can affect the surrounding population over large distances. VOCs are often released as "fugitive emissions" through pump seals, agitator glands, flange leaks, accidental spillage, or when handling open containers during batch charging or discharging operations. The effects of fugitive emissions, however, are usually confined to the process plant boundaries.

1.3.2 Water Pollutants

Water pollutants from the process industries include oxygen-demanding wastes, characterized by high biological oxygen demand (BOD) and chemical oxygen demand (COD), suspended particulate matter (TSS, or Total Suspended Solids), toxic metals, pesticides, etc. Management systems must ensure that treated effluents comply with applicable emission standards to avoid adverse effects on aquatic or marine life.

1.3.3 Solid Wastes

Solid or semisolid waste generated by the process industries includes by-products, residues, substandard products, contaminated packaging materials, etc., which cannot be reprocessed economically into useful products. Unless disposed of in an environmentally acceptable manner, they cause significant damage to the environment and affect the quality of soil, groundwater, and river or ocean beds.

1.4 OTHER HAZARDS

1.4.1 Electricity

Inadvertent contact with a live electrical system can easily injure or kill a person. It is a well-known hazard in our day-to-day life, both at home and in the workplace. People get used to the safe operation of electrical equipment and can forget the importance of safe work practices. Shocks or burns from sparks damage body tissue and can cause severe injuries or fatalities. An "Arc Flash" can occur when working with medium-to-high voltage electric circuits. This phenomenon causes white-hot metallic particles to be ejected that can cause extremely serious burns and injuries. However, a discussion of the calculation methods and safety precautions for this hazard are beyond the scope of this book that focuses primarily on process hazards. In process plants, the ignition of flammable gases by electric sparks may cause fires and explosions. Hazardous areas must be classified, and only appropriate electrical equipment recommended for service in hazardous areas should be used to minimize such hazards. This aspect has been discussed further in Chapter 3.

1.4.2 Hazards in Maintenance Work

Routine or unscheduled maintenance is essential for ensuring the safety and integrity of process plants. However, many maintenance-related accidents have occurred in the past. Unfortunately, these accidents continue to occur at a depressing frequency because of misunderstandings and habitual neglect of essential precautions. These are especially important while handing over the plant from production to maintenance, and vice versa. Several case histories are described in Chapter 2.

1.5 CLASSIFICATION CATEGORIES AND LABELING OF HAZARDOUS CHEMICALS

Until recently, countries have had unique systems for the classification and labeling of chemical products, resulting in many conflicting rules and inconsistent practices. In several major chemical-producing countries, inconsistent systems have coexisted, an expensive and cumbersome headache for government regulators. It has also been costly and confusing both for companies (who must comply with all prevailing regulations) and for workers, who need to understand process safety requirements for labeling of hazardous chemical packages or containers, and for selecting the proper PPE.

Furthermore, packaging and labeling for the export of hazardous chemicals often presented vexatious problems. In many cases, the exporters had to comply with inconsistent packaging and labeling rules in the exporting and importing countries.

In 1992, to address this lack of uniform standards, the United Nations Conference on Environment and Development (UNCED), often called the "Earth Summit", developed the Globally Harmonized System of Classification and Labeling of Chemicals (GHS), including safety data sheets (SDS) and easily understandable symbols. The GHS defines and classifies chemical product hazards and specifies health and safety information on labels and SDS. The goal is to ensure (1) the adoption of uniform rules for classifying hazards, and (2) the use of the same format and content for labels and SDS around the world.

1.5.1 Globally Harmonized System (GHS)

The GHS is the result of an international mandate and was developed by an international team of hazard communication experts. Details of this system are available in the UN documents.

The voluntary GHS standard has already been implemented by legislation in many countries and classifies hazards into the following categories:

- Physical Hazards
  - Explosives
  - Flammable gases
  - Flammable aerosols
  - Oxidizing gases
  - Gases under pressure
  - Flammable liquids
  - Flammable solids
  - Self-reactive substances and mixtures
  - Pyrophoric liquids
  - Pyrophoric solids
  - Self-heating substances and mixtures
  - Substances and mixtures that, in contact with water, emit flammable gases
  - Oxidizing liquids
  - Oxidizing solids
  - Organic peroxides
  - Corrosive to metals
- Health Hazards
  - Acute toxicity
  - Skin corrosion/irritation
  - Severe eye damage/eye irritation
  - Respiratory or skin sensitization
  - Germ cell mutagenicity
  - Carcinogenicity
  - Reproductive toxicity
  - Target organ systemic toxicity – single exposure
  - Target organ systemic toxicity – repeated exposure
  - Aspiration toxicity
- Environmental Hazards
  - Hazards to the aquatic environment
    - Acute aquatic toxicity
    - Chronic aquatic toxicity

Readers interested in the details of this method of classification are referred to the GHS literature on the Internet.

1.5.2 Adoption of GHS by Countries

In the European Union countries, GHS has been implemented by *Regulation (EC) No 1272/2008 on classification, labeling and packaging of substances and mixtures (the "CLP Regulation")*. According to the new rules, the deadline for substance classification was December 1, 2010; for mixtures, the deadline was June 1, 2015. *(EC) 1272/2008* aligns previous EU legislation with GHS hazards.

In the United Kingdom, GHS is adopted through the *Chemicals (Hazard Information and Packaging for Supply) Regulations 2009.*

The United States has implemented GHS through the revised *Hazard Communication Standard* (*HCS*) laws and the Occupational Safety and Health Administration (OSHA) rules. In the USA, all hazardous substances and mixtures need to be classified and labeled following the new HCS after June 1, 2015.

Japan has implemented the GHS through its *Industrial Safety and Health Law* (*ISHL*) regulatory framework. The *JIS standards* (*Japan Industrial Standards*) JIS Z7252 and Z7253 are the prevailing standards for classification, labeling, and standard data sheets (SDS/MSDS).

The People's Republic of China (SAC) has implemented, under its *National Standard Series GB 30000-2013, 28* compulsory new standards for chemical classification consistent with the GHS as of November 1, 2014.

South Korea implemented a newly revised *Standard for Classification and Labelling of Chemical Substances and MSDS* from July 1, 2013.

Since 2013, Singapore has implemented its *Standard Specification (SS 586:2014) for Hazard Communication for Hazardous and Dangerous Goods.*

India has not officially adopted the GHS for chemicals as yet. In July 2011, the Government Ministry of Environment and Forests published draft *Hazardous Substances (Classification, Packaging, and Labeling) Rules, 2011*, fully aligned with the GHS. Even though the rules had not officially been adopted as of the last quarter of 2018, the new 16-section draft SDS requirements can help prepare SDS for the Indian market. The classification rules are *Manufacture, Storage, and Import of Hazardous Chemicals, Rules 2000*, amended through 2018. For the packaging and labeling of hazardous chemicals, the *Central Motor Vehicle Rules of 1989, amended through 2018*, apply.

1.6 PROVISION OF HAZARD INFORMATION

It is essential to provide information and appropriate training on specific hazardous chemicals to anyone exposed directly or indirectly. Numerous governmental laws and regulations specify that suppliers of hazardous chemicals have the responsibility to provide such information. Industry associations and professional societies play a crucial role in developing appropriate codes and guidelines, devising the proper formats for passing on the information, and education and training.

The details required to be provided to a supplier would depend on the underlying purpose and the background of those who would need them. A supplier should generally be aware of how their materials will be used and must ensure that each SDS addresses the hazards of all uses that can reasonably be anticipated. For example, SDS containing a detailed information on physicochemical properties, safety, and health-related data are essential in manufacturing and R&D environments for preparing in-house safety instructions for plant personnel.

Firefighting personnel, emergency responders, transporters, and members of the public need critical safety information that they can understand easily and quickly. For this purpose, product labels and safety cards containing hazard category symbols, placards, and "Dos and Don'ts" instructions as bulleted points are more appropriate than a full SDS.

1.6.1 Safety Data Sheets (SDS)

The SDS is a vital document that provides information on a chemical substance or mixture's hazards and includes advice on precautions during handling. The SDS is sometimes referred to as the material safety data sheet (MSDS).

In most countries, suppliers of chemicals must provide SDS to their customers. Suppliers include manufacturers, importers, distributors, wholesalers, and retailers. MSDS for products are also available on the web sites of manufacturing companies and large retail suppliers. Such data should routinely be validated or checked to ensure they remain current.

The SDS or MSDS usually provides information under the following 16 headings:

1. Chemical product and company identification (including trade or common name of the chemical)
2. Hazards identification (composition/information in a way that unambiguously identifies them for conducting a hazard evaluation)
3. First-aid measures
4. Firefighting measures
5. Accidental release measures
6. Handling and storage
7. Exposure controls/personal protection
8. Physical and chemical properties
9. Stability and reactivity
10. Toxicological information
11. Ecological information
12. Disposal considerations
13. Transport information
14. Regulatory information
15. Other information.

REFERENCE

1. Mannan, S.: *Lees' Loss Prevention in the Process Industries* (4th Ed., Butterworth-Heinemann, Oxford, 2012).

2 Overview of Some Major Accidents in the World

In this chapter, we have reviewed some major, well-documented accidents in the chemical and allied industries. We describe clearly how such catastrophic accidents occur, their consequences, and how to avoid them. In this review, the selected accidents cover commonly occurring phenomena such as fire, explosion, and release of toxic substances to the atmosphere. Generally, more than one accident has been included under each type to give some idea about the range of circumstances, causes, and consequences.

Criteria followed in selecting the accidents are as follows: (1) they are well known, generally by name; (2) the consequences are enormous; and (3) the published findings contain sufficient details to be of educational value.

We have provided a summary of the conclusions and lessons learned for ready reference at the end of this chapter. Human memory is fallible, and readers should find this summary helpful.

Figures on fatalities and injuries included under various accidents are based on published references and private communications, and some of them may be disputable. These should be regarded as order-of-magnitude values, adequate for the objectives of this book.

2.1 CLEVELAND, OHIO[1,2]

Date of accident:	October 20, 1944
Location:	Cleveland (Ohio), USA
Nature of accident:	LNG leakage and fire/explosion
Fatalities and injuries:	128 killed, 200–400 injured, substantial property loss

2.1.1 Brief Description of Facility and Process

East Ohio Gas Co. constructed the US's first (and most likely the first global) commercial liquefied natural gas (LNG) liquefaction, storage, and vaporization facility in 1941 at Cleveland, Ohio, for natural gas supply to industrial and domestic consumers. The facility was located in the center of the city to facilitate gas distribution. It had four LNG tanks: the first three were spherical (17.4 m in diameter), and the fourth, added last in 1943, was of semi-toroidal construction (21 m tank diameter and 13 m tank height). Figure 2.1 shows the semi-toroidal inner tank. The material of construction used for the inner container of all the tanks reportedly was 3.5% Ni-steel with carbon < 0.09%, and the outer container was of carbon steel.

DOI: 10.1201/9781003107873-2

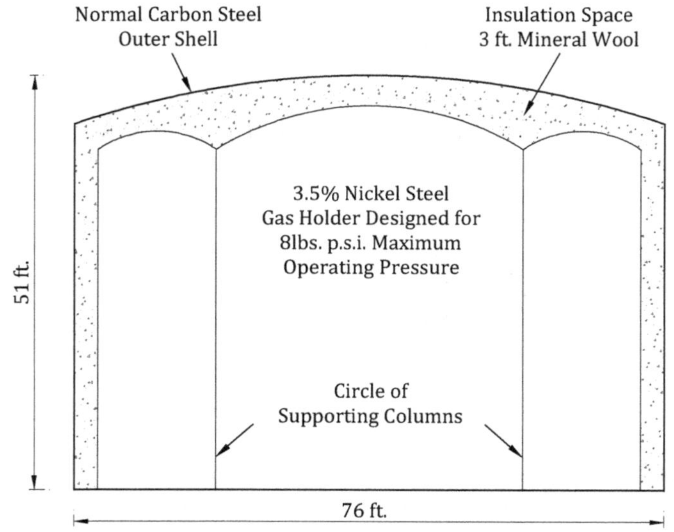

FIGURE 2.1 Cross section of semi-toroidal construction, Tank 4.

The internal space between two containers had 1 m-thick cork insulation: the bottom one-third is made of solid cork and the top space granular cork.

LNG is a liquefied natural gas consisting of methane with 5%–10% impurities (nitrogen and C_2 to C_4 hydrocarbons). The atmospheric boiling point is around −160°C, and the vapors are extremely flammable. The plant had a liquefaction capacity of 113,300 sm^3/day and a re-gas capacity of 85,000 sm^3/day. The liquefaction process used pre-chilling by ethylene followed by two-stage expansion. The LNG was stored at 8 psig (55 kPag) pressure at −156°C. The re-gas process required steam heating.

2.1.2 The Accident

In October 1944, all the tanks were filled in the early afternoon, and Tank 1 (spherical) was being topped up. The liquefaction section had stopped operation. At 2.40 p.m., rumbling sounds were heard and, immediately after that, "clouds of vapor" were seen close to the semi-toroidal Tank No. 4. An estimated 3,800 m^3 (1,710 tons) of LNG stored in Tank 4 was released and flowed along the stormwater drain running along the main street. Tank 4 was at a higher elevation, and LNG flowed towards Tank 3, which also failed, releasing reportedly 1,700 m^3 (765 tons), most likely owing to the failure of its carbon steel supporting columns contacting very low-temperature LNG.

An investigation report concluded that damage to the supporting columns caused Tank 3 to fail; a fire ignited by shrapnel from Tank 4 contributed to this structural failure. However, brittle failure of the carbon steel columns in contact with extremely cold LNG is a far more likely explanation. Vaporized LNG was soon ignited by an unknown fire source, most likely on the street, and vapors around tanks and all along the storm-water drain were engulfed immediately in an enormous fire. An engineer present in the site reportedly estimated the maximum flame height to be 85 m. However, it is unlikely that such a flame height can be estimated reliably by visual observation alone. An open-drain fire and an ignition followed by an explosion, likely vapor cloud explosion, (VCE), were also reported in the underground sewer.

2.1.3 Causes, Circumstances, and Consequences

The U.S. Bureau of Mines investigated the accident in some detail but could not reach a definitive conclusion as to the cause.

However, most experts later determined that use of the wrong material of construction for Tank 4 caused "low-temperature embrittlement" failure of the tank. The recommended material of construction for LNG vessels was not finalized until later. Designers had been aware of the low-temperature stress limitation of 3.5% nickel steel but believed erroneously that this limitation was not sufficient to make it unsuitable. Today, authorities agree that 9% nickel steel has the required ductility suitable for LNG duties.

Some doubts were expressed later by some investigators who concluded that owing to the acute shortage of nickel during World War II, even 3.5% Ni-steel was not used for Tank 4 construction.

The general consensus is that, because the accident occurred during World War II, wartime material shortages caused an improper compromise in materials selection.

The Cleveland fire and explosions caused very significant damage to the facility (which was never restarted) and to the mostly wooden buildings located up to 400 m away. This accident resulted in the death of around 128 persons, and 200–400 people were injured. However, Tank Nos. 1 and 2 showed no visual damage.

2.1.4 Lessons/Recommendations

- It is vital to use the right materials of construction and ensure quality control in material procurement and fabrication.
  - 9% Ni steel (which universally is accepted now) should be specified for the conventional LNG tank inner shells.
  - LNG tank design should follow US standards API 625,[3] API 620 Appendix Q,[4] and NFPA-59A,[5] or EU/UK standard BS/EN 5555[6] and EN 1473.[7] For a full containment tank (concrete outer shell), prestressed, posttension reinforced concrete should be used, as specified in the US Standard ACI 376-1.[8]
  - For full-containment tanks, the inner surface of the tank shell's concrete should be provided with a low-temperature steel liner up to a height of 5 m to serve as a vapor barrier in case of a minor leak in the inner shell.
  - For land-based membrane tanks (a recent development), the membrane material should be stainless steel, as per tank designer specification and European Standard EN-14620.[9]
- LNG plants should be located far from densely populated areas that have a mix of homes and businesses.

- Facility layouts should comply with applicable standards such as NFPA 59A or EN 1473. A formal quantitative risk assessment is mandatory for LNG installations, following guidance in NFPA-59A.

2.2 FEYZIN, FRANCE[1,10–12]

Date of accident:	January 4, 1966
Location:	Feyzin Refinery, France
Nature of accident:	Propane leakage and fire/BLEVE
Fatalities and injuries:	18 killed, 81 injured, substantial property loss

2.2.1 Brief Description of Facility and Process

The LPG storage tanks were in the ELF Feyzin Refinery's site near Lyon, south of France. A highway was running alongside this site. The LPG tank farm consisted of the following:

- Four spherical pressure vessels (1,200 m^3 each) for propane
- Four spherical pressure vessels (2,000 m^3 each) for butane
- Two horizontal bullet pressure vessels (150 m^3 each) for propane and butane.

The eight LPG storage vessels were located in a 114.5 m × 55 m bund wall with a central subdivision forming approximately square bund halves. These contained two propane and two butane spheres each. The bund walls were 0.5 m high, while the one in between was 0.25 m high.

The LPG storage spheres were about 450 m away from the nearest refinery unit and about 300 m from the village houses. The shortest distance between an LPG sphere and the motorway was 42.4 m, and the spacing between individual spheres varied from 11.3 to 17.2 m.

Each sphere had fixed water sprays, both at the top and at mid-height, plus a single spray directed towards the bottom connections.

There was a three-way valve on top of each sphere, beneath two identical pressure relief valves, so that one was always in service with the other isolated. The propane spheres had relief valve settings of 18.0 barg, and the butane spheres were set to 7.5 barg.

All spheres had fireproof steel supports.

Propane is a highly flammable gas at ambient temperature with a normal boiling point of −43°C. In Feyzin, propane was stored in spherical tanks as a liquefied gas at about 0°C. The produced propane batches were contained in the propane tanks. Each product batch was sampled, and the bottom layer of water was drained at regular intervals. The sampling line was a branch with a valve positioned between two 2″ purge valves on the propane sphere's bottom drain line used to drain any water and denser oil residues.

2.2.2 The Accident

On the day of the accident, an operator was draining water from the propane sphere by opening valve A first and then valve B (Figure 2.2). When this operation was nearly complete, he closed the upper valve A and then cracked it open again. Valve B was

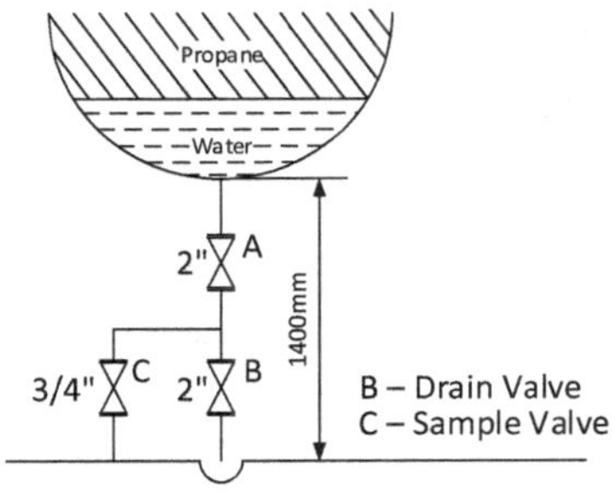

FIGURE 2.2 Drain valves underneath propane tank at Feyzin.

then in the open position. There was no flow, and he opened valve A further. A blockage (propane hydrate or ice) cleared, and propane gushed out. The operator was unable to close the upper valve A. He then tried to close the lower valve B, but it was too late as, by then, valve B was also frozen open.

The entire inventory of propane in the sphere (approximately 450 tons) spewed out, forming a vast vapor cloud. Since propane is denser than air, this cloud spread 150 m onto the motorway. The police had stopped highway traffic following an alarm; however, a car approaching from a side road ignited the gas about 25 minutes after the release had started. The fire flashed back to the sphere, which was surrounded immediately by flames.

The sphere sprinklers were started, but the available water flow was inadequate. The firemen seemed to have used most of the available water for cooling neighboring spheres to stop the fire from spreading, in the belief that the relief valve would protect the vessel that was on fire.

About 90 minutes after the fire started, the vessel suffered a disastrous boiling liquid, expanding vapor explosion (BLEVE), and burst. This phenomenon has been discussed in detail in Chapter 8. Flying debris broke the legs of an adjacent sphere, which fell over. Its relief valve discharged liquid into the fire and, 45 minutes later, this sphere also burst. Altogether, five spheres and two other pressure vessels burst, and an additional three were damaged. The fire also spread to adjacent gasoline and fuel oil storage tanks.

Out of the 18 people killed in this accident, 11 were firemen.

2.2.3 Causes, Circumstances, and Consequences

The Feyzin accident was caused by the water freezing at the propane–water interface at the outlet of valve A, located just 1.4 m below the bottom of the sphere. The operator had failed to follow established procedures. The result was a BLEVE. Subsequent fires led to 18 deaths (including 11 firefighters). Property damage included the destruction of five of the spheres, two horizontal cylindrical tanks, four floating roof jet fuel and gasoline tanks, as well as damage to many homes and other constructions off-site.

2.2.4 Lessons/Recommendations

Had the vessel been about 5 m higher above the ground and had a remotely actuated valve been in place, the flow could reliably have been stopped promptly. Locating valves below such vessels and not on a line at some safe distance away from the vessel's shadow is a practice strongly to be discouraged.

Lessons learned from this accident have resulted in the adoption of several design standards for pressurized tanks used for liquefied flammable gas storage:

- New LPG installations and existing installations (wherever feasible) should use an earth-covered (mounded) vessel using clean sand for the cover (see Chapter 8). Earth-covered cylindrical pressurized tanks with capacities up to about 4,000 m^3 are constructed nowadays in many refineries and storage terminals. Such tanks are claimed to be "zero BLEVE" tanks.
- For existing installations and all other above-ground pressurized LPG storage facilities where the tanks are not mounded, provide fireproof insulation. Vermiculite concrete or other suitable, well-proven material would act as an affirmative barrier for heat input. No water sprays on LPG vessels would then be required.
- The main product line should slope away from the vessel. All valves and fittings should be relocated outside the tank's shadow. Only one connection for draining water should be provided. It should be robust, firmly supported, and fully welded up to the first remotely operated fire-safe isolation valve. After that, the second valve on the drain line can be manually operated. The size of the second valve on the drain line should be restricted to 20 mm (3/4 inch).
- Unless the tank is mounded or not covered with fireproof insulation, provide water sprays at 10 L/min/m^2 of the tank surface area. Fit a remotely operated emergency pressure release valve (on the top of the vessel and connected to the flare) so that vessel pressure can be reduced to one-fifth of the design stress in about 10 minutes. This time can be relaxed to 30 minutes if the vessel is insulated, and to 1 hour if the floor beneath is also sloped to capture any spillage in a collection pit.
- Install combustible gas detectors around the installation to provide early warning of a leak.
- All LPG and other above-ground pressurized vessels must have their pressure relief valve and downstream release vent/flare line designed and sized adequately, following American Petroleum Institute (API) standards 520,521,526 and 527 (or equivalent). In addition, such installation should follow API 576 "Inspection of Pressure Relieving Devices", a Recommended Practice developed and published by API that describes inspection and repair practices of automatic pressure-relieving devices commonly used in the oil and petrochemical industries.

2.3 FLIXBOROUGH, UK[1,11,13,14]

Date of accident:	Saturday, June 1, 1974
Location:	Flixborough Works of Nypro Ltd., UK (Cyclohexane plant)
Nature of accident:	Cyclohexane leakage and vapor cloud explosion
Fatalities and injuries:	28 killed, 56 injured, extensive damage on-/off-site

2.3.1 Brief Description of Facility and Process

Nypro Ltd., in Flixborough UK, had a large caprolactam plant. The Flixborough works were designed to produce 70,000 tons/annum of caprolactam, an essential raw material for nylon production. Until 1972, the plant had a capacity of 20,000 tons/annum of caprolactam. It used cyclohexanone as feedstock that was produced by hydrogenation of phenol. In 1972, additional capacity was built, in Phase 2, that enabled cyclohexanone manufacture via cyclohexane oxidation. Cyclohexane is a highly flammable liquid with a flash point of –20°C and a normal boiling point of 81°C.

The plant consisted of six reactors in series. The plant consisted of six reactors in series, arranged so that each successive reactor was at a lower elevation than the one preceding to allow the cyclohexane to flow by gravity from one reactor to the next as shown in Figure 2.3a. The connection between any two reactors was through a 28″ diameter pipe with bellows at the two ends. In these reactors, cyclohexane was oxidized catalytically to cyclohexanone and cyclohexanol using injected air. The feed to the reactors, starting with Reactor No. 1, was a mixture of fresh cyclohexane and recycled unreacted cyclohexane. The reactants flowed from one reactor to the next by gravity. The final reactor product still contained approximately 94% cyclohexane; this passed through an after-reactor (to complete the reaction). Finally, a distillation section separated cyclohexanone and cyclohexanol from unreacted cyclohexane. Between the after-reactor and the distillation section, there was also a chain of mixers and separators for caustic/water washing to remove acidic impurities formed in the reaction. The reactors were operated at a pressure of 8.8 kg/cm^{2}g and a temperature of 155°C. Each reactor contained about 20 tons of cyclohexane.

2.3.2 The Accident

On March 27, a cyclohexane leak was observed on Reactor No. 5, owing to a vertical crack in the sidewall. The facility operator decided to conduct a thorough inspection of the reactor during the next scheduled downtime. The next day, this crack had already extended to a full 2 m in length; the installation was then closed, and Reactor No. 5 was withdrawn for inspection.

A bypass between Reactors 4 and 6 was built, as shown in Figure 2.3a, without even a preliminary safety study. Since the required 28″ diameter piping was not in stock, available 20″ diameter piping was used for bypass connection. Based on a sketch produced on the shop floor, a three-piece, 20″ diameter dogleg-shaped bypass connection was quickly fabricated in the plant workshop and fitted between Reactor Nos. 4 and 6. It was connected directly with the 28″ bellows fitted at both the reactors.

On April 1st, following a leak test, the unit resumed operations with the 20-inch (508-mm)-diameter elbow pipe connecting the two 28-inch-diameter expansion bellows, on Reactors 4 and 6, via a plate and flange (see Figure 2.3b). The entire assembly was supported by temporary scaffolding placed so as not to interfere with pipe movement.

Until May 29, this stop-gap installation operated without any unusual problems being reported.

On May 29, the installation was shut down after discovering a cyclohexane leak. Between 29th and 31st May, the unit was started and shut down a few times when leaks were detected. Finally, by 31st May, after several leak repairs and supposedly successful pressure tests, the unit was cleared for a start-up.

Start-up began on the morning of Saturday, 1st June. As the production rate was increased gradually, the 20″ bypass pipeline shown in Figure 2.3b ruptured late in

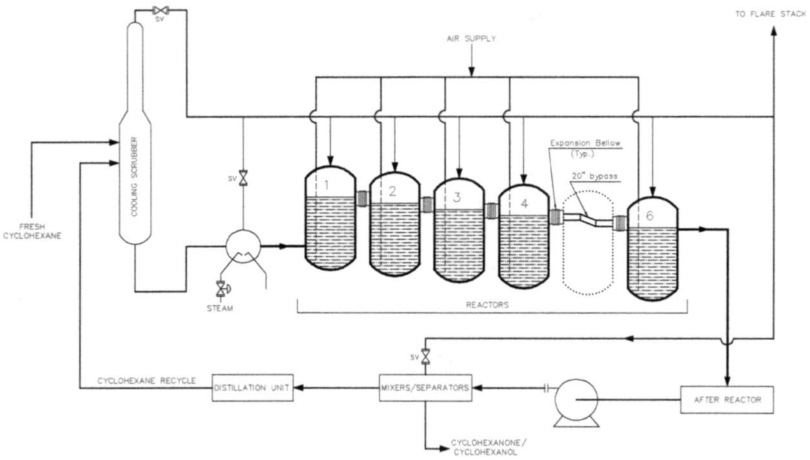

FIGURE 2.3a Flow diagram of cyclohexane oxidation plant.

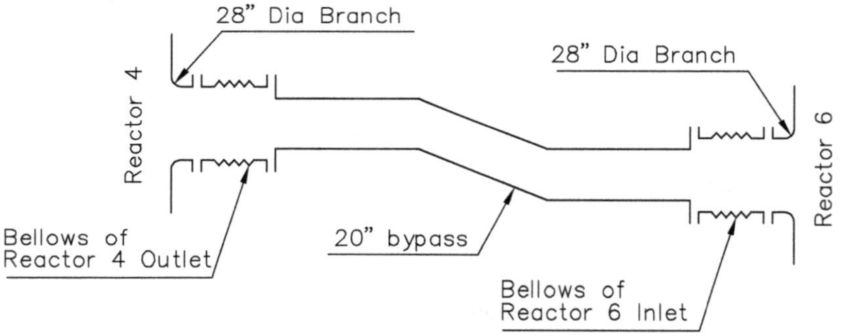

FIGURE 2.3b Sketch of temporary bypass assembly for Flixborough reactors.

the afternoon. Between 30 and 40 tons of cyclohexane escaped as vapor and mist from an estimated inventory of 120 tons in the five reactors and one after-reactor. The vapor ignited immediately and resulted in a massive VCE.

2.3.3 Causes, Circumstances, and Consequences

The report of the official Court of Inquiry,[14] and many papers that appeared in the open literature following the Flixborough accident, leads to the following main conclusions on the cause of this accident:

- The disastrous outcome was set when (1) one of the reactors in the cyclohexane oxidation train was removed, owing to a leak, and (2) the gap between the flanking reactors was bridged by an ill-designed and inadequately supported bypass assembly that retained the original bellows at each end.

Because of the bellows at each end, the bypass line was free to rotate or "squirm" when the pressure was raised, causing the bellows to fail.
- The design of highly stressed piping requires specialized engineering knowledge; this was not understood at the Nypro plant. As a result, there was no proper design study or consideration of the need for supports, no safety testing, no reference to relevant engineering standards (such as BS 3351:1971[15]), and no reference to the bellows manufacturer's "Designer's Guide".
- When the modifications mentioned above were being carried out, there was no qualified and credentialed Works Engineer with sufficient authority at the site. As a result, there was unseemly haste in getting the process restarted with minimal delay.
- Thus, the Flixborough accident was the result of deplorable management and engineering failures.

The released VCE was noticeable up to 50 km away and practically leveled the entire site. In many plant sectors, intense fires ensued, with flames flaring 70–100 m high. The instantaneous jump in pressure at the epicenter of the explosion, calculated at over 2 bar, destroyed all stationary fire protection equipment, further complicating emergency and rescue services.

Twenty-eight people died, and 36 others were injured within the site. Had the accident occurred on a weekday, the number of casualties would undoubtedly have been much higher. Outside the works, injuries and property damage was widespread, but, fortunately, no one else was killed.

At Flixborough, the total inventory of cyclohexane, naphtha, benzene, and toluene was 1,650 tons at the time of the accident, far above the licensed quantities; the entire stock took two and a half days to burn following the explosion.

The explosion demolished the control room, close to the plants, killing all the 18 operators and destroying the entire control system and records.

2.3.4 Lessons/Recommendations

A seemingly simple change or modification in a plant without due consideration by qualified, competent, and experienced engineers can have disastrous consequences.

The Flixborough accident, together with those at Seveso (Section 2.4) and Bhopal (Section 2.9), occurred in the same decade and negatively impacted public perceptions concerning the chemical industry. These events led to a spate of new regulations for the control of major accident hazards. A step change followed in the chemical engineering profession, which added new emphasis, priorities, and safety practice standards.

The following recommendations are based on lessons learned after the Flixborough accident:

- Top management must ensure professional and accountability-based management in production and supporting functions, including process safety engineering. Site safety systems and procedures should include "management of change" (MOC) rules and cover all modifications. An authorized and qualified manager should formally approve any proposal for alterations in existing plants/processes before its implementation, and this requirement should apply to temporary modifications as well. Approval should follow only after the request has been duly examined and vetted from the workability, engineering design, and safety points of view by a nominated team of professional engineers.

- As a matter of policy, the control room should be located away from hazardous plants, sufficiently away from the operational area, and outside the blast pressure range capable of causing structural damage. Otherwise, it should be protected by blast-proof walls and roofs that would enable operators to evacuate or remain safely inside in an emergency or after a safe shutdown.
- In operations involving manufacture, storage, or handling of hazardous chemicals, management should critically examine the inventory levels and avoid taking undue risks by limiting the inventory of dangerous substances. Although this did not increase the severity of the accident, investigators made adverse comments concerning the licensing system's ineffectiveness.

2.4 SEVESO, ITALY[1,11,16–18]

Date of accident:	July 10, 1976
Location:	The factory of Industrie Chimiche Meda S.A. (ICMESA) at Seveso, Italy
Nature of accident:	Runaway reaction leading to the release of highly toxic dioxin into the atmosphere
Fatalities and injuries:	No human fatality, but many people developed chloracne, an unpleasant skin disease, and an area of 17 km^2 became uninhabitable

2.4.1 Brief Description of Facility and Process

The plant, located in Meda, 20 km north of Milan (Lombardy, Italy), was owned by an international chemical group that manufactured perfumes and pharmaceutical products. Department B of the plant manufactured 2,4,5-trichlorophenol (TCP) that was subsequently transported to another group site to produce herbicides and antiseptics. The TCP production unit included three 10,000-L reactors and various platforms, columns, condensers, pumps, etc., spread out over 230 m^2.

Dioxin is 2,3,7,8-tetrachloro-dibenzo-p-dioxin (TCDD). It is known to be an ultra-toxic substance. An LD_{50} (oral) value of 0.6 μg/kg of body weight has been quoted for guinea pigs and 22 μg/kg for rats (as compared to an LD_{50} of 6,400 μg/kg for sodium cyanide). A mild symptom of TCDD poisoning is chloracne, an acne-like skin condition caused by chemical exposure. TCDD is a stable solid, is almost insoluble in water, and is resistant to destruction except by incineration at high temperatures near 1,000°C.

Outside of the Seveso context, dioxin is a little-known chemical. We now provide some details on the process and the plant used at Seveso to explain how the accident occurred. The plant was being used for the manufacture of TCP from 1,2,4,5-tetrachlorobenzene (TCB). The main chemical reactions are shown in Figure 2.4a. A batch process was used, and a schematic diagram of the reactor unit is shown in Figure 2.4b[1].

TCP was being produced in two steps. In the first step (Reaction 1), TCB reacted with sodium hydroxide at a temperature of 150°C–160°C in a reaction medium that

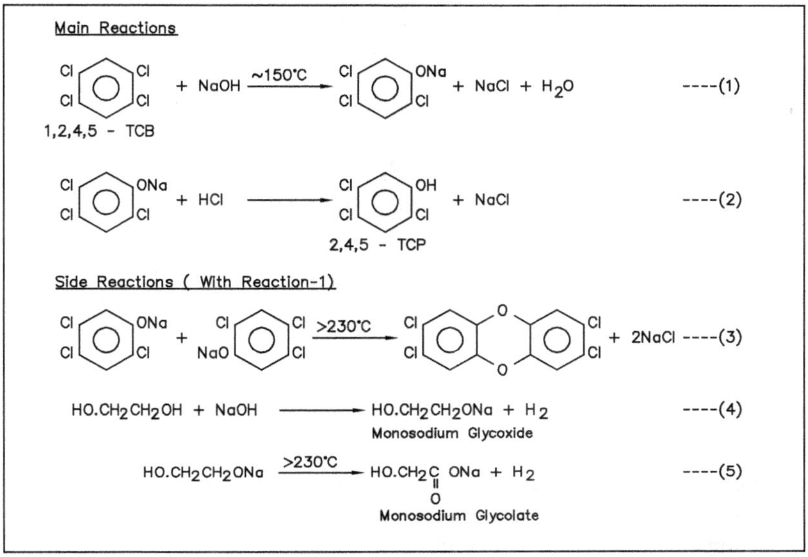

FIGURE 2.4a Reaction scheme for 2,4,5-TCP.

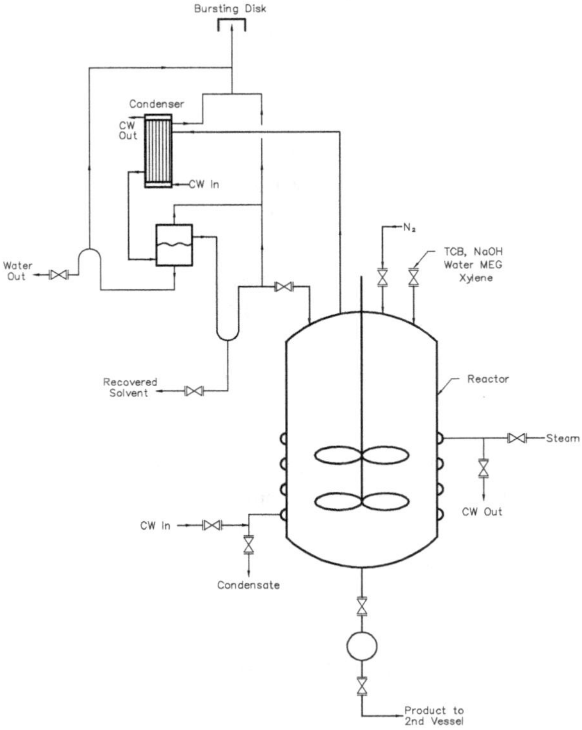

FIGURE 2.4b Schematic diagram of Seveso reactor.

comprised ethylene glycol and xylene. The purpose of xylene was to remove water by azeotropic distillation. The distillate was condensed, water was separated and removed, and xylene returned to the reactor as reflux.

After completion of Reaction 1, as indicated by no further water formation, the batch was heated to about 170°C to remove xylene and about 50% of the ethylene glycol. This temperature was above that of the process utilities that usually were available. Accordingly, it was decided to utilize turbine exhaust steam from the power plant on site, which was at around the temperature of the external heating coil in the reactor. This exhaust steam was at 12 bar and 190°C. The batch was then cooled quickly to 50°C–60°C by adding cold water, thereby terminating undesirable reactions. This procedure for Reaction 1 has been followed at Seveso since 1971.

After cooling by water addition, the batch was transferred by nitrogen pressuring to a second vessel; there, it was treated with hydrochloric acid to convert sodium trichorophenate to TCP, according to Reaction 2. TCP was then purified by distillation, and the residue left after distillation was incinerated. Exhaust steam (from a steam turbine at the site) was used to heat the reactor at 12 barg and 190°C.

Reaction 1 (primary reaction) was accompanied by several side reactions, of which the important ones are shown in Figure 2.4a. These side reactions have been reviewed by Wilson[18] based on the work of Milnes[17] and Theopfanous.[41] The formation of dioxin is the result of Reaction 3, which is favored at temperatures above 230°C and is an exothermic reaction. The other significant exothermic reaction is the decomposition of sodium-2-hydroxy ethoxide (Reaction 5), which starts at 230°C and proceeds uncontrollably to about 410°C.

2.4.2 The Accident

The batch involved in the accident was started at 16:00 on 9 July, and Reaction 1 was completed as usual. By 05:00 of 10 July (Saturday), all the xylene and about 15% of ethylene glycol (instead of the usual 50%) had been removed. The operation had to be stopped and the plant shut down for the weekend to comply with current Italian law that prohibited operations during weekends. The steam supply valve was closed, and the agitator turned off. No cooling was done, and the batch was left at 158°C.

The site was closed on the weekend, and the electrical load fell sharply, resulting in the exhaust steam temperature rising to around 300°C and causing the reactor wall to approach this temperature. The residual heat in the jacket then heated the upper layer of the mixture next to the wall by radiation. Also, the liquid in contact with the wall of the reactor was heated by conduction and convection. This hot liquid presumably flowed along the reactor wall by free convection. Thus, in the absence of agitation, a hot layer is formed on the top of the reaction mass, which is asserted to have set off the runaway reactions. The temperature increase from the runaway reaction caused the rupture of the bursting disk on the reactor.

2.4.3 Causes, Circumstances, and Consequences

The bursting disk rupture released a white cloud that traveled downwind in a southeastern direction from the site. Deposition from this cloud, which contained just 2 kg or so of highly toxic TCDD (dioxin), made approximately 17 km^2 of the area uninhabitable, of which 3.7 km^2 was rendered entirely off-limits.

Within days of release, about 3,300 animals, mostly poultry and rabbits, were found dead. Emergency slaughtering of animals was carried out to prevent TCDD from entering the food chain. By 1978, over 80,000 animals had been slaughtered. Fifteen children were quickly hospitalized for skin inflammation. By the end of August, Zone A had been evacuated completely and fenced. 1,600 people of all ages had been examined, and 447 were found to suffer from skin lesions or chloracne.

Although the Seveso accident did not result in any human fatality, it resulted in the death or slaughter of a vast number of animals who were feared to have been affected by TCDD toxicity.

2.4.4 Lessons/Recommendations

The Seveso accident was the result of several unacceptable deviations from the standard operating procedure (SOP):

- The operation was interrupted sooner than the time required for the batch to reach a state of completion when water could be added to terminate the reactions. The temperature of the partially completed batch increased significantly beyond the normal operating temperature during the weekend when the plant was left unattended.
- The agitator was shut down, which facilitated the formation of a sufficiently hot liquid layer at the top of the reaction mass, allowing the runaway reactions to occur. Had the agitation continued, the mixed mean temperature of the batch would have been much lower.
- The operating procedure was grossly inadequate as it did not consider any of the above situations and, consequently, made no mention of how to deal with them.
- Very likely, the operating personnel did not know enough about these physical and chemical phenomena to anticipate the consequence of these deviations. Even if some of the senior staff were knowledgeable, that knowledge was not disseminated properly.

The following recommendations are based on the lessons learned from the Seveso accident:

1. Methods for exploring unexpected hazards in chemical plants, such as a HAZOP or What-If or any similar study, must be carried out starting at the initial design stage, before commissioning, during operation, or before implementing any contemplated change (see Chapter 12).
2. Operating procedures must be comprehensive with clear instructions for any eventualities.
3. When working with highly toxic chemicals, sound knowledge of all known hazards must be imparted to plant personnel by formal training.
4. The on-site and off-site emergency plans should specify the necessary precautions in case of accidental releases.

5. The location of the plant must consider all off-site effects.
6. Any future development planning around a hazardous chemical site should seek prior approval by senior management.

The Seveso accident is regarded as a major chemical industry disaster, with impacts exceeded only in the Bhopal accident. A direct result of this accident is the EC Directive (82/501/EEC) on Control of Major Accident Hazards, generally known as the Seveso I Directive.

2.5 QATAR, PERSIAN GULF[1]

Date of accident:	April 3, 1977
Location:	Umm Said, Qatar, Persian Gulf
Nature of accident:	Leakage of liquefied propane from the storage tank, followed by fire
Fatalities and injuries:	Seven killed, over a dozen injured, and extensive damage to the plant

2.5.1 Brief Description of Facility and Process

The Gas Processing Plant at the Umm Said, Qatar, produced NGL, propane, and butane. Refrigerated liquid propane and butane were stored in two single-walled, atmospheric storage tanks of 260,000 barrels (41,300 m^3) and 125,000 barrels (19,900 m^3).

Chilled propane was stored at −42°C and liquid butane at about 0°C, both at atmospheric pressure. Unconfirmed reports mention that the material used in tank construction was made of carbon steel (not low-temperature carbon steel, LTCS). The tanks were hydrotested with seawater.

2.5.2 The Accident

On April 3, 1977, the propane tank containing 236,000 barrels (37,500 m^3) of refrigerated liquid propane suffered a massive failure. The wave of liquid propane swept over the containment dikes and inundated a 51,000 barrel/day (8,100 m^3/day) process area before igniting. The refrigerated butane tank was also destroyed, as was most of the process area. The fire burned out of control for 2 days and was extinguished after 8 days.

Reportedly, the tank weld that failed had been repaired following a weld failure incident a year earlier, when 14,000 barrels (2,200 m^3) of liquid propane were released. In that instance, a massive vapor cloud traveled 500 ft but, fortunately, did not ignite.

2.5.3 Causes, Circumstances, and Consequences

From the available information, it appears that either or both of the factors listed below caused the accident:

1. Weld failure was caused by corrosion because of microbiological sulfate-reducing bacteria that remained inside the tank, following hydrotesting of the tank with seawater.
2. Weld failure of carbon steel in low-temperature service.

The accident reportedly resulted in seven deaths, over a dozen (mostly plant personnel) injuries, and extensive damage to the plant.

2.5.4 Lessons/Recommendations

Many lessons were learned from studies of the Qatar accident. The most important recommendations are as follows:

1. A hydrotest should be carried out using pH-controlled fresh water. If testing is done with seawater (a common practice for facilities in coastal areas and gas storage tanks in ports), the following should be ensured:
   a. The chloride level in seawater is below 30ppm.
   b. Immediately after the hydrotest, the tank's inside wall is cleaned thoroughly with fresh water at least three times.
2. It was a grievous error to use standard carbon steel construction for the tank; the recommended material is LTCS as per ASTM 537 (low manganese content), following API Standard 620, Appendix Q,[4] or NFPA 58[19] or equivalent EN standard.
3. Refrigerated hydrocarbon tanks should preferably be of full containment design, as defined in NFPA-58 or EN.BS 7777 or equivalent standards.

2.6 CARACAS, VENEZUELA

Date of accident:	December 19, 1982
Location:	The electrical power plant at Tacoa, near Caracas, Venezuela
Nature of accident:	Fire and boil over in a fuel oil tank
Fatalities and injuries:	Over 150 people were killed, including 40 firefighters, and severe injuries to people on the nearby beach.

2.6.1 Brief Description of Facilities and Process

The Tacoa Power Plant is an electrical power station in Tacoa, which is owned by C.A. La Electricidad de Caracas, Venezuela; it was built in 1978. The plant had two cone-roof tanks, each 55 m in diameter and 17 m in height, for No. 6 fuel oil storage. No. 6 fuel oil was fuel oil mixed with 6%–20% heavy naphtha as per the power plant fuel specification. The tanks were served by a 100-mm water main to enable deluging the roof and the shell to prevent overheating. It also had a semifixed foam system. The tanks were located on a steep hill, presumably because of site topography.

2.6.2 The Accident

At about 06:00 on 19th December, while it was still dark, two men went to the top of Tank No. 8 to gauge the level, and a third person waited in a vehicle. Approximately 2 minutes later, a huge explosion blew off the top of the tank. The oil in the tank caught fire, with dense, black smoke spreading hundreds of feet into the air. Both men were killed.

Typically, No. 6 fuel oil does not ignite easily. However, the company's specification allowed admixing of 5%–20% low-boiling fractions (heavy naphtha). The flash point of the oil was 71°C. It had been specified that the oil would not be heated beyond 65°C; however, the actual temperature was 82°C, and the tank's high-temperature alarm had sounded some 6 hours before the accident.

The source of ignition is not known. The men who had gone to the tank top did not carry any flashlight with them, and it has been suspected that they lighted a match to be able to read the level.

The destruction of the tank top rendered the roof water deluge system ineffective, and it is not known whether the foam system could have been operated. Moreover, the tank was located on a steep hill, and this hindered the firefighting effort. Further, as the tank was under a full top-surface fire, it was decided to let the oil in the tank burn itself out, and water sprinklers were fully diverted to keep adjacent tanks cool.

About 8 hours after the fire started, a violent boil over was due to the existence of a water layer at the bottom of the tank. Burning oil (together with steam) erupted from the tank like a volcano and surrounded the second tank (Tank No. 9) and flowed down the hill into the sea.

2.6.3 Causes, Circumstances, and Consequences

The mixture of hydrocarbons with wide ranges of boiling points supported the development of a heat wave and rollover below the burning surface. External cooling efforts had been suspended at the tank that was on fire. Therefore, the liquid temperature rose rapidly to over 100°C and propagated towards the bottom of the tank. When the wave reached the water layer at the bottom of the tank, the water vaporized almost instantaneously, causing the violent expulsion (boilover) of the unburned and burning liquid hundreds of feet from the tank.

The use of a fuel specification that allowed 5%–20% low-boiling fractions (heavy naphtha) in the fuel oil resulted in stratified layers in the tank with widely varying boiling points: a light hydrocarbon fraction on the top, the heavier hydrocarbon layer in the middle, and the water layer at the bottom. These conditions enabled the boil over phenomenon to occur. The occurrence of a full top-surface fire, and a combination of other potent factors, caused a disastrous result.

The accident killed over 150 persons, including 40 firefighters, many civil defense workers, and onlookers. About 400 civilians enjoying the nearby beach suffered serious to moderate burn injuries, and approximately 500 homes were also severely damaged in the fire.

2.6.4 Lessons/Recommendations

- Organizations operating storage tanks containing mixtures of light hydrocarbons with heavy fuel oil (or Bunker C) must update their emergency response plans to acknowledge the much greater risk posed by stratified layers of combustible liquids of widely varying density in atmospheric tanks. Mitigation plans and actions must be specified to prevent boilovers and protect emergency responders.

- The requirements spelled out in NFPA's flammable and combustible liquids code handbook must be observed.
- Firefighters must be well-trained in handling oil and gas fires under different scenarios.
- During firefighting operations, crowding by onlookers not directly involved in firefighting must be prevented.

2.7 MEXICO CITY[1,11,16]

Date of accident:	November 19, 1984
Location:	LPG installation of PEMEX (Petroleos Mexicanos) at San Juan Ixhuatepec, Mexico City
Nature of accident:	LPG leakage leading to a major fire, vapor cloud explosion, and BLEVE
Fatalities and injuries:	At least 500 people were killed; more than 7,000 injured; extensive damage on-/off-site

2.7.1 Brief Description of Facility and Process

The site was an LPG storage terminal owned by PEMEX fed by pipelined LPG from three different refineries for supply to wholesale distributors. The storage capacity was 16,000 m^3, consisting of two spheres of 2,400 m^3 each, four spheres of 1,600 m^3 each, and 48 cylindrical tanks from 45 to 270 m^3 capacity. The daily throughput was 5,000 m^3. The LPG consisted of 80% butane and 20% propane. The refineries supplied the facility with LPG daily.

A general layout of the plant is shown in Figure 2.5. Southwest of the spheres, there was a water pond and, beyond the pond, a ground-level flare. Adjoining the PEMEX terminal toward the east, there was a distribution depot owned by Unigas and, further east, an LPG bottling plant with distribution facilities. The Unigas site had 67 trucks at the time of the accident.

2.7.2 The Accident

Early on 18th November, the plant was being fed from a refinery 400 km away. A fall in pressure was registered in the control room and at a pipeline pumping station 40 km away. This drop was caused by a leak in an 8″ diameter pipeline located between a sphere and a series of cylinders adjacent to the spheres. Unfortunately, the operators could not diagnose the cause of the pressure drop.

The release of LPG continued for about 5–10 minutes and soon caught fire. The isolation valve provided upstream of the ruptured pipeline was to be operated manually but could not be approached quickly, as it was engulfed in the fire. The massive leak continued, and a large gas cloud, 150 m × 2 m high, traveled downwind in a southwesterly direction and ignited on reaching the flare, resulting in a flash fire (deflagration). The cloud burned and dissipated, but a flame at the rupture point continued as the leakage was not stopped. This flame heated the nearest sphere that suffered a BLEVE with a fireball of about 300 m diameter, causing further damage

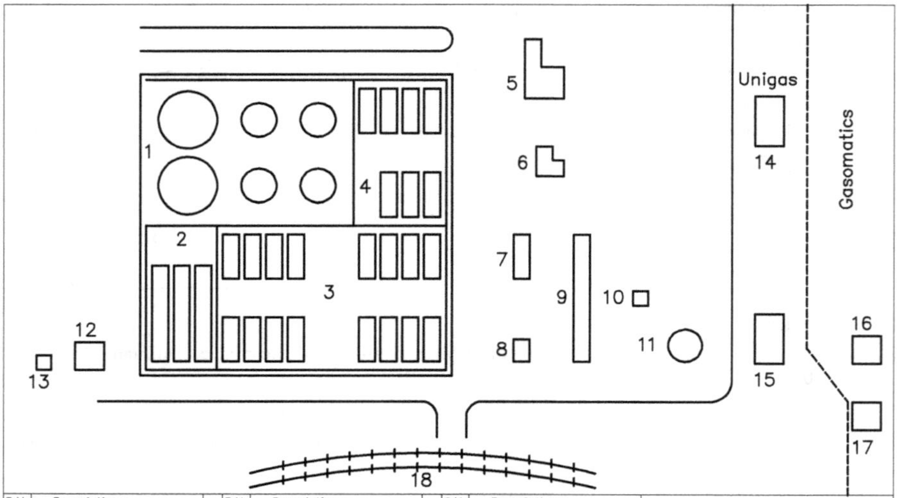

S.N.	Descriptions	S.N.	Descriptions	S.N.	Descriptions
1	LPG Spheres	8	Fire Pumps	15	LPG Storage, Unigas
2		9	Road Car Loading	16	Bottling Plant, Gasomatics
3	LPG Bullets	10	Gas Bottles Stores	17	LPG Storage, Gasomatics
4		11	Water Tower	18	Rail Car Loading
5	Pipe Valve Manifold	12	Pond		
6	Control Room	13	Flare Pit		
7	Pump House	14	Bottling Plant, Unigas		

FIGURE 2.5 Sketch plan of PEMEX site in Mexico City.

and additional BLEVEs. Altogether, four spheres and 15 horizontal cylindrical tanks (called bullets) underwent a BLEVE during the next 1½ hours, and some parts of the tanks were propelled up to 1,200 m from the plant.

2.7.3 Causes, Circumstances, and Consequences

No root cause was provided officially by Pemex for the rupture of the 8″ pipeline, and no investigation report is available in the public domain. However, according to some third-party reports, the causes of the pipe rupture could have been:

i. Overfilling of the sphere, coupled with the failure of the relief valve
ii. A surge in pressure in the LPG feed line during the start of pumping
iii. High localized corrosion.

The absence of online flow monitoring and an automatic shutoff system at the first isolation valve in the LPG terminal allowed a leak at full flow for up to 10 minutes. The terminal's firewater system was destroyed in the initial blast caused by the deflagration. The water spray systems were inadequate to cool the first sphere on which a jet fire from the ruptured pipe impinged, and this deficiency caused a BLEVE. After successive BLEVEs, burning LPG rained down nearly 150 m around the site on residential homes. Reportedly, there were 19 explosions (mostly BLEVE) at several spheres and bullets.

There were 500–600 deaths and over 5,000 injuries, mostly to members of the public. Practically all the houses within a 300 m radius were severely damaged. When the plant was built, the nearest buildings were 360 m away, but homes were subsequently built on the intervening ground. Eventually, the closest houses were only 130 m from the plant. The explosion flung large and small metal fragments up to 1.2 km from the site.

2.7.4 Lessons/Recommendations

- The very high number of casualties from this accident was caused by the facility being very close to a residential area, as happened in the Bhopal, India disaster (see Section 2.8).
- There were no gas detectors, and the LPG leak was not discovered immediately.
- There was no automatic or even remote-operated fire-safe isolation or shut-off valve system at the site boundary. The control room was not located at a safe distance for enabling reliable emergency shutdown.
- Owing to the absence of adequate sprinklers on the spheres, a BLEVE could not be prevented.
- It appears that no process hazard analysis (PHA) was carried out either before freezing the design or before commissioning.

The approach to the PEMEX LPG terminal was highly congested, and this interfered with the emergency response.

The recommendations arising out of the accident are:

- Site selection for facilities with high potential for accidents, with significant off-site consequences, shall be away from residential areas and shall be based on a PHA, followed by risk analysis.
- Pressurized LPG storage terminals and new facilities with storage capacities below 10,000 tons should be mounded in compliance with applicable regulatory requirements.
- Larger LPG storage terminals (new facilities of storage capacities 10,000 tons and above) should be refrigerated overground tanks, operating at near atmospheric pressure and conforming to applicable codes, e.g., US NFPA, EN, or OISD (India).
- Existing above-ground pressurized LPG storage terminals should be provided with "fire-safe insulation" for a fire of a minimum of 2-hour duration engulfing the vessel.
- The facility must have a disaster management plan to deal with on-site and off-site emergencies, and it should periodically be well-rehearsed.

2.8 BHOPAL, INDIA[1,20–22]

Date of accident:	December 3, 1984
Location:	Insecticide factory of Union Carbide India Limited at Bhopal, India
Nature of accident:	Release of highly toxic methyl isocyanate (MIC) to the atmosphere
Fatalities and injuries:	More than 5,000 people killed; approximately 200,000 injured

2.8.1 Brief Description of Facilities and Process

Union Carbide India Limited (UCIL) was operating a plant at Bhopal, Madhya Pradesh, to manufacture pesticides. The main intermediate methyl isocyanate (MIC) was used to manufacture the pesticide Sevin (a carbamate). Since December 1960, UCIL had been importing MIC in 200-L stainless steel drums from Union Carbide's plant in West Virginia, USA. They were marketing Sevin after adding diluents etc. Subsequently, UCC and UCIL started manufacturing MIC at Bhopal and stored it in underground tanks.

MIC was being manufactured by UCIL using monomethylamine (MMA) and phosgene as feedstocks. MMA and chlorine were procured raw materials, and phosgene was manufactured from carbon monoxide and chlorine. The two primary reactions are:

$$\underset{\text{Phosgene}}{COCl_2} + \underset{\text{Methylamine}}{CH_3NH_2} = \underset{\text{Methylcarbamoyl Chloride}}{CH_3NHCOCl} + \underset{\text{Hydrogen Chloride}}{HCl} \tag{2.1}$$

$$\underset{\text{Methylcarbamoyl Chloride}}{CH_3NHCOCl} = \underset{\text{Methyl Isocyanate}}{CH_3NCO} + \underset{\text{Hydrogen Chloride}}{HCl} \tag{2.2}$$

In the first reaction, MMA reacts with excess phosgene in the vapor phase to yield methylcarbamoyl chloride (MCC). After quenching with chloroform, the product is stripped of the excess phosgene, which is recycled.

In the second reaction, MCC is pyrolyzed to obtain crude MIC that is distilled to separate it from MCC, chloroform, and residues.

The distilled MIC was transferred to one of the three type 304 stainless steel storage tanks. Each tank had a capacity of 57 m^3 or approximately 40 tons of MIC. The normal boiling point of MIC is 39°C, and the storage condition was specified to be between 0°C and 5°C. The temperature was maintained by circulating the MIC through an external heat exchanger cooled by chloroform that, in turn, was chilled in a 30-ton refrigeration unit.

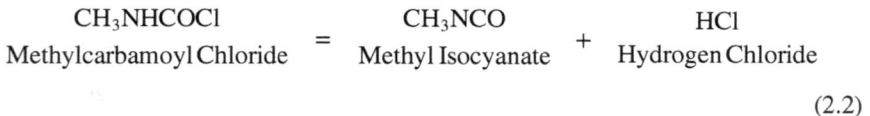

A simplified flow diagram of one of the storage tanks, with details relevant to the accident, is shown in Figure 2.6. The tanks were mounded and covered with a concrete deck for protection and insulation.

The MIC storage tanks were designed for a pressure of 2.8 barg. Each tank was fitted with a pressure relief valve and a bursting disk upstream of the relief valve. The tank's pressure was controlled by supplying high-purity nitrogen from a nitrogen header through a control valve. Venting from the tank was controlled by a similar control valve discharging to the process vent header (PVH). A pressure controller manipulated both control valves. Blowdown from the pressure relief valve was connected to a separate relief valve vent header (RVVH). About a year before the accident, the RVVH and PVH were connected through a "jumper line" to allow gas to be routed to a scrubber to enable repairs at one of the vent headers.

The PVH and RVVH were connected to a vent gas scrubber, where any MIC in the escaping gas would be destroyed by caustic before any discharge through a 30-m atmospheric stack. Alternatively, the treated gas from the scrubber could be diverted to a flare stack of similar height. There was also a direct route from the vent headers to the flare stack.

The RVVH of the MIC tanks was also connected, upstream of the scrubber, to several pressure safety valve lines, including a quench filter and phosgene stripping still filter. Each of these lines was provided with an isolation valve and a bleeder valve. Details of these additional lines are not available, and whatever has been shown on these lines in Figure 2.6 is based on Lees,[1] who cites a report by Bhushan and Subramanian.[21]

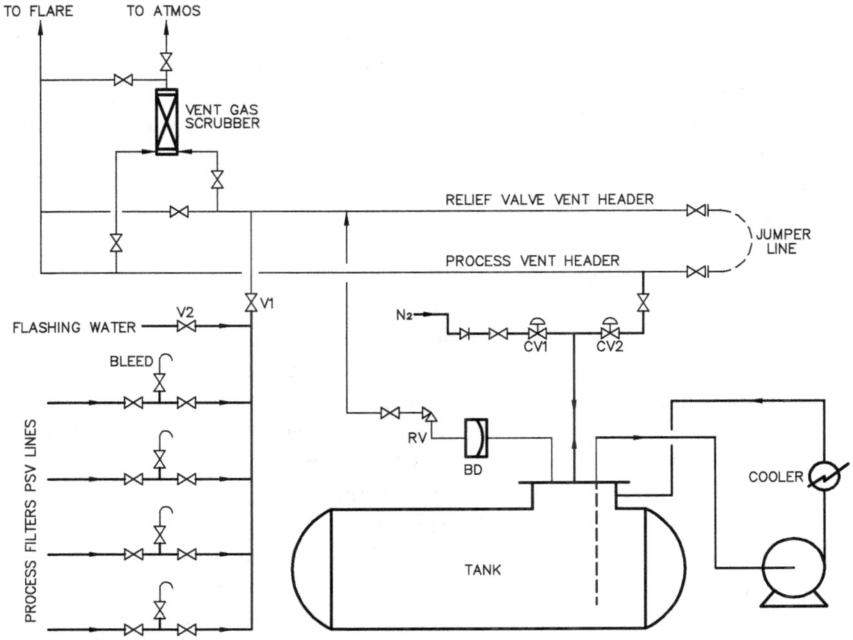

FIGURE 2.6 Flow diagram of MIC storage system.

2.8.2 The Accident

Early on the morning of December 3, 1984, a relief valve lifted on a storage tank (Tank No. 610) containing MIC at UCIL's plant in Bhopal, India. Some 30 tons of highly toxic MIC gas escaped to the atmosphere and downwind as a cloud, heavier than air, and as an "unending thick mist". The wind direction was such that the MIC cloud engulfed an enormous population living near the plant, killing and injuring many thousands.

2.8.3 Causes, Circumstances, and Consequences

The accident's direct cause was water ingress into MIC Tank 610 sometime on the evening of December 2, 1984. The quantity was at least 500 kg. (The paper "Investigation of large magnitude events Bhopal as a case study[23]", presented in the IChemE (UK) Conference in 1988, mentions about 500 kg).

Water reacts exothermically with MIC to produce methylamine and carbon dioxide. The heat generated could cause a runaway reaction. During normal operation, MIC fumes and other escaping gases first pass through a pipe of 2″ diameter to a vent gas scrubber (with a circulating alkali solution, if needed) to neutralize MIC in the escaping gases before discharge to the atmosphere. However, this system was grossly undersized and, further, not maintained properly.

The MIC reaction with water begins slowly, but if there is no agitation – as in tank 610 – it soon accelerates. Consequently, upon water entry, the pressure in tank 610 started rising rapidly. After an hour or so, the tank pressure rose above the relief valve setting, releasing highly toxic MIC gas. The relief valve vent from the MIC tanks was led to a 4″ diameter RVVH and ultimately to the grossly undersized vent gas scrubber for neutralization.

The detailed sequence of events was described in an investigation report. At 23.00, on December 2, 1984, the control room operator noticed that the pressure in Tank 610 was 10 psig or 0.7 barg. By 00:15, the field operator reported leakage of MIC near the scrubber and in the process filter area. The control room operator observed the pressure in Tank 610 to be 2.1 barg and still rising rapidly. Within 15 minutes, the pressure reached 3.8 barg, the top of the scale. By then, the extremely toxic MIC gas was gushing out of the stack. This release continued until about 02:00 when the safety valve re-seated after a release of about 30 tons of MIC.

Extensive subsequent investigations were conducted 20,21,22,23, and an analysis of the tank residue was performed. Also, additional experiments were performed to explain the formation of various by-product compounds. The major components in the residue were MIC trimer, dimethylisocyanurate (DMI), trimethylbiuret (TMB), tetramethylbiuret (TRMB), and several other complex organic compounds. The exothermic reactions caused the formation of these by-products; it was concluded that about 450–900 kg of water had entered the storage tank. Also, the tank's temperature had exceeded 200°C.

However, exactly how the water entered the MIC storage tank (Tank 610) was not established because of divergent findings in two subsequent detailed investigations: one by a committee of third-party Indian experts (commissioned by the Government of India[24]) and the other by UCC's experts, along with UCIL.[20]

Two hypotheses have been offered. The first suggests that the introduction of water was a deliberate act of sabotage (UCC Report[20]). The second (Government of India's appointed team[24]) concludes that leakage of water occurred during the plant's water flushing operation. The second possibility, considered in some detail by Lees[1] and Shrivastava,[22] appears to have been based on statements made by the plant operators: water leaked through Valve V1 (see Figure 2.6) into the RVVH and passed through the jumper line into the PVH. From the PVH, the water entered Tank 610 through blowdown valve CV2, which had either been left in the open position or did not seal properly. The fact that tank pressurization occurred gradually before the accident lends credibility to the hypothesis that water leaked into the tank through CV2.

Sometime after the accident, several eminent scientists contradicted claims by UCC that MIC and water do not react exothermically[22]. It is generally accepted today that the primary cause of the catastrophic increase in tank pressure was the extensive heat released from chemical reactions caused by water leakage into the tank. Of course, it is well known that failure to install blinds as a positive means of isolation before undertaking maintenance work – as happened in this case – is one of the major contributing factors for such accidents (see also the Piper Alpha accident, Section 2.9).

The majority of the Bhopal inhabitants who were affected were impoverished laborers living in huts and sleeping on floors, a common practice in tropical countries. They were fast asleep at night when the MIC release occurred. The highly toxic MIC gas affected a vast number of poor people. Initially, about 2,000 people were killed within a few days, and this figure rose to 5,295 over the next few months (submitted by the Government of India to the Supreme Court of India). Another 15,000 or so were reportedly injured severely. The government stated that 4,294 people suffered devastating health problems (eyes, throats, lungs). Of those still alive in 2018, most were continuing to suffer grievously.

This event caused extreme shock, panic, and prolonged anxiety among the local population and shook the global chemical industry throughout the world to its roots. Bhopal is recognized as the worst disaster, by far, in the history of the chemical industry in the world. Since then, most of the changes worldwide in laws and regulations have been made to address severe shortcomings in the safety performance of the chemical process industries.

2.8.4 Lessons/Recommendations

The lessons learned following the Bhopal accident are as follows:

Catastrophic accidents such as Bhopal result from repetitive oversights and sustained neglect by plant management. The Bhopal tragedy resulted from the grossly negligent safety culture in the ranks of Corporate Management, compounded by the lackadaisical attitude of both the internal and external auditing teams. The company and the government both failed to ensure that such a highly hazardous industry would not be located close to residential areas. A list of that went wrong in Bhopal is as follows:

- In June 1984, the refrigeration unit for cooling the MIC storage tanks was shut down, and MIC storage temperature in subsequent months was reportedly between 15°C and 20°C, as opposed to a required range of 0°C to 5°C.

The temperature alarm was disconnected and not reset properly. Therefore, no automatic warning based on tank temperature was available.

- In October 1984, MIC refining continued operation at a higher-than-normal temperature, contrary to the SOP, resulting in higher-than-normal chloroform content in MIC.
- In the same month, the MIC production unit was shut down, and the scrubber circulation pump was set to a standby position with a manual start.
- The flare stack was rendered inoperative during maintenance work (date unknown).
- On 30th November and 1st December, Tank 610 could not be pressurized. The bursting disk was tested and found to be intact. It is likely that the control valve meant for venting the tank was leaking.
- During the evening shift of 30th December, cleaning four process filter PSV lines by water flushing was commenced. Valve V1 that isolates RVVH from the MIC tanks was closed but not blinded off, leaving open the possibility of water leaking through during flushing. Isolation valves and bleeder valves on the four lines were opened, and the water supply was turned on by opening Valve V2.
- The flushing operation began at 22:30. After noticing that water was not coming out of the bleeder lines, the operator shut off the water supply; however, his supervisor, who was new on the job, ordered him to resume the operation.

The recommendations arising from this accident are as follows:

- Large amounts of highly toxic substances should never be stored in tanks. In other similar facilities, hazardous intermediate products, such as MIC, are consumed immediately after production. At Bhopal, the MIC was used to feed a downstream pesticide-manufacturing plant that was not operational at the time of the accident. It is reprehensible that immediate steps were not taken to de-inventory the contents of the MIC storage tank. (The MIC inventory remained in the tank for a prolonged period).
- In hazardous chemical plants, operations must follow the SOPs strictly. The deplorable practice of ignoring higher-than-design MIC tank pressures that resulted from malfunctioning of the refrigeration system in Bhopal must be condemned.
- Formal permits to work must be issued by operations management, and maintenance work should be undertaken only after confirming that all permit conditions have been met. Use of blinds to ensure positive isolation is required to be specified clearly in work permits for work in hazardous environments. Site-specific process safety procedures must ensure that such vital safeguards are never ignored.
- The MOC procedure needs to be enforced rigorously. In the Bhopal accident, the RVVH and PVH were connected through a "jumper line" to allow gas to be routed to a scrubber if repairs had to be done on one of the vent headers, in gross contravention of safety rules. The risk of reverse flow (possibly during water flushing) was, therefore, created. The MOC procedure

must inexorably be followed to prevent new, unforeseen hazards in chemical plants. Deceptively innocuous changes in the Bhopal plant operations and maintenance practices led to a total disaster, as has also happened in many other cases (see Flixborough Section 2.1).

- Top management must imbue a strong safety culture in plants and ensure that shop-floor operators and supervisors, and contract personnel working in the plant, are thoroughly trained, particularly on safety matters. In Bhopal, the supervisor of the water flushing operation was not well trained and experienced. A fully alert and qualified supervisor would not have ordered the flushing operation to continue without first investigating why water was not coming out through the bleeder lines.
- For facilities where toxic hazards exist, on-site and off-site emergency plans should be prepared before any production commences. All site personnel must be made conversant, through regular drills, with all emergency response procedures. Further, all members of the public who could be affected by such major events must be apprised of the precautions to be taken in such circumstances, no matter how unlikely the accident might be.
- If the affected Bhopal victims had simply been trained to seek higher ground, such as the higher floors of nearby high-rise buildings at heights above the dense MIC cloud, many lives would have been saved. In Bhopal, no such plan was in existence. Large numbers of people lived in slums around the plant that did not know what to do in the event of such a massive toxic gas release. Instead of running cross-wind, many ran along the path of the toxic cloud that soon engulfed them.
- Even worse, local hospitals were utterly unprepared to deal with the disastrous medical emergency that ensued and had no prior knowledge or training concerning the toxicity of MIC. The medical staff was never apprised about the short- and long-term health hazards of MIC, and there was utter confusion about how to treat the enormous influx of panicky patients.
- As in Mexico City, minimum safe distances for construction around hazardous facilities must be enforced. Construction of homes, offices, shops, hotels, etc., must be forbidden close to plants processing or storing highly toxic and hazardous substances. This requirement must be ensured by the proper disclosures to the local authorities by the company well before plant licensing.

2.9 OFFSHORE OIL RIG PIPER ALPHA, NORTH SEA[1,25,26]

Date of accident:	July 6, 1988
Location:	Piper Alpha oil production platform in the North Sea
Nature of accident:	An explosion followed by a major fire
Fatalities and injuries:	165 people killed

2.9.1 Brief Description of Facility and Process

The Piper Alpha oil platform was owned by a consortium of Occidental Petroleum (Caledonia) Ltd., Texaco Britain Ltd., International Thomson plc, and Texas Petroleum Ltd. It was located in the Piper field, some 110 miles northeast of Aberdeen, Scotland. Piper was connected through pipelines to the Flotta terminal at Orkney Isles 128 miles away, and also to three other platforms in the field at the following distances:

Platform	Distance from Piper Alpha (miles)
MCP-01	34
Claymore	22
Tartan	12

The Piper Alpha platform had two main decks: the production deck at 84 ft level and the deck support frame at 68 ft level. Figure 2.7a shows the west elevation, and Figure 2.7b is a simplified process flow diagram.

The production deck consisted of four modules, Modules A, B, C, and D:

Module A: wellheads
Module B: production separators and main oil line pumps (for export)
Module C: gas compression plant
Module D: power generation and utility facilities

The deck support frame at 68 ft level held riser terminations and pig traps for the Tartan and MCP-01 gas pipelines under the Module B condensate injection pumps,

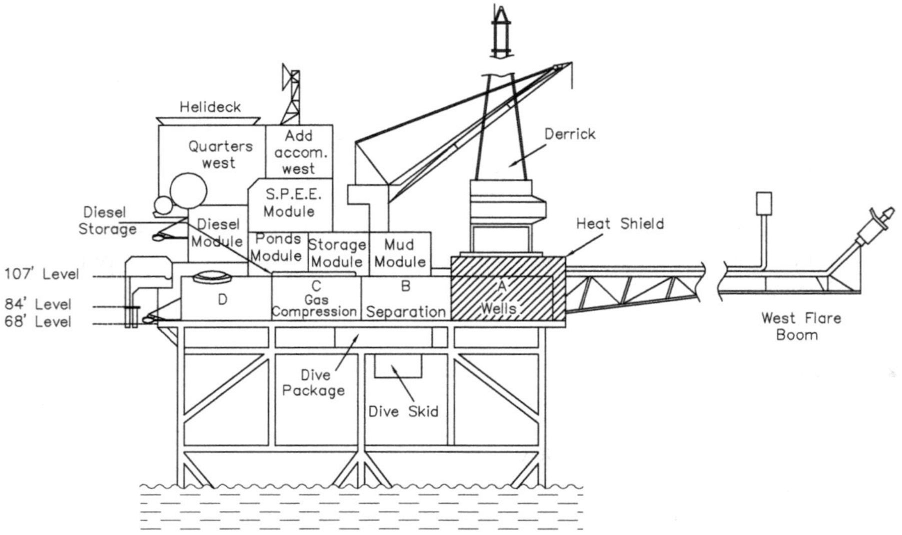

FIGURE 2.7a Piper Alpha – west elevation.

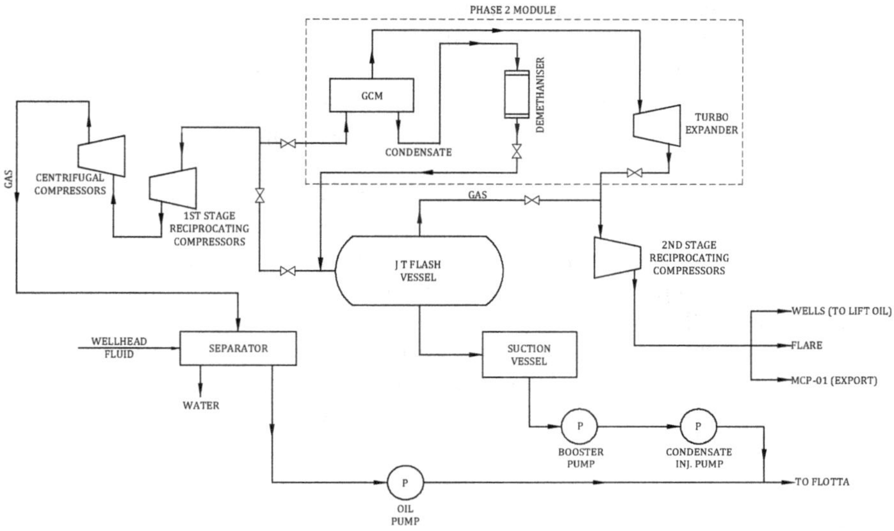

FIGURE 2.7b Simplified flow diagram of the Piper Alpha production process.

and the Joule–Thomson (JT) flash drum under Module C. Under Module A was the flare knockout drum, and under Module D, the Claymore gas riser termination and the pig trap.

Between the 107 and 174 ft levels, above Module D, was the housing area, above which was the helicopter deck. The control room was on a mezzanine level in the upper part of Module D. Also, above Module D was Sub-Module D containing foam storage, foam pumps, offices, and stores.

Originally, the platform was designed and used for oil exploration and production. Later, owing to the UK's high gas demand, the owners decided to add a gas processing module to the platform to produce both oil and gas. Unfortunately, this modification resulted in the gas processing module being located close to the living quarters. Notably, the gas compression unit was located next to the control room in a manner that flouted generally accepted safety criteria for platform layout.

The reservoir fluid from the production wells was passed to two parallel separators operating at a pressure of 155 psia, and separated by gravity into oil, water, and gas streams (see Figure 2.7b).

From the separators, the oil was pumped down the oil export pipeline to the Flotta terminal by two booster pumps followed by the main oil line pumps. Water from the separators was passed to a plate skimmer, followed by a hydrocyclone to remove any oil before being dumped.

The gas from the separators was passed through a condensate knockout drum to three centrifugal compressors, where it was compressed to 675 psia. It was then boosted to 1,465 psia by the first stage of two reciprocating compressors and further processed in one of two operating modes, namely, the Phase 1 and Phase 2 modes.

In Phase 1 mode, the gas from the first-stage reciprocating compressor system was passed through a JT valve to the JT flash drum. The JT expansion knocked out some of the condensate, which was removed from the JT flash drum. Gas from the JT flash drum outlet was passed to the second-stage inlet, where it was compressed to 1,735 psia. The high-pressure gas from the second stage went three ways: to serve as lift gas, or to MCP-01 as export gas, or to flare.

In Phase 2 mode, the gas from the first stage of the reciprocating compressors was passed to the gas conservation module (GCM), where it was dried using molecular sieve beds. It was then let down through a turbo-expander to a pressure of about 635 psia and returned to the Phase 1 plant at the outlet of the JT flash drum. The condensate formed in the GCM was passed to a distillation column, where methane was stripped off, and the stripped condensate was passed to the JT flash drum.

The JT flash drum condensate was passed via a suction vessel to two parallel booster pumps and two reciprocating condensate injection pumps to join the main oil line at 1,100 psia.

2.9.2 The Accident

The accident occurred at the condensate injection pumps that inject condensate into the main oil export line. Figure 2.7c is a simplified flow diagram.

On July 6, Pump B was in operation, and Pump A was down for preventive maintenance, for which a maintenance permit had been issued. In the early hours of the night shift, the running Pump B suddenly tripped and could not be restarted. A stoppage of a few minutes of pumping would have led to the entire platform shutting down. The shift in-charge immediately sought information from maintenance on the status of Pump A and requested its start-up as soon as possible. He was informed that maintenance had not started work on Pump A and he rescinded the work permit immediately. When Pump A was started, a gas alarm sounded almost immediately, followed by an explosion. This event was followed by a series of small and, eventually, massive explosions.

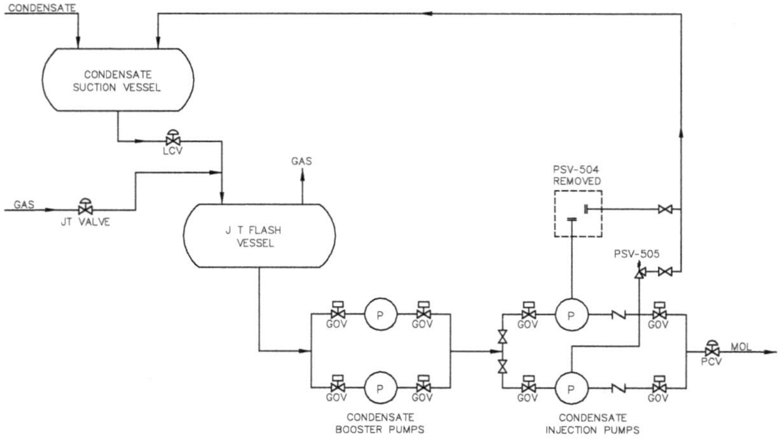

FIGURE 2.7c Simplified flow diagram of condensate injection pump unit at Piper Alpha.

2.9.3 Causes, Circumstances, and Consequences

Before issuing the permit to take Pump A out for routine maintenance, the shift operation in-charge had issued another permit earlier on the same day for fixing the leaky safety valve PSV-504 on the same pump. After receiving the pump maintenance permit, the shift instrument technician removed the PSV-504 in the morning for testing and recertification, closing the opening by fitting blind flanges. However, the accident investigation report established that the joint fitted with the blind flange was neither properly tightened nor pressure-tested for leaks. The location was not at an easily accessible height, and the fact that the pump was to be taken for maintenance shortly after that, in all likelihood, prompted the technician to take this shortcut.

Another lapse was as follows: before removing the PSV, Pump A had been isolated (presumably by mechanical technicians) only by closing the gas-operated valves (GOVs) on the suction and delivery lines. Blinds had not been inserted to ensure positive isolation of Pump A, according to SOP.

The maintenance team was not able to restore the PSV that evening. The permit to work was suspended, and the team went off duty, intending to put the PSV back on the next day.

At about 21:50, Pump B tripped, and the operators failed to restart it. The permit-to-work system exchanged between the operations (process) and maintenance teams failed when Pump A was started up. This operation caused an immediate spillage and spreading of light condensate in the pump area from the improperly tightened PSV-504 connection fitted with the blind flange. At 22:00, there was an explosion (the first explosion), when the initial leak of 45 kg of condensate was ignited in a highly congested area of Module B. That, in turn, resulted in an explosion (deflagration) that destroyed the firewalls between Modules B and C, and also between Modules C and D. These firewalls were not designed to withstand an explosion.

The fragments hit and ruptured several small-bore pipeworks in the adjacent crude oil metering skid and a 4" condensate line passing through the firewalls, spilling a large quantity of crude oil. This leak resulted almost immediately in a massive fireball originating from the west side of Module B, and also a large oil pool fire at the west end of that module. The large oil pool fire gave rise to a massive smoke plume, which enveloped the platform at and above the production deck at 84 ft level. The adjacent living quarters were quickly engulfed by smoke. The pool fire extended to the deck below at 68 ft level, where – after 20 minutes – the gas riser from the pipeline connection between the Piper and Tartan platforms ruptured. This rupture resulted in a massive jet flame, which enveloped much of the platform.

At about 22:50, the MCP-01 gas riser at Piper was ruptured by a second, violent explosion, the shock waves from which were felt a mile away. Debris was strewn 800 m from the platform. This explosion disabled the central communication system, which was at the center of Piper. The other platforms were, therefore, unable to communicate with Piper. They had been made aware that there was a fire at Piper but did not understand its scale and, unfortunately, continued normal operations and pumping for some time, thereby adding fuel to the fire at Piper.

Besides, the firewater deluge system did not work. The auto-start was in manual mode, contravening operating instructions, and therefore did not start immediately after the fire. The explosion that followed knocked out the main power supplies and, perhaps, also damaged the water pumps and the water mains.

Most of the people on the platform were in the living quarters, trapped by smoke. The emergency rescue system failed. Escape routes to the lifeboats were blocked, and the staff was told that rescue helicopters were expected to arrive. However, the helicopter deck was already unusable. Some 62 workers escaped, mainly by climbing down knotted ropes or jumping into the sea; unfortunately, 165 perished.

2.9.4 Lessons/Recommendations

The Piper Alpha accident is the worst offshore accident to date, in terms of human life lost, and has taught us many lessons:

- This accident resulted from poor plant layout and inadequate safety distances between modules. It was designed originally as an oil rig and subsequently modified to accommodate the gas module.
- A disastrous failure of the "permit to work" system occurred (lessons from other accidents, such as Bhopal, had not been heeded!).
- Poor maintenance practices were followed: failure to insert a blind at the GOVs and abandoning flange joints that had not properly been blinded. When combined with a failure to transmit critical information at shift handover – both essential requirements under the permit to work system – these failures led to condensate from a pump spilling out from the removed PSV blind joint, causing the disaster.
- The auto-start of the firewater deluge system was deliberately put in manual mode (reportedly for enabling off-duty swimming by staff), making it utterly ineffective – the manual start-switch could not be reached owing to the raging fire.

In this connection, it would be appropriate to quote the observation made in Paragraph 14.52 of the official inquiry report[25]:

> Senior management were too easily satisfied that the permit to work system was being operated correctly, relying on the absence of any feedback of problems, as indicating that all was well. They failed to provide the training required to ensure that an effective permit to work system was operated in practice. In the face of a known problem with the deluge system, they did not become personally involved in probing the extent of the problem and what should be done to resolve it as soon as possible. They adopted a superficial response when the issue of safety was raised by others. They failed to ensure that emergency training was being provided as they intended. Platform personnel and management were not prepared for a major emergency as they should have been.

From a review of all the case studies in this chapter, it should be evident that the kinds of inadequacies described above are not limited to Piper Alpha.

The primary recommendations arising out of the accident specific to offshore installations are as follows:

1. *Isolation of hydrocarbon inventories*: Considerable effort must be directed towards the effective isolation of pipeline hydrocarbon inventories from fixed installations. Sub-sea isolation valves must be provided, and emergency shutdown systems (ESDs) must be protected in such a way that they would be able to survive major fires, explosions, or other serious accidents. Note that ESDs must have their own instrumentation and shutdown logic processors, typically PLCs. All shutdown instrumentation and the PLCs must run independently of normal process control instrumentation and the digital control system (DCS) used for routine operations. The overall reliability and survivability of these systems have now become critical design issues.
2. *Limitation and disposal of HC inventories*: Minimize hydrocarbon inventories and provide high-pressure water jet equipment for rapid depressurization systems on board. Dump systems for the rapid de-inventorying of hydrocarbon liquids must also be considered.
3. *Mitigation of explosion effects*: Offshore platforms must have reliable remedial measures, such as strong blast walls and additional relief panels. Analyses of explosions in the modules have revealed that, in many situations, significant overpressures can be generated. Considerable care must be taken concerning the equipment layout to reduce potential flame acceleration and generation of overpressures.
4. As detailed in earlier case histories, most onshore installation recommendations are entirely applicable for offshore facilities.

2.10 BHARAT PETROLEUM REFINERY, BOMBAY, INDIA[27]

Date of accident:	November 9, 1988
Location:	Oil refinery of Bharat Petroleum Corporation Limited (BPCL) in Bombay, India
Nature of accident:	Fire in the aromatics tank farm
Fatalities and injuries:	35 people killed, and 23 injured

2.10.1 Description of Facility and Process

Bharat Petroleum Corporation Ltd (BPCL) is a public sector oil refining company in India. Originally, it was built and operated by Burmah-Shell Refineries in the Mahul suburb of Mumbai (Bombay). It was nationalized by the Government of India in 1976, renamed, and debottlenecked to a much larger refining capacity.

The Mahul refinery started at 2.2 million tons/year (50,000 BPD) refinery in 1955. Over the years, it has been expanded dramatically to 12 million tons/year. It is a large, integrated coastal refinery with many storage tanks for crude oil and intermediate and final petroleum products. A broad sketch of the tank farm area is shown in Figure 2.8 (courtesy of BPCL). There were 20 atmospheric tanks (both fixed and floating roof) in the tank farm. These were reportedly designed as per API 610 and contained light naphtha feedstock, benzene, and toluene, with an aggregate capacity of 20,000 tons. There were pipelines around the tank farm and several pumping stations on the western periphery to supply products to storage tanks.

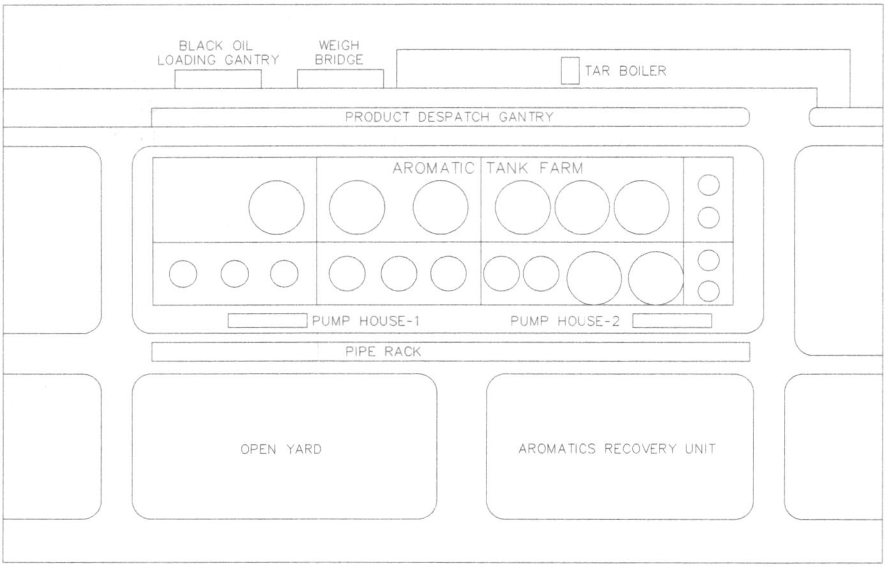

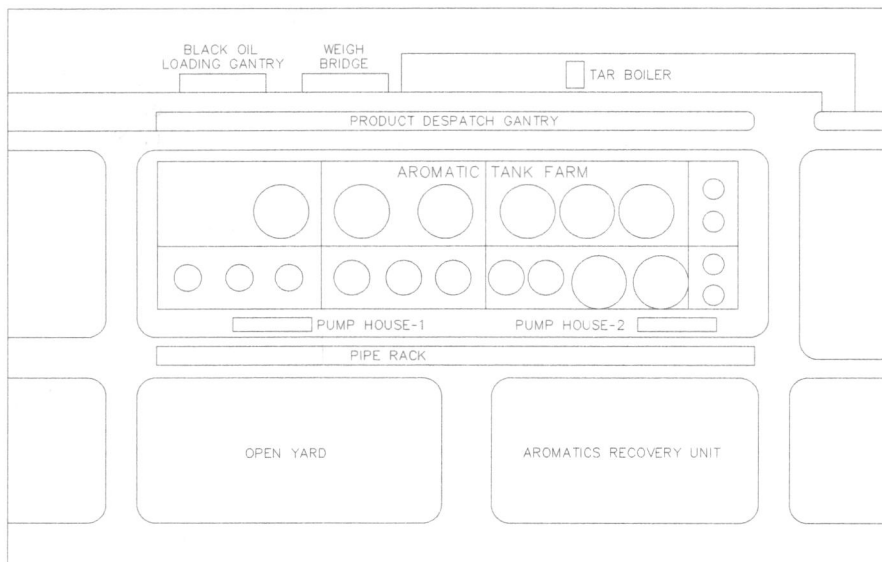

FIGURE 2.8 Plan of the aromatics tank farm at BPCL (courtesy of BPCL).

2.10.2 The Accident

The refinery had a major shutdown in October–November 1988 when several equipment modifications and tie-ins of new equipment were completed. On 9th November, the aromatics recovery unit had just come on stream.

The accident started at 11.55 a.m. with Tank No. 354, a domed-roof tank of 3.5 m diameter and 8 m height. The tank contained light naphtha under a nitrogen blanket at 0.25 kg/cm^{2}g or 3.6 psig. Immediately before the accident, the tank was being filled with light naphtha. The tank was overfilled when a recently repaired level indicator malfunctioned. The resulting overpressure ruptured the roof/shell joint at two segments with a bang, and the escaping liquid formed a large vapor cloud that spread rapidly over an extended area.

The cloud ignited, causing a VCE and a flash fire, followed by a large-scale fire that continued over a wide area. The fire covered Tank 354, the bitumen stacking area near the weighbridges, the product dispatch facility (PDF) gantry, and the storm-water channel near Pump-House 11. In a span of 8–10 hours, the fire spread to four other tanks (360, 362, 363, and 357). The fire lasted for 48 hours.

2.10.3 Causes, Circumstances, and Consequences

According to an investigation report, the ignition's likely cause was a static electricity discharge resulting from the free fall of flammable liquid from the pressure safety valve outlet on Tank 354. However, experienced experts familiar with the incident attributed the cause to be vapors reaching an open-flame tar boiler operating over half a kilometer away. The tar boiler was found wholly incinerated. The VCE was preceded momentarily by a flash fire. Immediately after that, a large-scale fire followed that engulfed the plant areas mentioned above.

The damage caused by the fire has been summarized as follows:

- Fifty-eight persons received burn injuries, of which 35 were fatal. The victims were mostly contractors engaged in product dispatch operations. Several workers in the bitumen loading area were scorched, and the charred body of the driver of a car was found on the road between the PDF gantry and the aromatics tank farm.
- Five storage tanks were damaged, and 2,000 tons of stored products was lost.
- The aromatics pump house and the associated piping connecting the tank farm and the pump house were destroyed. About 30% of the canopy in the PDF gantry center was blown up, apparently by the VCE. The tar boiler was burned down, and several lorries parked in the area for loading bitumen drums were damaged.
- The total monitory loss was estimated to be Rs.3.2 crores ($25 million).

2.10.4 Lessons/Recommendations

The lessons to be learned from this accident are as follows:

- The tank was designed as per API 620 for low pressure (0.5–15 psig). In such tanks, best practices require that two independent-level gauging systems should be provided to minimize the risk of overfilling.
- In this case, it appears that there was only one locally mounted level indication instrument, and monitoring of this level was solely dependent on the vigilance of the operator. A critical safety control system was not installed as there was no high-level trip for the pump (risk criteria are discussed in Chapter 13).
- Extensive and periodic operator training is an essential requirement for safe operations. An alert operator would have been aware of the possibility of flawed tank-level readings and would not have continued pumping for long without investigating.
- Finally, a critical-level control instrument should be checked independently following repairs; there is no indication that this was done. Besides, redundant instrumentation (with one-out-of-two voting logic) for such critical measurements must always be provided.

The primary recommendations arising out of the accident are as follows:

- In all storage tanks for hazardous chemicals, at least two independent digital-level instruments providing readings at the control room must be installed. Also, at least one of these should provide a local indication.
- When storing volatile flammable and highly toxic chemicals, a high-level trip for the feed pumping system must be provided.

- The level indicators and level safety trips must be checked regularly. After periodic maintenance, they must be re-checked independently before restart.
- Process and maintenance operators must be trained rigorously for the installed instrumentation and process control systems.

2.11 PETROCHEMICAL COMPLEX, PHILLIPS PETROLEUM, PASADENA, USA[1,11,28]

Date of accident:	October 23, 1989
Location:	Phillips 66 Company's chemical complex at Pasadena, Texas (USA), near Houston
Nature of accident:	Leakage of ethylene from polyethylene plant resulting in a massive vapor cloud explosion
Fatalities and injuries:	23 killed, more than 314 injured, extensive property damage

2.11.1 Brief Description of Facility and Processes

Phillips 66 Company's Houston Chemical Complex (HCC) facility was located at the Houston Ship Channel in Pasadena, Texas, USA. It is a major petrochemical complex and has two high-density polyethylene (HDPE) plants with a combined capacity of 680,000 tons annually. The HDPE plants use Phillips' proprietary catalyst and "loop reactor" process. The accident occurred in one of the two polyethylene plants at the site. The process polymerizes ethylene, dissolved in isobutene, at 700 psig. Particles of polyethylene polymer settle out and are removed via settling legs. The configuration of a typical leg is shown in Figure 2.9. Each leg is connected to the reactor loop through an air-operated ball valve, which normally is kept open so that polythene product particles can settle into the leg.

In addition to its regular employees at the site, the company employed approximately 600 daily contract workers for routine maintenance and construction of new facilities.

2.11.2 The Accident

While cleaning a 10-inch-diameter "settling leg" of the high-pressure (700 psig[28]) reactor loop, a supposedly-closed valve vented a highly flammable mixture of ethylene, isobutane, hydrogen, and hexane. A huge vapor cloud formed and ignited 60–90 seconds later, resulting in a massive VCE.

2.11.3 Causes, Circumstances, and Consequences

The primary cause of the accident was the release of hydrocarbons through a large pneumatically operated isolation valve. This valve had erroneously been assumed to be closed by the maintenance team. Unfortunately, it was open because of incorrectly connected air-supply lines to it.

Contractors were doing maintenance work to clear a blockage in one of the settling legs when the accident happened. The usual isolation procedure was to close the

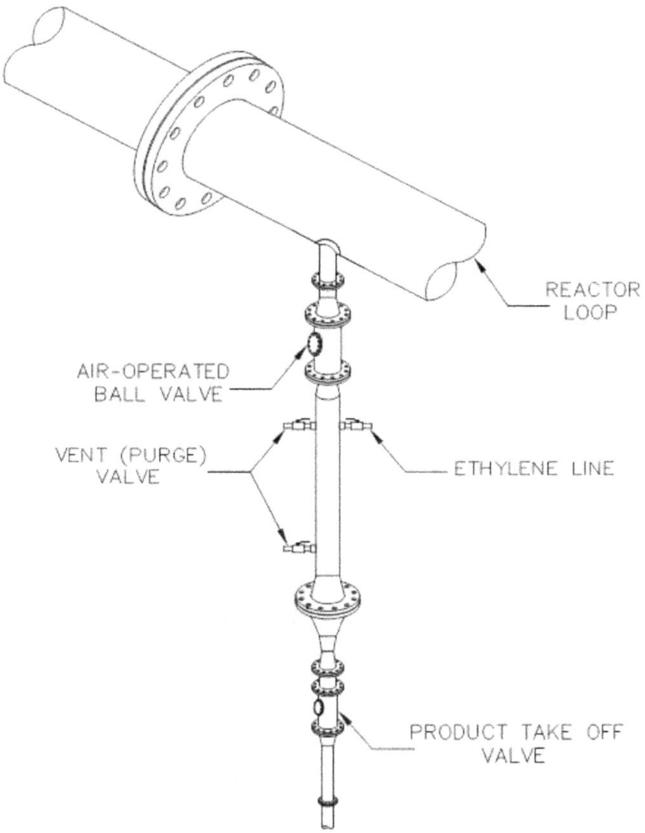

FIGURE 2.9 Arrangement of settling leg at Phillips loop reactor.

pneumatically operated ball valve (Figure 2.9) and disconnect the pneumatic lines from the valve to avoid accidental opening during maintenance. No other method of isolation, such as a slip plate or blind flange, was used.

The maintenance team partially disassembled the leg and removed part of the blockage when the release occurred. The mass of the gas mixture released was estimated to be about 39 tons of ethylene, isobutane, hydrogen, and hexene and resulted in a massive VCE.

It was subsequently established that the pneumatically operated ball valve was, in fact, open at the time of the release. There were two air hoses connected to the valve: one to open the valve and the other to close it. In this instance, the air-supply hoses were incorrectly connected, causing the valve to be open instead of closed.

The firefighting water system at the plant was part of the process water system. When the first explosion occurred, some fire hydrants were sheared off at the ground level by the blast. The result was inadequate water pressure for firefighting. The shut-off valves, which could have been used to prevent water loss from ruptured lines in the plant, were out of reach in the burning wreckage. No remotely operated fail-safe isolation valves existed in the combined plant/firefighting water system. Also, the regular-service fire-water pumps were disabled by the fire that had destroyed their electrical power cables. Of the three backup diesel-operated fire pumps, one had been taken out of service, and one ran out of fuel in about an hour. Hoses were used to bring in firefighting water from remote sources: settling ponds, a cooling tower, a water main at a neighboring plant, and even the nearby Houston Ship Channel.

The explosion and fire covered a 600 ft×800 ft area, resulting in 23 deaths (all workers) and injuring 314 persons (185 Phillips 66 Company's employees and 129 contract employees).

The explosion also both destroyed polyethylene plants and affected other plants as well at the site, causing $715.5 million worth of damage, plus additional business disruption losses estimated at $700 million.

2.11.4 Lessons/Recommendations

Major lessons to be learned from this accident are as follows:

- It appears that no adequate PHA, followed by HAZOP, was conducted. The possibility of failures caused by incorrect connection of pneumatic hoses for air-operated valves was not contemplated and hence was not addressed.
- The permit-to-work system and its enforcement were faulty. There appears to have been no adequate method of valve isolation and checking before the permit was issued.
- Plant operators lacked proper training concerning these potential hazards; however, they were entrusted with contract labor supervision.
- Careful monitoring by senior plant personnel was lacking when contractors were performing critical maintenance jobs.

The recommendations arising out of the accident are as follows:

- PHA and HAZOP are mandatory for process plants and must be done under experts with plant management's full participation.
- Work permits must be comprehensive and prepared after careful study of PHA and HAZOP reports.
- Critical issues such as proper air-supply connections to pneumatically operated valves should be duly addressed by providing different sizing/threading and color of air inlet/outlet hoses. Such connections must be tested prior to unit start-up.
- Operator training must include familiarization with visible and latent hazards.
- Adequate oversight by senior operations management staff must be ensured in all major maintenance activities.

These are generally applicable lessons for ensuring safe plant maintenance at all process plants. Unfortunately, well-established process safety procedures are forgotten all too often, and extremely unsafe shortcuts are taken. As a result, such accidents continue to occur at a depressingly high frequency.

2.12 LPG IMPORT TERMINAL HINDUSTAN PETROLEUM, VISHAKHAPATNAM, INDIA[29]

Date of accident:	September 14, 1997
Location:	LPG terminal of HPCL Refinery at Vishakhapatnam, India
Nature of accident:	Leakage of LPG from the manifold area below a sphere containing LPG, followed by a massive vapor cloud explosion
Fatalities and injuries:	57 people killed, all on-site

2.12.1 Brief Description of the Facility and the Process

Hindustan Petroleum Corporation Ltd (HPCL) is a Government-owned Oil Refining Company that came into being in 1977 when the Government of India nationalized two private coastal refining companies set up in India. The first was by ESSO in the Mahul suburb of Mumbai (Bombay), and the second was by Caltex in Vishakhapatnam (Vizag) on the east coast of India. Both entities were made part of HPCL. HPCL now has four large refineries in India, including a lubricating oil refinery. The Visakhapatnam east coast refinery, where the accident took place, now processes 8.3 million tons (186,000 BPD) of crude oil a year.

There was a large pressurized LPG terminal in the refinery that also served as an LPG import terminal. The storage facility consisted of eight spheres of 1,200-tonne capacity each and other spheres of smaller size. The smaller size spheres were used for decanting LPG. Adjoining the LPG terminal, there was an LPG bottling plant, also owned by HPCL.

The terminal received LPG from the refinery and also via imports. Imported LPG was unloaded into storage tanks from ships berthed at Vishakhapatnam Port through an 8-inch-diameter, 4.2 km long pipeline. After every unloading operation, the pipeline's LPG inventory was transferred to a storage tank by pumping water through the pipeline (the water was retained in the pipeline).

LPG was unloaded only on weekends because of an agreement with the Indian Navy when the nearby Naval Workshop was closed. The LPG unloading pipeline was thus kept filled with water when idle; water was displaced by LPG only during unloading. During unloading, LPG initially pushed the water to a spherical tank for water decanting. This tank received all 133 tons of water from the 4.1 km unloading line. Following this, the arriving LPG was routed to a designated LPG spherical tank. Next, the collected water was drained through the bottom drain line at the decanting sphere via a 2-inch valve. After the entire shipload of LPG was unloaded, and part of it was pumped to the adjoining bottling plant or was loaded into road tankers or tank wagons for distribution to other bottling plants.

2.12.2 The Accident

The accident happened on September 14, 1997 (Sunday), at about 06:30 am. During the unloading from the ship, LPG leaked in the manifold area under a small decanting sphere tank for a significant period. The resulting vapor cloud traveled a considerable distance and finally was ignited, resulting in a massive VCE. As a result of this explosion, several LPG sphere relief valve vents caught fire, and several product tank roofs were blown off, giving rise to open roof fires over those tanks.

2.12.3 Causes, Circumstances, and Consequences

HPCL and various governmental agencies set up several investigation committees. Inexplicably, none of their reports was released to the public domain. As a result, several possible scenarios about the primary cause were propounded by independent safety experts, in India and abroad, and others who visited the site soon after the accident.

The likely causes of the accident are as follows:

In the immediately preceding LPG unloading operation, 100% displacement of LPG in the unloading pipeline with water was not accomplished because a sufficient amount of water was not available. As a result, a substantial LPG inventory remained in the front portion of the pipeline (i.e., the refinery side of the unloading line). This deviation from the SOP was not communicated to the next shift's operating staff.

On September 14, the plant operator/supervisor asked the LPG ship to unload, assuming that (as usual) the ship would be pumping LPG only after first pushing the water in the unloading line to the decanting sphere. However, in reality, it was pumping LPG and hardly any water to the decanting sphere. Thus, after the normal time for water decanting had elapsed, the operator diverted the LPG delivery line to the LPG spheres and opened the 2-inch water drain valve at the decanting spheres. Very soon, LPG gushed out with full force. The operator tried to close the valve but failed as it was frozen cold by the flashing LPG. Unfortunately, the operator reportedly suffered severe cold burns and, eventually, died. The leak continued for a considerable time, forming a vast vapor cloud that spread over a large area and, ultimately, ignited.

A second theory also supports LPG filling of the decanting sphere in the same manner but contends that violations of SOPs caused the leak. The operator deviated from standard practice: (1) open the decanting sphere's drain valve, (2) retreat for some time instead of remaining present throughout the decanting process, and (3) return after the decanting was complete to close the valve.

A third theory blames the maloperation of a Hammer-Blind Valve provided on the 8" line for positive isolation at the LPG terminal end.

The massive vapor cloud reportedly found an ignition source when it reached the canteen, resulting in a huge VCE. The blast damaged the upper mountings of many spheres and devastated the LPG terminal when large jet flames from relief valves burned uncontrollably for many hours. The blast also damaged several petroleum storage tanks. Numerous buildings, including the fire station, administrative building, canteen, the control room, and many other structures, were destroyed.

The accident reportedly killed 57 people and injured 70. Fatalities would likely have been much higher had the accident occurred during normal working hours instead of Sunday morning.

2.12.4 Lessons/Recommendations

Many important lessons were learned from this incident:

- The LPG ship unloading procedure should have been preceded by a comprehensive PHA, HAZID, and HAZOP. These would very likely have identified the hazards and need for critical safeguards.
- Operator training was inadequate, and SOPs were not comprehensive and deeply flawed.
- Ship unloading is a hazardous activity that requires constant monitoring and supervision.
- Mock drills did not contemplate a massive leak of the type that occurred.
- The fire station should have been located away from the operating area to prevent damage during the fire and explosion in the plant area.

The primary recommendations arising out of the accident are as follows:

- PHA, HAZID, and HAZOP must be done, and all recommendations must be incorporated in the operating instructions for all LPG storage and similar hazardous installations.
- All LPG port terminals must have "gas detectors" in unloading jetty, storage, and operating areas because imported LPG cargos are supplied free of odorizing mercaptan.
- Automatic shutoff controls must be provided at the ship unloading line both at the Jetty and at the terminal ends.
- The LPG unloading line should be designed (1) for a high surge pressure, or (2) for a predetermined closing time of the auto shutoff valve to avoid high surge pressures (similar to "water hammer").
- During any manual draining operation, operators must remain vigilant.
- For ensuring reliable intershift communications, logbooks must be completed diligently by operators during every shift. The quality of the logbook recordings should be checked by management regularly.
- Hammer-Blind Valves must not be used in pipelines intended for carrying LPG or any other hazardous material.
- Following this accident, HPCL scrapped the pressurized LPG import terminal at Vishakhapatnam. In a subsequent joint venture with Total of France, HPCL constructed underground cavern storage for LPG imports (a much safer solution).

2.13 GRANDE PAROISSE, AMMONIUM NITRATE FACILITY TOULOUSE, FRANCE[30,31]

Date of accident:	September 21, 2001
Location:	AZF fertilizer factory in Toulouse, France, belonging to the Grande Paroisse branch of the total group
Nature of accident:	The explosion of about 200–300 tons of ammonium nitrate in a warehouse
Fatalities and injuries:	31 people killed and about 2,500 injured

2.13.1 Brief Description of Facility and Process

The AZF fertilizer factory was located in an industrial zone, south of Toulouse, about 3 km from the center of the town. Created in 1924 as ONIA, it belonged to the Grande Paroisse group since 1991. The company was the leading French producer of fertilizers and ranked third in Europe. The main processes in Toulouse were nitrogenous fertilizer and industrial nitrates, and chlorine compound synthesis. The plant manufactured ammonia and transformed it into ammonium nitrate (AN), a part of which was then used to manufacture fertilizer and industrial nitrates. They also manufactured melamine (a raw material used in the manufacture of resins and adhesives) and chlorinated products used for water treatment applications.

The site included several large hazardous material storage facilities:

1. Two cryogenic ammonia tanks (5,000 and 1,000 tons)
2. A 315-ton pressurized ammonia storage tank
3. Two 56-ton liquid chlorine tankers
4. 1,500 tons of oxidants
5. 15,000 tons of solid bulk AN
6. 15,000 tons of AN in sacks
7. 1,200 tons of AN in hot liquor solution
8. 2,500 tons of methanol.

The site was governed by the EU's SEVESO 2 directive owing to the presence of ammonia, chlorine, toxic or combustible substances, AN, and nitrate-based fertilizers within the scope of French legislation regarding classified facilities.

The AZF plant was authorized under the terms of legislation governing classified facilities. Finally, several hazard studies have been conducted since 1982 and were updated every 5 years. Some of them were completed in 2000, and the latest one in 2001 before the explosion. In these studies, dozens of accidental scenarios were examined, although the detonation of AN was disregarded based on available feedback. The contingency plan thus did not foresee the scenario that transpired.

2.13.2 The Accident

On September 21, 2001, at 10:17 a.m., a severe explosion (detonation) occurred in Shed 221. The blast, felt several kilometers away, corresponded to a magnitude of 3.4 on the Richter scale. There was significant dust fallout, and a crater was observed outside the plant. A large part of the AZF plant's 70-hectare area was devastated, and debris littered the site. 31 people were killed.

2.13.3 Causes, Circumstances, and Consequences of the Accident

On the eve of the explosion, 15–20 tons of additive-containing AN was brought into shed 221. That morning, ancillary materials (from the AN packaging and the manufacturing shops) were also transferred into the shed. A bin from another storage zone was brought in less than 30 minutes before the explosion. Several inquiries and expert evaluations were conducted: (1) a judicial inquiry, (2) an administrative inquiry conducted by the French Ministry of the Environment with the participation of industrial hazard experts, (3) an ATOFINA internal inquiry, and (4) an inquiry by the CHSCT (the plant committee for hygiene, safety, and working conditions).

In May 2006, the final report was presented by the expert investigators. They attributed the disaster to an unfortunate admixing of a few dozen kilograms of sodium dichloroisocyanurate with 500 kg of spilled AN, 20 minutes before the detonation.

It is worth recalling that, since the very beginning of the industrial manufacturing of nitrogenous fertilizers early in the 20th century, AN has been involved in several major industrial accidents while being stored or dispatched. Some of the important ones are the 1921 explosion at Oppau, Germany, that killed more than 500 people, and the 1947 explosion in Texas City, Texas, that killed some 600 people.

Despite the production of coated nitrate particles to avoid nitrate agglomeration, several AN detonations have still occurred globally. These events confirm (1) the complex characteristics (e.g., chemical composition, particle distribution, density, humidity) associated with these categories of products, and (2) their detonation potential under circumstances that promote instability: mixing, hazardous reactions with other materials or pollutants, temperature, and containment.

The strength of the explosion was estimated to be equivalent to 30–40 tons of TNT. The experts made various measurements and observations. They estimated that the detonation gave rise to an overpressure between 140 mbar (the threshold for lethal impacts at distances between 280 and 350 m) and 50 mbar (the threshold for irreversible effects on human health at distances between 680 and 860 m).

The accident resulted in many casualties: 21 deaths at the AZF site, one at SNPE, and nine outside the site (two of whom were in the hospital) who were killed immediately by the explosion or succumbed in the days that followed. More than 30 people were seriously injured. Casualties included a student killed at a college located 500 m from the epicenter and several others who were injured when a concrete structure collapsed. Two people also died in a vehicle maintenance establishment located 380 m away, and one person died in the EDF (electric power company) building located 450 m from the epicenter.

A large cloud of dust from the detonation and red smoke drifted to the northwest. The appearance of the smoke is linked to the emergency shutdown of the nitric acid-manufacturing installation. Before dissipating rapidly, the cloud that contained ammonia and nitrogen oxides sickened witnesses who complained of eye and throat irritations.

Thousands of people were hospitalized: 8,042 people were examined as part of a legal, medical investigation. The explosion also caused significant damage to chemical companies located on the chemical platform outside the AZF plant, on the other side of the Garonne; these are also governed by the SEVESO 2 directive: SNPE and Isochem (a subsidiary of SNPE). Two plants located on the SNPE plant grounds were subjected to colossal damage (Raiso and Air Liquide). The Tolochimie plant (part of the SNPE group), also governed by the Seveso-2 directive and located to the south of the AZF plant, was only slightly damaged.

2.13.4 Lessons/Recommendations

Hazard studies were conducted at the facility every 5 years; the last was done in the year of the explosion. These studies ignored the possibility of an AN explosion hazard completely, a most glaring omission: catastrophic explosions in facilities storing AN were known to have occurred on several past occasions.

AN, a primary ingredient in mixed fertilizers, is potentially a high explosive, especially when mixed with organic ingredients. AN, mixed with about 5% fuel oil, is an important commercial blasting agent in many mining operations. AN solutions in water containing emulsified oil are used in commercial slurry explosives formulations. Accordingly, AN needs to be handled with the utmost care. Heating solid AN contaminated with any organic/combustible material must, therefore, be avoided.

Recommendations concerning storage and handling of AN and site location are as follows:

1. Hazard studies (HAZOP, HAZID, HAZAN) must be comprehensive and not omit any hazard that might expose the plant and all off-site facilities to risks when AN is present in significant quantities.
2. Case histories of similar accidents in other facilities must be studied thoroughly, and safety systems and procedures must be adopted to minimize hazards in the storage or handling of AN.
3. Decisions on the location of AN plants and bulk storage facilities must be made most carefully; they must involve company management, local authorities, and safety experts. Such facilities must always be located at safe distances away from population centers.
4. The utmost care must be taken to avoid the contamination of AN with unknown impurities.
5. Installation of blast-proof walls should seriously be considered around AN storage.
6. Disaster management plans must be drawn up that consider all possible explosion scenarios, even if the probability of such explosions is considered to be extremely low.

2.14 SPACE SHUTTLE COLUMBIA, NASA FLORIDA[32,33]

Date of accident:	February 1, 2003
Location:	Re-entry of the shuttle to earth atmosphere
Nature of accident:	Explosion and disintegration
Fatalities and injuries:	All seven astronauts killed, and the shuttle destroyed

2.14.1 Brief Description of Space Program and the Shuttle

The National Aeronautical and Space Administration (NASA)'s Space Shuttle Program started with its first launch on April 12, 1981, and consisted of 135 missions; the last one was on July 21, 2011. NASA's space shuttle fleet consisted of five shuttles:

Atlantis, Columbia, Challenger, Discovery, and Endeavor. The shuttle program helped the construction of the International Space Station and inspired generations. However, two of the shuttles, Challenger and Columbia, perished in catastrophic accidents in 1986 and 2003, respectively.

Though not related directly to the chemical process industries, the events of these critical missions have provided essential lessons for all industries engaged in hazardous activities.

2.14.2 The Accident

The space shuttle Columbia's last unfortunate mission, STS-107, was launched on January 16, 2003, on a 16-day voyage from NASA's facilities at the Kennedy Space Center (KSC) in Florida. The accident took place during re-entry on February 1, 2003. Just 16 minutes before the scheduled touchdown, Columbia was destroyed.

2.14.3 Causes, Circumstances, and Consequences

Columbia had been launched on this mission for the 28th time. After approximately 82 seconds into the launch, a large piece of insulating foam (approximately 20 inches long and 2 to 6 inches wide) broke off from the external fuel tank. This foam impacted Columbia's left-wing underside at a speed of about 500 mph relative to the Columbia vehicle.

The foam impact was discovered on the second day of the mission during a review of launch videos. Considerable discussion and analysis occurred during the rest of the mission, focusing on whether any of the delicate tiles that make up the shuttle's thermal protection system (TPS) could have been damaged. NASA's management discounted the significance of the impact, and no inspection for damage was made. Unfortunately, no contingency plans for dealing with TPS damage were formulated, even though it might have been possible to launch a rescue mission up to the 5th day after Columbia's launch.

The foam's impact had created a hole in the heat-resistant reinforced carbon-carbon (RCC) panels on the wing's leading edge. During re-entry, superheated air entered through this breach into the wing behind the RCC panels' cavity. The breach widened rapidly, destroying the insulation protecting the support structure for the wing's leading edge. Subsequent melting of the thin aluminum spars within the wing resulted in its catastrophic failure. Columbia, tumbling out of control at speed over 10,000 mph, was torn apart. All astronauts on board died.

2.14.4 Lessons/Recommendations

Lessons from this tragic incident for process safety management are as follows:

1. Over time, flouting of instructions regarding safety procedures became routine, ultimately leading to this catastrophic consequence. RCC foam separation, from very small to large sizes as in the case of Columbia, was commonly accepted at NASA.

2. Even in a leading space research organization such as NASA, crucial operating safety instructions for the space shuttle were flouted repeatedly and with impunity.
3. Overconfidence is known to cloud the judgment of even the highest management levels and compromises rational decision making. The previous Challenger disaster had occurred 17 years earlier and was caused by failed O-rings that were frozen in the rocket engine during a freak winter event. Before that launch, the designers had warned that the frozen O-rings could very well result in outright failure, but their concerns were ignored. In both disasters, the recommendations of technical expert teams were overruled inexplicably by NASA's Management. Deferring the Challenger launch (to fix the O-ring problem) and sending a rescue mission after discovering the damage caused by the dislodged foam from Columbia were both entirely feasible and could have averted the mishaps.
4. Not learning lessons from experience is a serious obstacle to progress in process safety. After the Challenger disaster, one NASA Senior Manager stated that "*the space shuttle would always have to be treated as an R&D project and considering it 'operational' was inappropriate and sent the wrong signals to the public*".

The primary recommendations from the incident are as follows:

- Irrespective of management seniority, we must learn from past accidents.
- SOPs must be followed strictly, and the considered advice from domain experts must never be discounted or dismissed.
- The safety of men, machines, and property must always be prioritized over schedules and budgets.
- The possibility of accidents, however remote, must never be discounted. Overconfidence is insidious but extremely dangerous in mission-critical projects.

2.15 LNG LIQUEFACTION FACILITY, SKIKDA, ALGERIA[34]

Date of accident:	January 19, 2004
Location:	Skikda LNG facility at Skikda, Algeria
Nature of accident:	Explosion and fire
Fatalities and injuries:	27 workers killed and at least 70 injured

2.15.1 Brief Description of Facility and the Process

The Skikda LNG facility (known as GL1K) was located approximately 500 km east of the capital, Algiers. The facility contained six LNG trains, built in phases from 1971 to 1981, that processed natural gas from the Hassi R'Mel fields located approximately 500 km south. The gas composition was methane (83%), ethane (7.1%),

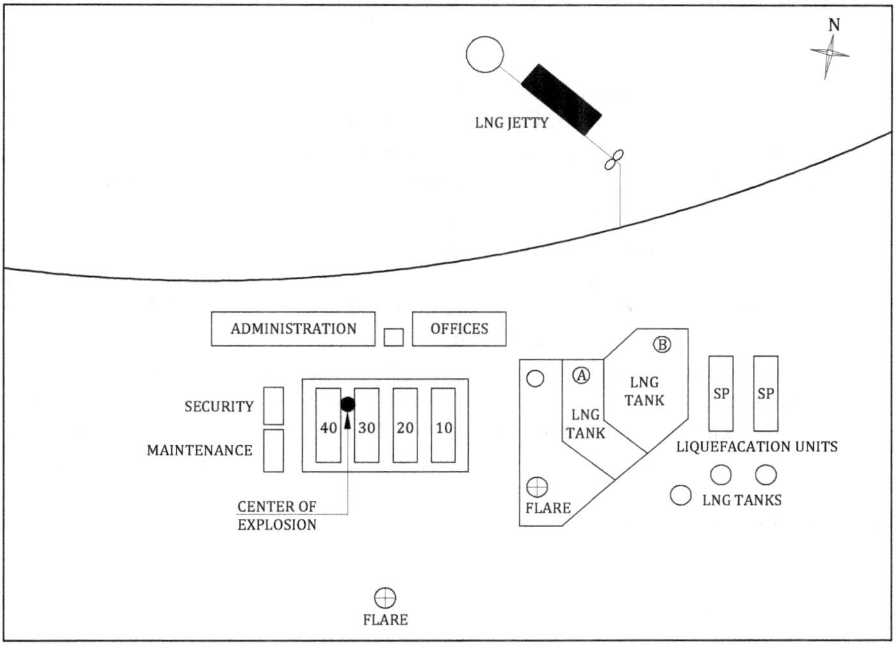

FIGURE 2.10 Broad layout of LNG complex at Skikda before explosion.

propane (2.2%), butane (1.0%), and nitrogen (5.8%). The plant separated ethane, propane, butane, and LNG. The ethane product was sent to a petrochemical plant located on the eastern side of GL1K. Propane and butane were used in the domestic market or exported, and LNG was exported.

Out of the six parallel trains (see Figure 2.10), Trains 40, 30, 20, and 10 were about 30 m apart. Trains 5 and 6 were located remotely to the east. This older facility, unlike modern-day LNG liquefaction facilities, had refrigeration compressors powered by steam turbines. Each train had an independent liquefaction plant and a boiler to generate steam to drive the refrigeration cycle compressor.

2.15.2 The Accident

Just before the accident, Train 10 had been shut down for major maintenance. Train 6 was shut down for minor maintenance. Train 40, where the accident happened, had been operating steadily for 6 days, following routine maintenance.

At about 18:40 on January 19, 2004, the day of the accident, a hydrocarbon leak occurred in the semi-confined area between Liquefaction Train 40's control room, boiler, and the liquefaction area. A small explosion followed this leak. Almost immediately, a massive explosion followed that killed many plant personnel and seriously damaged a part of the plant.

2.15.3 Causes, Circumstances, and Consequences

A series of cascading events appear to have caused a massive explosion and fire. Investigation reports concluded that a leak of heavier hydrocarbons occurred in the semi-confined area between Train 40's control room, boiler, and the liquefaction area. The boiler's air intake fan ingested a hydrocarbon/air mixture (instead of air) that was within flammable limits. This mixture caused a small explosion in the firebox enclosure. The explosion breached the boiler and provided an ignition source for the external accumulation of combustible gas, leading to a second, massive explosion.

The accident killed 27 people, and more than 70 were injured. Victims were located mainly in the maintenance and fire control buildings, Train 40 control room, and Train 30. Most of the deaths and injuries resulted from the explosion and flying debris impacts, rather than from the fire itself. The proximity of the Train 40 control room to the administrative, maintenance, and security/fire control buildings contributed to many injuries and fatalities. Trains 40, 30, and 20 were virtually destroyed, although the effect diminished with distance. Damage to Train 20 was not as severe as in Train 40, and damage to Train 10 was minimal. Damage outside the plant areas was limited to broken windows.

2.15.4 Lessons/Recommendations

Several primary lessons from this accident are as follows:

- The proximity of the Train 40 and Train 30 to control rooms, the security/fire building, and administrative and maintenance buildings (all of which housed many people) was a significant contributor to the number of injuries and fatalities. The layout of the facility did not consider all relevant hazards and consequences. Since this was an old plant, adequate hazard identification, HAZOP, Consequence Analysis, and Risk Assessments may not have been carried out.
- Boilers and fired equipment are hazard sources, especially in a facility handling light hydrocarbons. This location was close to the liquefaction unit (refrigeration compressor), increasing the hazard considerably.
- The gas detection system appeared to be either nonexistent or dysfunctional.
- The facility was not modernized to follow current safety practices, as was the case for other similar facilities at the time.

The principal recommendations arising out of the Skikda accident are as follows:

- Hazardous facility layouts should consider measures to reduce the risk (both on-site and off-site) to acceptable levels.
- The control room should be far away from any plant area and located near the exits to ensure that the primary operating staff can take timely actions according to a disaster control plan. For existing older plants, where this may be difficult to achieve, the control room building should be made fire and blast-proof.

- Hazard control steps (hazard identification, HAZOP, consequence analysis, LOPA, quantitative risk analysis, etc.) should be initiated from the design stage and be continued until the facility begins operation. Safety audits should then be carried out routinely to ensure reliable implementation of the Process Safety Management system.
- Facilities with potential hazards that could result in severe consequences must adopt continuous safety improvement programs in process technology, process controls, best operating practices, and best equipment maintenance.

2.16 BP REFINERY, TEXAS CITY, TEXAS, USA[35,36]

Date of accident:	March 23, 2005
Location:	BP Refinery at Texas City, Texas, USA
Nature of accident:	Vapor cloud explosion and fire
Fatalities and injuries:	15 people killed and over 170 injured

2.16.1 Brief Description of Facility and Process

Amoco's Texas City refinery, the third-largest in the USA, was acquired by BP as part of its merger with Amoco in 1999. In 2000, the refinery's daily crude throughput capacity was 437,000 barrels (69,500 m^3) per day. The refinery was built in 1934 but reportedly had not been well maintained for several years.[36] The consulting firm Telos had examined conditions at the plant and released a report in January 2005 that enumerated numerous safety-related problems or issues, including "broken alarms, thinned pipe, chunks of concrete falling, bolts dropping 60 ft". The staff had previously been reported to be overcome by fumes. The refinery had seen a succession of five refinery managers in the six years since BP inherited it after its 1999 merger with Amoco.[36]

The explosion occurred at the isomerization (ISOM) unit of the refinery. The ISOM unit converts low octane, paraffinic light hydrocarbons into isomeric paraffins. These isoparaffins have higher octane ratings and are then blended profitably into the gasoline pool. The ISOM distillation unit comprised a raffinate splitter column with a blowdown drum and a "vent stack".

The raffinate splitter was a 164-ft-tall fractionating column with 70 trays at a 2 ft spacing, a feed surge drum, a fired heater reboiler, and a fin-fan overhead condenser. The overhead product was a light raffinate distillate from the column. The bottom product was pumped to a 45,000 barrel (7,200 m^3) heavy raffinate storage tank.

The relief valve blowdown from the splitter overhead did not go to a flare system; instead, it entered a 10-ft-diameter (62 m^3 volume) vertical drum with a 113-ft-tall high stack ("vent stack"). The stack was also not connected to a flare system, and any flashed vapors would, therefore, escape directly to the atmosphere. The drum was provided at the bottom with a liquid outflow line leading, via a gooseneck, to the site's closed sewer system. (It should be noted that this anachronistic and unsafe design was commonplace many years ago and has since been abandoned in favor of adequately designed relief systems.)

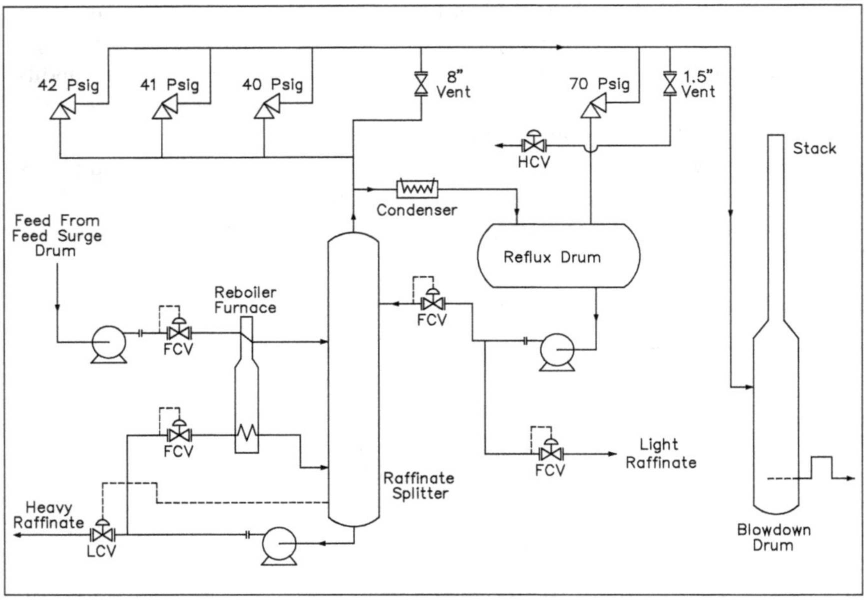

FIGURE 2.11 Flow diagram of raffinate splitter and blowdown system.

A flow diagram of the raffinate splitter and the blowdown "vent stack" system is shown in Figure 2.11.

2.16.2 The Accident

On March 23, 2005, at 13:20, an explosion and fire occurred at the ISOM unit. Liquid hydrocarbon liquid spilled to the ground from the overfilled blowdown stack. This liquid vaporized partially very rapidly, and unvaporized liquid spread along the ground. Eventually, a nearby ignition source caused a massive VCE that resulted in multiple casualties and caused substantial equipment destruction, as detailed below.

2.16.3 Causes, Circumstances, and Consequences of the Accident

On the day of the accident, the raffinate splitter in the ISOM unit was being started up after a planned, month-long temporary outage in another part of the ISOM and ARU units. Charging of the feed to the splitter started at 02:03 during the night shift of March 22/23. At 02:44, the reboiler flow control valve was opened to establish reboiler circulation. The column's liquid level continued to rise until the high-level alarm, set at 72%, activated at 03:05.

The operator acknowledged the alarm and reduced the feed rate, while the alarm remained "ON" in the "acknowledged" condition. The redundant hard-wired high-level alarm, set at 78%, failed to operate, and the level continued to rise to 100% by 03:16 (100% level was equivalent to a height of 10.25 ft in the bottom of the 164-ft-tall tower). At 03:20, the operator closed the column feed and the reboiler recirculation and decided to leave the remainder of the start-up to the day shift. The column pressure at that time was 4 psig (slightly higher than the normal value of 14.7 psig, owing to the presence of residual nitrogen in the system after inerting with nitrogen that had been carried out before start-up).

At 04:59, the operator left the site before the day shift operator arrived, and after making shift relief, not with the day shift supervisor, but with the supervisor of the Naphtha Desulphurization Unit (NDU)/Aromatics Recovery Unit No. 2. The day shift operators arrived at around 06:00 and the day shift supervisor at 07:13. No prejob safety review was conducted, nor a walk-through of the detailed operating procedure.

At 09:21, the field operators relieved the column pressure from 4 psig to atmospheric pressure by briefly opening the 8" chain-operated vent valve at the top of the column. After that, at 09:41, the day shift operator started the reboiler circulation and re-introduced feed to the column, unaware of the prior high-level alarms. At approximately 10:00, the reboiler fired heater's burners were lit before establishing heavy raffinate rundown, contrary to the established start-up procedure. Shortly afterward, this day shift supervisor left the site to attend to a personal family matter. Unfortunately, before leaving, he did not ensure that the proper procedures were being followed.

The splitter bottom temperature continued to rise at approximately 75°F/h, far exceeding the 50°F/h limit specified in the start-up procedure. Throughout this period, feed into the tower continued, and the heavy raffinate rundown remained closed. No liquids had been taken out of the splitter despite the continuous feed input.

By 12:40, the splitter pressure had steadily climbed to 33 psig, greatly exceeding the normal value of about 20 psig. Also, the base temperature reached 302°F, greatly exceeding the normal range of 280°F–290°F. At this point, the operators opened the overhead 8″ chain-operated vent valve, for the second time, for about 10 minutes and brought the pressure down to 22.6 psig.

The feed to the column was continuing at the normal rate. Eventually, at approximately 12:45, the column became 100% full of liquid (137 ft from the bottom) when the normal bottom liquid level range was just 6–7 ft. This absurd situation was not diagnosed by the operator, despite the enormous resultant increase in column base pressure. At 13:15, the base pressure peaked at 63 psig, and the three relief valves, with pressure set-points of 40, 41, and 42 psig, opened. The column's liquid contents were discharged into the "vent stack" that also continued to fill up to the very top. The vent stack level continued to rise because its bottom drain was too small to accommodate the extremely high rate at which it was being filled by the relief overflow from the splitter column.

At about 13:20, a mixture of vapor and liquid emerged "like a geyser" from the top of the vent stack. This liquid flowed down the outside of the drum and the stack, forming an enormous liquid pool on the surrounding ground. This volatile material vaporized partially and formed a ground-level vapor cloud ignited by an unknown nearby source, resulting in a massive explosion.

The explosion severely damaged the ISOM unit and destroyed a contractor's trailers located in the adjacent Ultracracker unit on the ISOM unit's west side. Many of those killed or injured were contractors congregated in or around a temporary trailer used to support turnaround work at the Ultracracker.

The explosion also caused several secondary hydrocarbon releases and fires that lasted for about 2 hours.

2.16.4 Lessons/Recommendations

The primary lessons from this accident are as follows:

- The refinery was of vintage origin (1934). Maintenance practices at the site appeared to have deteriorated to a state where accidents would be "about to happen", a conclusion drawn in the publicly-available Telo Consultants' report. None of the previously identified and badly needed upgrading work had been initiated until the date of the incident.
- The work culture in the organization was indifferent. Operators did not follow SOPs, and supervisors were absent at critical times. Knowledgeable supervision is most important during the start-up and shutdown phases of petroleum refinery operations, and especially so after turnarounds.
- According to the U.S. Government's Baker Report, the leadership required to develop and cultivate safety (especially process safety) was nonexistent. The refinery also had been through as many as five managers in the 6 years (1999–2005) since BP acquired it, following its 1999 merger with Amoco.
- The ISOM plant was started up without checking all safety-critical equipment and control systems hardware. The hard-wired high/low-level alarm was reported not to be working even before the turnaround. Inexplicably, the alarm was not repaired, and the night shift operator did not record the failure of this critical measurement for the following day shift to take corrective action.
- Neither the ISOM unit operators nor their supervisors seemed to have had adequate training and background to appreciate the hazards involved. Feeding the fractionating column for an abnormally long period without removing any product, which led to column overfilling and, ultimately, the accident, betrays extreme inattention to crucial details, lack of knowledge, and operating skills.
- There is no reason or excuse for contractors engaged in turnaround work to park temporary trailers in process areas. Inadequate management attention to contractors' work has resulted in many past accidents, and this issue deserves special attention.

The primary recommendations following this incident are as follows:

- Oil refineries and petrochemical plants handle flammable and explosive chemicals. Managers heading these organizations must be well qualified and well trained, and they, in turn, must imbue a good work and safety culture and discipline among the staff and workers under them.
- Plant- and unit-level managers must be well trained in process safety and maintenance procedures. They should be thoroughly familiar with all process safety management systems: MOC, hazard identification and analysis, LOPA, HAZOP, risk analysis, etc. They also must ensure that these are duly followed for all work carried out under their jurisdiction.

- Parking of nonessential contractor trailers (temporary offices), trucks, and vehicles must never be allowed in process areas. If contract workers are needed, they must be trained beforehand and must follow all site safety and operational rules.
- Strict observance of safety systems, e.g., work permit systems, checking of key safety instruments (periodically during normal operations and after repair and shutdown work), is imperative.

2.17 IMPERIAL SUGAR, PORT WENTWORTH, GEORGIA, USA[37]

Date of accident:	February 7, 2008
Location:	Port Wentworth, Georgia, USA
Nature of accident:	Dust explosion
Fatalities and injuries:	14 people killed and 36 injured

2.17.1 Brief Description of Facility and Process

Imperial Sugar Company, headquartered in Sugar Land, Texas, was incorporated in 1924. The company purchased the Port Wentworth facility from Savannah Foods and Industries, Inc. in December 1997. At the time of the accident, Imperial Sugar operated the Port Wentworth facility, a sugar manufacturing and packaging facility in Gramercy, Louisiana, and a warehousing operation in Ludlow, Kentucky.

The sugar manufacturing facilities refined raw sugar into granulated sugar. Some granulated sugar was used to make powdered sugar, specialty sugars, and liquid sugar products. Sugar products were packaged into bulk tank loads, 100-pound bags, and small boxes and bags. The company produced more than 1.3 million tons of sugar annually, making it one of the largest sugar refiners in the United States.

2.17.2 The Accident

At about 7:15 p.m. on February 7, 2008, a sugar dust explosion occurred at an enclosed steel conveyor belt in the granulated sugar storage silo at Port Wentworth, Georgia. Seconds later, massive secondary dust explosions propagated throughout the entire granulated and powdered sugar packing buildings, bulk sugar loading buildings, and parts of the raw sugar refinery. Security cameras (located to the north, south, and west of the facility) captured the sudden, violent fireball eruptions out of the penthouse on top of the silos, the west bucket elevator structure, and surrounding buildings.

2.17.3 Causes, Circumstances, and Consequences

Subsequent investigations revealed that sugar and corn-starch conveying equipment pieces were not designed or maintained to minimize sugar dust releases into the work area. Inadequate housekeeping practices resulted in significant combustible sugar and sugar dust residues on the floors and elevated surfaces throughout the packing buildings. Airborne combustible sugar dust accumulated, above the minimum explosive concentration, inside the newly enclosed steel belt assembly under silos 1 and 2. An overheated bearing in the steel belt conveyor most likely caused a primary dust explosion that led to massive secondary dust explosions and fires throughout the packing buildings.

Over the years, the facility had experienced granulated sugar and powdered sugar fires caused by overheated bearings or electrical devices in the packing building. However, none of these incidents resulted in a significant incident or fire before the February 2008 incident. The company management and the managers and workers at both Port Wentworth, Georgia, and Gramercy, Louisiana, did not recognize the significant hazard posed by sugar dust, despite the continuing history of near-misses.

The enclosure installed on the steel conveyor belt under silos 1 and 2 created a confined, unventilated space where sugar dust could readily accumulate above the minimum explosive concentration. The enclosed steel conveyor belt was not equipped with explosion vents to safely vent a combustible dust explosion outside the building.

Company management and supervisory personnel had reviewed and distributed a *Combustible Dust National Emphasis Program* in October 2007. However, they did not promptly and routinely remove all significant accumulations of sugar and sugar dust throughout the packing buildings and in the silo penthouse. In May 2007, a risk assessment for the Port Wentworth facility was performed by Zurich Services Corporation. The report submitted to Imperial Sugar management did not address the combustible dust hazards adequately.

Eight workers died at the scene, including four who were trapped by falling debris and collapsing floors. Two of these fatally injured workers had reportedly re-entered the building to rescue their co-workers but themselves failed to escape. Nineteen of the 36 workers suffered severe burn injuries.

The accident caused substantial damage to the entire granulated and powdered sugar packing buildings, bulk sugar loading buildings, and parts of the raw sugar refinery. Three-inch-thick concrete floors heaved and buckled from the explosive force of the secondary dust explosions as they moved through the four-story building on the south and east sides of the silos. The wooden plank roof on the palletizer building was shattered and blown into the adjacent bulk loading area for sugar rail cars.

2.17.4 Lessons/Recommendations

The lessons learned after this accident are as follows:

- Near-misses invariably precede most catastrophic accidents; management must understand these and take corrective actions urgently. One company report highlighted the hazard of sugar dust, but this was not acted on by the management. A second report, unfortunately, did not address the combustible dust hazards adequately.
- Dust explosions are critical hazards in process plants. If the plant managers had been trained in process safety and hazard analysis, they would have implemented ventilation plans to maintain the plant's process areas free of sugar dust in the air and accumulations on the ground.

The primary recommendations following this accident are as follows:

- Plant managers and supervisors must be trained well in process safety engineering. In particular, for those working in dust-prone areas, the training must cover the precautions that must be taken to eliminate dusty environments and thus avoid explosions resulting from accumulations of dust.
- There must be formal, written operating procedures, reviewed and revised as necessary from time to time, based on hazard studies, safety audits, etc. There must be regular checks on the quality of housekeeping for plants that are dust-prone or have other similar process hazards.
- Housekeeping checks must be performed regularly, and process and engineering modifications made for reducing to a minimum, if not eliminating, the dust nuisance.
- In the manufacture or handling of hazardous chemicals, there must be a definitive, site-specific disaster management plan that must incorporate periodic mock drills and be updated from time to time.

2.18 INDIAN OIL CORPORATION PRODUCT TANK FARM, JAIPUR, RAJASTHAN, INDIA[38,39]

Date of accident:	October 29, 2009
Location:	Jaipur City, India
Nature of accident:	Unconfined vapor cloud explosion (detonation)
Fatalities and injuries:	Eleven people were killed. The number of people injured is unknown

2.18.1 Description of Facility and Process

Indian Oil Corporation Limited (IOCL), established in 1964, is the largest oil company in India, with combined 80 million tons of annual refining capacity. It operates 11 of 23 Indian refineries at various locations and is also the largest oil product distributor in India. Under its marketing division, the government-owned company has many product tank farms (also called isolated storage) throughout the country. These tank farms receive products (gasoline, diesel, and kerosene) from refineries through pipelines or railcars for distribution throughout the country.

The huge tank farm in Jaipur, shown in Figure 2.12a, had a storage capacity of 110,000 m^3. It received liquid products from the Koyali refinery in the state of Gujarat. The Jaipur tank farm was located in a relatively large plot of 105 acres (42 hectares). Two other refineries of IOCL in north India (Mathura and Panipat) receive pipelined crude oil from the Vadinar Port in Gujarat, located around 1,200 km away. A crude oil stabilization unit and crude oil booster station for the crude pipelines were also installed within the tank farm, and their operation was under IOCL's pipeline division. This facility was closed down after the accident in 2009.

The IOCL tank farm also supplied the products by pipelines to the marketing terminal of BPCL, located around 10 km away.

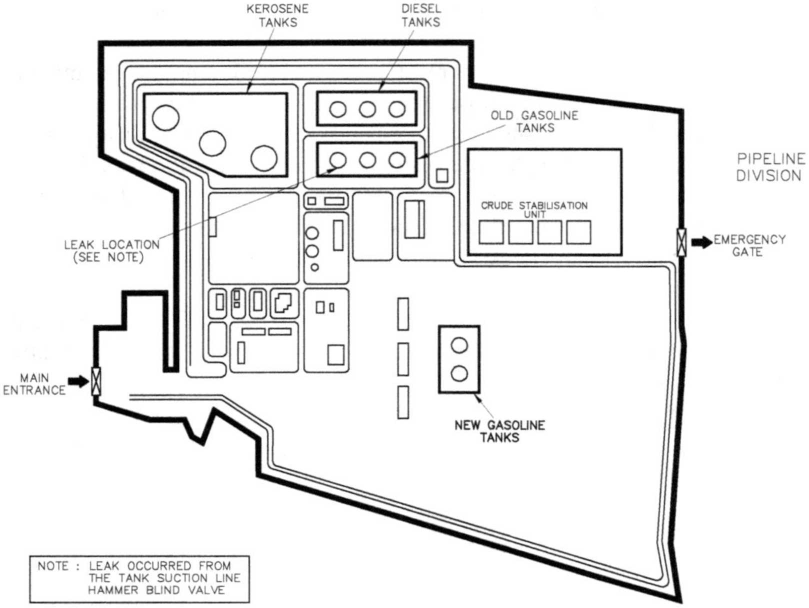

FIGURE 2.12a IOCL Jaipur tank farm.

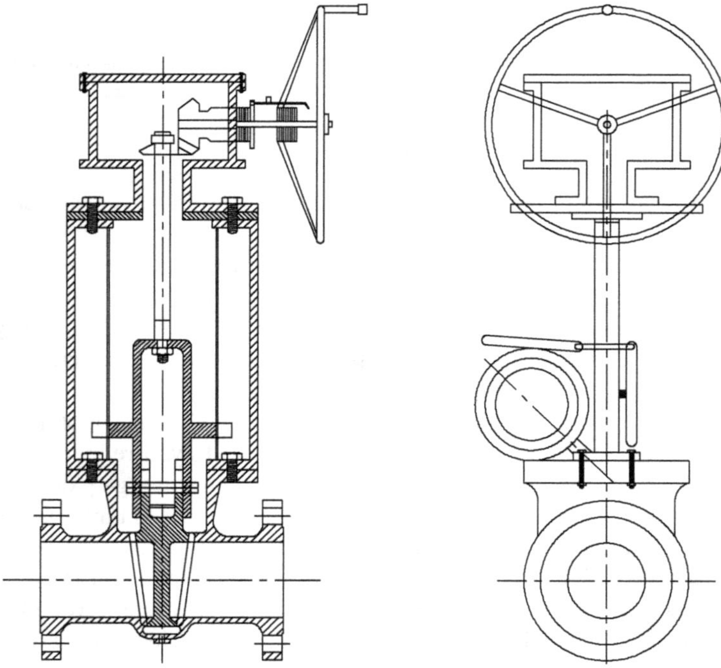

FIGURE 2.12b Hammer-Blind Valve.

2.18.2 The Accident

On the day of the accident (October 29, 2009), in the early evening, when the IOC tank farm was being lined up to supply gasoline by pipeline transfer to BPCL terminal, there was a major loss of containment at the gasoline storage tank being readied for transfer. A dike, provided as a secondary containment around the tank, also failed to retain the spilled gasoline. This spill caused a high, continuous gasoline release rate that kept evaporating to form a massive vapor cloud. After a lengthy interval of about 75 minutes, the cloud was ignited by an unknown source, resulting in a massive VCE, with the overpressure being felt up to 4 km away.

2.18.3 Causes, Circumstances, and Consequences

The cause of the accident was the maloperation of two isolation valves (one motor-operated gate valve and a Hammer-Blind Valve, as shown in Figure 2.12b), on the 10-inch line of a gasoline tank that was being lined up for gasoline transfer to BPCL's tank farm. Gasoline gushed out of the tank like a vertical fountain (reportedly reaching a height of about 12 ft) from the bonnet-less top of the Hammer-Blind Valve.

Inexplicably, the water drain valve on the dike was in a fully open position, causing an outright failure of the secondary containment that continued for almost 75 minutes. A vast amount of gasoline from the dike poured out through the open dike water drain valves into the storm-water drain, spreading throughout the tank farm, evaporating quickly, and generating colossal quantities of gasoline vapors. After a significant time lapse since the initiation of the spill, the vast vapor cloud found a source of ignition, and a devastating blast occurred.

The most likely cause was an ignition source in the pipeline area, near the pipeline division's control room, and was likely a nearby motor bicycle (subsequently found completely charred). The ignition caused a confined or partially confined explosion within the building that initiated a detonation wave as it vented from the building. The directional indicators pointed to a source in the Pipeline Division area in the northeast corner of the site.[42]

IOCL's tank farm was planning to pump around 1,600 m^3 gasoline to BPCL's tank farm in the evening. The line-up of the tank was supposed to be done by two operators. However, during line-up, the younger of the operators, who was competent, was absent. The other operator, not as highly regarded, was left alone to carry out the task. The supervisor advised this operator only to perform the line-up until the second operator returned. The time was around 5:50 p.m.

After about 15–20 minutes, the supervisor heard the sole operator (who was on the other side of the tank) shouting for help, saying that there was an enormous gasoline leak and that fumes had overcome him. The supervisor rushed to the tank area, jumped over the dike, and ran to the other side of the tank only to find the operator lying unconscious. The supervisor lifted him and tried to carry him out of the dike away from the leak, but, in the process, he was overcome with fast-spreading, dense gasoline vapors.

The supervisor then dropped the operator inside the dike. In attempting to jump over the dike, he stumbled and fell outside the dike in a semiconscious state. His fall was witnessed by a BPCL manager who had just arrived on the other side of the plant road about 100 m away. Being near the gate, the manager immediately informed plant security, who alerted others via walkie-talkie and announced an emergency.

On being alerted by the walkie-talkie, the second operator, who happened to be taking a break, rushed to the gasoline tank area and entered the dike to control the situation and rescue the other operator. He, too, was immediately overcome by gasoline vapors and, drenched in gasoline, fainted inside the dike. Unfortunately, both the operators perished in the subsequent blast; their remains were identified only by items such as key rings.

This leak of epic proportions continued with gasoline vapors spreading rapidly over the 105-acre plot for 75 minutes until the VCE took place.

During a subsequent investigation, it was found that the emergency shutdown (ESD) system in the control room was dysfunctional. *It was found that the ESD had remained nonfunctional for several prior years, despite an OISD (Oil Industry Safety Directorate, India) Audit Report that had been communicated to the highest levels of IOCL's Pipeline Division.*

A local stop/start-switch for a motor-operated valve (MOV) was installed very close to the tank, and, since it was within the dike, it could not be approached for closing the MOV without personal protective equipment (PPE). However, no PPE was found in the tank farm. IOCL's senior managers, including the General Manager, reached the site but remained helpless bystanders outside the gate. By the time they arrived, gasoline vapors had reached the gate. Anyone daring to enter the tank farm without the proper PPE would likely have choked. Unfortunately, no PPE was available.

Finally, one of the managers in the tank farm drove to the BPCL tank terminal (his car was manually pushed more than 100 m away from the gate and then started). When the PPEs were finally brought, they could not be deployed in the absence of trained emergency responders/firefighting staff. Finally, a volunteer worker agreed; however, just as he was ready to go in, the VCE took place. The explosion mechanism was later investigated jointly by Fire and Blast Investigation Group (FABIG), UK, and GL Noble Denton, UK.[42] It was established that, despite the relatively open area, a VCE had occurred. This open-flame VCE could have been initiated as a deflagration. However, it soon progressed to a DDT (deflagration-to-detonation) transition with an unprecedentedly high blast overpressure of 2 barg.

The incident caused the death of 11 people, six of whom were company personnel (including one contractor) and the other five were from a neighboring industry. At the time, three of the senior managers of the pipeline division were in a meeting behind the plot. They realized the danger only much later when the gasoline vapors reached them and ran towards the rear emergency gate. To their horror and misfortune, they found the gate locked. The only other escape route available to them was the main gate at the farthest corner that was unapproachable without PPE. They were trapped and most likely kept trying unsuccessfully to jump over the wall until the VCE took place.

The blast, which destroyed the facilities and building within the terminal, resulted in 9 of the 11 tanks catching fire in immediate succession, with their roofs blown off by the blast. The fire lasted 11 days. The blast also caused extensive damage to the buildings, structures, and some of the neighboring industries. The effects of the blast extended almost 4 km from the terminal and resulted in glass panes being shattered. Besides, there was extensive damage to cars in a showroom close-by.

2.18.4 Lessons/Recommendations

The Jaipur IOC tank farm accident is a rare example of a large VCE in a mostly open area where a flash fire would generally have been expected. A detailed investigation carried out by GL-Noble Denton (assisted by the FABIG, UK) stated as follows:

- Overpressures exceeding 2 barg were generated across almost the entire site; this is not consistent with the event being caused by an explosion in one area of the site that would produce a decaying blast wave propagating across the site.
- The VCE could not have been caused by a deflagration alone, as evidenced by the high, area-wide overpressures and directional indicators in the open areas.
- The overpressure damage and the directional indicators showed that the flammable vapor cloud covered almost the entire site, an area approximately four times that in the December 11–13 Buncefield accident in Hemel Hempstead, UK.
- As with the Buncefield incident, the overpressure and directional indicator evidence was consistent with a detonation propagating through most of the cloud.
- The directional indicators point to a detonation source in the Pipeline Division area, in the northeast corner of the site.
- Unlike Buncefield, the possibility of the detonation occurring because of flame acceleration in trees does not appear consistent with the evidence.
- The most likely cause of the detonation is a flame in either the pipeline area's control room or the pipeline pump house. This flame most likely resulted in a confined or partially confined explosion or deflagration, progressing to a "DDT transition", as it vented from the building. In drawing this conclusion, it would seem necessary for some of the directional evidence to be affected by a lack of symmetry in the vapor cloud, which did not seem unreasonable.
- The exact source of the transition to detonation cannot be determined owing to the limited evidence from the Pipeline Division's area, mainly because restoration work was ongoing before the site visit in February 2010. However, had the CCTV records that were in police custody at the time of writing been analyzed, they would very likely have allowed firm conclusions to be reached.

Recommendations to avoid such accidents are as follows:

- Management of large organizations should have uniform standards for process safety at all their hazardous facilities. The same safety culture should prevail in oil tank farms as in petroleum refineries.
- Company management must take safety audits seriously and ensure full implementation of their recommendations.
- Strict operating discipline should be enforced to ensure qualified operator presence during essential operations at all facilities.
- Facilities should replace all obsolete and unsafe items (such as Hammer-Blind Valves) when positive isolation is required.

- All hazardous facilities should maintain the required types of PPE in good condition, and these should readily be available. All hazardous facilities must have access to trained emergency responders.
- There must be a site-specific disaster management plan for on-site and off-site emergencies. Emergency gates/exits must be readily accessible, and security personnel must be nearby.
- The layout of hazardous facilities must be consistent with formal hazard identification/HAZOP studies.

2.19 BP DEEPWATER HORIZON OFFSHORE RIG[40]

Date of accident:	April 20, 2010.
Location:	Macondo well, Mississippi Canyon, Block 252, in the Central Gulf of Mexico, USA
Nature of accident:	Vapor cloud explosion and fire
Fatalities and injuries:	Eleven people killed and 17 injured

2.19.1 Description of Facility and Process

On March 19, 2008, BP acquired the lease on Mississippi Canyon, Block 252, in the Central Gulf of Mexico, at the Minerals Management Service's (MMS) lease sale 206. The 10-year lease started on June 1, 2008. BP, Anadarko Petroleum, and MOEX Offshore shared ownership (65%, 25%, and 10%, respectively), with BP as the lease operator. The Macondo well is located in this area. The MMS approved the first exploration plan for the lease on April 6, 2009, and a revised exploration plan on April 16, 2009.

The Macondo well is located approximately 48 miles from the nearest shoreline, 114 miles from Port Fourchon, Louisiana. This well was an infrastructure-led development, meaning that the "exploration well" was designed to be completed later to serve as a "production well" if sufficient hydrocarbons were found. BP's primary objective for the Macondo well was to evaluate a Miocene geological formation for commercial hydrocarbon-bearing sands. Although the original well plan was to drill to an estimated total depth of 19,650 ft, the actual total depth was 18,360 ft.

Initial drilling of the Macondo well began with Transocean's leased semisubmersible rig, the *Marianas*, on October 6, 2009, and continued until November 8, 2009, when the *Marianas* was secured and evacuated for Hurricane Ida. The *Marianas* was subsequently de-moored and removed owing to hurricane damage that required dock repairs. After the repairs, the rig went off-contract.

Transocean's *Deepwater Horizon* rig had been under contract to BP in the Gulf of Mexico for approximately nine years. During this time, it had drilled approximately 30 wells, two-thirds of which were exploration wells. The rig was chosen to finish the Macondo work, well after completing its previous project (the Kodiak "appraisal well"). On January 31, 2010, *Deepwater Horizon* arrived on-site. Drilling activities recommenced on February 6, 2010.

As is typical of exploratory wells in the Gulf of Mexico, the well-encountered pore pressures and fracture gradients differed from the design throughout the drilling process. The mud weights and well casing setting depths were also quite different from the original design.

2.19.2 The Accident

On the evening of April 20, 2010, a well-controlled event for temporary abandonment of the Macondo well failed. *Temporary well abandonment refers to the procedures that a rig crew uses to secure a well so that a rig can safely remove its blowout preventer (BOP) and riser from the well and leave the well site.* Control of the well was lost, allowing hydrocarbons to escape from the well into Transocean's *Deepwater Horizon* rig, thus entering the drilling riser and reaching the rig, resulting in explosions and subsequent fires (Figure 2.13).

From shortly before the explosions and until May 20, 2010, when all ROV intervention ceased, several efforts were made to seal the well. Eventually, many weeks after the initial accident, the well was permanently plugged with cement and "killed" on September 19, 2010.

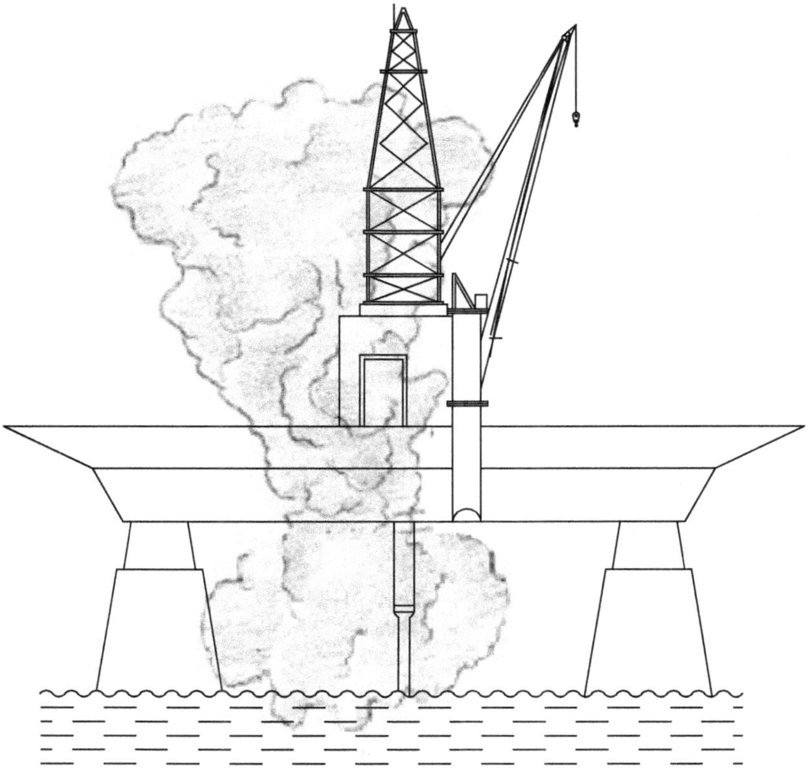

FIGURE 2.13 Deepwater horizon rig on fire.

2.19.3 Causes, Circumstances, and Consequences

The accident occurred on April 20, 2010, and was caused by a well-integrity failure followed by a loss of hydrostatic control. Unfortunately, this was also followed by a failure of the BOP – widely considered to be an ultimate safeguard – so that flow from the well could not be sealed off at the ocean floor. These concomitant failures allowed the release and subsequent ignition of hydrocarbons, leading to a catastrophic fire and several explosions that killed 11 employees and injured many others.

The daily uncontrolled release of thousands of barrels of oil from the well into the Gulf of Mexico continued for many weeks. This massive release resulted in extensive pollution of the sea and heavy bituminous deposits along the coastline in the States of Louisiana and Mississippi. Innumerable aquatic and bird species perished, cleanup costs were exorbitant, the local economy suffered severe damage, and BP's reputation suffered lasting damage.

BP's internal investigation report mentioned significant technical failures that caused the following events:

1. The annulus cement barrier did not isolate the hydrocarbons.
2. The shoe track barriers did not isolate the hydrocarbons.
3. The negative-pressure test was accepted, even though well integrity had not been established.
4. The influx of hydrocarbons was not recognized until hydrocarbons were already in the well's riser.
5. Well-controlled response actions failed to regain control of the well.
6. Diversion to the mud gas separator resulted in gas venting onto the rig.
7. The fire and gas system did not prevent hydrocarbon ignition.
8. The BOP failed disastrously and did not seal the well.

Eleven people lost their lives, and 17 others were injured. The fire, which was fed by hydrocarbons from the well, continued for 36 hours until the enormous rig sank. Hydrocarbons continued to flow from the reservoir through the wellbore and the failed BOP on the seabed for 87 days, causing an oil spill of national and international significance.

The Deepwater Horizon accident represented the most massive marine oil spill in the history of the petroleum industry. It was estimated to be anywhere from 8% to 31% larger in volume than the previous largest oil spill (*Ixtoc I*, also in the Gulf of Mexico). The U.S. government estimated the total discharge into the sea at 4.9 million barrels. After several failed efforts to contain the flow, BP succeeded in cementing the well and declared it sealed on September 19, 2010. However, reports in early 2012 indicated that the well site was still leaking oil.

2.19.4 Lessons/Recommendations

Many crucial lessons were learned from this accident:

BP had initiated a substantial cost-cutting exercise at the *Deepwater Horizon* rig. For highly hazardous, deepwater exploratory drilling, the best and safest equipment must be employed. The *Deepwater Horizon* exploration rig used a cheaper liner of "long string design". In contrast, other major operators such as Exxon-Mobil, Shell, and Chevron use the safer, improved "Liner Design". BP had retained the older design as it was considered too costly to change to the modern version.

For critical projects, such cost-cutting is often a prelude to disaster. It was reported that extreme pressure was exerted on the senior management at the *Deepwater Horizon* site to expedite task completion and move on to the next site.

The well's cementing contractor Halliburton's recommendation to use 21 "centralizers" to ensure the cement liner's proper positioning was not heeded by BP, who opted for using just 6 of them. For testing well integrity, cement regulators also found that BP decided against the recommended 9-to-12-hour procedure known as a "cement bond log".

BP also failed to fully circulate drilling mud, a 12-hour procedure that could have helped detect gas pockets that later shot up the well and exploded on the drilling rig. Such failure to adhere to the standard and well-proven operating procedures often results in severe consequences, as happened at Macondo.

Based on a thorough review of the factual record, the following recommendations are appropriate:

- Comprehensive risk assessment and safety management is mandatory for high-quality, deep well design and operation.
- Last-minute changes to plans can easily defeat the purposes of carefully laid-down operating procedures and must not be allowed without extensive safety and engineering reviews.
- Observing and responding appropriately to critical indicators (of well-controlled response and similar hazardous processes and activities) are vital for safeguarding life, property, and the fragile marine environment.
- Emergency bridge response training of all personnel must be comprehensive.
- Rig layouts must thoroughly be reviewed by experts, based on the cumulative experience gained from all offshore accidents.

2.20 SUMMARY AND CONCLUSIONS

The 19 case histories described above illustrate how catastrophic accidents have been caused by inadequate attention to process safety engineering and discipline in the design, operation, instrumentation, control, and maintenance of plants. We find that a few common factors contribute to many accidents, but the lessons learned tend soon to be forgotten, and accidents continue to occur with a distressing frequency.

A major cause of casualties and injuries in process plant accidents has been the proximity of control rooms to process equipment. In the days of pneumatic control systems and instrumentation, the concept was that a control room, which has to be well manned, should be located centrally in the processing area to enable easy visual monitoring and access to critical equipment. However, for many decades now, improvements in data communications (fiber-optic cable, wireless transmitters, etc.) and the development of very accurate sensors with distributed DCSs have enabled reliable transmission of data and signals to remote locations. *It is no longer necessary to locate control rooms close to hazardous operating units or to require the operating personnel physically to visit the operating areas routinely.* Some of the most significant technological advances are as follows:

1. High-resolution color TV cameras that can be rotated as needed to provide excellent remote viewing of well-lit areas, even at night
2. "Sniffers" that detect the presence of hydrocarbons and chemicals in the air
3. Thermal radiation sensors and flame detectors that can sense abnormal conditions
4. Noise and vibration monitors for machinery that substitute for human judgment
5. Robots and aerial drones to monitor hazardous areas where personnel exposure might be risky
6. Motorized valves for remote actuation of valves that traditionally have been operated manually.
7. Use of advanced technologies such as artificial intelligence, advanced analytics, edge computing, and digital twin modeling should be encouraged. These would enable (a) rapid detection of abnormal situations, (b) rapid development of options for corrective actions, (c) automated execution of validated corrections, and (d) avoidance of computational, communication, or logical errors by ensuring controlled duplication of critical applications.

The causes of large accidents in the process industries can be summarized as follows:

- Inadequate safety culture and lack of top-management awareness and sensitivity to safety issues
- Ill-defined responsibility/accountability down the line on matters related to safety
- Deficient formal operating procedures or unauthorized deviations and shortcuts being tolerated when the work is performed
- Design or technological deficiencies in automation: (1) insufficient or improper automation instrumentation and process control system design, (2) configuration or tuning of control loops, and (3) absence of ESD systems or improper design logic
- Use of improper materials of construction
- Unnecessarily high inventories of hazardous materials
- Inadequate safeguards against human error
- Insufficient process safety and hazard recognition training of operators and supervisors
- No prior hazard identification or risk assessment, leading to ill-preparedness to face deviations, upsets, or emergencies
- Lack/disregard of process safety review systems and MOC procedures for maintenance permits; improper modifications of plants and processes
- Failure to check credentials and performance of contract personnel
- Unauthorized and dangerous shortcuts, etc.

- Deficiencies in plant location and control room layout, which contribute to large numbers of fatalities and injuries in the event of an accident
- Lack of any credible safety audit or accident investigation systems that risk losing opportunities for taking timely corrective actions.

There are many such lessons to be learned. These points have been re-emphasized in the various chapters of this book. The authors express the hope that readers will appreciate these points, remember them, and act accordingly to help achieve the ultimate goals of profitable and accident-free operations.

REFERENCES

1. Mannan, S.: *Lees' Loss Prevention in the Process Industries* (4th Ed., Butterworth-Heinemann, Oxford, 2012).
2. Lemoff, T. C.: *LNG Incident, Cleveland* (NFPA, October 20, 1944).
3. API-625: *Tank Systems for Refrigerated Liquefied Gas Storage.*
4. API-620: *Design and Construction of Large, Welded, Low-Pressure Storage Tanks.*
5. NFPA-59A: *Standard for the Production, Storage, and Handling of Liquefied Natural Gas (LNG).*
6. EN/BS-7777: *Flat-Bottomed, Vertical, Cylindrical Storage Tanks for Low-Temperature Service. Guide to the General Provisions Applying for Design, Construction, Installation, and Operation.*
7. EN 1473: *Installation and Equipment for Liquefied Natural Gas. Design of Onshore Installations.*
8. ACI 376-11: *Code Requirement for Design & Construction of Concrete Structures for Containment of Refrigerated Liquefied Gases.*
9. EN 14620: *Design and Manufacture of Site Built, Vertical, Cylindrical, Flat-Bottom Steel Tanks.*
10. Burn, A.: Aker Solutions & Hailwood Mar LUBW Germany, *Loss Prevention Bulletin, Issue 251* (IChemE October, 2016).
11. Kumar, R. L.: *Petroleum Laws, An Encyclopedic Work on the Petroleum Act, Rules and Allied Laws along with Important Control Orders* (Law Publishers (India) Private Limited, Allahabad, 1996).
12. CENELEC: *Technical Report CLC/TR 50404, Electrostatics – Code of Practice for the Avoidance of Hazards Due to Static Electricity* (June, 2003).
13. *Loss Prevention Bulletin*, Issue 251 (October, 2016).
14. Parke, R. J.: *The Flixborough Disaster: Report of the Court of Inquiry* (Her Majesty's Stationery Office, London, 1975).
15. BS 3351: *Specification Piping Systems for the Petroleum Industry.*
16. Marshall, V. C.: *Major Chemical Hazards* (Ellis Horwood, Chichester, 1987).
17. Milnes, M.H.: Formation of 2,3,7,8-Tetrachloro-Dibenzo-Dioxin by Thermal Decomposition of Sodium 2,4,5-Trichlorophenate, *Nature*, 232, pp. 395–396 (1971).
18. Wilson, D. C.: *Lessons from Seveso, Chemistry in Britain* (July, 1982).
19. NFPA 58: *Liquid Petroleum Gas Code.*
20. Loss Prevention Bulletin (No. 063): *Bhopal: The Company's Report (based on Union Carbide Corporation's Report – March 1985)* (IIChE, June, 1985).
21. Bhushan, B. and Subramanian, A.: *Bhopal: What Really Happened?* (Business India, No. 102, February 25–March 10, 1985).
22. Shrivastava, P.: *Bhopal: Anatomy of a Crisis* (Ballinger, Cambridge, MA, 1987).
23. Kelkar, A. S.: *Investigation of Large Magnitude Events. Bhopal as a Case Study* (Paper presented in IChemE Conference, London, England, 1988).

24. *Report of Varadarajan Committee, Constituted by the Government of India.*
25. The Hon Lord Cullen: *The Public Inquiry into the Piper Alpha Disaster, Volumes 1 and 2* (October, 1990).
26. Drysdale, D. D. and Sylvester-Evans, R.: *The Explosion and Fire on the Piper Alpha Platform, A Case Study* (July 6, 1988).
27. Ganesan, G. K.: *Light Naphtha Tank Fire at BPCL* (Paper presented at the Oil Industry Safety Seminar, New Delhi, October 15, 1990).
28. Federal Emergency Management Agency: *Report Phillips Petroleum Chemical Plant Explosion and Fire Pasadena, Texas.*
29. Ghosal, S.: *Analysis of Unusual Events in LPG Installations.*
30. *Toulouse Chemical Factory Explosion* (Wikipedia).
31. Report of French Ministry of Sustainable Development – DGPR/SRT/BARPI: *Explosion in AZF Fertilizer Plant, Toulouse.*
32. *Space Shuttle Columbia Disaster* (Wikipedia).
33. Columbia Case History: *Building Process Safety Culture: Tools to Enhance Process Safety Performance.*
34. Federal Energy Regulatory Commission and U. S. Department of Energy: *Report on the U.S. Government Team Site Inspection of the Sonatrach Skikda LNG Plant in Skikda*, Algeria (March 12–16, 2004).
35. B.P.: *Final Investigation Report on BP Texas City Refinery Explosion* (December 9, 2005).
36. *Texas City Refinery Explosion – Background* (Wikipedia).
37. U.S. Chemical Safety and Hazard Investigation Board Investigation: *Sugar Dust Explosion and Fire* (Report No. 2008-05-I-GA, September, 2009).
38. *Report of Investigation Committee constituted by MoPNG Govt. of India*, Indian Oil Terminal Fire at Jaipur India.
39. *Report of GL Noble Delton: Characteristics of the Vapor Cloud Explosion Incident at the IOC Terminal in Jaipur.*
40. *BP-Deepwater Horizon Accident Investigation Report* September 8, 2010 (b) The Bureau of Ocean Energy Management, Regulation and Enforcement Regarding the April 20, 2010, Macondo Well Blowout September 14, 2011.
41. Theophanous T.G.: A Physico-Chemical Mechanism for the Ignition of the Seveso Accident, *Nature*, 291, pp. 640–642 (1981).
42. Johnson, D. M.: *GL Noble Denton, Characteristics of the Vapor Cloud Explosion Incident at IOC Terminal at Jaipur, Report Number: 11510* (October 29, 2009).

3 Fundamentals of Fire Processes

Fire is a chemical reaction process between combustible substances and oxygen in the air, releasing heat and emitting light and smoke. Controlled fires as a source of heat and power are of immense benefit in homes and industry, while uncontrolled fires can cause untold misery through the loss of life and property.

In chemical process plants, fuels are stored, used, and handled in large quantities. It is, therefore, of the utmost importance that we manage them safely to avoid accidental fires. This chapter aims to enable the readers to understand methods to achieve this objective by providing a comprehensive background on the fundamentals of fire processes.

3.1 HOW FIRE STARTS

The three essential requirements for a fire to start are fuel, oxygen, and heat. Fire technologists often express these requirements in the form of the fire triangle, as shown in Figure 3.1.

If one of the three requirements in the fire triangle is missing, the fire does not start. Similarly, if one of them is removed, the fire is extinguished.

Subsequently, fire technologists refined our understanding of fire phenomena further, and the fire triangle was changed to a fire tetrahedron to reflect a fourth element, chain chemical reaction.

The fire tetrahedron (see Figure 3.2) is a four-sided representation of the four factors necessary for fire: (1) fuel (any substance that can undergo combustion), (2) heat

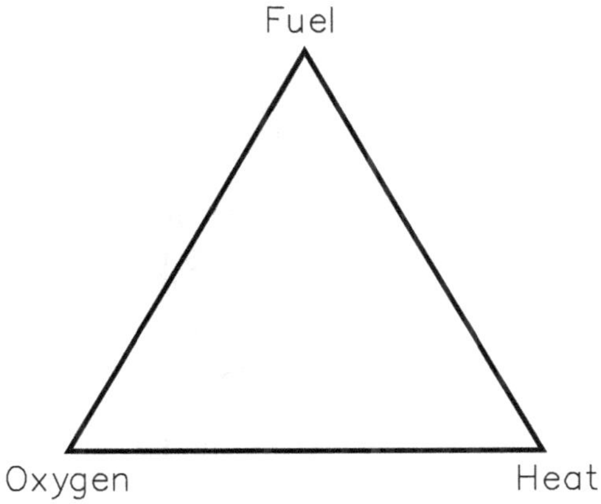

FIGURE 3.1 Fire triangle.

DOI: 10.1201/9781003107873-3

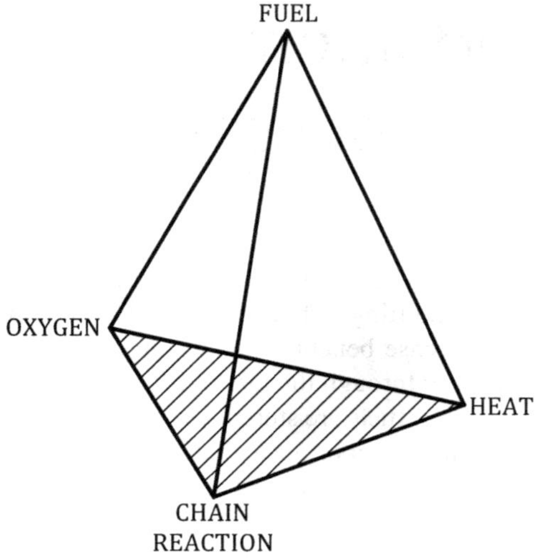

FIGURE 3.2 Fire tetrahedron.

(sufficient energy to release vapor from the fuel and cause ignition), (3) oxidizing agent (air containing oxygen), and (4) uninhibited chemical chain reaction.

The role of heat is to ignite the fuel/oxygen mixture (initiate the oxidation process), such as by contact with a hot object or an electric spark.

Fuels may be solids (e.g., wood or charcoal), liquids (e.g., gasoline or kerosene), or gases (e.g., LPG or natural gas). Gaseous fuels or vapors of liquid fuels combine with oxygen in the gas phase, giving rise to flames. A *flame* is the visible portion of the volume within which the oxidation of fuels in gaseous form occurs. It may be noted that some flames, such as those with a hydrogen-rich fuel, are exceedingly hot but may not be visible, especially under well-lit conditions or in the daytime. Unfortunately, and as confirmed by the personal experience of one of the authors, hydrogen fires from leaks have been known to occur spontaneously with no heat or identifiable ignition source at all[1].

For solid fuels, thermal decomposition, or *pyrolysis*, at the surface layer of the solid is necessary to yield flammable, volatile products of a low molecular weight that can sustain flame combustion. Pyrolysis may not yield volatile products at a rate higher than some minimum value required for sustaining the flame. The combustion then proceeds at the surface of the solid, possibly producing smoke but no flame. Such non-flaming combustion is known as *smoldering*.

A fuel/air mixture can be ignited if the mixture's fuel concentration is within a flammable range. Below the low end, this range is called the *lower flammability limit* (LFL), and the mixture is too lean to ignite. Similarly, above the high end of this limit, called the *upper flammability limit* (UFL), the mixture is too rich to ignite. The concentration range between these two limits constitutes the *flammability range*. For example, for methane, the LFL is 5 vol%, the UFL is 15 vol%, and the flammability range is from 5 to 15 vol%.

3.1.1 Flammability Limits

3.1.1.1 Pure Fuels

- LFL and UFL values for selected fuel gases/vapors in the air at atmospheric pressure are given in Table 3.1. These values are based on Lees[2] and Strehlow[3], all compiled mainly from Zabetakis.[4]

When data are not readily available, these limits can roughly be estimated by one or more of the following methods:

- Table 3.1 shows that for hydrocarbons and alcohols, the LFL is approximately one-half of the stoichiometric composition (C_{st}), and the UFL is between 2 and 3 times the stoichiometric composition (see Table 3.1). This approximation is not valid for acetylene, however.
- Empirically, for many hydrocarbon/air mixtures, it has been observed that the LFL (volume%) multiplied by the fuel's net heating value (kJ/g-mole) is approximately 4,340.[3] Since LFL values are readily available, this relationship can also be used to estimate the fuel's net heating value.
- For liquid fuels of known flash point, the LFL is equal to the concentration of the vapor in air at the flash point temperature, which can be calculated from the vapor pressure of the liquid at the flash point temperature.

TABLE 3.1
Flammability Limits (% Volume) in Fuel/Air Mixtures at 1 atm

Fuel	LFL	UFL	C_{st}	LFL/C_{st}	UFL/C_{st}
Methane	5.0	15.0	9.50	0.53	1.58
Ethane	3.0	12.4	5.66	0.53	2.19
Propane	2.1	9.5	4.03	0.52	2.36
n-Butane	1.8	8.4	3.13	0.58	2.68
n-Hexane	1.2	7.4	2.16	0.55	3.42
Ethylene	2.7	36.0	6.54	0.41	5.50
Propylene	2.4	11.0	4.46	0.54	2.47
Acetylene	2.5	100.0	7.75	0.32	12.90
Benzene	1.3	7.9	2.72	0.48	2.90
Acetone	2.6	13.0	4.99	0.52	2.61
Methanol	6.7	36.0	12.28	0.55	2.90
Ethanol	3.2	19.0	6.54	0.49	2.90
Diethyl ether	1.9	36.0	3.38	0.56	10.65
Hydrogen	4.0	75.0	29.58	0.14	2.54
Carbon monoxide	12.5	74.0	29.58	0.42	2.50

C_{st} is the % volume of fuel in the fuel/air mixture for 100% stoichiometric combustion.

Example 3.1

Calculate the LFL and UFL for propane and ethanol based on stoichiometric equations for combustion.

$$\text{Propane: } C_3H_8 + 5O_2 = 3CO_2 + 4H_2O$$

$$\text{Ethanol: } C_2H_5OH + 3O_2 = 2CO_2 + 3H_2O$$

Stoichiometric compositions of fuel/air mixture, based on 1 mole of propane/ethanol, are as follows (air contains approximately 21% oxygen and 79% nitrogen by volume):

	Propane			Ethanol	
	Moles	Mole%		Moles	Mole%
Propane	1.0	4.0	Ethanol	1.0	6.5
Oxygen	5.0	20.2	Oxygen	3.0	19.6
Nitrogen	18.8	75.8	Nitrogen	11.3	73.9
Total	24.8	100.0	Total	15.3	100.0

For propane/air mixture at 50% of the stoichiometric value of 4%, the LFL is (0.5) (4.0) = 2.0%. Similarly, at 250% (midway between 200% and 300%), the UFL is (2.5) (4.0) = 10.0%. (The experimental values are 2.1% and 9.5%, respectively.)

Similarly, for ethanol/air mixture, LFL = (0.5) (6.5) = 3.3%, and UFL = (3.0) (6.5) = 19.5% (The experimental values are 3.2% and 19.0%, respectively.)

For a fuel having a molecular formula $C_mH_nO_p$, the stoichiometric value of C_{st} for complete combustion can be calculated as follows:

$$\text{Moles of air per mole of fuel: } n_{air} = \left(m + n/4 - p/2\right)\left(100/21\right), \text{ and } C_{st} = 100/\left(n_{air} + 1\right)$$

3.1.1.2 Dependence of LFL and UFL on Pressure and Temperature

Above atmospheric pressure, an increase in pressure widens the flammability range by raising the UFL and reducing the LFL. The increase in the UFL is very significant. However, the drop in the LFL is relatively small. The following empirical expression for UFL, as a function of pressure, is provided by Crowl and Louvar[5] based on the work of Zabetakis.[6]

$$\text{UFL}_p = \text{UFL} + 20.6 \left(\log_{10}P + 1\right) \tag{3.1}$$

where

P = pressure, MPa,

UFL_p = upper flammability limit at pressure P,

UFL = upper flammability limit at 1 atm.

For example, the UFL for methane is 15%. From Equation (3.1), the value of UFL_p for methane at 200 atm (20.26 MPa) is calculated as follows:

$$UFL_p = 15 + (20.6)\left[\log_{10}(20.26) + 1\right] = 62.5\%$$

In the subatmospheric region, a decrease in pressure narrows the flammability range by raising the LFL and reducing the UFL until the two coincide. At this point, the mixture becomes nonflammable.

The flammability range widens with an increase in temperature. Drysdale[7] quotes the following empirical correlations based on the work of Zabitakis[6]:

$$LFL_T = LFL_{25}\left[1 - (0.00078)(T - 25)\right] \tag{3.2}$$

$$UFL_T = UFL_{25}\left[1 + (0.000721)(T - 25)\right] \tag{3.3}$$

where T = temperature, °C.

Example 3.2

The experimental values for LFL25 and UFL25 for ethane at 25°C and 1 atm are 3.0% and 12.4%, respectively. Calculate these values at 1 atm and 500°C. Also, calculate the value of UFL at 60 atm and 25°C.

From Equations (3.2) and (3.3), LFL and UFL at 500°C are calculated as follows:

$$LFL_T = 3\left[1 - (0.00078)(500 - 25)\right] = 1.89\%, \text{ and}$$

$$UFL_T = 12.4\left[1 + (0.000721)(500 - 25)\right] = 16.65\% \text{ respectively.}$$

For the effect of pressure, using Equation (3.1), at 60 atm or 6.08 MPa,

$$UFL_p = 12.4 + (20.6)\left[\log_{10}(6.08) + 1\right] = 49.1\%.$$

3.1.1.3 Mixture of Fuels in Air

For a mixture of flammable gases/vapors, the flammability limits in air at 25°C and atmospheric pressure are estimated by Le Chatelier's principle[5]:

$$100/LFL_{mix} = y_1/LFL_1 + y_2/LFL_2 + \ldots + y_n/LFL_n \tag{3.4}$$

$$100/UFL_{mix} = y_1/UFL_1 + y_2/UFL_2 + \ldots + y_n/UFL_n \tag{3.5}$$

where

y = volume percent, and subscripts "mix" for the mixture, and (1, 2, … n) are the components.

When the combustibles are less than 100%, the values of y in Equations (3.4) and (3.5) should be calculated on a total combustible basis.

Example 3.3

Calculate the LFL and UFL of the following gas mixtures: (i) 84% methane, 10% ethane, and 6% propane, and (ii) 2% methane, 6% ethane, 0.5% ethylene, and 91.5% air. The values of LFL and UFL for constituent gases can be taken from Table 3.1.

Case (i): In this case, the volumetric concentrations add up to 100%.

	Y (vol%)	LFL (vol%)	UFL (vol%)	*y*/LFL	*y*/UFL
Methane	84.0	5.0	15.0	16.80	5.60
Ethane	10.0	3.0	12.4	3.33	0.81
Propane	6.0	2.1	9.5	2.86	0.63
Total	100.0			22.99	7.04

Using Equations (3.4) and (3.5), LFL of the mixture is 100/22.99, or 4.35%, and UFL of the mixture is 100/7.04, or 14.2%.

Case (ii): In this case, volumetric concentrations of constituent combustible gases add up to less than 100%. Therefore, y values are to be calculated on the total combustible basis.

	Conc. vol%	*Y* vol%	LFL vol%	UFL vol%	*y*/LFL	*y*/UFL
Methane	2.0	23.53	5.0	15.0	4.71	1.57
Ethane	6.0	70.59	3.0	12.4	23.53	5.69
Ethylene	0.5	5.88	2.7	36.0	2.18	0.16
Total combustibles	8.5	100.00				
Air	91.5					
Total	100.0				30.42	7.42

Using Equations (3.4) and (3.5), the LFL of the mixture is 100/30.42, or 3.3%, and the UFL of the mixture is 100/7.42, or 13.5%. Since the mixture contains 8.5% of total combustibles, it is already in the flammable region, and additional fresh air is not required.

TABLE 3.2
Flammability Range in Oxygen at Ordinary Temperatures and Pressures

	LFL (vol%)	UFL (vol%)
Methane	5.1	61
Ethane	3.0	66
Propane	2.3	55
Butane	1.8	49
Ethylene	3.0	80
Propylene	2.1	53
Hydrogen	4.0	94
Carbon monoxide	15.5	94
Ammonia	15.0	79

3.1.1.4 Flammability Range in Oxygen

In chemical process plants, occasions can arise when a fuel gas/vapor has to be handled in an atmosphere of pure oxygen. The LFL in oxygen is about equal to that in air; however, the UFL is much greater than in air. Thus, the flammability range in oxygen is far wider than in air. Flammability data for some common gases, based on the U.S. Bureau of Mines Bulletin No. 503 (1952), are given in Table 3.2.

3.1.1.5 Effect of Addition of Inert Gases

The addition of inert gas (such as nitrogen, carbon dioxide, or steam) to a flammable mixture reduces oxygen volume percent in the mixture, which results in a narrowing of the flammability range.

This "limiting oxygen concentration (LOC)", also known as the "minimum oxygen concentration (MOC)", is defined as the limiting concentration of oxygen below which combustion is not possible, independent of fuel concentration. It is expressed in volume percent of oxygen. The LOC varies with pressure and temperature. It is also dependent on the type of inert (nonflammable) gas.

As the inert gas concentration increases, a stage is reached when the LFL and the UFL coincide, and the mixture becomes nonflammable. Figure 3.3 shows the effect of nitrogen, water vapor, and carbon dioxide on the flammability of methane/air mixture at 25°C and 1 atm.[2,7]

From Figure 3.3, a mixture of methane and air can be made nonflammable by adding nitrogen or carbon dioxide to the mixture, to the extent of 38% or 24% by volume, respectively. Table 3.3 shows the minimum concentration of nitrogen and carbon dioxide to suppress the flammability of selected gases in air. Carbon dioxide can be seen to be more effective than nitrogen in rendering the mixture nonflammable.

Expressing the flammability data in the form of a ternary diagram is useful for determining the degree of inerting necessary before equipment can be used for handling flammable materials. The requirement is explained in Example 3.4.

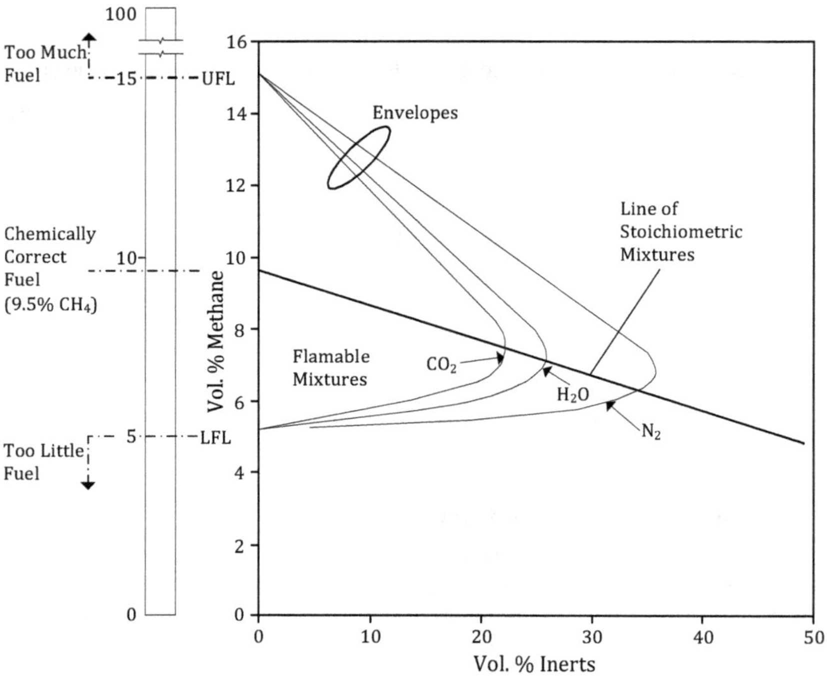

FIGURE 3.3 Effect of inert gases on flammability of methane in air.

TABLE 3.3
Minimum Inert Gas Concentration for Suppression of Flammability of Selected Substances in Air[2]

	Nitrogen (vol%)	Carbon Dioxide (vol%)
Methane	38	24
Ethane	46	33
Propane	43	30
n-Butane	41	28
Ethylene	50	41
Propylene	43	30
Benzene	45	32

Example 3.4

A vessel has been constructed for storage of liquefied butane under pressure at ambient temperature. It is to be commissioned by filling liquefied butane from another tank. The vessel has been checked for mechanical integrity and contains air at atmospheric pressure. The LFL for butane in air is equal to that in oxygen

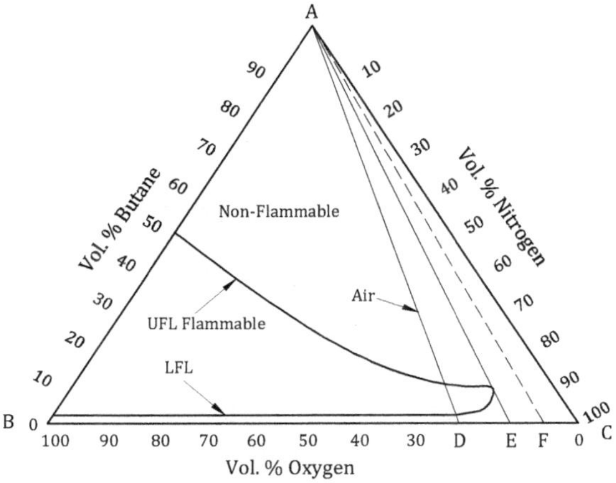

FIGURE 3.4 Flammability diagram of n-butane/oxygen/nitrogen system.

(1.8% by volume). The UFLs are 8.4 vol% in air and 49 vol% in oxygen. The minimum oxygen percentage below which no mixture is flammable is 12 vol%.

Based on the data provided, the flammability diagram of the butane vapor/oxygen/nitrogen system is first drawn on a triangular graph paper (see Figure 3.4).

In the flammability diagram (Figure 3.4), point A represents 100% butane vapor, point B represents 100% oxygen, and point C represents 100% nitrogen. Point D on the line BC represents the composition of air (21% oxygen by volume). The line DA represents mixtures of air and butane vapor. This line intersects the UFL line at 8.4% and the LFL line at 1.8%, the two compositions representing the UFL and LFL, respectively, of butane in air. Also, the UFL line intersects the line AB at 48% butane, and the LFL line intersects the line AB at 1.8%; these two concentrations represent the UFL and LFL values, respectively, for butane in mixture with oxygen.

The line AE forms a tangent to the envelope of the flammability region. This line passes through 12% oxygen, the MOC necessary for butane to burn.

If butane vapor is introduced directly into a vessel containing air, assuming uniform mixing, the butane concentration in the vessel will increase, following the line DA as butane is added. The mixture, therefore, will pass through the flammable region. For safety, the air in the vessel should be first purged with nitrogen until the oxygen concentration corresponds to that at point E. Butane vapor may then be introduced safely into the vessel. This procedure will ensure avoiding the flammable region. In practice, to provide a safety margin, purging with nitrogen is continued until the concentration of oxygen is lowered to about 5%, represented by point F. Further addition of butane will follow the line FA, with a sufficient safety margin against entering the flammable region.

The typical sequential steps for commissioning a tank to store flammable liquefied gases include (1) purging the air with an inert gas (at atmospheric pressure) to bring down the oxygen concentration to a safe value, (2) displacing the inert gas with vaporized liquid, and finally, (3) introducing the liquid itself into the tank. This procedure is necessary to avoid a brittle failure by a sudden cooldown of flashing liquid, leading to severe vessel chilling immediately following the liquid's introduction in refrigerated tanks/vessels.

Example 3.5

A low-temperature carbon steel tank is to be commissioned to store liquefied propane at atmospheric pressure and −43°C. The tank has been purged with nitrogen, and liquid propane feeding at the tank bottom has started. Pressure in the tank is being maintained at 40 mmHg gauge by controlled venting to a safe place. The intermediate step of displacing nitrogen with propane vapor has been neglected. To estimate the temperature at the tank bottom during the initial period of filling, calculate the saturated liquid propane temperature, assuming that the propane concentrations are 10, 20, 50, and 90 (vol%) in the vapor space immediately above the liquid.

The Antoine equation gives the vapor pressure of propane:

$$\log_{10} P = 6.82973 - 813.2/(t + 248) \tag{3.6}$$

where

P = vapor pressure, mmHg,
t = temperature, °C.

The total pressure in the tank is (40 + 760) or 800 mmHg. At 10% propane concentration, the partial pressure for propane in the vapor space is (0.1)(800) = 80 mmHg. Substituting for $P = 80$ in Equation (3.6), the saturated liquid temperature is calculated to be −83°C. Similar calculations for other propane concentrations give the following results:

Propane concentration (vol%)	10	20	50	90
Partial pressure of propane (mmHg)	80	160	400	720
Temperature (°C)	−82.9	−72.2	−55.6	−43.3

Thus, the localized temperature at the tank bottom could go down to as low as −70 to −80°C during the initial period, compared to the recommended design temperature of about −50°C for steel storage tanks. Such a low temperature can easily lead to a brittle fracture of the tank, which is an obvious and severe safety hazard.

3.1.2 Flash Point

For liquid fuels, the flash point is an important property used to determine the degree of flammability of hazardous liquids.

As temperature rises, vapor pressure increases; therefore, the vapor space fuel concentration also increases. If the vapor mixture in equilibrium with the liquid contains less fuel vapor than required for the LFL, it will not ignite; it is below its flash point temperature. As the temperature of the liquid is increased further, the vapor mixture will enter the flammable range. At an even higher temperature, the vapor concentration will become too rich to sustain combustion. This temperature corresponds to the UFL.

TABLE 3.4
Flash Point Temperatures of Selected Liquids

	Flash Point (Closed Cup) (°C)	Flash Point (Open Cup) (°C)
Benzene	−11	
n-Butane	−60	
Styrene	32	38
Toluene	4	7
p-Xylene	25	31
Acetone	−18	−9
Ethanol	13	
Methanol	12	

Thus, the flash point of a flammable liquid is the temperature at which the liquid's vapor pressure is just sufficient to yield a concentration of vapor in the air that corresponds to the LFL.

In practice, the lower and upper flash point temperatures are measured in several types of standardized apparatus. The most common methods for measuring flash point are the Pensky-Martens closed-cup method (described in ASTM D 93-61) and the open-cup method (described in ASTM D 92-57). Results from these methods are generally fairly close. The flash points of selected liquids are shown in Table 3.4.

When the flashpoint of a fuel is not readily available, it can be estimated provided its LFL and vapor pressure are known, as illustrated in Example 3.6.

Example 3.6

The vapor pressure of benzene is given by the Antoine equation as follows:

$$P = 10^{[6.90564 - 1211.033/(220.790 + t)]}$$

where
P = vapor pressure of benzene, mmHg,
t = temperature, °C.

Calculate its flash point.

Assuming the atmospheric pressure to be 760 mmHg, the concentration of benzene in the vapor space above the liquid is calculated as follows:

Temperature, t (°C)	Vapor Pressure (mmHg)	Concentration (vol%)
−11	13.58	1.8
−15	10.49	1.4

The LFL for benzene in air is 1.4% by volume. From vapor pressure data, this concentration occurs in the vapor space at −15°C. Hence, the estimated flashpoint is −15°C, as compared to the experimental value of −11°C. Hence, the calculated values from this method should be taken as order-of-magnitude estimates only.

3.1.3 Fire Point

The fire point of a flammable liquid is the lowest temperature at which the liquid, when placed in an open container, will give off sufficient vapor to continue to burn after being ignited. The fire point is usually a few degrees above the open cup flash point temperature.

3.2 HEAT BALANCE IN FLAMES

In a flame, the fuel vapor undergoes rapid oxidation reactions with air. The heat release rate is equal to the fuel burning rate multiplied by its calorific value. Water is formed from the oxidation of the hydrogen in the fuel and remains in a gaseous form. The net (or lower) calorific value must, therefore, be used in this calculation.

About 25%–40% of the heat released in combustion is radiated from the flame. A part of the heat released is also used to vaporize liquid fuels before combustion can commence. The quantitative effects of radiated heat and hot product gases are covered in subsequent chapters.

3.3 TYPES OF FLAMES

3.3.1 Premixed and Diffusion Flames

Combustion air may mix with the gaseous fuel before ignition, or it may diffuse into the flame as combustion proceeds. Depending upon the type of air mixing, the flames are classified into two types:

- *Premixed flame*: Here, the fuel and the air are mixed before ignition. The simplest example is the laboratory Bunsen burner with an open-air inlet. Here, the rate of combustion is determined by the reaction kinetics. In chemical plants, premixed flames are encountered in flash fires (considered later) resulting from the accidental release of flammable gases.
- *Diffusion flame*: In this type of flame, the fuel and the air are initially unmixed. Air is drawn into the fuel (and into the flame after ignition) from the surrounding atmosphere by diffusion. The simplest example of a diffusion flame is the Bunsen burner with the air inlet port closed. Here, the burning rate is limited by the diffusion rate of air and is slower than that in a premixed flame. Diffusion flames are encountered in the burning of gas jets and also combustible liquids and solids.

3.3.2 Pool Fire

When a flammable liquid spills onto the ground and the vapor above it is ignited, the result is a near-cylindrical flame on the surface of a liquid pool. This fire is called a pool fire. A pool fire has a diffusion flame, and the air for combustion is entrained into the flame from the surrounding atmosphere. When the burning liquid pool is contained within an enclosure, for example, by a dike, the resulting flame is a confined pool fire. In the absence of containment, the liquid fuel pool spreads as the burning continues, and the process becomes an unconfined pool fire.

A fire on the top of a liquid storage tank (tank fire) is an example of a confined pool fire. A fire in a trench is also a form of a pool fire. A pool fire may also occur when an immiscible liquid fuel, such as motor spirit, is spilled over water. As a rule of thumb, the flame height above a pool fire will be approximately 1.5 times the pool diameter. However, this will vary depending on the material in the pool and the environmental conditions.

3.3.3 Jet Fire

When a combustible gas or vapor issues from containment under pressure through a small opening and is ignited, the result is a plume-shaped flame known as a jet fire. The fires from ruptured pipes or holes in tanks are examples of jet fires. These can do extensive damage when they impinge on adjacent tanks and pipe racks. The fire on top of a flare stack is also an example of a jet fire. Here, air for combustion may be premixed and drawn into the flame as combustion proceeds. Jet fires typically can be extinguished only by removing the fuel source.

3.3.4 Vapor Cloud Fire

If a flammable gas or vapor accidentally leaks into the atmosphere, and if it does not find an ignition source in the immediate vicinity, it continues to travel in the wind's direction. As it travels, it disperses in the air and gets diluted to progressively lower concentrations. A flammable gas cloud may be formed, and the concentration at the outer boundary of such a cloud is equal to the LFL.

After some time, and at some remote location, if an ignition source is encountered while the gas is still flammable, the gas will ignite, and the resulting flame will travel backward through the premixed cloud at high speed. Such a momentary flash is known as a flash or vapor cloud fire.

Initially, the flash fire is a deflagration. The flame front will start to travel at subsonic speed and gradually accelerate until it reaches the fuel emission source. The acceleration of the flame front generally will increase in case of confinement and obstacles in its path. As it progresses, if the flame front velocity surpasses sonic velocity, the fire becomes a vapor cloud explosion, a phenomenon known as the "deflagration-to-detonation" transition.

On the other hand, if the flame speed remains subsonic, the flash fire will cease upon reaching the source of emission. However, the emitted gas/vapor will continue to burn as a jet fire until the leakage stops or as a pool fire until the spilled liquid has been consumed completely.

3.3.5 Fireball

This type of fire occurs if a large mass of flammable vapors is released suddenly into the atmosphere and is ignited immediately. A typical example is the sudden release of a vapor and mist mixture, following a boiling liquid, expanding vapor explosion (BLEVE). In this case, the flammable material burns on the fireball's surface, and the interior is too rich to burn. The initial momentum of the release and the hot flame's high buoyancy cause the burning mass to rise like a fireball that expands and rises until burnout.

3.4 IGNITION

3.4.1 Requirements and Characteristics of Ignition Sources

While an ignition source is an essential requirement for starting a fire, not all ignition sources can successfully ignite flammable mixtures. Some of the basic requirements are ignition energy, ignition temperature, and flame size.

The ignition energy requirement depends on the nature of the chemical or mixture, and the minimum value occurs near the stoichiometric mixture composition. The values of minimum ignition energy for some selected fuel/air mixtures are given in Table 3.5. Minimum ignition energy is a crucial consideration in ignition probability assessment in situations where there is potential for a spark or static discharge.

The auto-ignition temperature is the minimum temperature at which a flammable gas/air mixture ignites spontaneously. The lower the auto-ignition temperature, the easier it is to ignite the mixture. The auto-ignition temperature for a given fuel/air mixture varies with the composition, volume, pressure, and presence of impurities. Quoted values do not usually mention the conditions under which measurements were made; therefore, due allowances should be made when using such data. Table 3.5 gives the auto-ignition temperatures for some selected fuel/air mixtures.

TABLE 3.5
Minimum Ignition Energy and Auto-Ignition Temperature for Selected Fuel/Air Mixtures[8]

	Minimum Ignition Energy (mJ)	Auto-Ignition T (°C)
Carbon disulfide	0.01–0.02	
Hydrogen	0.019	
Acetylene	0.02	
Methane	0.29	537
Ethane	0.24	515
Propane	0.25	466
n-Butane	0.25	405
Ethylene	0.12	
Benzene	0.22	

Another essential factor for successful ignition is establishing a critical flame size that would enable flame propagation until all of the flammable mixture has been consumed. The term "quenching diameter", defined as the maximum diameter at which flame propagation is suppressed, is often used. For a slot-like aperture, the "critical slot width" is useful in designing flame arresters. Critical slot widths for common hydrocarbons are generally 1–2 mm. Lees[2] mentions that the quenching diameter is about 1.5 times the critical slot width.

Some potential sources of ignition in process plants are as follows:

- Flames
- Hot work
- Electrical equipment
- Static electricity
- Hot surfaces
- Hot particles
- Friction and impact
- Reactive, unstable, and pyrophoric materials
- Vehicles with the engine running
- Lightning
- Smoking.

These items are generally self-explanatory. However, some discussion on hot work, electrical equipment, and static electricity is appropriate, as these items account for an appreciable proportion of ignition accidents in process plant operations.

While it is crucial to control potential ignition sources, ignition sources remain plentiful despite our best efforts. It would be unusual for a released flammable vapor to avoid coming into contact with an ignition source. Therefore, while ignition sources must be controlled as carefully as possible, this is no substitute for preventing flammable releases.

3.4.2 Hot Work

Hot work, such as welding, cutting, or grinding, has been responsible for many fires in process plants. Such maintenance operations should be carried out only after the work has been duly authorized by process personnel (issuance of a work permit). Before any hot work permit is issued, it must first be verified by physical inspection, using appropriate instrumentation as necessary, that the work area is free of any flammable gas. Prior purging with air may be necessary to achieve this condition. All potential sources of flammable liquids or vapors in the area must be positively closed, locked, and tagged to prevent inadvertent opening. The work area must also be monitored frequently to ensure the complete absence of flammable vapors. Monitoring is doubly important when work is started early in the day, as heating of equipment by the sun can cause pockets of trapped flammables to be released.

3.4.3 Electrical Equipment

Electrical equipment is widely used in process plants and may easily be a source of ignition unless close control is exercised. Such controls include hazardous area classification and safeguarding of equipment located in hazardous areas.

Several definitions exist for area classification, such as the Institute of Petroleum U.K. Classification, the NFPA 70 Classification, or Classification by the International Electrochemical Commission (IEC). IEC classification is in terms of zones, defined in Table 3.6.

For petroleum products, the vapor space of a storage tank or a gasoline tanker is classified as Zone 0. Guidelines are available for determining the extent of Zones 0, 1, and 2 for a range of hazardous equipment and operations, such as storage tanks, pumping areas, road and rail tanker loading areas, drum filling, etc. Such area classifications are mandatory for petroleum installations, and the extent of the zones depends on the area jurisdiction.

Once the hazardous area classification has been made, the next step is to select the right type of electrical equipment for use in that area. Commonly used categories are intrinsically safe, flameproof, or increased safety equipment. Equipment selection requires specialized knowledge, and only the equipment that has been tested and certified by an approved authority should be used. The process engineer's responsibility is to specify usage conditions, such as temperature, nature of flammable gas or vapor, and the probability of leakage.

3.4.4 Static Electricity

Static electricity is a ubiquitous source of ignition in process plants. Many severe fires and explosions have been caused by static electricity igniting flammable clouds. Also, there have been instances where investigators have failed to pinpoint the actual source of ignition and have mistakenly attributed the cause to static electricity – even though the evidence was weak.[2] Hence, process engineers need to develop a good understanding of ignition by static electricity. With this objective in mind, Chapter 4 has been devoted entirely to the issue of static electricity.

TABLE 3.6
Electrical Classification of Hazardous Areas[2]

Zone 0	A zone in which an explosive gas/air mixture is continuously present, or present for long periods
Zone 1	A zone in which an explosive gas/air mixture is likely to occur normally
Zone 2	A zone in which an explosive gas/air mixture is not likely to occur; if it occurs, it will persist for a short time only
Nonhazardous	An area in which an explosive gas/air mixture is not expected to be present in quantities high enough to require special precautions for the construction and use of electrical apparatus

3.5 EFFECT OF THERMAL RADIATION

Fires produce gaseous combustion products and heat. Combustion products can cause adverse health effects from smoke, carbon monoxide, and other toxic gases formed during the combustion process. The combustion products carry 60%–75% of the heat generated in the flame as sensible heat, thereby heating up whatever comes in their path and contributing to fire spread.

About 25%–40% of the heat generated in the flame is dissipated as radiation from the flame. Humans near the flame absorb this radiation and can suffer pain, burns, or even fatalities. The effect on plant and machinery, buildings, etc. includes ignition of any flammable materials in the vicinity, failure of equipment and structures caused by loss of structural strength, and consequent domino effects.

3.5.1 Effect on the Human Body[2,9]

The effect of thermal radiation on the human body depends mainly on the three factors: type of clothing, the radiation intensity, and the duration of exposure.

In general, exposure to thermal radiation causes the bare skin temperature to rise progressively. Pain ensues at about 45°C, followed by burn injuries of increasing severity. For skin, the effect depends on the type of clothing and whether the clothing itself ignites.

For correlating the effect of radiation intensity for a given exposure duration, the terms "thermal dose" and "thermal load" are used. The thermal dose is the product of radiation intensity (kW/m^2) and exposure duration (seconds). It represents the quantity of radiation received in kJ/m^2. The thermal load is a dimensional empirical variable, defined as:

$$\mathrm{TL} = (t)\left(I^{4/3}\right) \tag{3.7}$$

where

TL = thermal load, kJ/m^2

I = intensity, kW/m^2

t = exposure duration, seconds

API RP 521[6] lists the radiation intensity values versus time to pain for exposed bare skin. These are shown in Columns 1 and 2 of Table 3.7. The thermal dose and thermal load values calculated from these data are shown in this table in Columns 3 and 4, respectively. Column 4 shows the variation of thermal load with intensity, with an average value of 118. Thus, a guideline value of 118 for thermal load could be used for estimating the time to experience pain in the intensity range of 2–20 kW/m^2.

The average solar radiation intensity on a clear, hot summer day in the tropics can reach about 1 kW/m^2. No pain may be felt even after extended exposure; however, a severe sunburn may, in fact, result.

The severity of injuries caused to the skin by heat radiation is classified as first-, second-, or third-degree burns. These are defined as follows[10]:

TABLE 3.7
Time to Experience Pain on Exposure to Thermal Radiation

Intensity (kW/m^2)	Time (seconds)	Thermal Dose (kJ/m^2)	Thermal Load, seconds $(kW/m^2)^{4/3}$
1.74	60	104	125
2.33	40	93	123
2.90	30	87	124
4.73	16	76	126
6.94	9	62	118
9.46	6	57	119
11.67	4	47	105
19.87	2	40	107

- A first-degree burn is superficial and is characterized by red, dry, and painful skin.
- A second-degree burn extends through the epidermis (0.07–0.12 mm). This type of burn is characterized by blister formation and wet skin, which is also red.
- A third-degree burn extends to the dermis (thickness 1–2 mm) in which hair roots and free nerve extremities are present. The burned skin lacks sensation, is dry, and has a white, yellow, or black color. Such burns are excruciatingly painful owing to the exposed nerves, and treatment can last many weeks or months, often requiring skin grafts.

Empirical equations (probit functions) are used to express the probabilities of first- and second-degree burns and fatalities. The probit function is defined as follows:

$$P = \frac{1}{2\pi}\int_{-\infty}^{u-5} e^{-u^2/2}\,du \tag{3.8}$$

As defined above, the probit is a random variable with a mean of 5 and a variance of 1. Table 3.8 shows the relationship between % probability and the probit function.

For hydrocarbon fires, the TNO Green Book[9] gives the following equations for first- and second-degree burns and also for fatalities:

$$\text{First-degree burns } Y = -12.03 + (3.0186)\ln(\text{TL}) \tag{3.9a}$$

$$\text{Second-degree burns } Y = -15.34 + (3.0186)\ln(\text{TL}) \tag{3.9b}$$

$$\text{Fatality } Y = -12.80 + (2.56)\ln(\text{TL}) \tag{3.9c}$$

TABLE 3.8
Relationship between Percentage and Probit[9]

Percentage	Probit									
%	0	1	2	3	4	5	6	7	8	9
0	–	2.67	2.95	3.12	3.25	3.36	3.45	3.52	3.59	3.66
10	3.72	3.77	3.82	3.87	3.92	3.96	4.01	4.05	4.08	4.12
20	4.16	4.19	4.23	4.26	4.29	4.33	4.36	4.39	4.42	4.45
30	4.48	4.50	4.53	4.56	4.59	4.61	4.64	4.67	4.69	4.72
40	4.75	4.77	4.80	4.82	4.85	4.87	4.90	4.92	4.95	4.97
50	5.00	5.03	5.05	5.08	5.10	5.13	5.15	5.18	5.20	5.23
60	5.25	5.28	5.31	5.33	5.36	5.39	5.41	5.44	5.47	5.50
70	5.52	5.55	5.58	5.61	5.64	5.67	5.71	5.74	5.77	5.81
80	5.84	5.88	5.92	5.95	5.99	6.04	6.08	6.13	6.18	6.23
90	6.28	6.34	6.41	6.48	6.55	6.64	6.75	6.88	7.05	7.33

Probability values for Table 3.8 above may be approximated fairly accurately by the following function:

'% Probability = A/(1 + exp(B – C X Probit))^(1/D)

Where:

A = 103.72719, B = 6.6665734, C = 1.4167812, D = 0.69653309

Example 3.7

(a) A person doing maintenance work in a plant is suddenly exposed to thermal radiation from fire in adjoining equipment. The intensity of radiation is 8 kW/m^2, and the layout is such that he would need about 20 seconds (including average reaction time) to escape to a safe distance. Estimate the likely level of injury.

From Equation (3.7), thermal load (TL) = (20) ($8^{4/3}$) = 318 seconds $(kW/m^2)^{4/3}$. Since this thermal load is much higher than the threshold of 118 (Table 3.7), the injury would be more severe than just pain. Using Equation (3.9a) for the first-degree pain:

Probit $Y = -12.03 + 3.0186 \ln(318) = 5.33$. This corresponds to a 63% probability of first-degree burns.

(b) A person is working in the open when he is suddenly exposed to thermal radiation of 25 kW/m^2 from a high-intensity fireball. He would require at least 20 seconds to run indoors. What is the probability of fatal injury?

$$\text{From Equation (3.7), Thermal Load (TL)} = (20)(25)^{4/3} = 1{,}462$$

$$\text{Using Equation (3.9c), Probit } Y = -12.8 + 2.56 \ln(1{,}462) = 5.86$$

This value of the probit function corresponds to an 80% probability of fatality.

3.5.2 Effect on Plant and Machinery

Data on damage to plant equipment and materials by thermal radiation are limited. Lees[2] quotes data from several sources on radiation intensity design limits, but corresponding values for exposure duration are not provided. Some of these values

TABLE 3.9
Thermal Radiation Intensity vs. Effect on Plant and Materials[8]

Intensity (kW/m^2)	Effect
12.5–15	Piloted ignition of wood
25	Spontaneous (non-piloted) ignition of wood
12	Plastic melts
18–20	Cable insulation degrades
38	Damage to process equipment
100	Steel structural elements fail[12]

are given in Table 3.9. These limits should be used only as broad order-of-magnitude criteria for the lower limit of safe values in hazard assessment studies.

3.6 FIRE PREVENTION SYSTEMS

Fire prevention systems are intended to reduce the likelihood of fires occurring; these are discussed in the following sections.

3.6.1 Good Housekeeping

Poor housekeeping is a significant contributor to fire accidents. Many devastating fires are known to have been caused by accumulations of oil/solvent-soaked waste rags, jute bags, etc., lying unattended in drains or other areas that contacted powders or aqueous solutions of oxidizing salts.

3.6.2 Control of Flammable Materials

The layout of warehouses, drum stores, or bulk storage areas, and how they are operated and maintained, has a significant effect on the likelihood of fires. Drum storage areas contain materials of different flammability categories and may even contain drums with mutually incompatible materials. Proper segregation and labeling are, therefore, of paramount importance. Compartmentalization of warehouses for different categories of materials using fireproof walls and doors should also be considered.

3.6.3 Control of Sources of Ignition

Areas where flammable materials are stored, processed, or handled, are likely to have occasional leaks of a magnitude and frequency depending on the nature of the operation and housekeeping practices. Prior electrical area classification and deploying electrical equipment conforming to such classifications can minimize the incidence of fires.

Other practical steps to control ignition are maintaining properly earthed electrical systems and preventing static charge accumulation for flammable liquids and combustible dusts.

3.6.4 Fire Hazards Awareness

Awareness starts at the top of the management ladder and extends to people on the shop floor who work in hazardous environments. At the management level, commitment to safe working methods and procedures, installing safety systems, and training/periodic re-training for all workers are paramount. The focus of such training should be to make all personnel well aware of the fire hazards in their work environment, the importance of complying with specified precautionary measures, proper and prompt emergency response, and correct identification of fire situations.

3.6.5 Monitoring

Regular monitoring of housekeeping standards, plant maintenance operations to ensure integrity, routine testing of detection systems, ignition control, etc., are all essential for minimizing the risk of accidents and fires.

3.7 FIRE PROTECTION SYSTEMS

Fire protection systems are intended to minimize damage once a fire has started. These include both passive and active fire protection systems. Routine testing of fire protection equipment, sensors, sprinkler and deluge systems, and local/remote alarms must be mandated so that these systems are not found to be non-functional when they are needed most urgently. As discussed in Chapter 2, neglect of such precautions has resulted in several relatively minor incidents escalating to major disasters.

3.7.1 Passive Fire Protection

Passive fire protection includes measures to limit fire spread and contain the damage. Typical examples are (1) dikes around storage tanks and (2) buildings divided into compartments separated by walls and doors with appropriate fire resistance.

Passive fire protection also includes some forms of fire insulation or fireproofing. The main emphasis in fireproofing is generally to protect supporting structures. In some cases, this is extended to cover critical equipment and vessels containing flammable materials that could feed a fire.

The advantage of fireproofing is that it can "buy time" in which other active fire-fighting resources can be put into operation. A criterion usually specified for fireproof coating is that it should hold the substrate below 1,000°F or 538°C for 90 minutes while its surface is exposed continuously at 1,800°F or 982°C,[2] a difficult requirement to meet, as such coatings generally fail much earlier.

Fireproofing suffers from the disadvantage that, over time, corrosion may occur underneath the coating that might be would be difficult to detect early enough. Given these disadvantages, it is desirable that even where fireproofing has been provided, other active fire protection systems are reliably and quickly made operational after a fire is detected.

3.7.2 Active Fire Protection

Active fire protection involves automated detection of a flammable material leak at the first instance and immediate detection, next immediate detection of fire, followed by cooling of surrounding facilities to prevent escalation, and fire extinguishing.

3.7.2.1 Detection of Flammable Material[11]

Many industrial processes produce flammable gases and vapors, which can burn when mixed with air, sometimes violently. Typical examples include the following:

- Removal of flammable materials from tanks and pipes in preparation for entry, line breaking, cleaning, or hot work such as welding
- Evaporation of flammable solvents in drying ovens
- Spraying, spreading, and coating of articles with paint, adhesives, or other substances containing flammable solvents
- Manufacture of flammable gases
- Manufacture and mixing of flammable liquids
- Storage of flammable substances
- Solvent extraction processes
- Combustion of gas or oil
- Combined heat and power plants
- Heat treatment furnaces in which flammable atmospheres are used
- Battery charging.

Flammable gas detectors can make valuable contributions to the safety of these processes. They can be used to trigger alarms if a specified concentration of the gas or vapor is exceeded, providing an early warning of a problem and ensuring people's safety. However, a detector does not prevent leaks from occurring or indicate what action should be taken. It is not a substitute for safe working practices and maintenance.

A *fixed* flammable material detector is permanently installed in a chosen location to provide continuous monitoring of plant and equipment. It is used for early warning of leaks from a plant containing flammable gases or vapors, or monitoring concentrations of such gases and vapors within the plant. Fixed detectors are particularly useful where there is a possibility of a leak into an enclosed or partially enclosed space where flammable gases could accumulate.

A *portable* detector is usually a small hand-held device that can be used for testing an atmosphere in a confined space before entry. These detectors are useful for tracking leaks or providing early warnings of flammable gas or vapor when hot work is carried out in a hazardous area.

A *transportable* detector is not hand-carried but can still be moved readily from one place to another. A primary purpose is to monitor an area while a fixed detector is undergoing maintenance.

Portable and transportable detectors are always "point detectors". Fixed gas detectors can be point detectors or open-path detectors.

Point detectors measure the concentration of the gas at the sampling point of the instrument. The unit of measurement can be:

- % volume ratio
- % lower explosion limit (LEL) for a flammable gas
- ppm or mg/m^3 for low-level concentrations (primarily used for toxic gases).

Open-path detectors, also called beam detectors, typically consist of a radiation source and a physically separate, remote detector. The detector measures the average concentration of gas along the path of the beam. The unit of measurement is concentration multiplied by path length:

(% LEL) (m), or
(ppm) (m).

Systems can be designed with path lengths of 100 m or more. However, it is impossible to distinguish whether a reading is due to a high concentration along a small part of the beam or a lower concentration distributed over a longer length. Also, they are not specific to a particular gas; for example, steam or water vapor can produce false readings and alarms.

There are various types of sensors for flammable gas detectors:

- Catalytic (pellistor)
- Infrared
- Thermal conductivity
- Flame ionization.

3.7.2.2 Detection of Fire

Fires can be detected by sensing heat, radiation, smoke, and other products of combustion. Some basic principles are described below. However, the choice of detectors, their locations, and spacing should be decided based on field conditions and supplier recommendations.

- Flame Detectors

  Detection of a flame is done by sensing infrared or ultraviolet emissions from the flame. These detectors generally respond rapidly, infrared detectors in about 10 seconds, and UV detectors virtually instantaneously.[2] Also, flame detectors are available that are very sensitive even to small flames. A practicable design could be expected to be triggered by a fire of just 3 kW intensity at a 20 m distance.[12] Usually, it is necessary for the detector to "see" the flame. However, some reflections at IR wavelengths may be picked up by an IR detector.

  Flame detectors should discriminate between a flame's emissions and the continuous background emission, such as from the sun or process furnaces. They use the flickering characteristic of flame that is absent in the emission from a background source.[3]
- Heat Detectors

  Heat detectors sense the heat in the combustion gases rising from the flame. These detectors include temperature-measuring instruments, quartz bulbs, and devices that disintegrate, melt, or change characteristics when exposed to fire. Detectors are linked with systems to actuate alarms or set off water sprays. A simple detection system, used frequently on an air conduit, melts at a rated temperature of about 68°C. This causes the air pressure to fall and, in turn, actuate a drench valve on the water spray system.

Heat detectors are widely used in buildings. These detectors are located at the ceiling. The response of the detectors depends on the ceiling height and the spacing between detectors. Drysdale[7] has described a method for estimating the minimum size of flame that can be sensed by a detector of a specified temperature rating for given ceiling height and detector spacing.

- Smoke Detectors

  Smoke detectors are based on the ability of smoke particles to obscure or scatter light. These detectors are widely used in buildings but generally are less effective in open plants.
- Detection by People

  Despite the large-scale use of instrumental devices for detecting fires, it is worth mentioning that trained people can be excellent fire detectors. They will respond rapidly to an incipient fire, assess possible consequences, and take appropriate action. A fire and its location can be communicated to a control room by actuating a manual call point, which traditionally is of the break-glass type.

A combination of instrumental/manual detection and actuation of emergency systems by operators in the control room would usually provide a dependable system that is less prone to improper triggering. However, a system that relies heavily on manual detection must always be discouraged.

3.7.2.3 Cooling by Water

Water is used as a fire extinguishing agent, and for cooling equipment and vessels exposed to heat from neighboring fires. Systems for water delivery include fixed water spray systems, fixed water monitors, and mobile systems based on fire hoses. The use of water for cooling is considered in this section and for extinguishing in the next section.

The required flow rate of water for cooling tanks and vessels exposed to thermal radiation from nearby fires depends on the incident radiation intensity. The rate should also be high enough to allow a cohesive water film to form under windy conditions. The mechanism of heat intake by water is complex. It depends on the duration of exposure of the falling film to radiation, evaporation, mist formation, absorption, and scattering of radiation by water vapor and droplets.

For cooling storage tanks with water spray when a neighboring tank is on fire, NFPA 15 recommends a minimum water rate of 0.25 USgpm/ft^2 or 10.2 L/min/m^2 of the exposed surface.[13] Assuming simplistically that the incident radiation is entirely used up in raising the sensible heat of water, water usage of 10 L/min/m^2 at a typical radiation intensity of 25 kW/m^2 would increase water temperature by 36°C. In actual practice, the temperature rise will be less because there would be a 3%–5% evaporation of the water. Thus, 10 L/min/m^2 should be a safe figure, even in areas where water temperature in summer could be as high as 30°C.

3.7.2.4 Fire Extinguishing

3.7.2.4.1 Firefighting Media

Firefighting media include water, foam, dry powder, carbon dioxide, and other agents (e.g., Halotron, FM-200, DSPA). In earlier days, Halon was used quite commonly; however, since 1994, it is no longer produced and has been replaced by other agents mentioned above. Water extinguishes fire by direct cooling. It is preferable to apply water as a spray since droplets absorb heat more quickly than a jet. NFPA 15[13] recommends a general water spray rate from 0.15 to 0.5 USgpm/ft^2, or 6.1–20.4 L/min/m^2, for extinguishing fires involving ordinary combustible solids or liquids, while emphasizing that design density should ideally be based on test data.

Firefighting foam is an aggregate of air-filled bubbles formed from aqueous solutions and is lower in density than flammable liquids. It is used to provide a blanket between the fire and the flammable liquid and thus extinguish the flame. The foam also prevents re-ignition by suppressing the formation of flammable vapors.

Foam-making compounds are used to make foams of various expansion ratios (low, medium, and high expansion foams). Typical low expansion foam is made by introducing the foam compound into a water hose to yield a 3%–6% aqueous solution and then mixing the solution with air in an ejector nozzle for a 10:1 expansion. Several NFPA standards such as NFPA 11,[14] 11A,[15] 11C,[16], 16,[17] and 16A[18] describe various foam and foam/water systems. These give details of foam requirements and foam-making equipment and have been considered in Chapter 5.

Dry chemicals are of different types, and their mechanism of extinction varies accordingly. Some products act against fire by causing chain termination of the combustion reactions. Others, such as bicarbonates, undergo endothermic decomposition and produce carbon dioxide. Both mechanisms help extinction.

Extinguishment by carbon dioxide is achieved by reducing fuel and oxygen concentration in the fuel/air mixture. Also, carbon dioxide has a significantly higher heat capacity than other gases in the flame and can cool the flame below the minimum limiting temperature required for flame propagation. However, carbon dioxide is an asphyxiant and is also a toxic gas. Therefore, doors in unoccupied buildings protected by carbon dioxide should invariably be closed whenever the system is actuated.

Halons used a chain termination mechanism for combustion reactions. A typical Halon is a bromochlorodifluoromethane (Halon 1211). Once widely used, Halons have now been phased out in favor of ozone-friendly formulations.

Halotron is a proven clean fire extinguishing agent designed for streaming applications in portable and hand-held fire extinguishers. It is a widely distributed and used halocarbon-based clean fire extinguishing agent for such applications. Halotron has a very low ozone depletion potential (ODP) of 0.0098, as well as a low global warming potential (GWP) of 77 (based on HCFC-123).

FM200 is a synthetic chemical fire suppression gas and extinguishes fire by removing the free radicals or heat from the fire tetrahedron (oxygen, heat, fuel, and uninhibited chemical chain reaction). The typical concentration range for an FM200 system is between 7.9% and 8.5%. It is safe for people and electronic equipment. During firefighting, the gas is used in a concentration not to be harmful to human health and life. When using the FM200, the gas concentration of oxygen in the room can be reduced by 3%.

DSPA (dry sprinkler powder aerosol) extinguishing agents were developed to replace the Halon extinguishing systems. The DSPA compound will form an aerosol, which contains 70% gas and 30% solid particles, mostly potassium-based. An aerosol is a suspension of fine particles in a carrier gas. For DSPA, nitrogen is the carrier gas containing very fine potassium particles.

According to the US EPA (2012), a DSPA generator contains no Class I or Class II ODS, nor any compounds with measurable GWP. Compared to Halon 1301, HCFC Blend A (NAFS III®), HFC-227ea, and HFC-125, the impact of DSPA compounds is much lower.

3.7.2.4.2 Classification of Fires

Traditionally, fires have been categorized according to the preferred method of extinguishment into four classes: Classes A, B, C, and D,[19] besides electrical fires.

- Class "A" Fires

  These fires generally involve solid carbonaceous materials which gradually decompose during combustion to form flammable gases and glowing embers. Many solid organic materials, such as fibers and plastics, are in this class. By far, the most effective firefighting medium is water.
- Class "B" Fires

  Class "B" fires involve flammable liquids or melts. Depending on the situation, foam, dry powder, or Halotron can be used effectively. For tank or pool fires in areas enclosed by dikes, foams are widely used. However, it is difficult to form a foam blanket on a free-flowing liquid, and it is impossible to do so when the liquid is flowing over a vertical surface.

  If the flammable liquid is soluble in water, a small fire may be fought effectively with copious water sprays. Alcohol-resistant foams are particularly useful on fires involving polar solvents against which conventional protein or synthetic foams are ineffective.

  Since foams are water-based, these must not be used where contact with water can result in a hazard.
- Class "C" Fires

  This class of fire involves flammable gases. The only effective way to put out a gas fire is to cut off the supply of gas. Therefore, remote isolation valves must readily be accessible in a fire situation. Where this is not possible (e.g., with gas cylinders), water sprays may be used to cool surrounding stock and steelwork. Gas fires can be extinguished using Halotron, but there is a risk of re-ignition or even a gas/air explosion. Therefore, in many cases, the fire may best be left to burn itself out.
- Class "D" Fires

  These fires involve metals such as aluminum, magnesium, or sodium. They can be extinguished using special powders that fuse on contact with hot metal and seal off access to the air. However, some powders are dangerous to use for these fires. Therefore, the compatibility of powders in specific fire situations should be checked beforehand with media suppliers. Water or Halon must never be used in this class of fire.
- Electrical Fires

  Fires involving electrical equipment are fought mainly using portable carbon dioxide or dry chemical extinguishers. Continuous water jets conduct electricity and must never be used on electrical fires. Water sprays are used, however, to protect against fires in oil-cooled electrical equipment such as transformers.

3.7.2.5 Firefighting Plan

While the most appropriate extinguishing media should be specified, it should be appreciated that water is sometimes the only option available, for example, in remote tank farms. The selection of firefighting media and associated equipment for firefighting requires domain knowledge and significant expertise. Also, engineers operating a process plant are most knowledgeable about the hazards likely to arise from toxic emissions or water incompatibility in fire situations.

Therefore, a comprehensive firefighting plan should be prepared in consultation with suppliers of firefighting media and equipment and in-house/external fire brigades, as applicable. Design and operation of firefighting capabilities, including personnel training, should formally be tested routinely, using the principles and procedures described in this chapter.

REFERENCES

1. Spontaneous Ignition of Hydrogen: *Research Report RR615 by the Health and Safety Laboratory for the Health and Safety Executive* (2008).
2. Mannan, S.: *Lees' Loss Prevention in the Process Industries* (4th Ed., Butterworth-Heinemann, Oxford, 2012).
3. Strehlow, R. A.: *Combustion Fundamentals* (McGraw-Hill, New York, 1984).
4. Zabetakis, M. G.: *Flammability Characteristics of Combustible Gases and Vapors* (U.S. Bureau of Mines Bulletin 627, Pittsburgh, PA, 1965).
5. Crowl, D. A. and Louvar, J. F.: *Chemical Process Safety: Fundamentals with Applications* (Prentice Hall, Englewood Cliffs, NY, 1990).
6. Zabetakis, M. G.: *Fire and Explosion Hazards at Temperature and Pressure Extremes*, (AIChE Symposium Series 2, 1965).
7. Drysdale, D.: *An Introduction to Fire Dynamics* (2nd Ed., Wiley, New York, 1999).
8. Lees, F. P.: *Loss Prevention in the Process Industries* (2nd Ed., Butterworth-Heinemann, Oxford, 1996).
9. TNO: *Methods for the Determination of Possible Damage (Green Book)* (16th Ed., CPR, The Hague, 1992).
10. TNO: *Methods for the Calculation of Physical Effects (Yellow Book)* (14th Ed., CPR, The Hague, 1992).
11. Health and Safety Executive, UK: *The Selection and Use of Flammable Gas Detectors*.
12. Kletz, T.: *What Went Wrong* (3rd Ed., Gulf, Houston, TX, 1995).
13. NFPA 15: *Standard for Water Spray Fixed Systems for Fire Protection* (1996).
14. NFPA 11: *Standard for Low-Expansion Foam* (1994).
15. NFPA 11A: *Standard for Medium- and High-Expansion Foam Systems* (1994).
16. NFPA 11C: *Standard for Mobile Foam Apparatus* (1995).
17 NPA 16: *Standard for the Installation of Deluge Foam-Water Sprinkler and Foam-Water Spray Systems* (1995).
18. NFPA 16A: *Standard for the Installation of Closed-Head Foam-Water Sprinkler Systems* (1994).
19. ICI Fire Protection Guide No. 11: *Fire Prevention and Protection for Warehouses* (1984).

4 Static Electricity

Static electricity is a widespread source of ignition in process plants. Many extensive fires and explosions have been caused when static electricity ignited flammable gases and dusts. In our experience, the understanding of this hazard by process engineers is often rudimentary or nonexistent. Even during formal accident investigations, there have been instances when investigators have either failed to pinpoint the real source of ignition or falsely attributed the cause to static electricity.[1] This chapter on static electricity is designed to help readers gain a better quantitative understanding and appreciation of this critical hazard.

The following topics and issues have been considered in this chapter:

- Historical background and basic concepts of static electricity
- Mechanisms of static charge generation and their relevance in process plant operations
- Methodology, with illustrative examples, for the assessment of static charge accumulation to levels high enough to cause static discharges
- Types of discharge that occur under various situations and their associated energies, vis-à-vis minimum energy requirements for ignition of flammable mixtures
- Precautionary measures for control of static ignition hazards.

4.1 HISTORICAL BACKGROUND OF STATIC ELECTRICITY[2,3]

As far back as the 6th century B.C., the Greek philosopher Thales noticed that when rubbed briskly, amber (the hard sap of a kind of pine tree) attracts light bodies such as pieces of straw or paper. He was led to this observation by the Greeks' practice of spinning silk with an amber spindle; the rubbing of the spindle caused the silk to adhere to it. The Greek equivalent for the word amber is *elektron*, and it is believed that the term *electrification* arises from this Greek word.

No systematic study of the phenomenon mentioned above was known, however, until about 1600 A.D., when Dr. Gilbert, a physician to Queen Elizabeth I of England, carried out detailed investigations about this property of amber. He found that, besides amber, many other substances could similarly be electrified, for example, by rubbing glass with silk.

During his investigation, Dr. Gilbert tried but failed to electrify metals by rubbing and concluded that this was impossible. However, more than 100 years later, in 1734, Du Fay found that metal could be charged by rubbing with fur or silk, but only if it were held with a glass or amber handle, not manually. Du Fay's experiments followed Gray's discovery in 1729 that electric charges could be transmitted through the human body, water, and metals. These are examples of *conductors*; glass and amber

DOI: 10.1201/9781003107873-4

are examples of *insulators*. The glass or amber handles in Du Fay's experiment isolated the electrified metal from the earth.

4.2 BASIC CONCEPTS OF STATIC ELECTRICITY

The process of electrification (e.g., by rubbing glass with silk) is caused by the movement of electrons. In an uncharged body, atoms are electrically neutral, i.e., the total negative charge on the electrons is equal to the total positive charge on the atomic nuclei. Electrons are free to move, for example, during rubbing or when a body is placed in an electric field. An excess of electrons in a body, relative to its neutral state, means that the body is negatively charged. Similarly, a deficit of electrons relative to a neutral state means that the body is positively charged. The magnitude of the charge is expressed in coulombs (C).

When the properties of flowing (current) electricity were discovered, the term *static* came into use to distinguish an electric charge that was at rest from one that was in motion. Today the term *static electricity* is used to describe phenomena from an electric charge, regardless of whether the charge is at rest or in motion.[3] *Streaming current* or *charging current* (expressed in amperes) is defined as the flow rate of charge into a given system per unit of time. It is the current caused by the rate of flow of charged material into a system; it should be distinguished from the rate of static charge generation within a batch system, such as during the rubbing of solids or stirring of liquids.

Static electricity involves high potentials, generally of the order of kilovolts, and very low currents, generally in the range of milliamperes (mA). Therefore, static electricity effects are associated with an electric field, but unlike current electricity, static electricity has no significant magnetic field effects.

Earth is a near-infinite "reservoir" of electrons and is conventionally referred to as having zero potential. Therefore, if a charged conductor is electrically connected to the earth, its potential becomes zero. Earthing also removes all charge from a conductor unless there is another charged body in its vicinity that influences by induction.

Bodies with unlike charges attract each other, and those with like charges repel. Based on Coulomb's law, if two bodies (that are small compared to the distance between them), having charges Q_1 and Q_2 coulombs, are placed in vacuum at a distance of r (m) between them, then the force F (N) between them is given by:

$$F = Q_1Q_2/\left(4\pi\varepsilon_o r^2\right),$$

where

$$\varepsilon_o = \text{the constant permittivity of space } \left(8.85 \times 10^{-12} \text{ Farad (F) / meter}\right).$$

If the two bodies are placed in a medium with a dielectric constant of ε (instead of vacuum), the force is given by:

$$F = Q_1Q_2/\left(4\pi\varepsilon\varepsilon_o r^2\right) \tag{4.1}$$

The dielectric constant ε of the medium is dimensionless and is also known as relative permittivity.

The region surrounding an electrostatic charge, where its influence (i.e., the force of attraction or repulsion) extends, is called its electrostatic field. This field is usually visualized as electrostatic lines of force originating from positive charges and terminating at negative charges (or at infinity). Arrows are shown on the lines of force to indicate the direction of a positive charge. The intensity E of an electrostatic field (also called field strength) at a given location is defined as the force exerted on a positive unit charge placed at that location. Field strength is expressed as Newton/coulomb, or V/m, the latter showing that the field strength also represents the potential gradient ($E=-dV/dr$). Here, the negative sign indicates that the potential decreases as the field intensity increases.

The theoretical derivation of field strength distributions is exceedingly complex owing to many factors. A uniform electric field is an ideal case in which the field lines are parallel to one another, for example, between the plates of a large, parallel plate air capacitor. In practical situations, the fields are rarely uniform, and this needs to be considered while considering the criteria for static discharge (see Section 4.6.1).

As charge accumulates in a body, its potential rises. The charge required to raise the potential by 1 V is defined as the electrical capacitance of the body. This concept is similar to that used in heating an object: the heat required to raise its temperature by 1°C is the body's thermal capacitance. The unit of capacitance is the Farad, which is equal to 1 Coulomb/Volt. The potential of a body can also be raised by the presence of charge on other nearby bodies, as discussed in later sections.

The insulating property of a material that affects its ability to accumulate and dissipate static charges depends upon its conductivity/resistivity and dielectric constant. The former has been considered in Section 4.3. The dielectric constant is a dimensionless parameter, defined as the ratio of the permittivity of the material to the permittivity of vacuum. Metals have an infinite dielectric constant, while gases and vapors have a dielectric constant close to unity. High dielectric constant liquids are generally conductive, and low dielectric constant liquids are nonconductive.

4.3 CONDUCTORS AND INSULATORS

The ability of a material to acquire, accumulate, and dissipate static charges depends on its conducting/insulating properties, that is, on the material's electrical conductivity.

4.3.1 LIQUIDS

Liquids are classified as having low, medium, or high conductivity. Low-conductivity liquids are also known as "static accumulating" liquids, nonconductive liquids, or insulating liquids. Medium-conductivity liquids are also known as "semiconductive" liquids, and high-conductivity liquids as "conductive liquids".

The conductivity is expressed in terms of picoSiemens/m (1 pS/m $=10^{-12}$ Siemens/m $=10^{-12}$ mho/m). The conductivity ranges for the three categories of

TABLE 4.1
Classification of Liquids Based on Electrical Conductivity[3,5]

	Conductivity (pS/m)	Examples
Low conductivity	$<10^2$	Paraffins (hexane, heptane), aromatics (toluene, xylene), kerosene
Medium conductivity	10^2–10^4	Fuel oils, gasoline containing up to 5% ethanol as an additive
High conductivity	$>10^4$	Crude oil, alcohols, ketones, water

liquids are given in Table 4.1. The petroleum industry generally uses 50 pS/m for the upper limit of conductivity (instead of 100 pS/m) for low-conductivity liquids, and 1,000 pS/m (instead of 10,000 pS/m) for the lower limit of conductivity for high-conductivity liquids.[4]

The conductivity of a liquid is sensitive to its temperature and purity. Therefore, the conductivity ranges for various categories in Table 4.1 should be regarded as order-of-magnitude values only. For example, for benzene or hexane, the conductivity of high-purity samples could be well below 0.01 pS/m. However, the values for typical commercial products are higher by 1 or 2 orders of magnitude.

4.3.2 Solids

Solid materials can transmit electric currents through their volume or over their surfaces. They are, therefore, characterized in terms of both volume resistivity and surface resistivity. Volume resistivity is the resistance of a body of unit length and unit cross-sectional area, and is expressed in ohm.m. Surface resistivity is defined as the resistance across a surface element having a unit length and unit width, and is expressed in ohm/square or ohm. In effect, the surface resistivity is the resistance between two opposite sides of a square and is independent of the size of the square or its dimensional units.

Solids are classified as conductive, dissipative (semiconductive), or insulating (nonconductive). The demarcation is complex and varies according to the object in question (material container, liner, clothes, footwear, gloves, floor, etc.). The definitions include volume resistivity, surface resistivity, and leakage resistance. CENELEC[6] gives definitions for most of these cases, and for powders, it continues with the classification based on volume resistivity, as follows:

1. Low-resistivity powders, e.g., metals, with volume resistivities up to about 10^6 ohm.m
2. Medium-resistivity powders, e.g., many natural organic powders, with volume resistivities in the range of 10^6–10^{10} ohm.m
3. High-resistivity powders, e.g., polymers, some synthetic organic powders, and dry natural organic powders, with volume resistivities of 10^{10} ohm.m or higher.

The surface resistivity of solids is sensitive to the humidity of the surroundings. At high humidities, the surface of many materials that are not hydrophobic adsorbs enough moisture to ensure a surface conductivity that is high enough to prevent the accumulation of static electricity. The same materials could also become good insulators at relative humidities below 30%.

4.4 GENERATION OF ELECTROSTATIC CHARGE[5,7,8]

4.4.1 Mechanisms of Charge Generation

There are three primary mechanisms of electrostatic charging:

- Relative movement at material interfaces (contact charging)
- Induction
- Charge transfer.

4.4.1.1 Relative Movement at Material Interfaces

This charging phenomenon occurs when solids or liquids move relative to materials in contact with them, e.g., the flow of a liquid through a pipeline or the settling of droplets in a continuous phase of another fluid.

Electric charges are always present at material interfaces. When there is no relative movement, the charges on either side of the interface are generally equal and opposite so that, as a whole, the interface is neutral. Relative movement of the materials at an interface causes separation of the charges on either side of the interface, each material carrying equal and opposite charges. An essential condition for charging by this process is that at least one of the materials has low electrical conductivity.

When movement is continuous, a stream of charge is carried with the flowing material, causing a continuous electric current, known as "streaming current" or charging current. For example, if a liquid flows through a 50-mm-diameter pipeline at a velocity of 4 m/s, and acquires an average charge density of 50 $\mu C/m^3$, then the streaming current (charge density times the volumetric flow rate) is 0.4 μA.

Typical processes that give rise to charging by this method are as follows:

- The flow of low-conductivity liquids through pipes, valves, filters, etc.
- Pouring liquids from buckets
- Settling of droplets or solids in liquids
- Stirring of two-phase liquids or solid–liquid mixtures
- Pouring solids from containers
- Grinding/sieving
- Pneumatic conveying of solids.

4.4.1.2 Induction

Electrostatic induction occurs between two bodies when they are not in contact with each other. An already-charged primary body raises the electric potential of its surroundings since it has an electric field around it. If an insulated secondary body made of conducting material is brought into this field, there is an instantaneous redistribution of electrons in the secondary body (polarization). In this state, the secondary body has no net charge on it, but it has a potential equal to the primary's average potential. This potential varies proportionally with the charge in the primary and decreases as the distance between them increases.

If the secondary body, polarized as above, is earthed momentarily, its potential reduces to zero. However, it retains a charge, called the induced charge, which is of a sign opposite to that in the primary. The induced charge keeps the secondary at the earth's potential despite the charge on the primary. If the secondary body is moved away from the primary electric field's influence, the induced charge in the secondary will raise its potential. Therefore, the secondary body will have both a potential and a charge until the charge is dissipated away.

Typical examples of the phenomenon of acquiring high potentials by induction are as follows:

- A person working ungrounded in the vicinity of a reactor where a powder is being charged from a bag
- An earthed metallic level probe in the vapor space of a container being filled with a charged liquid.

4.4.1.3 Charge Transfer

When a charged object contacts an uncharged body, the charge is shared between them to the extent that their conductivities allow. An example of this is filling a charged liquid into a container when the liquid shares its charge with the container.

4.4.2 Quantitative Relationships for Charge Generation

4.4.2.1 Charge Generation on Liquids

Electrostatic charging of liquid by flow through pipes is highly variable. Based on BS-5958[9] and a detailed paper by Walmsley et al.,[10] the following empirical equations are suggested for order-of-magnitude estimates of streaming current and charge density for single-phase turbulent flow of low-conductivity liquids through industrial pipes:

$$I_s = K_1 v^2 d^2 \left(1 - e^{-\frac{t}{\tau}}\right) \tag{4.1a}$$

$$Q_v = 5v \left(1 - e^{-\frac{t}{\tau}}\right) \tag{4.1b}$$

where

K_1 = empirical constant (4 in Walmsley's paper)
I_s = streaming current, μA
Q_v = charge density, μC/m³
d = diameter of the pipe, m
v = velocity of the liquid in the pipe, m/s
t = residence time of the liquid in the pipe, seconds
τ = time constant or relaxation time of the liquid, seconds.

The exponential decay term in Equation (4.1a) or (4.1b) approaches zero if the residence time of the liquid in the pipe exceeds about 3 times the relaxation time. Hence, the streaming current or the charge density of the liquid becomes stationary. For liquids having a conductivity less than about 2 pS/m, this condition is achieved if the residence time exceeds about 100 seconds.

To limit static charge generation, API Recommended Practice 2003[7] recommends the maximum allowable velocity in a pipe as 7 m/s. The presence of a second phase (e.g., immiscible liquid or entrained solids) in the flowing liquid significantly increases the streaming current. Therefore, in those situations, it is considered necessary to restrict the maximum velocity to 1 m/s when vessels are being filled, besides allowing sufficient waiting time for charge relaxation before any object (such as a temperature or level probe) is lowered into the vessel. The velocity restriction of 1 m/s also applies if there is any water at the bottom of a tank that may be stirred during tank filling.

The disintegration of a liquid jet into small droplets during splash filling of a vessel can produce a highly charged spray or mist, irrespective of the liquid's conductivity. In general, the more conducting the liquid, the greater the charge generation. For example, a water jet produces more charge than an oil jet.[6]

Installation of a fine filter in a pipeline results in a high charging current. Typical charge densities in the liquid leaving such a filter can range from 10 to 5000 μC/m^3. However, an accurate calculation is not achievable since even identical filters can give widely differing results.[9]

4.4.2.2 Charge Generation in Powders

In NFPA's definition, powders include pellets, granules, and dust particles. Pellets have diameters greater than 2 mm, granules between 420 μm and 2 mm, and dusts 420 μm or less. Aggregates of pellets and granules will often contain significant amounts of dust.

Powders can be charged by contact and separation between particles or between particles and other surfaces such as bags or pipe walls. Charging may also occur when particles break apart. Contact and separation processes between particles result in bipolar charging; the charge on coarser particles is predominantly of a sign opposite to the sign of the charge on finer particles.

Charging between particles and boundary surfaces, on the other hand, is unipolar; the charges on the particles and the boundary surface are opposite in sign. Interparticle charging and particle–boundary charging usually coincide. The effect overall is, therefore, complicated and can change radically if some relevant parameter is changed. Such parameters include materials of construction, particle size distribution, flow velocity, material density, and atmospheric humidity. A more detailed discussion is available.[5]

The charge generation in powders is usually expressed as a charge-to-mass ratio, or mass charge density, μC/kg. Table 4.2 shows data on charge levels for medium-resistivity powders emerging from various operations. These values are for suspended powders before compaction.

The onset of discharges limits the maximum charge density on powder. For spherical particles, an empirical equation[5] shows that the maximum charge-to-mass ratio, μC/kg, is $24.7/(\rho\ r^{1.3})$, where ρ is the density of powder particles, kg/m^3, and r is the

TABLE 4.2
Typical Charge Levels on Medium-Resistivity Powders Emerging from Various Operations[6]

Operation	Mass Charge Density (μC/kg)
Sieving	10^{-3}–10^{-5}
Pouring	10^{-1}–10^{-3}
Grinding	1–10^{-1}
Micronizing	10^{2}–10^{-1}
Pneumatic conveying	10^{3}–10^{-1}

radius of the particle, m. Thus, for a powder with a particle density of 1,300 kg/m^3, and a particle radius of 100 μ, the maximum charge-to-mass ratio is 3,000 μC/kg, or three times the peak value given in Table 4.2 for pneumatic conveying. For finer particles, e.g., those with a radius of 100 μ, the estimated charge-to-mass ratio is 24,000 μC/kg, or 24 times the peak value given in Table 4.2.

4.5 ACCUMULATION OF ELECTROSTATIC CHARGE

The conditions under which electrostatic charges accumulate to produce hazardous potentials on the different categories of accumulators can be examined in the following sections.

4.5.1 Accumulation in Liquids

The tendency of a liquid to accumulate charge is indicated by its charge relaxation time, τ, which is defined as follows:

$$\tau = \frac{\varepsilon \varepsilon_o}{\gamma_c} \tag{4.2}$$

where

ε = dielectric constant or relative permittivity of the liquid
ε_o = permittivity of space (see Section 4.2.5)
γ_c = conductivity of the liquid, S/m.

The significance of relaxation time can be appreciated by considering a simple example of the loss of charge from a batch of liquid in an earthed conducting vessel. The rate of loss of charge from liquids is determined mainly by conduction, and the effect of convection and diffusion is negligible. Under this condition, the rate of loss of charge is given by:

$$-\frac{dQ}{dt} = \frac{Q\gamma_c}{\varepsilon \varepsilon_o} = \frac{Q}{\tau} \tag{4.3}$$

where
Q is the charge in the liquid at time t.

By integrating Equation (4.3) under the condition that $Q=Q_o$ at $t=0$, we get:

$$Q = Q_o e^{-t/\tau} \quad (4.4)$$

It follows from Equation (4.4) that at $t=\tau$, Q is 37% of Q_o. Hence, the relaxation time is the time required for the charge level in the liquid to come down to 37% of its original value. It is evident from Equation (4.4) that the level of relaxation depends on the properties of the liquid only and is independent of the geometry of the vessel.

Expressed in terms of charge density, Equation (4.4) can be written as:

$$Q_v = Q_{vo} e^{-t/\tau} \quad (4.5)$$

where
τ = charge relaxation time, seconds
t=time, seconds
Q_v=volumetric charge density, μC/m^3
Q_{vo}=initial volumetric charge density, μC/m^3.

Equations (4.2)–(4.5) are applicable for liquids with conductivities above about 2 pS/m (i.e., those liquids that follow the exponential decay or Ohmic relaxation law). For lower conductivities, the retained charge follows a hyperbolic relaxation law, which is given as follows[5,7]:

$$Q_v = \frac{Q_{vo}}{1 + \mu Q_{vo} t / (\varepsilon \varepsilon_o)} \quad (4.6)$$

where
τ=time, seconds
Q_v=volumetric charge density, μC/m^3
Q_{vo}=initial volumetric charge density, μC/m^3
μ=charge carrier mobility, about 10^{-8} m^2/(V.s).

The choice of the relaxation model affects the retained charge markedly, as can be seen from Table 4.3. In this table, the charge density after 100 seconds has been calculated for different values of the initial charge density, using Equation (4.5) for the exponential decay model and Equation (4.6) for hyperbolic relaxation. A value of 30 seconds for relaxation time, τ has been assumed for use in the exponential decay model.

Table 4.3 shows that for hyperbolic relaxation, the retained charge density is roughly independent of the initial charge density. The use of a residence time of 100 s in these calculations is based on the BS-5958[9] recommendation between the filter and a receiving vessel for low-conductivity liquids.

TABLE 4.3
Charge Density After 100 Seconds for the Hyperbolic and Exponential Decay Models

	Q_v (μC/m³)	
Q_{vo} (μC/m³)	Hyperbolic	Exponential
5,000	17.6	178.4
2,000	17.5	71.3
1,000	17.4	35.7
500	17.1	17.8
100	15.0	3.6

We now consider a typical flow system comprising an earthed (or grounded) vessel into which there is a continuous input of a charged liquid (conductivity more than 2 pS/m). The vessel's loss of charge consists of charge carried away by the flowing liquid and the relaxation charge loss. The terms in Figure 4.1 are as follows:

I_s = rate of charge inflow with the liquid entering the vessel (streaming current)
I_{out} = rate of charge outflow with the liquid leaving the vessel
F = volumetric flow rate of liquid
V_c = volume of the vessel
Q = total charge in the vessel at time t.

Assuming that the liquid in the vessel is well mixed, resulting in a uniform charge density in the vessel, we have:

$$I_{out} = \frac{FQ}{V_c} \tag{4.7}$$

Hence, the charge balance equation can be written as follows:

$$\frac{dQ}{dt} = I_s - \frac{FQ}{V_c} - \frac{Q}{\tau} \tag{4.8}$$

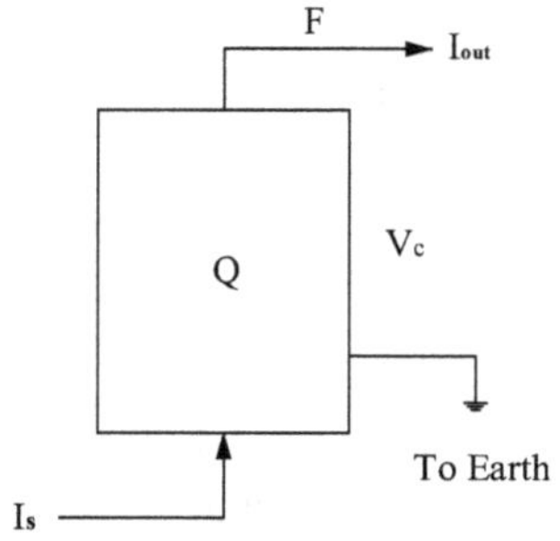

FIGURE 4.1 Charge accumulation in a flow system.

By integrating Equation (4.8) under the initial condition that $Q = Q_o$ at $t = 0$, we get:

$$Q = \frac{I_s}{P} + \left(Q_o - \frac{I_s}{P}\right)e^{-Pt} \tag{4.9a}$$

$$P = \frac{F}{V_c} + \frac{1}{\tau} \tag{4.9b}$$

When the tank is being filled from an empty condition, $F = 0$ (no liquid outflow) and $Q_o = 0$, whence,

$$Q = I_s\tau\left(1 - e^{-\frac{t}{\tau}}\right) \tag{4.10}$$

Charge accumulation in the liquid will increase its electrical potential, thereby leading to the possibility of a hazardous discharge. Calculation of the potential is quite complex, and models are available only for simple geometrical shapes. The analytical procedure for calculating the potential in a grounded cylindrical vessel containing a liquid of uniform charge density has been provided by Asano.[8] For hazard analysis, it is necessary to calculate the maximum potential, which occurs at the center of the liquid–vapor interface at a given filling ratio. The calculation can be done readily using a dimensionless equation provided by Asano,[8] which is as follows:

$$\Phi^*_{\max} = \sum_{n=1}^{\infty} \frac{1}{q_n{}^3 J_1(q_n)} \frac{1 - \dfrac{1}{\cosh(q_n \alpha H)}}{\varepsilon + \dfrac{\tanh(q_n \alpha H)}{\tanh\{q_n H(1-\alpha)\}}} \tag{4.11}$$

where

$$\Phi^*_{\max} = \frac{\varepsilon_o \Phi_{\max}}{2Q_v\, b^2}, \quad H = \frac{d+p}{b}, \quad \alpha = \frac{d}{d+p}$$

In Equation (4.11):

$\Phi^*_{\max}$ = maximum dimensionless potential at a given filling ratio
$\Phi_{\max}$ = corresponding maximum potential, V
H = ratio of total height to the radius of the vessel
α = ratio of liquid height to the total height of the vessel
d = height of liquid, m
p = height of vapor space, m
b = radius of the vessel, m
ε = dielectric constant of the liquid
ε_o = permittivity of space
Q_v = charge density in the liquid, C/m^3.

TABLE 4.4
Calculated Values of Φ^*_{max} for Different Values of α and H

	Φ^*_{max} (at $\varepsilon=2$)				
H	$\alpha=0.2$	$\alpha=0.3$	$\alpha=0.5$	$\alpha=0.7$	$\alpha=0.9$
0.5	0.0011	0.0022	0.0051	0.0069	0.0045
1.0	0.0042	0.0084	0.0169	0.0211	0.0142
1.5	0.0085	0.0157	0.0271	0.0314	0.0223
2.0	0.0134	0.0222	0.0334	0.0368	0.0227
3.0	0.0223	0.0315	0.0391	0.0406	0.0337
4.0	0.0290	0.0366	0.0408	0.0414	0.0370
5.0	0.0336	0.0391	0.0414	0.0416	0.0387

Note that $J_0(q_n)$ and $J_1(q_n)$ are Bessel functions of order 0 and 1, respectively, and q_n are the roots of the equation $J_0(q_n)=0$.

For convenience, the calculated values of Φ^*_{max} for different α and H values are given in Table 4.4. Asano[8] has also presented similar results in graphical form and has also shown that Φ^*_{max} reaches a maximum value of 0.0417 as H tends to infinity. In these calculations, the dielectric constant of the liquid has been assumed as 2.

Example 4.1

Toluene is being pumped through a 300-mm-diameter, 500-m-long pipeline at an average velocity of 4 m/s. Calculate the streaming current. The properties of toluene are as follows:

Density $=871$ kg/m^3
Viscosity $=0.0006275$ kg/(m.s)
Relative dielectric constant $=2.4$
Conductivity $=1$ pS/m

Reynolds number, Re $=(0.3)$ (4) $(871)/0.0006275=1.67\times10^6$, which far exceeds the threshold value of 2,300 for laminar flow. Hence, the flow is turbulent.

Since the conductivity is less than 2 pS/m, an average relaxation time of 30 seconds can be assumed.

The residence time of the liquid flowing through a 500 m length of the pipeline $=500/4=125$ seconds.

Since the residence time is more than 100 seconds (i.e., three times the relaxation time), the pipeline can be regarded as long. Therefore, from Equation (4.1a), with $K_1=10$,

$$I_s=(10)\left(4^2\right)\left(0.3^2\right)=14.4\,\mu\text{A}$$

Example 4.2

A 4-m-diameter and 6-m-high cylindrical process vessel is to be half-filled by pumping toluene from a storage tank through a 100-mm-diameter, 50-m-long pipeline connected to the vessel bottom. The filling rate is 25 L/s. The vessel and the filling line are made of conducting material and are duly earthed. Toluene conductivity is 1 pS/m, and its dielectric constant is 2.4. The dynamic viscosity of toluene is 0.0006275 kg/(m.s). Calculate the maximum potential acquired by the liquid.

Volume of toluene to be filled into the vessel = (0.785) (4^2) (3) = 37.7 m^3.

Filling time = 37.7/0.025 = 1,507 seconds

Velocity of toluene in the filling line = 0.025/0.785/0.1^2 = 3.18 m/s

The Reynolds number in the filling line = (0.1) (3.18) (871)/0.0006275 = 4.42×10^5, which exceeds the 2,300 threshold value for laminar flow. Hence, the flow is turbulent.

The residence time of the liquid in the filling line = 50/3.2 = 15.6 seconds

The relaxation time of toluene = 30 seconds (see Example 4.1).

Since the liquid's residence time in the filling line is less than the relaxation time, the exponential decay term is not negligible. Hence, from Equation (4.1a) with $K_1 = 10$

Streaming current $I_s = (10)\ (0.1)^2\ (3.2)^2\ [1 - e^{-(15.6/30)}] = 0.415\ \mu A$

Since the filling time is far more than 3 times the relaxation time, the exponential term in Equation (4.10) will be negligible.

Hence, accumulated charge $Q = I_s\,\tau = (0.415)\ (30) = 12.45\ \mu C$

Charge density in the liquid = 12.45/37.7 = 0.33 $\mu C/m^3$

From Table 4.4, at $\alpha = 0.5$ and $H = 3$, $\Phi^*_{max} = 0.0391$

Hence, from Equation (4.11), the maximum potential is:

$$\Phi_{max} = (2)\left(0.33\times10^{-6}\right)\left(2^2\right)(0.0391)/\left(8.8\times10^{-12}\right) = 11{,}730\,\text{V}.$$

The above calculation is for $\varepsilon = 2$. Use of Equation (4.10) at $\varepsilon = 2.4$ gives:

$$\Phi^*_{max} = 0.0345, \quad \text{and} \quad \Phi_{max} = 10{,}300\,\text{V}.$$

4.5.2 Accumulation on Insulated Conductors

Electrostatic charge accumulates on an insulated conductor whenever it is subjected to a charging current. For example, when a stream of charged liquid flows into an insulated metal container, the container accumulates charge and thereby acquires a potential.

The process of accumulation is usually represented by an equivalent electrical circuit as shown in Figure 4.2. The input of charge is the charging current, the output of charge is the leakage current to earth, and the difference between the input and the output is accumulated in the conductor. The unsteady state equation for charge balance is as follows:

$$C\frac{d\Phi_t}{dt} = I_s - \frac{\Phi_t}{R} \tag{4.12}$$

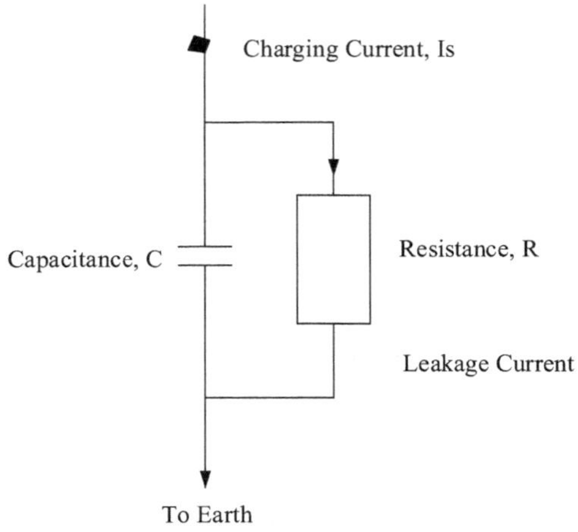

FIGURE 4.2 Equivalent circuit for an electrostatic charging of a conductor.[9]

where

t = time, s
I_s = charging current, A
Φ_t = electrical potential of the conductor, V
C = capacitance of the conductor, F
R = resistance between the conductor and the earth, ohm.

By integrating Equation (4.12) under the condition that at $t=0$, $\Phi_t=0$, we get:

$$\Phi_t = I_s R\left(1 - e^{-t/RC}\right) \tag{4.13}$$

The maximum potential is reached when t greatly exceeds RC (the exponential term approaches zero). Hence,

$$\Phi_{\max} = I_s R \tag{4.14}$$

Thus, when subjected to a charging current, the maximum potential acquired by an insulated conductor is determined by the charging current and the resistance between the conductor and the earth. The capacitance of the conductor affects the rate at which the maximum potential is reached. The product of *R* and *C* is known as the time constant of the system, expressed in seconds.

To ensure that the potential acquired by a conductor does not reach a level at which an incendive discharge might occur, the resistance to earth must not exceed a threshold value. Equation (4.14) is useful for determining this maximum resistance. Thus, if the maximum charging current in a given situation is 10 μA, and the

TABLE 4.5
Capacitance of Some Common Conductors[9]

	Capacitance (pF)
Small metal items (scoop, hose nozzle)	10–20
Small containers (bucket, 50 L drum)	10–100
Medium containers (250–500 L)	50–300
Major plant items (reaction vessels) immediately surrounded by an earthed structure	100–1000
Human body	100–300

threshold potential to avoid an incendiary discharge is 1,000 V, then the resistance to earth must be below 10^8 ohm.

The energy stored in an insulated conductor is given by

$$W = \frac{1}{2}C\Phi^2 = \frac{1}{2}Q\Phi = \frac{1}{2}\left(\frac{Q^2}{C}\right) \tag{4.15}$$

where

W = energy, J
C = capacitance, F
Φ = potential, V
Q = accumulated charge, C.

Calculation of the capacitance, particularly of large objects such as liquid containers, is a challenging exercise as it depends on several factors related to the surroundings. Actual measurement using a bridge is desirable. However, considering the inaccuracies involved in other factors such as charge density, rough estimates should be adequate for most hazard assessment work. Values of capacitance in picoFarads (pF) for some common conductors are given in Table 4.5.

(Note: 1 pF = 10^{-12} F).

Example 4.3

Rework Example 4.2, assuming that the earth connection is broken and the tank becomes ungrounded.

In this case, the charge dissipated from the liquid will appear on the outside surface and will equal the charge accumulated in the liquid (the tank will act as a Faraday pail). At an assumed capacitance of 1,000 pF for the tank (see Table 4.5), a charge of 12.6 μC will raise the tank's potential to $(12.6 \times 10^{-6})/10^{-9} = 12{,}600$ V, relative to ground. Hence, the maximum liquid potential would be 11.7 kV relative to the tank wall and (11.7 + 12.6), or 24.3 kV relative to the ground.

Based on what has been discussed so far, the maximum potential is reached as a steady-state value when the charging current equals the rate of dissipation to earth. In some cases (e.g., in powder handling processes), the charge is lost through static discharges at lower voltages before this steady condition is reached. The phenomenon of static discharge is dealt with in Section 4.6.

4.5.3 Accumulation on Lined/Coated Containers

Case (i): Conductive Liquid

Let us consider a case where a conductive liquid is filled into a metallic tank with an inner liner or coating for corrosion resistance, based on Britton's[11] method. Several factors could give rise to a charging current, such as splash filling from a nozzle or the presence of a second dispersed phase in the liquid.

If the liner is thick and is also made of a high-resistivity material, the incoming charge would accumulate in the liquid. The accumulated charge, Q_t, at time t would be:

$$Q_t = Q_v F t = I_s t \tag{4.16}$$

where

t = filling time, s
Q_t = accumulated charge in the liquid, μC
Q_v = incoming charge density, μC/m^3
I_s = streaming current, μA
F = volumetric rate of filling, m^3/s.

Since the liquid is conductive, the potential would always be uniform throughout the liquid. The liquid/liner/tank wall combination can be considered a parallel plate capacitor, with the liner acting as a dielectric. The capacitance of the liquid at time t, C_t, can be estimated as follows:

$$C_t = \frac{\varepsilon\varepsilon_o A_w}{a} = \frac{\varepsilon\varepsilon_o\left(\pi b^3 + 2Ft\right)}{ab} \tag{4.17}$$

where

t = time, seconds
C_t = capacitance, F
ε = dielectric constant of the liner
A_w = area of the wetted wall, m^2
a = liner thickness, m
b = radius of the tank, m.

The liquid potential, Φ_t (V) at time t is, therefore:

$$\Phi_t = \frac{Q_v F t a b}{\varepsilon\varepsilon_o\left(\pi b^3 + 2Ft\right)} \tag{4.18}$$

Example 4.4

Isopropyl acetate, a conductive liquid having a conductivity of about 30,000 pS/m, is being charged from a top nozzle into a 1.5-m-diameter, 4 m^3 vertical cylindrical reaction vessel. The vessel is made of carbon steel and is lined inside with a 2-mm-thick layer of polyethylene (dielectric constant = 2) for corrosion protection. The liquid charging rate is 4 L/s; the total volume of liquid to be charged is 2 m^3. The estimated streaming current is 3 μA. Calculate the potential developed in the liquid as a function of time, assuming that the steel tank and the filling system are duly earthed.

Volumetric flow rate, $F = 0.004$ m^3/s
Inlet charge density, $Q_v = 3/0.004 = 750$ $\mu C/m^3$
Dielectric constant of liner, $\varepsilon = 2$
Permittivity of space, $\varepsilon_o = 8.85 \times 10^{-12}$ C/(V.m)
Liner thickness, $a = 0.002$ m
Vessel radius, $b = 0.75$ m

Substituting these values in Equations (4.18) and (4.19a and b), we get the values of capacitance and liquid potential at various values of time until completion of charging at 500 seconds, as shown below:

Charging time (seconds)	100	200	300	400	500
Capacitance (pF)	25,100	34,500	44,000	53,400	62,800
Potential (V)	12,000	17,400	20,500	22,500	23,900

This is a hazardous situation since conductive liquids are prone to produce incendiary sparks at potentials above a threshold of about 1,000 V. Therefore, replacing the liner with a corrosion-resistant coating of 50 to 100 μ thickness would be desirable because the liquid potential would no longer exceed the threshold potential.

Sometimes, it is possible to use a liner made of semiconducting or antistatic material (volume resistivity 10^8–10^{10} ohm.m) instead of a high-resistivity material. In such cases, charge from the liquid is lost through the liner according to Ohm's law, and the potential drop across the liner is given by:

$$\Phi = I_s R \tag{4.19a}$$

$$R = \frac{\rho\, a}{A_w} \tag{4.19b}$$

where

Φ = potential of the liquid, V
R = liner resistance, ohm.

For a given liner, the conservative situation would correspond to an initial phase of the filling operation when the wetted area, A_w, is the minimum; therefore, the resistance is a maximum.

Example 4.5

Rework Example 4.4 assuming that polyethylene is replaced as the liner by a material of volume resistivity $= 10^{10}$ ohm.m, and the thickness of the liner remains unchanged.

A_w at the start of filling $= \pi\ (0.75)^2 = 1.77\ m^2$

$R = 10^{10}\ (2 \times 10^{-3})/1.77 = 1.13 \times 10^7$ ohm

Hence, $\Phi = (3 \times 10^{-6})\ (1.13 \times 10^7) = 34$ V, which is negligible.

Case (ii): Nonconductive Liquid

No simple model for estimating tank potential is available in this case. The lining material should be selected assuming that the lining introduces a negligible additional impediment to charge dissipation from the liquid.[5] Accordingly, the relaxation time of the liner ($\tau = \varepsilon\ \varepsilon_o\ \rho$) is less than or equal to that of the liquid. For example, a liner material of dielectric constant equal to 4 and a volume resistivity of 10^{10} ohm.m will have a relaxation time of 0.36 seconds, which satisfies the above criterion (a low-conductivity liquid, such as toluene, has a relaxation time of about 20 seconds).

In Example 4.5, if such a material is used as a liner (charging current $= 3$ μA, $A_w = 1.8\ m^2$, and lining thickness $= 2$ mm), then the maximum resistance of the liner is estimated to be $10^{10}\ (2 \times 10^{-3})/1.8$, or approximately 1.11×10^7 ohm. Therefore, for a charging current of 3 μA, the maximum potential drop in the liner would be:

$(3 \times 10^{-6})\ (1.11 \times 10^7)$, or 33.3 V, which is negligible.

For a nonconducting material of volume resistivity equal to 10^{12} ohm.m (such as glass), the liner's relaxation time increases to 36 seconds, and the maximum resistance for the same reactor conditions would be 100 times higher. The maximum potential drop would thus increase to 3,000 V. The potential gradient, in this case, would correspond to 1.5×10^6 V/m, which is close to the breakdown voltage of 10^7 V/m for common liner materials.

4.5.4 Accumulation on Powders

The relaxation time of the powder in bulk (i.e., when settled in a drum or silo) is:

$$\tau = \rho_v \varepsilon \varepsilon_o \tag{4.20}$$

where

$\tau =$ relaxation time, seconds

$\rho_v =$ volume resistivity of the powder in bulk, ohm.m

$\varepsilon =$ bulk dielectric constant of the powder in bulk

$\varepsilon_o =$ permittivity of space, F/m.

A bulked powder is a mixture of particles with interparticulate voids filled with gas (mostly air). Therefore, the bulk properties of a powder heap (such as density, dielectric constant, or volume resistivity) depend not only on the properties of the solid particles but also on the packing density. Packing density, in turn, depends on the average particle size/shape and the standard deviation of the particle size distribution. The actual packing density might differ from the one obtained during volume resistivity measurements by the standard method. Considering the inherent variability in the values of these parameters, the relaxation time estimated using Equation (4.20) should be regarded only as a guide for the charge accumulation tendency.

The dielectric constant for air is about 1, and that for solid particulates varies over a narrow range (usually between 2 and 4). The volume resistivity of powder particles, on the other hand, varies widely from less than 10^6 ohm.m for conductive powders to greater than 10^{10} ohm.m for nonconductive powders (and up to 10^{16} ohm.m for thermoplastic resins). Hence, the relaxation time of a powder heap is determined mainly by the conductivity of the powder and is relatively insensitive to its dielectric constant.

Accumulation of charge is, therefore, considered under categories, depending on the conductivity of the powder, as follows:

- Conductive (Low-Resistivity) Powders

  Conductive powders, such as aluminum, become charged during flow, but this charge is lost almost immediately as the powder is transferred into grounded, conductive containers.

  However, a charge may accumulate on a conductive powder if it is filled into a nonconductive container or a container with a nonconductive lining. If a grounded metal rod is brought near such a charged powder, the charge accumulated on the powder may be dissipated through a spark, which could ignite the powder. Sometimes, earthed metallic rods are placed inside such containers to facilitate charge dissipation from the powder, but this should be done before the commencement of powder filling.
- Semiconductive (Medium-Resistivity) Powders

  Semiconductive powders (volume resistivity between 10^6 and 10^{10} ohm.m) are believed not to produce bulking brush discharges while also being too resistive to produce sparks. Bulk discharges are possible, but the effective energy of such discharges is usually less than 1 mJ. Hence, ignitions are not expected.
- Nonconductive (High-Resistivity) Powders

  Nonconductive powders lose charge at a slow rate owing to their high resistivity. They tend to accumulate charge even in properly grounded equipment such as powder silos. Relaxation time could increase, from 0.2 seconds for a powder with a dielectric constant of 2 and volume resistivity equal to 10^{10} ohm.m, to more than an hour with a dielectric constant of 3 and volume resistivity equal to 10^{14} ohm.m. Nonconductive powders do not give rise to spark discharges, but other discharges (corona, brush, bulking brush, and propagating brush, as discussed below in Section 4.6) are not unlikely.

4.6 ELECTROSTATIC DISCHARGE

Five types of electrostatic discharge (discussed below) need to be considered for assessing ignition hazards[5]:

1. Spark discharge
2. Corona discharge
3. Brush discharge
4. Propagating brush discharge (PBD)
5. Bulking brush discharge (also known as cone discharge).

4.6.1 Spark Discharge

A spark discharge occurs between two conductors (liquid or solid) separated from each other by a gap and at different potentials. It is characterized by a well-defined luminous discharge channel of high current density. Spark discharges are also called capacitor discharges since any system of two conductors isolated from each other is effectively a capacitor.

A spark discharge occurs between conductors when the electrical field strength between them exceeds a level, known as breakdown strength, that depends on the gap width. As a guide, the breakdown strength of air between flat or large radius surfaces 10 mm or more apart is about 3,000 kV/m at atmospheric pressure and increases as the gap decreases.[9] Electric fields that produce sparks are usually nonuniform, however, and a value of 3,000 kV/m is therefore attained somewhere in the gap, even when the average field exceeds about 500 kV/m. Therefore, a potential of about 1,000 V is regarded[9] as sufficient to cause a discharge across a gap of about 2 mm.

Britton[5] provides a discussion on the effect of gap width and the nature of the gas present in the gap. He also gives a usable criterion, based on

a. The capacitance of the conductor
b. The diameter of the earthed electrode
c. The nature of the gas between the conductor and the electrode.

FIGURE 4.3a Spark discharge.[5]

TABLE 4.6
Variation of Minimum Ignition Voltage and Corresponding Ignition Energy (mJ) with Capacitance and Electrode Diameter

	Capacitance (pF)		15 mm diameter	1.5 mm diameter	0.5 mm diameter	Points
Hydrogen (28 vol%	146	V	2,300	1,900	1,700	1,000
in air)		mJ	0.39	0.26	0.21	0.073
	30	V	2,500	2,300	–	1,500
		mJ	0.094	0.079	–	0.034
Methane (8.5 vol%)	146	V	8,000	5,500	5,350	3,200
in air		mJ	4.67	2.21	2.09	0.75

Finally, Britton gives experimental data on the minimum ignition voltage variation with conductor capacitance and electrode diameter for hydrogen and methane. Some of these data are reproduced in Table 4.6. Calculated values of ignition energy for various combinations of voltage and capacitance are also shown in the table. "Points" in the table refers to steel gramophone needles.

This table shows that both the ignition voltage and the ignition energy decrease with the capacitance of the charged conductor and the size of the earthed electrode. In each case, the ignition energy is several times greater than the minimum ignition energy (MIE) of the gas (0.016 mJ for hydrogen and 0.21 mJ for methane, see Section 4.7).

Typical examples of spark discharge (and possible consequential ignition) are as follows:

- A metal can that is floating on a charged low-conductivity liquid discharging to the side of an earthed tank-truck compartment
- A conducting liquid that is being filled into a drum with nonconducting internal lining discharging to an earthed metallic filling nozzle
- A person wearing woolen clothing and insulating shoes tries to grab an earthed railing in an area where a flammable hydrocarbon is being filled into small containers open to the atmosphere.

Spark discharges can be excluded by earthing/bonding of all conductors (Section 4.9).

4.6.2 Corona Discharge[5,10]

Corona discharges occur when an earthed conductor (usually called an electrode), with a diameter less than 5 mm, and having a sharp, pointed end, is moved towards a charged surface, such as nonconductive liquid, plastic sheet, or powder. This type of discharge occurs when the electric field at the sharp surface is very high (above 3,000 kV/m).

Corona discharges are characterized by a hissing sound and a faint glow. The energy dissipated through the corona discharge is less than 0.2 mJ and, therefore,

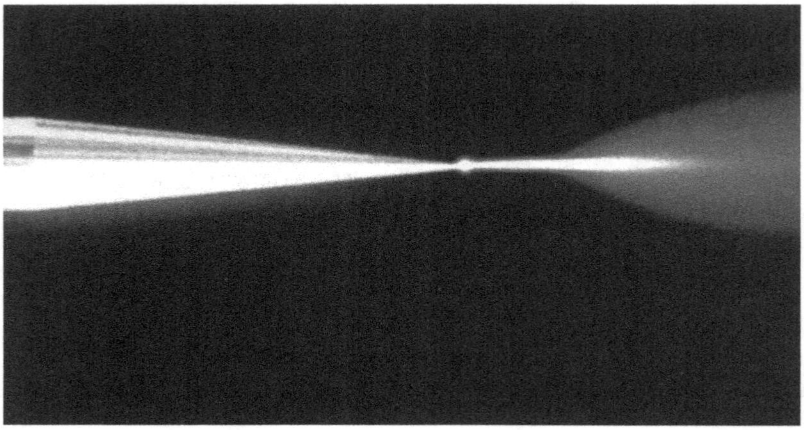

FIGURE 4.3b Corona discharge.[5]

insufficient to ignite most gases and vapors; exceptions include hydrogen, acetylene, or carbon disulfide.

4.6.3 Brush Discharge[5,10]

Brush discharges occur between grounded, blunt conducting bodies – diameter greater than 10 mm – and isolated, charged nonconducting surfaces (e.g., between a person's finger and a plastic surface or between a metallic dip-leg and the surface of

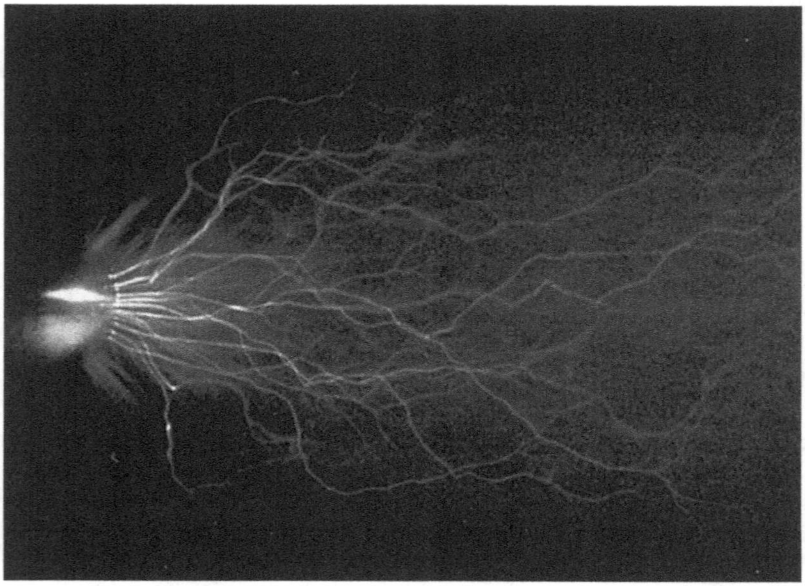

FIGURE 4.3c Brush discharge.[5]

a nonconducting liquid in a tank). Brush discharges are characterized by a hot, conducting plasma channel from the conductor and terminating in a brush-like network of smaller channels reaching the insulating surface.

Under atmospheric conditions, the threshold potential difference between the conductor and the charged insulating surface is of the order of 20–25 kV for a brush discharge to occur.

Unlike in a spark, where almost the entire charge of the charged conductor is dissipated in one discharge, the energy transferred in a brush discharge is a small fraction of the charge on the insulating material. This is because the mobility of the charges along the surface, or through the volume of the nonconducting material, is too low compared with a spark's duration.

The incendiary potential of brush discharges depends on several factors. This is a complex subject; interested readers may wish to consult references.[5,10] For hazard assessment, ensuring that the accumulated potential for nonconducting liquids does not exceed 20–25 kV is sufficient. Where this is not practically feasible, flammable gas mixtures must be diluted using inert gas. This approach is referred to as "inerting".

Typical activities that exhibit a potential for brush discharges are as follows:

- The approach of earthed metallic tools, or a human fingertip, to highly charged insulating surfaces, e.g., plastic pipes used for conveying liquids or dusts, plastic bags, intermediate bulk containers
- Discharging of solids from plastic bags in the vicinity of metal fittings (e.g., above an access port of a reaction vessel)
- Feeding nonconducting liquids at high rates into a tank having earthed internal fittings or measuring probes
- Lowering of a conductive sampling can onto a highly charged liquid surface.

4.6.4 Propagating Brush Discharge[5,10,12]

PBD can occur when the two surfaces of an insulating sheet (or layer) of high-resistivity material, and high dielectric strength, are charged to a high surface charge density but of opposing polarity.

The necessary conditions for a PBD are as follows:

a. The thickness of the insulating layer is less than around 8 mm
b. The surface charge density is at least 250 $\mu C/m^2$
c. The charges on the two surfaces are of opposite polarity.

The insulating sheet is often backed by a metal plate, as with an inner plastic lining or coating on a metallic container; however, a backup plate is not an essential condition for PBD.

For isolated insulators, the surface charge density is limited by the breakdown strength of the surrounding medium. If the insulator is in direct contact with a grounded metallic substrate, much larger charge densities can be accommodated on the exposed surface until electrical breakdown occurs spontaneously through the insulating layer to the ground.

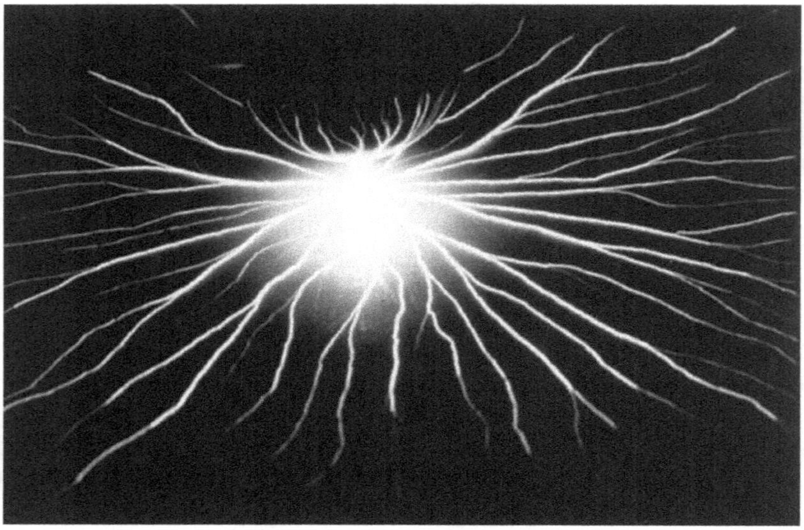

FIGURE 4.3d Propagating brush discharge.[5]

Even when the surface charge density is insufficient for a spontaneous PBD to occur, a PBD can be initiated by an approaching grounded metal electrode whenever the external field is enhanced sufficiently to produce a brush discharge adjacent to the area of the sheet. This eliminates the charge from a small patch of the sheet and produces intense radial fields over the surface. These fields lead to further breakdowns that eliminate the charge from other areas, and thus, the discharge propagates. It may continue until the entire surface charge has been removed.

The energy released in a PBD is high (1 J or more), depending on the charged sheet's thickness and surface charge density. It can readily ignite flammable gases, vapors, and dusts. Typical industrial operations where sufficient surface charge densities could be built up to cause PBD are as follows:

- High-velocity pneumatic transfer of powders through an insulating pipe or a conductive pipe with an insulating internal coating/liner
- The continuous impact of powder particles onto an insulating surface, e.g., a dust deflector plate in a cyclone separator
- Filling of large containers made from insulating materials, e.g., flexible intermediate bulk containers (FIBC).

4.6.5 Bulking Brush Discharge[5,10,13]

Bulking brush discharges, also known as cone discharges, occur on the surface of a highly charged heap of bulk powder or granules during the filling of large silos. As the charged, dispersed powder settles and "bulks" in a silo, there is a huge increase in volumetric charge density. As a result, high-intensity fields are produced. These lead to the ionization of the air at the dump cone's surface and the formation of the

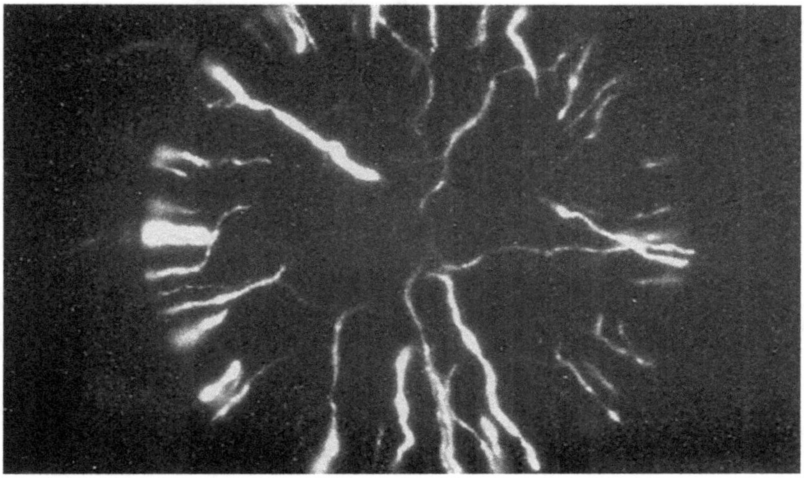

FIGURE 4.3e Bulking brush discharge.[5]

highly conductive, hot discharge channels propagating from the wall to the center of the dump cone.

Based on experimental data on vertical, cylindrical silos, the maximum observed charge transfer varies with the diameter of the silo as follows[8]:

$$Q = 2.3D^{2.8} \tag{4.21}$$

where

Q = maximum value of charge transfer, μC
D = diameter of the silo, m.

Thus, for a silo of 3 m diameter and a bin diameter of 1 m, the maximum predicted charge transfers are about 50 and 2 μC, respectively. Methods for estimating the energy of bulking brush discharges do not appear to be sufficiently developed. However, hybrid mixtures and dusts having ignition energies less than 20 mJ should always be considered at risk from such discharges.

4.7 IGNITION OF FLAMMABLE VAPORS AND DUSTS BY ELECTROSTATIC DISCHARGE

The Minimum Ignition Energy (MIE) for a few selected flammable gas mixtures and vapors in air is shown in Table 4.7.

MIE values for flammable gases or vapors in oxygen are much lower, which is about 1% of the values in air.

For powders, the MIE values vary with particle size. The smaller the diameter, the lower the MIE. The values quoted in Table 4.7 are for the most easily ignited size range, i.e., below 75 μ (or −200 mesh size).

TABLE 4.7
Minimum Ignition Energies of Gases and Dusts

Gases and Vapors in Air[5]

	mJ		mJ
Acetaldehyde	0.13	Ethylene oxide	0.065
Acetone	0.19	n-Heptane	0.24
Acetylene	0.017	Hydrogen	0.016
Acrylonitrile	0.16	Isooctane	0.25
Benzene	0.20	Methane	0.21
n-Butane	0.25	Methanol	0.14
Carbon disulfide	0.009	Methyl ethyl ketone	0.21
Cyclohexane	0.22	Propane	0.25
Diethyl ether	0.19	Propylene	0.18
Ethane	0.23	Styrene	0.18
Ethanol	0.23	Toluene	0.24
Ethyl acrylate	0.18	Vinyl acetate	0.16
Ethylene	0.084	Xylene	0.2

Dusts and Powders in Air[9]

	mJ		mJ
Aluminum	10	Polystyrene	15
Epoxy resin	9	Rice	50
Nylon	20	Sulfur	15
Polyethylene	30	Wheat flour	50
Polypropylene	30		

4.7.1 Hybrid Mixtures[9,10]

Situations often arise in process plants when flammable dust is suspended in an atmosphere containing flammable gas or vapor. A typical example is loading ingredients in powder form into a reaction vessel containing flammable solvents. Such suspensions of dust in an atmosphere containing a proportion of flammable gas or vapor are known as "hybrid mixtures".

Hybrid mixtures pose particularly severe problems because of high charge densities from powder handling operations and low ignition energies of flammable vapors. The MIE of a hybrid mixture is challenging to assess. However, a conservative estimate can be made by assuming that the MIE of the mixture is at or near the MIE of the gas alone. As a result, there can be a violent explosion even if the flammable vapor concentration is below the lower flammability limit (LFL).

4.8 HAZARDS FROM PEOPLE AND CLOTHING

The human body is an excellent electrical conductor, and this fact has been responsible for numerous instances of static discharge. Persons can accumulate a significant charge on their bodies if they wear synthetic clothing and insulating shoes or walking on an insulated floor. The problem is aggravated in the winter, as dry weather increases the rate of static generation. A person insulated from the ground can also pick up charge by momentarily touching a charged object or by induction from a nearby charge generating operation.

Many of us who wear warm synthetic clothing in winter have experienced a shock by touching a water tap or other earthed device. The spark energy of the discharge that makes us "feel" such shocks needs to be only about 1 mJ. It follows from this observation that the potential acquired from wearing synthetic clothing is about 3 kV, assuming an average capacitance of 200 pF for the human body.

It must be understood that an electrostatic spark discharge from a human body can readily ignite a flammable gas or vapor whose MIE is in the region of 0.1–0.2 mJ. Therefore, in refineries and process plants, safety management systems often require operators to discharge themselves by touching an earthed metallic plate before undertaking any job that might expose them to flammable gases or vapors. The same precautions apply when filling gasoline into the fuel tank of an automobile.

At low values of relative humidity, walking across a carpet or getting up from an upholstered chair often results in body voltages up to about 10 kV, corresponding to a stored energy of 10 mJ at an average body capacitance of 200 pF. In fact, during normal industrial activities, the potential of the human body could reach much higher values, up to about 50 kV (stored energy = 250 mJ)[5] (e.g., while standing close to a freshly piled stack of thermoplastic resin). Assuming that just 20% of such stored energy effectively causes ignition, the need for reliable earthing of operators becomes readily apparent.

Wherever significant static generation is envisaged, footwear and flooring materials should both be either antistatic or conductive. The antistatic property enables static dissipation at an acceptable rate, while the conductive property allows the charge to flow to earth. The combined resistance of footwear and flooring should be less than 10^6 ohm.

4.9 EARTHING AND BONDING

Earthing, also referred to as grounding, is a process of providing an electrical pathway between a conductor and the earth so that the conductor is at zero potential. Bonding is the process of connecting two or more conductive objects through a conductor to reach the same electrical potential, but not necessarily at the earth's zero potential.

In process plant environments with potentially flammable atmospheres, there are many conductors that, if not suitably earthed, can become charged to a hazardous level. Some of these are necessary components: plant structures, reaction vessels, pipes, valves, storage tanks, and drums. Others may be present by accident or as a result of carelessness or poor housekeeping, e.g., lengths of redundant wire or metallic cans floating on nonconductive liquids.

The primary consideration in earthing is to prevent the build-up of potential on a conductor to a hazardous level. This is achieved by ensuring that the resistance to ground is low enough to enable dissipation of charge as fast as the rate of charge accumulation in the conductor. Electrostatic charging currents rarely exceed 10^{-4} A.

A potential of at least 300 V is generally considered sufficient to initiate an incendiary discharge in normal industrial operations. However, in explosives manufacturing plants, anything voltage above 100 V is considered to be hazardous. Using 100 V for static dissipation, the maximum resistance to ground is usually specified as $100/10^{-4} = 10^6$ ohm. In those operations where the charging current does not exceed 10^{-6} A, the same criterion gives a maximum earthing resistance of 10^8 ohm.

In this connection, it needs to be mentioned that earthing connections are necessary not only to avoid static hazards but also for personnel protection against lightning and shocks from electrical systems. For these requirements, an earthing resistance of 1–10 ohm is considered adequate. A ceiling of 10 ohm for resistance is also convenient for monitoring and maintenance; any higher value indicates that an intended metallic path is not established reliably, possibly because of corrosion or loose connections. In practice, a maximum value of 10 ohm is considered adequate for all wholly metallic systems.

With metallic connections, the maximum bonding resistance is usually specified as 10 ohm even when the earth resistance is higher.[10]

Ensuring earthing/bonding continuity across pipe joints deserves special mention. If a nonconductive gasket is used, there may be no continuity across the connection. This can be corrected by using flexible graphite-filled or other conductive gaskets. Alternatively, jumper cables can be used. Auditing to ensure electrical continuity should be done routinely after maintenance or painting work.

4.10 EXAMPLES OF STATIC IGNITION

Walmsley[10] and Kletz[14] have described several accidents caused by static ignition. Four of them are mentioned below to illustrate the theoretical principles discussed in the preceding sections.

4.10.1 Draining Flammable Liquids into Buckets[14]

Acetone was routinely drained into a metal bucket. One day, the operator hung the bucket on the drain valve instead of placing it on the metal surface below the valve. The acetone caught fire during the draining operation.

During normal operation, the static charge acquired by the acetone flowed to earth via the grounded metal surface on which the bucket was placed. This route was not available when the bucket was hung on the drain valve. Also, the bucket's handle was covered with plastic; this prevented the charge from flowing to earth via the grounded drainpipe. As a result, the static charge accumulated on the acetone and the bucket, and a spark passed between the bucket and the drain valve, igniting the acetone.

4.10.2 Removing Synthetic Clothing from Body[14]

A driver drove his car to a filling station, removed the cap from the end of the filler pipe, and held it in his hand while an attendant filled the car with gasoline. The driver took off his woolen pullover and threw it into the car. After the filling was complete, the driver was about to replace the cap on the end of the filler pipe when a spark jumped from the cap to the pipe, and a flame appeared at the end of the pipe. It was soon extinguished. The flame could not travel back into the gasoline tank as the mixture of gasoline and air was too rich to be flammable.

The driver acquired a static charge while removing the woolen pullover, and the pullover acquired an equal and opposite charge. The driver was wearing nonconducting shoes, and therefore, the charge on his body could not escape to earth. The lesson from this accident is that nonconducting garments should not be removed from the body in areas where a flammable atmosphere might be present.

4.10.3 Charging High-Resistivity Flakes/Powders[9]

An operator was manually charging epoxy resin flakes from bags, via an earthed metal funnel, into a dissolver vessel filled partially with xylene. The contents of the vessel were being stirred continuously to facilitate dissolution. After the contents from several bags had been added, an explosion occurred inside the vessel.

A static charge can be acquired by flakes (which could be mixed with dusts) while charging from plastic or paper bags. A static charge could also be generated in the dissolver while stirring the two-phase suspension of solids in the liquid. The flakes' high resistivity would not have allowed any significant charge dissipation during passage through the chute, despite the chute being earthed. Also, xylene's low conductivity would have allowed little dissipation of charge from the suspension despite the dissolver itself being earthed.

Brush discharge from the charged suspension to either the stirrer or the vessel wall likely ignited the air–xylene mixture. Also, the ignition sensitivity would have been enhanced by resin dust (hybrid mixture).

Such operations are commonplace in the process industries, and operators must be trained and aware of potential hazards. Steps to prevent ignition would include inerting the vapor space (e.g., by nitrogen purging), increasing the liquid's conductivity by adding a polar solvent or an antistatic agent, and limiting the speed of the agitator.

4.10.4 Filling Polyethylene Granules into a Silo[10]

A degassing silo was being used regularly with polyethylene granules from a pneumatic conveyor. Ethylene gas emitted from the stored granules was removed by blowing purge air at the bottom of the silo. The purge air blower broke down, and an explosion and fire occurred inside the silo.

Owing to the purge air blower's failure, the ethylene concentration in the air inside the silo rose above the LFL. The volume resistivity of polyethylene is greater than 10^{14} ohm.m. Therefore, granules acquired charge during pneumatic transportation, and charge dissipation from the granule pile was extremely slow.

Bulk surface discharges have frequently been observed in silos receiving pneumatically conveyed granulate. This phenomenon likely caused the ignition of the flammable air–ethylene mixture. As a precautionary measure, the formation of a flammable mixture in such systems should be prevented, for example, by installing an automatic emergency nitrogen purging system that would activate after an air blower failure.

4.11 SUMMARY OF COMMON PRECAUTIONARY MEASURES FOR STATIC HAZARDS

Static hazards can arise in many unexpected ways in the diverse operations of chemical plants. In-depth examinations and hazard assessment studies enable identifying the potential causes of accidents, enabling the development of precautionary measures. Some common measures are summarized below:

a. While transferring low-conductivity liquids through pipes, charge generation is controlled by maintaining liquid velocity below a threshold value. For liquids with a conductivity below 50 pS/m, the velocity in the pipe should not exceed 1 m/s unless the inlet has been submerged in the liquid in the receiver. If a second, immiscible phase is present, this low velocity should be maintained even after the inlet has been submerged.
b. Splash filling into a tank must be avoided if a flammable atmosphere may be present in the tank using either a bottom inlet or a dip pipe that extends to the tank bottom. For top feeding, the feed pipe can also be bent to direct the liquid towards the vessel wall at a low velocity to avoid jet impingement. This practice would allow the liquid to flow slowly down the wall of the vessel by gravity.
c. Liquid–liquid mixing or mixing of liquids with particulate matter is particularly vulnerable to static ignition. The vessel, the agitator, and the interconnecting piping should be in good electrical contact and also with the earth. An antistatic agent may be added to raise the conductivity of low-conductivity liquids, provided that such an additive is compatible with the product or process.
d. A fine particle filter in a pipeline carrying a liquid can significantly increase electrostatic charge generation in the liquid. A minimum residence time of 100 seconds between the filter and the receiving vessel is recommended to ensure charge relaxation to a safe level.[6,9,12] When this is not feasible, the receiving vessel's vapor space should be "inerted" properly, e.g., using nitrogen.
e. Many accidents are known to have occurred during the transfer of flammable liquids through hoses. Hoses can be conductive, semiconductive, or insulating. Depending on the type of hose and operating conditions, precautions to be taken would vary. Comprehensive guidance is provided by BS 5958,[9] CENELEC,[6] and Walmsley.[10] Also, ISO 8031 and ISO 8330 have provided new definitions for hoses.
f. When filling low-conductivity liquids into a vessel, sufficient time should be allowed for dissipation of the charge before any attempt is made to insert any level gauge or sampling device into the vessel. This waiting time could be as high as 30 minutes, depending on process conditions.

g. During washing of tanks with high-velocity water or steam jets (e.g., for tanks in ships), a charged mist of fine droplets is formed. The formation of water slugs adds to the problem. Incendiary discharges are common unless due precautions are taken. Safe procedures should be developed, with specialist assistance and advice, before undertaking such cleaning operations.

h. Transport contractors often use tanker trucks for multiple products. Filling tanks that contain flammable vapors from previous loads should be avoided. "Switch loading" of a low volatility liquid, such as gas oil, into a tanker that previously contained a volatile product (such as gasoline) has caused many accidents. Grounding a tanker truck will not prevent the ignition of vapor by a static discharge: while grounding will only prevent a discharge from the tank vessel to earth, it will not prevent a discharge from the liquid in the tank to the tank wall or the filling arm.

i. The manual addition of powders from drums/sacks to flammable liquids is commonplace in chemical plant batch operations.[15] Electrostatic charges generated during this operation can accumulate and cause high potentials on the loading chute, the receiving vessel, the powder stream, and the personnel carrying out the operation. In addition to ensuring that bonding and earthing measures are adequate, providing an inert atmosphere in the receiving vessel may also be necessary.

REFERENCES

1. Mannan, S.: *Lees' Loss Prevention in the Process Industries* (4th Ed., Butterworth-Heinemann, Oxford, 2012).
2. Nelkon, M and Parker, P.: *Advanced Level Physics* (4th Ed., Heinemann, London, 1982).
3. NFPA 77: *Recommended Practice on Static Electricity, 2007 Edition.*
4. Reppermund, J. and Britton, L. G.: *Hazards of Static Accumulating Flammable Liquids* (Paper presented at SCHC Spring Meeting, Houston, TX, 2009).
5. Britton, L. G.: *Avoiding Static Ignition Hazards in Chemical Operations* (Center for Chemical Process Safety (CCPS), AIChE, New York, 1999).
6. CENELEC: *Technical Report CLC/TR 50404, Electrostatics – Code of Practice for the Avoidance of Hazards due to Static Electricity* (June, 2003).
7. API Recommended Practice 2003: *Protection against Ignitions Arising Out of Static, Lightning and Stray Currents* (September, 1998).
8. Asano, K.: Electrostatic Potential and Field in a Cylindrical Tank Containing Charged Liquid, *Proceedings of the Institution of Electrical Engineers*, 124, pp. 1277–1281 (1977).
9. BS 595: British Standard: *Code of Practice for Control of Undesirable Static Electricity, Part 1: General Considerations and Part 2: Recommendations for Particular Industrial Situations* (1991).
10. Walmsley, H. L., and Kalisvaart, P.: The Avoidance of Electrostatic Hazards in the Petroleum Industry, *Journal of Electrostatics*, 27, pp. 1–193 (1991).
11. Britton, L. G.: Static Ignition Hazards of Conductive Liquids during Container Filling, *Process Safety Progress*, 29, pp. 98–102 (2010).
12. Klinkenberg, A. and Van der Minne, J. L.: *Electrostatics in the Petroleum Industry, A Royal Dutch/Shell Research and Development Report* (Elsevier, Amsterdam, 1958).

13. NFPA 704: *Standard System for the Identification of the Hazards of Materials for Emergency Response*, National Fire Protection Association, Boston, MA (2007).
14. Kletz, T.: *What Went Wrong* (3rd Ed., Gulf, Houston, TX, 1995).
15. ICI Fire Protection Guide No. 11: *Fire Prevention and Protection for Warehouses* (1984).

5 Pool Fire

This chapter presents available methods for estimating the intensity of thermal radiation at various distances from a pool fire. The combustion process depends on numerous factors, including the fuel characteristics, the air supply, degree of mixing, and several other ambient conditions. A theoretical treatment of this process would be quite elaborate and considered inappropriate for hazard assessment work. Accordingly, only widely used empirical or semiempirical methods have been covered in the discussions below. Readers interested in theoretical aspects may find references listed at the end of this chapter to be useful.

Pool fires include fires from spillage of liquid fuels at ground level and fires in storage tanks with the top cover removed (tank fires). Together, these scenarios cover the majority of major accidental fires in chemical plants and storage installations. Mention has also been made of fires on water, although data on such fires are limited.

5.1 SIZE AND SHAPE OF FLAMES

In almost all the models below, the flame is visualized as a cylinder of diameter D and length L. The flame cylinder is assumed to be vertical when the air is still, and tilted downwind when the wind is blowing (see Figure 5.1). Normal oscillations on the surface of the flame are ignored.

5.1.1 Confined Pool Fire on Land

A confined pool fire is one in which the burning liquid pool is contained within a fixed area, such as an area surrounded by dikes. The diameter of the flame is assumed to be equal to the diameter of the pool.

5.1.1.1 Pool Diameter

For a confined pool of noncircular surface area, the pool diameter is expressed in terms of the equivalent diameter:

$$D_e = \frac{4A_p}{P_p} \tag{5.1}$$

where

D_e = equivalent diameter of the pool, m
A_p = surface area of the pool, m^2
P_p = perimeter of the pool, m.

For a square pool, the equivalent diameter is equal to the length of the square's side. Similarly, for a vertical cylindrical vessel, it equals the vessel diameter.

DOI: 10.1201/9781003107873-5

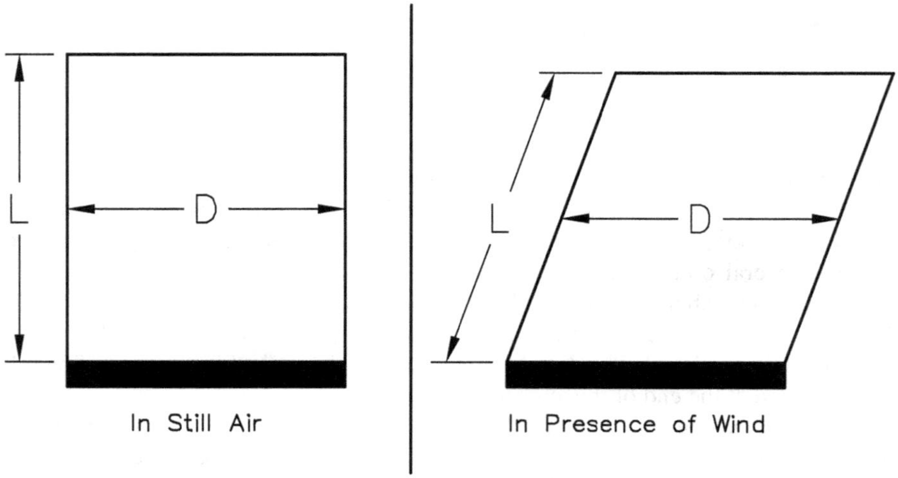

FIGURE 5.1 Typical flame geometry in case of a pool fire.

If the volume of the spilled liquid and the depth of the pool are known, the equivalent diameter can be calculated as:

$$D_e = \sqrt{\frac{4V}{\pi\delta}} \tag{5.2}$$

where

V = volume of liquid, m^3

δ = depth of the pool, m.

5.1.1.2 Burning Rate

The burning rate of liquid fuels, kg/(s.m^2), is an important parameter that needs to be estimated for calculating the flame height and the duration of the fire. It is usually expressed as the *mass burning rate*, defined as the rate of burning the pool's surface area. Sometimes, the burning rate is also expressed as a *regression rate*, defined as the rate of decrease of the liquid depth, cm/min. The regression rate is useful for quick estimates of a fire's duration if the depth of the pool is known. The regression rate is the volumetric loss of liquid per unit time per unit surface area of the pool. It can be calculated by dividing the mass burning rate by the liquid density, in dimensionally consistent units.

Data on mass burning and regression rates for some common flammable liquids are shown in Tables 5.1a and b. In the absence of published data, the regression rate (and hence the mass burning rate) can be estimated using Equation (5.3), due to Burgess and Hertzberg[1]:

$$R = 0.0076\frac{\Delta H_c}{\Delta H_v} \tag{5.3}$$

TABLE 5.1a
Heats of Combustion and Vaporization at Atmospheric Pressure[2]

	ΔH_c (MJ/kg)	ΔH_v (kJ/kg)
LNG (CH_4)	55.51	440.8
LPG (Propane)	50.33	428.7
n-Butane	49.5	378.8
Hexane	48.31	333.4
Heptane	48.06	325.0
Methanol	22.65	1,085.7
Ethanol	29.67	864.1
Benzene	41.83	338.9
p-Xylene	42.87	312.5
Acetone	30.81	495.1
Diethyl ether	36.74	357.5
Gasoline[a]	47.0	794
Kerosene[a]	46.2	1,211
Fuel oil (heavy)[a]	44.0	1,520
Crude oil[a]	44.2	1,455

[a] Approximate values.

where

R = regression rate, cm/min
ΔH_c = heat of combustion, kJ/kg
ΔH_v = latent heat of vaporization of the liquid, kJ/kg.

As an example for toluene, $\Delta H_c = 40{,}600$ kJ/kg and $\Delta H_v = 416$ kJ/kg. From Equation (5.3), the regression rate is (0.0076) (40,600)/416 = 0.74 cm/min. At a density of 871 kg/m^3, the mass burning rate is (0.74/100) (871/60) = 0.107 kg/(s.m^2).

The heats of combustion and vaporization for various substances obtained from the literature are shown in Table 5.1a.

Based on the data given in Table 5.1a, regression rates and mass burning rates have been provided for various substances in Table 5.1b.

5.1.1.3 Flame Height

5.1.1.3.1 Still Air Conditions

A widely used correlation for the assessment of flame length with still air conditions is due to Thomas[4]:

$$\frac{L}{D} = 42\left(\frac{m}{\rho_a\sqrt{gD}}\right)^{0.61} \tag{5.4}$$

where

L = flame height, m
D = diameter of the flame, m

TABLE 5.1b
Mass Burning Rate and Regression Rate for Liquid Fuels[3]

Liquid	Density (kg/m^3)	Mass Burning Rate ($kg/s/m^2$)	Regression Rate (cm/min)
CH_4	300	0.0479	0.957
Propane	506	0.0753	0.892
Butane	585	0.0968	0.993
Hexane	665	0.1221	1.101
Heptane	690	0.1293	1.124
Methanol	801	0.0212	0.159
Ethanol	796	0.0346	0.261
Benzene	883	0.1381	0.938
p-Xylene	865	0.1504	1.043
Acetone	799	0.0630	0.473
Diethyl ether	721	0.0938	0.781
Gasoline[a]	740	0.0555	0.45
Kerosene[a]	820	0.0396	0.29
Heavy fuel[a]	940–1,000	0.0356	0.22
Crude oil[a]	830–880	0.0342	0.15–0.33

[a] Approximate for the listed wide-boiling petroleum mixtures.

m = mass burning rate of the burning pool, kg/(s.m^2)
ρ_a = density of ambient air, kg/m^3
g = gravitational acceleration = 9.81 m/s^2.

Thomas' equation was developed from dimensional analysis and experimental data for burning cribs of wood (spruce) sticks arranged in a square cross section. The cribs were placed on a weighing platform so that the burning rate could be measured continuously. During steady burning, it was mainly the volatile matter from the wood that burned.

The density of air is found simply as follows:

$$\rho_a = (PM)/(zRT) \tag{5.5}$$

where

ρ_a = density of ambient air, kg/m^3
P = atmospheric pressure = 1 atm
M = molecular weight of air = 29.9647 kg/kgmol
z = compressibility factor (very close to 1 for ambient conditions)
R = gas constant = 0.082057 (m^3 atm)/(kgmol K)
T = ambient air temperature, K.

Another useful correlation for the prediction of flame height in still air is due to Heskestad[5]:

$$\frac{L}{D} = -1.02 + 15.6N^{1/5} \tag{5.6}$$

$$N = \left(\frac{C_p T_a}{g \rho_a^2 \left(\frac{\Delta H_c}{r_s} \right)^3} \right) \frac{Q^2}{D^5} \tag{5.7}$$

where

C_p = specific heat of air, kJ/(kg K)
T_a = ambient temperature, K
ΔH_c = heat of combustion of the burning liquid fuel, kJ/kg
r_s = stoichiometric mass ratio of air to fuel
Q = heat release rate, kJ/s
[other terms are as defined for Equation (5.4)].

The heat release rate Q, kJ/s, of the pool area is found as the product of mass burning rate, kg/s/m^2; the surface area of the pool, m^2; and the heat of combustion, kJ/kg. For a pure liquid fuel of molecular formula $C_\alpha H_\beta O_\gamma$ burning in air, the stoichiometric mass ratio of fuel to air for complete combustion is given by[13]:

$$r_s = 137.87 \frac{(\alpha + \beta/4 - \gamma/2)}{(12\alpha + \beta + 16\gamma)} \tag{5.8}$$

Heskestad's correlation was developed to predict the mean height of buoyancy-controlled, turbulent diffusion flames and is backed by experimental data. The correlation is claimed to be valid over a wide range of the parameter N (Equation 5.7) between 10^{-5} and 10^5, although the experimental data are limited to values of $N < 10^{-2}$ (or $L/D < 5$). Pool fire scenarios rarely involve L/D ratios greater than 5.

Table 5.2 gives the calculated values of L/D in still air for a hydrocarbon (such as hexane) and an oxygenated compound (such as ethanol) using Thomas' correlation and Heskestad's correlation. It is seen that the values of flame length from Heskestad's correlation are about 30% higher than those given by Thomas' correlation. For a given liquid, the L/D ratio decreases as the diameter increases. Also, at a given diameter, the L/D ratio decreases with the liquid mass burning rate.

5.1.1.3.2 In the Presence of Wind

Thomas' correlation for flame length in the presence of wind is as follows[3]

$$\frac{L}{D} = 55 \left(\frac{m}{\rho_a \sqrt{gD}} \right)^{0.67} U_*^{-0.21} \tag{5.9}$$

$$U_* = \frac{U}{U_c} \tag{5.9a}$$

TABLE 5.2
Calculated Values of Flame Height in Still Air for Hexane and Ethanol

	Hexane			Ethanol		
Thomas' Correlation						
Flame diameter, D, m	5	10	20	5	10	20
Ambient temperature, T_a, K	298	298	298	298	298	298
Air density, ρ_a, kg/m^3	1.186	1.186	1.186	1.186	1.186	1.186
Mass burning rate, m, kg/s/m^2	0.074	0.074	0.074	0.015	0.015	0.015
L/D	2.36	1.91	1.55	0.89	0.72	0.58
Flame length, L, m	11.8	19.1	30.9	4.45	7.21	11.7
Heskestad's correlation						
Mass burning rate, m, kg/s	1.452	5.809	23.236	0.294	1.178	4.710
Heat of combustion, ΔH_c, kJ/kg	44,700	44,700	44,700	26,800	26,800	26,800
A	6	6	6	2	2	2
B	14	14	14	6	6	6
Γ	0	0	0	1	1	1
Stoichiometric mass ratio, r_s	15.23	15.23	15.23	8.99	8.99	8.99
L/D	3.0	2.5	2.1	0.7	0.5	0.3
Flame length, L, m	15	25	42	3.5	5	6

$$U_c = \left(\frac{gmD}{\rho_a} \right)^{1/3} \tag{5.9b}$$

where

U = wind speed, m/s

U_c = characteristic wind velocity, m/s

U_* = dimensionless wind velocity; other terms are as defined for Equation (5.4).

Thomas' correlation for the angle of tilt of the flame, based on experimental data for wooden cribs, is:

$$\cos\theta = \frac{0.7}{U_*^{0.49}} \tag{5.10}$$

where

θ = angle of tilt from the vertical in the downwind direction

U_* = dimensionless wind velocity.

Another equation for flame tilt is due to Welker and Sliepcevich[7]:

$$F = \frac{\tan\theta}{\cos\theta} = 3.3\,\mathrm{Re}^{0.07}\,\mathrm{Fr}^{0.8}\left(\frac{\rho_g}{\rho_a} \right)^{-0.6} \tag{5.11a}$$

TABLE 5.3
Effect of Wind Speed on Flame Length and Angle of Tilt for a 10-m-Diameter Hexane Pool Fire

	Wind Speed (m/s)			
	Still Air	2	4	5
Flame length, m	19.1	18.1	15.6	14.9
The angle of tilt, °				
Equation (5.10)	–	48	62	65
Equation (5.11)	–	20	42	49

$$\cos\theta = \sqrt{\frac{-1+\sqrt{1+4F^2}}{2F^2}} \tag{5.11b}$$

$$\text{Re} = \frac{DU\rho_a}{\mu_a} \tag{5.11c}$$

$$\text{Fr} = \frac{U^2}{Dg} \tag{5.11d}$$

where
μ_a = viscosity of ambient air, kg/(m.s)
ρ_g = density of fuel vapor, kg/m^3
and other terms are as defined in Equations (5.9) and (5.10).

Welker and Sliepcevich's correlation was developed for small flames with diameters ranging from 10 to 61 cm and wind speed varying from 0.3 to 0.6 m/s. Liquid fuels used in the tests were methanol, acetone, hexane, cyclohexane, and benzene.

To show the magnitude of the effect of wind speed on the flame length and angle of tilt, calculated values for a 10-m-diameter hexane pool fire are given in Table 5.3. In these calculations, flame length has been calculated using Thomas' correlation (Equations (5.4) for still air, and (5.9) in the presence of wind).

The results in Table 5.3 show that Thomas' equation gives a conservative estimate of the angle of tilt. For plant layouts in open sites, a liberal provision for flame deflection (45°–60° as a rule of thumb) might be made for protecting surrounding facilities against flame impingement or exposure to high thermal radiation.

Example 5.1

A storage tank, 10 m diameter, 15 m high, and having a maximum working volume of 1,100 m^3 contains hexane at atmospheric pressure and ambient temperature. The tank is located on a flat, impervious surface and is surrounded by a

32 m×32 m square dike, 1.2 m high (equivalent to 110% of the tank's working volume). The tank starts leaking, covering the entire diked area, and is ignited. Calculate the height of the flame, assuming (a) still air conditions and (b) at a wind speed of 4 m/s. Data: mass burning rate = 0.074 kg/(s.m^2), specific heat of air = 1.04 kJ/(kg.K), ambient temperature = 300 K, density of ambient air = 1.18 kg/m^3, and heat of combustion of hexane = 45,000 kJ/kg.

The diameter of the flame, D, is the equivalent diameter of the pool:

$$D = (4)(32)(32)/4/32 = 32\,\mathrm{m}$$

a. Still Air Conditions

Thomas' correlation, Equation (5.4):

$$L/D = 42\left[0.074/\left(1.18\sqrt{\{(9.81)(32)\}}\right)\right]^{0.61} = 1.34$$

Hence, $L = (1.34)\ (32) = 43$ m.

Heskestad's correlation, Equations (5.6), (5.7), and (5.8):

$$r_s = 137.87\ \left[(6+14/4)/((12)(6)+14)\right] = 15.23$$

$$Q = 0.074\ (0.785)\left(32^2\right)(45{,}000) = 2.68\times10^6\ \mathrm{kJ/s}\ \ \text{(see note below)}$$

$$N = \left[(1.04)(300)/\left\{(9.81)\left(1.18^2\right)\left(45{,}000/15.23\right)^3\right\}\right]\ \left(2.68\times10^6\right)^2/32^5 = 1.89\times10^{-4}$$

Therefore, $L/D = -1.02 + 15.6\left(1.89\times10^{-4}\right)^{0.2} = 1.79$, and $L = (32)(1.79) = 57.2$ m

Note: The cross-sectional area of the burning pool has been assumed to include that of the tank, although the flame may not extend to the inside of the tank. The model does not include any effects from the presence of the tank inside the pool fire flame.

b. In the Presence of Wind

$$U = 4\,\mathrm{m/s}$$

From Equation (5.9b), $U_c = \{(9.81)(0.074)(32)/(1.18)\}^{0.333} = 2.70$ m/s

From Equation (5.9a), $U_* = 4/2.70 = 1.48$

From Equation (5.9), $L/D = (55)\left[0.074/\left(1.18\{(9.81)(32)\}^{0.5}\right)\right]^{0.67}(1.48)^{-0.21}$

Therefore, $L/D = 1.15$ and $L = (1.15)(32) = 36.94$ m.

5.1.1.3.2.1 Channel Fire A channel fire – also called trench fire or slot fire – is a type of pool fire that may occur on a spill of flammable liquid that has entered a catchment area (e.g., a trench or a channel) having a high aspect ratio (ratio of length to width). The following correlations for channel fires are based on Mudan and Croce, in Lees.[3] These correlations are empirical and are based on experimental data for LNG. Aspect ratios for channels varied between 2 and 30. The wind direction was parallel to the axis of the channel. The height of the flame and the angle of tilt of the flame are given as follows:

$$\frac{H}{W} = 2.2, \quad \text{Fr}' \geq 0.25 \tag{5.12a}$$

$$\frac{H}{W} = 0.88\left(\text{Fr}'\right)^{-0.65}, \quad 0.1 < \text{Fr}' < 0.25 \tag{5.12b}$$

$$\frac{H}{W} = 4.0, \quad \text{Fr}' \leq 0.1 \tag{5.12c}$$

$$\text{Fr}' = \frac{u_w}{2\sqrt{gW}} \tag{5.12d}$$

$$\cos\theta = 0.56, \quad \text{Fr}' \geq 0.25 \tag{5.12e}$$

$$\cos\theta = 0.36\left(\text{Fr}'\right)^{-0.32}, \quad 0.042 < \text{Fr}' < 0.25 \tag{5.12f}$$

$$\cos\theta = 1, \quad \text{Fr}' \leq 0.042 \tag{5.12g}$$

In Equations (5.12a) through (5.12g):

H = height of the flame, m
W = width of the pool, m
Fr' = a modified Froude number
u_w = wind speed, m/s
θ = angle of tilt from vertical, degrees.

This model has not been experimentally verified for liquids other than LNG.

In a review paper, Moorhouse and Pritchard[8] have suggested an alternative approach that assumes that the trench fire is made up of a series of individual pool fires placed next to each other along the length of the channel. The diameter of each of these small pool fires can be taken as the channel width, allowing relationships for circular pools to be used to estimate the flame height. Small-scale experimental studies on slot fires have indicated that this approach is valid, although no large-scale verification is available to date.

Example 5.2

Gasoline is spilled into a 10-m-wide, 80-m-long channel. It ignites immediately, giving rise to a pool fire. Wind speed is negligible; therefore, the flame may be taken as vertical. Calculate the height of the flame. The ambient temperature is 298 K; the density of air is 1.19 kg/m^3. Mass burning rate for gasoline is 0.055 kg/s/m^2.

From Equation (5.12d), at negligible wind speed, Fr' approaches zero. Hence, from Equation (5.12c), the flame height = (4) (10) = 40 m.

In the alternative method, substituting values of mass burning rate, air density, and flame diameter of 10 m in Equation (5.4), the ratio of flame height to pool diameter is 1.59. Hence, the flame height now is approximately 16 m, which is just 40% of the value obtained by the first method.

It should be clear that a reliable estimate is not possible without further experimental work.

5.1.2 Unconfined Pool Fire on Land

A typical example of an unconfined pool fire on land is the burning of gasoline leaking out of a tank truck following an accidental loss of containment. Such fires usually are time-dependent, the pool diameter is increasing with time until the burning rate becomes equal to the rate of spillage. After the tanker has been emptied, the remaining liquid in the pool will continue to burn with a diminishing diameter.

The simplified model below for a time-dependent flame diameter is based mainly on Cline and Koenig.[9] Assumptions in the model are as follows:

a. The spillage occurs on a flat, horizontal, and impervious surface.
b. Vaporization of the liquid in the preignition period is negligible (this will not be valid for cryogenic liquid spills).
c. Friction between the spreading liquid and the solid's surface is negligible.
d. The rate of spillage from the tanker is not affected by flame impingement or heat radiation from the flame.

A schematic diagram of the pool fire model is shown in Figure 5.2. The fuel is fed into the pool from an overhead tank at atmospheric pressure. The feed rate diminishes with time, as it depends on the height of the liquid in the tank that drops as fuel is spilled.

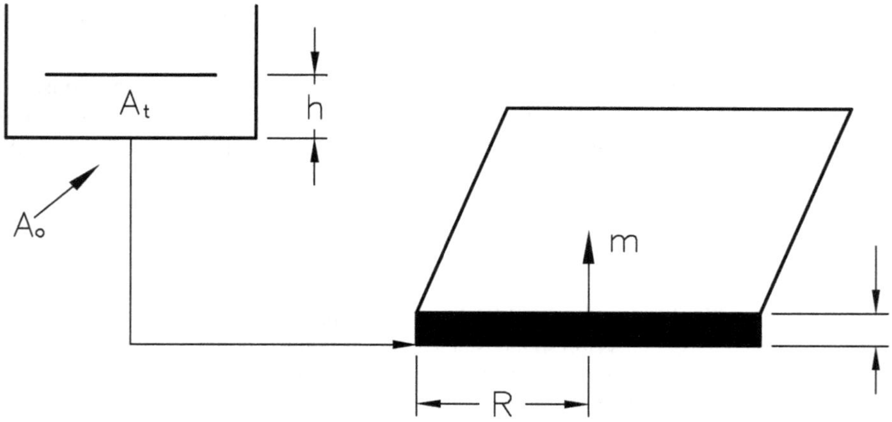

FIGURE 5.2 Schematic diagram of unconfined pool fire.

An unsteady-state material balance around the pool is given by Equation (5.13):

$$\rho\pi\delta\frac{dR^2}{dt} = \rho A_o V_t - m\pi R^2 \tag{5.13}$$

where
t = time, seconds
ρ = density of the liquid, kg/m^3
δ = depth of liquid in the pool, m
R = radius of the pool, m
A_o = cross-sectional area of the overhead tank outlet through which leakage occurs, m^2
V_t = the discharge velocity, m/s
A_o = leakage area, m^2
m = the mass burning rate, kg/s/m^2.

It is convenient to convert Equation (5.13) into the dimensionless form by defining two additional variables as follows:

$$\Phi = \frac{m\pi R^2}{\rho A_o V_o} \tag{5.13a}$$

$$\tau = \frac{mt}{\rho\delta} \tag{5.13b}$$

where
Φ = a dimensionless radius
m = the mass burning rate, kg/s/m^2
R = radius of the pool, m
ρ = density of the liquid, kg/m^3
A_o = cross-sectional area of the overhead tank outlet through which leakage occurs, m^2
V_o = velocity at tank outlet, m/s
τ = a dimensionless time
t = time, seconds
δ = depth of liquid in the pool, m.

The time-dependent discharge velocity, V_t, is related to the liquid head in the tank, as follows:

$$V_t = V_o - at \tag{5.14a}$$

$$V_o = C_o\sqrt{2gh_i} \tag{5.14b}$$

$$a = \frac{C_o^2 A_o g}{A_t} \tag{5.14c}$$

where

V_o = discharge velocity at the start of spillage, m/s
h_i = initial liquid level relative to the point of release, m
A_o = leakage area, m^2
A_t = cross-sectional area of the tank, m^2
C_o = discharge coefficient (dimensionless).

Equations (5.13) and (5.14) can be combined to give the pool spread equation in dimensionless form, as follows:

$$\frac{d\Phi}{d\tau} + \Phi = 1 - \beta\tau \tag{5.15a}$$

$$\beta = \frac{a\rho\delta}{mV_o} \tag{5.15b}$$

where

Φ = a dimensionless radius
τ = a dimensionless time.

Equation (5.15) is solved for two cases: (i) ignition at the start of spillage and (ii) ignition after a specified time following ignition.

Case (i): Ignition at $t = 0$

Integration of Equation (5.15) under the initial condition $\Phi = 0$ at $\tau = 0$ yields:

$$\Phi = 1 - \beta(\tau - 1) - (1 + \beta)e^{-\tau} \tag{5.16}$$

The maximum value of the pool radius and the time to reach this radius are given by:

$$\Phi_{\max} = 1 - \beta(\tau_{\max} - 1) - (1 + \beta)e^{-\tau_{\max}} \tag{5.17a}$$

$$\tau_{\max} = \ln\left(\frac{1 + \beta}{\beta}\right) \tag{5.17b}$$

Case (ii): Ignition at $t = t_i$

The first step is to determine the radius of the pool at the elapsed time (since the start of spillage) when ignition occurs, t_i. If we neglect vaporization during this preignition period ($0 \le t \le t_i$), a material balance can be expressed as follows:

$$\pi R^2\delta = A_o\int_0^t (V_o - at)\,dt = A_o\left(V_o t - \frac{at^2}{2}\right) \tag{5.18}$$

From Equation (5.18), and using definitions for Φ, τ, and β as above, it is possible to develop an equation for dimensionless radius Φ in the preignition period, as follows:

$$\Phi = (1 - \beta\tau)\tau \qquad 0 \le \tau \le \tau_i \tag{5.19}$$

$$\Phi_i = \left(1 - \beta\tau_i\right)\tau_i \tag{5.19a}$$

where

τ_i = dimensionless time corresponding to the time of ignition.

With the initial condition $\Phi = \Phi_i$ at $\tau = \tau_i$, Equation (5.15) can be integrated to obtain Φ in the postignition period, as follows:

$$\Phi = 1 - \beta(\tau - 1) - \left[1 - \Phi_i - \beta(\tau_i - 1)\right] e^{-(\tau - \tau_i)} \tag{5.20}$$

Equation (5.19) is valid for the preignition period ($0 \le \tau \le \tau_i$) and Equation (5.20) for the postignition period ($\tau_i \le \tau \le \tau_D$), where τ_D is the dimensionless time corresponding to the emptying time of the feed tank.

Once the maximum diameter of the burning pool is determined, the height of the flame can be calculated using the same equations as for a confined pool.

Example 5.3

A storage tank, 10 m diameter and 15 m high, containing hexane at atmospheric pressure and ambient temperature, leaks after a full-bore rupture of the bottom outlet of diameter = 150 mm. The tank is located on a flat, impervious surface and has no dike or containment around it. The initial liquid level is 12 m above the bottom outlet. Calculate the maximum diameter of the flame, assuming (a) ignition at time $t = 0$ and (b) ignition after time $t = 200$ seconds from the start of spillage.

Data: density of hexane = 650 kg/m^3, mass burning rate = 0.074 kg/s/m^2, discharge coefficient = 0.7, and depth of liquid pool = 20 mm.

Cross-sectional area of the storage tank, $A_t = \pi\ 5^2 = 78.5$ m^2

Cross-sectional area of the bottom outlet, $A_o = \pi\ 0.075^2 = 0.01766$ m^2

Using Equation (5.14) and $h_i = 12$ m, the values of V_o and α are calculated to be 10.74 m/s and 0.001082 m/s^2, respectively. The value of β is calculated from Equation (5.15):

$$\beta = (0.001082)(650)(0.020)/0.074/10.74 = 0.01769$$

a. For Ignition at $t = 0$

The pool radius at different times (up to a tank emptying time of 9,900 seconds) can be calculated using Equation (5.16). The results are as follows:

t (seconds)	100	200	712	2,000	4,000	6,000	9,900
T	0.569	1.134	4.053	11.385	22.769	34.154	56.353
Φ	0.432	0.672	0.928	0.816	0.615	0.414	0.021
R (m)	15.14	18.88	22.2	20.81	18.06	14.81	3.32

Hence, the maximum value of the pool radius is 22.2 m, which could also be calculated from Equation (5.17).

b. For Ignition at $t=200$ seconds

Calculation of the radius is done in two stages: for a time below 200 s using Equation (5.19), and Equation (5.20) otherwise. The results are as follows:

T (seconds)	100	150	200	300	2,000	6,000	9,900
T	0.569	0.847	1.138	1.708	11.385	34.154	56.353
Φ	0.566	0.847	1.127	1.061	0.816	0.414	0.021
R (m)	17.34	21.21	24.46	23.73	20.81	14.81	3.32

Hence, the maximum value of the pool radius is about 24.5 m, occurring just 3 minutes after ignition.

5.1.3 Pool Fire on Water

No model appears to have been developed for estimating flame diameter or flame height for pool fires on water. For spills enclosed by oil booms, models for confined pools on land could be used for estimation.

- For crude oils, gasoline, and other liquid petroleum products, the mass burning rates on water and land are nearly the same.
- For LPG, however, the burning rate on water is about twice that on land.
- For LNG, the burning rate on water is about three times that on land.[10]

5.1.4 Tank Fire

The term "tank fire" usually refers to a fire on top of a storage tank whose top has been completely blown off. Such fires are treated as pool fires. The differences from ground-level pool fires are as follows:

- In a tank fire, there is a vapor zone between the liquid surface and the top of the tank, whereas for a ground-level pool fire, the liquid surface coincides with the base of the flame.
- A cylindrical tank wall surrounds the vapor zone, and there can be entrainment of air. Sustained combustion is possible only after the vapor has reached the top of the tank. Therefore, flame height for a tank fire is measured from the top of the tank and not from the liquid surface, as is done for pool fires.

In modeling for tank fires, the flame diameter is usually taken to be equal to the diameter of the tank, and the flame height is calculated by Thomas' correlations (Equation (5.4) under still air conditions and Equation (5.9) in the presence of wind).

5.2 MODELING FOR RADIATION INTENSITY

In any fire, damage to surrounding bodies can be caused by direct flame contact or thermal radiation from the flame. The extent of radiation damage is determined by the intensity of thermal radiation on the receiver and the exposure duration, as explained in Chapter 3. This section deals with the procedure for determining the intensity of thermal radiation.

The intensity of radiation from a flame on a small area of a receiver is given by:

$$I = \mathrm{SEP} \times F \times \tau \tag{5.21}$$

where

I = intensity, kW/m^2
SEP = surface emissive power, kW/m^2
F = view factor, dimensionless
τ = transmissivity, dimensionless.

5.2.1 Surface Emissive Power of Flames

The surface emissive power (SEP) is the intensity of radiation at the surface of the flame. Where a fraction of the flame surface is covered by smoke, the SEP should be taken as the weighted average of:

i. The SEP for a clean surface
ii. The SEP for the soot.

Thus, if the SEP for the clean flame was 140 kW/m^2 (a typical value for hydrocarbon fires) and that for soot was 20%, and 20% of the flame surface was covered by soot, the weighted average SEP to be used in Equation (5.21) would be:

$$(120)(0.8) + (20)(0.2) = 116\,\mathrm{kW/m^2}.$$

Some judgment is required in selecting a proper value for SEP. Available experimental data are often limited and lack information on the prevailing conditions when the data were collected. Typical SEP values are 40–50 kW/m^2 for LPG pool fires, around 200 for LNG pool fires, and 60–130 for gasoline pool fires. While choosing a value for SEP, it should be ensured that for a clean flame, the energy emitted by radiation lies in the range of 25%–40% of the rate at which energy is released during the combustion process.

5.2.2 View Factor between a Flame and a Target

The view factor represents the fraction of the total radiant energy emitted by the flame surface that is received by the target or receiver. It is a function of:

i. The flame size
ii. The distance between the flame and the target
iii. The geometrical orientation of the target relative to the flame.

The view factor calculation involves three-dimensional numerical integration, although analytical expressions are available for simple flame sizes and geometries.[11–13]

Three models are available for the view factor. These are:

- Point source
- Solid flame
- Equivalent radiator.

The point source model assumes that the entire radiation would emanate from a point; this is unrealistic. The predicted values are subject to large errors in the near-field. This method, therefore, has not been considered any further in this discussion.

The solid flame model assumes the flame to be a solid cylinder with a uniformly radiating curved surface; this represents the best approximation of the physical situation. Applications are explained below for a pool fire at ground level and also for a tank fire.

The equivalent radiator model assumes the radiating surface to be a flat plate and, accordingly, the model is somewhat less accurate. The radiating surface is divided into several independent radiating areas, and the view factors for all these areas are added algebraically to obtain the combined view factor. A typical application is for calculating the view factors of channel fires.

The calculation procedure is given below for two cases:

- Pool fire and target at ground level
- Tank fire, with the target either at ground level or at an elevated position.

5.2.2.1 Case 1: Pool Fire and Target at Ground Level

- *Solid Flame Model*

  Figure 5.3 depicts the coordinate system for vertical as well as tilted flames relative to a target. View factor expressions are given below for tilted flames. The same expressions can be used for vertical flames, too, by setting the inclination (θ) equal to zero.

  A relatively simple equation, due to Stannard,[14] yields conservative results and is based on the solid angle subtended at a point:

$$A = \frac{1}{1+(X_r-1)\cos\theta} \tag{5.22a}$$

$$B = \frac{H_r}{X_r-\cos\theta} - \tan\theta \tag{5.22b}$$

$$F = \frac{2}{\pi}\sin^{-1}(A)\left[\sin\theta + \sin\left\{\tan^{-1}(B)\right\}\right] \tag{5.22c}$$

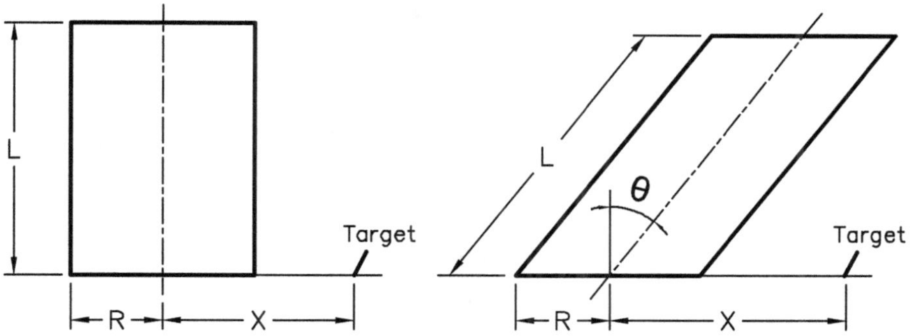

FIGURE 5.3 Coordinate system for vertical and tilted pool fire flames near a target.

where
- F = view factor
- H_r = dimensionless flame length (L/R)
- X_r = dimensionless distance of the target from the center of the flame (X/R)
- L = the flame length, m
- R = the flame radius, m
- θ = the angle of tilt of the flame from the vertical.

In the more accurate methods, equations are provided to calculate view factors for horizontal and vertical targets. The maximum value of the view factor is then calculated as the vector sum of these two values. In the geometrical orientation shown in Figure 5.3, the target is assumed to be downwind on a vertical plane passing through the target and the center of the cylindrical flame base. Several equations are available. Mudan's equations[15] are given in Equation 5.23(a–o). The definitions of H_r and X_r are the same as in Equation (5.22).

$$T_1 = \frac{H_r \cos\theta}{X_r - H_r \sin\theta} \tag{5.23a}$$

$$T_2 = \frac{H_r \cos\theta}{X_r - H_r \sin\theta} \tan^{-1}\left(\frac{X_r - 1}{X_r + 1}\right)^{1/2} \tag{5.23b}$$

$$T_3 = H_r^{\,2} + (X_r + 1)^2 - 2X_r(1 + H_r \sin\theta) \tag{5.23c}$$

$$T_4 = H_r^{\,2} + (X_r + 1)^2 - 2\,H_r(X_r + 1)\sin\theta \tag{5.23d}$$

$$T_5 = H_r^{\,2} + (X_r - 1)^2 - 2H_r(X_r - 1)\sin\theta \tag{5.23e}$$

$$T_6 = \tan^{-1}\left[\left(\frac{T_4}{T_5}\right)^{1/2}\left(\frac{X_r - 1}{X_r + 1}\right)^{1/2}\right] \tag{5.23f}$$

$$T_7 = \frac{\cos\theta}{\left(1+\left(X_r^2-1\right)\cos^2\theta\right)^{1/2}} \tag{5.23g}$$

$$T_8 = \frac{H_r X_r - \left(X_r^2-1\right)\sin\theta}{\left(X_r^2-1\right)^{1/2}\left(1+\left(X_r^2-1\right)\cos^2\theta\right)^{1/2}} \tag{5.23h}$$

$$T_9 = \frac{\left(X_r^2-1\right)\sin\theta}{\left(X_r^2-1\right)^{1/2}\left(1+\left(X_r^2-1\right)\cos^2\theta\right)^{1/2}} \tag{5.23i}$$

$$T_{10} = \tan^{-1}(T_8) + \tan^{-1}(T_9) \tag{5.23j}$$

$$T_{11} = \frac{H_r^2 + (X_r+1)^2 - 2(X_r + 1 + H_r X_r \sin\theta)}{(T_4)^{1/2}\,(T_5)^{1/2}} \tag{5.23k}$$

$$T_{12} = \frac{\sin\theta}{\left(1+\left(X_r^2-1\right)\cos^2\theta\right)^{1/2}} \tag{5.23l}$$

$$\pi F_v = -T_2 + \frac{T_1 \cdot T_3 \cdot T_6}{(T_4 \cdot T_5)^{1/2}} + T_7 \cdot T_{10} \tag{5.23m}$$

$$\pi F_h = \tan^{-1}\left(\frac{X_r+1}{X_r-1}\right)^{1/2} + T_{12} \cdot T_{10} - T_{11} \cdot T_6 \tag{5.23n}$$

$$F_{max} = \left(F_v^2 + F_h^2\right)^{1/2} \tag{5.23o}$$

We use Equations (5.23a–o) to show the calculated values of F_{max}, for assumed values of H_r, X_r, and θ, in Table 5.4. A positive value of θ indicates that the target is located downwind, and a negative value of θ indicates that

TABLE 5.4
Maximum View Factor (F_{max}) Using Mudan's Equations 5.23(a–o)

	$H_r=4$			$H_r=6$		
X_r	$\theta=0$	$\theta=30°$	$\theta=-30°$	$\theta=0$	$\theta=30°$	$\theta=-30°$
2	0.292	0.591	0.177	0.297	0.460	0.181
5	0.0860	0.135	0.0525	0.101	0.167	0.0600
10	0.0252	0.0310	0.0171	0.0343	0.0485	0.0219
15	0.0115	0.0125	0.00832	0.0164	0.0200	0.0112

the target is located upwind. A value of θ equal to zero indicates that the flame is vertical (negligible wind speed).

An example using the following inputs illustrates the procedure:

$$H_r = 4, \ X_r = 5, \ \theta = 30°$$

In this case, the intermediate results for Equation (5.23a–l) are as follows:

T_1	T_2	T_3	T_4	T_5	T_6
1.1547	0.79065	22.0	28.0	16.0	0.82390

T_7	T_8	T_9	T_{10}	T_{11}	T_{12}
0.19868	0.37463	0.56195	0.87042	0.94491	0.11471

Substituting these into Equations (5.23m–o), we get:

$$F_v = 0.118, \quad F_h = 0.0660, \quad \text{and} \quad F_{max} = 0.135$$

Similarly, we have calculated the values shown in Table 5.4 for the following cases:

$$H_r = 4 \text{ and } 6$$

$$X_r = 2, \ 5, \ 10, \ 15, \ \text{and}$$

$$\theta = 0, \ 30, \text{and} - 30°$$

These results show that the length of the flame, its diameter, the distance of the target from the center of the flame (at base), and the angle of the flame from the vertical all have a pronounced effect on the maximum view factor and, consequently, on the intensity of the radiation received at the target. Therefore, it is highly recommended that the location of adjacent equipment and minimum allowable safe distances for people from the fire must be determined most carefully and conservatively. The methods described above should provide useful guidance in such decision making.

It has been our experience for many existing facilities that such minimum safe distances have not been observed, and the resulting severe crowding of major equipment has led to a rapid escalation in economic losses and personnel injuries. Also, Chapter 2 describes many such past cases. For example, a recent fire in a highly congested tank farm area spread to adjacent tanks (seven tanks were engulfed) at International Terminal Co.'s facilities near the Houston Ship Channel in Deer Park, Texas.[16]

This accident resulted in an enormous fire and loss of containment, causing spills and burning of many millions of gallons of aromatic compounds, including benzene (a known carcinogen). Over 60,000 gallons of contaminant-laden wastewater was reported to have been recovered from the waterway using booms designed to contain such spills. The Ship Channel, a major waterway for transporting petroleum products and petrochemicals, was shut down by authorities for an extended period, resulting in severe economic losses to all neighboring plants. This accident caused massive supply chain disruptions lasting many months throughout the area's refining and petrochemical operations. This event also resulted in sharp spikes in refined product prices locally. Extensive litigation was launched against the owners, both by the regulatory authorities and by neighboring facilities that suffered disruptions and economic losses because of the fire and spill.

- *Equivalent Radiator Model*

  Figure 5.4 shows a vertical flame represented by an equivalent flat radiator of total area $A_1 + A_2$. The target is vertical and located at a distance X from the flame. The view factor between the flame and the target will be a maximum when $b_1 = b_2$. This model is used typically for a channel fire of height H and length $(b_1 + b_2)$.

  Since the target is considered to be vertical, only the view factor, F_v, is relevant. The following equation, based on the work of Morgan and

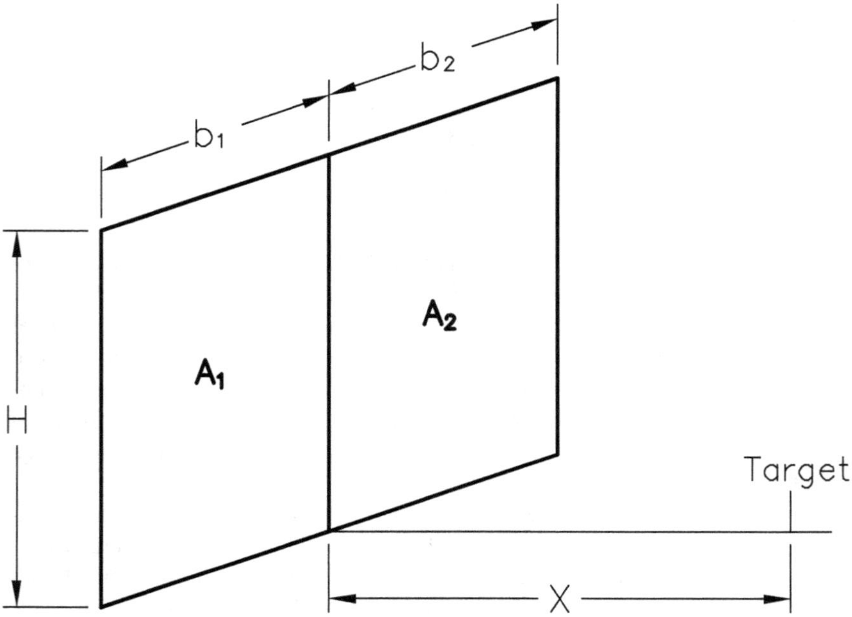

FIGURE 5.4 An equivalent flat radiator with a vertical target.

Hamilton, which appears in the 1992 edition of the Yellow Book,[17] can be used for this calculation:

$$H_r = \frac{H}{b} \quad X_r = \frac{X}{b} \tag{5.24a}$$

$$A = \frac{1}{\left(H_r^2 + X_r^2\right)^{1/2}} \tag{5.24b}$$

$$B = \frac{H_r}{\left(1 + X_r^2\right)^{1/2}} \tag{5.24c}$$

$$F_v = \frac{1}{\pi}\left[H_r A \tan^{-1}(A) + \frac{B}{H_r} \tan^{-1}(B)\right] \tag{5.24d}$$

In Equations (5.24a–d), areas A_1 and A_2 are equal ($b_1 = b_2 = b$) and F_v is twice the value for each section. For a cylindrical vertical flame, the equivalent radiator model can also be used, with b equal to the radius of the flame.

Example 5.4

A person is standing vertically in front of a vertical cylindrical flame at ground level. The flame diameter is 10 m, the length is 20 m, and the distance between the person and the flame's axis is 15 m. Calculate the view factor using (i) Stannard's equation, (ii) Mudan's equation, and (iii) the equivalent radiator model.

Using the given data, $H_r = 20/5 = 4$ and $X_r = 15/5 = 3$

i. Substituting $H_r = 4$, $X_r = 3$, and $\theta = 0$ in Equations (5.22a) through (5.22c), $F = 0.194$
ii. Substituting values of H_r, X_r, and θ in Equations (5.23a) through (5.23o), $F_v = 0.179$
iii. Substituting values of H_r and X_r in Equations (5.24a) through (5.24d), $F_v = 0.141$

5.2.2.2 Case 2: Tank Fire with Target at Ground Level/Elevated Position

Figure 5.5 shows a tank fire of flame length L_1 located on a tank of diameter D and height L_2. Target 1 is at ground level, Target 2 at the tank's height, and Target 3 is midway between the flame and ground level. For the three configurations shown in the figure 5.5, the view factors are as follows:

Target 1	$F_v[(L_1 + L_2), X] - F_v(L_2, X)$
Target 2	$F_v(L_1, X)$
Target 3	$2\,F_v(L_1/2, X)$

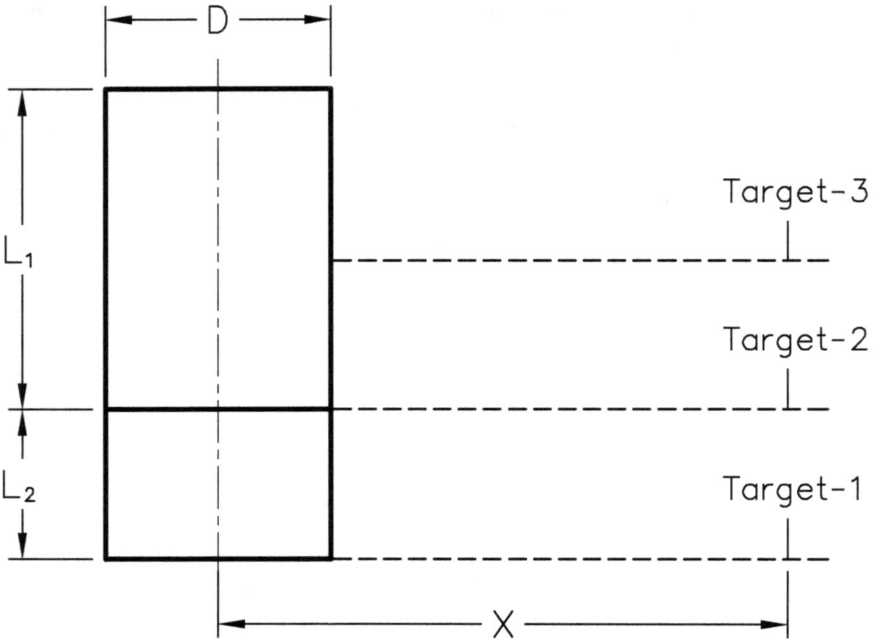

FIGURE 5.5 Flame/target configuration in case of a tank fire in still air.

F_v (L, X) stands for view factor on a vertical surface for a flame of height H and a target at a distance X from the flame's axis. View factor calculations for the above three cases are shown in Example 5.5.

Example 5.5

A storage tank containing a petroleum product suffers significant damage during tank filling. A tank fire erupts, covering the entire tank top. The tank diameter is 10 m, and the height is 8 m. The height of the cylindrical flame is 20 m. Wind speed is negligible; therefore, the flame can be regarded as vertical. The horizontal distance from the axis of the flame for each of the targets is 15 m. Calculate the view factor for each of the three targets. Use Mudan's equations[15] (5.23a–o) above for the view factor.

i. Location: Target 1 (Ground Level)
   For a flame of diameter = 10 m, and height = 28 m, $H_r = 28/5 = 5.6$.
   Dimensionless distance $X_r = 15/5 = 3$.
   Using Mudan's equations (5.23a) through (5.23o), $F_v = 0.1884$.
   For a hypothetical flame equal in size to the storage tank, $H_r = 8/5 = 1.6$.
   Dimensionless distance $X_r = 15/5 = 3$.
   Using Mudan's equations, $F_v = 0.1247$.
   Hence, the view factor for Target 1 = 0.1884 − 0.1247 = 0.0637.

ii. Location: Target 2
For a flame of diameter = 10 m, and height = 20 m, $H_r = 20/5 = 4$.
Dimensionless distance $X_r = 15/5 = 3$.
Using Mudan's equations, $F_v = 0.1792$ (same as in Example 5.4).

iii. Location: Target 3
For a flame of diameter = 10 m, and height = 10 m, $H_r = 10/5 = 2$.
Dimensionless distance $X_r = 15/5 = 3$.
Using Mudan's equations, $F_v = 0.141$.
Hence, the view factor for Target 3 = (2) (0.141) = 0.282.

5.2.3 Atmospheric Transmissivity

The atmospheric transmissivity modifies the radiation emitted from the flame, as it is only partially absorbed by the air between the flame and the target. The transmissivity, or the fraction transmitted, depends on the air's absorbing properties compared to the emission spectrum of the fire. Water vapor and carbon dioxide are the main absorbing components in the wavelength of heat radiation relevant to common industrial fires. Of these, water vapor has a more significant effect.

The TNO Yellow Book[10] provides a simple equation for the calculation of atmospheric transmissivity as follows:

$$\tau = 2.02(P_w X)^{-0.09} \tag{5.25}$$

where
τ = atmospheric transmissivity
P_w = partial pressure of water vapor in the air, Pa
X = the distance between the surface of the flame and the target, m.

This equation has been recommended for use when $10^4 < (P_w\ X) < 10^5$ N/m. The partial pressure of water, P_w, can be calculated by multiplying its water vapor pressure at ambient temperature with the fractional relative humidity of the ambient air. The vapor pressure of water, P_{ws}, can be calculated using the Antoine equation as follows:

$$\log_{10} P_{ws} = 8.10765 - \frac{1750.286}{235 + t} \tag{5.26}$$

where
P_{ws} = water vapor pressure, mmHg
t = the ambient temperature, °C.

Accordingly, for use in Equation (5.26), $P_w = (133.32)\ (P_{ws})\ (\%RH/100)$, where %RH denotes the percent relative humidity in the atmosphere, and P_{ws} is found from Equation (5.26) at ambient temperature, t °C. Here, the constant 133.32 converts vapor pressure from mmHg to Pa in Equation (5.26).

Example 5.6

Calculate the transmissivity of thermal radiation from a flame to a target at a distance of 25 m. Atmospheric temperature is 30°C, and relative humidity is 60%.

From Equation (5.28), the vapor pressure of water at 30°C is 31.8 mmHg. Therefore, the partial pressure of water vapor is:

$$P_w = (31.8)(0.6) = 19.1\ \text{mmHg} = (19.1)(133.32) = 2{,}546\ \text{Pa}.$$

At X=25 m, $P_w X$=(2,546) (25)=63,655 N/m, which is between 10^4 and 10^5.

Hence, from Equation (5.27), the transmissivity τ=(2.02) $(63{,}650)^{-0.09}$=0.746.

5.2.4 Assessment of Safety Distance

So far, we have examined detailed procedures for:

a. Calculation of the flame size
b. The SEP
c. The view factor
d. The atmospheric transmissivity.

The next step is to combine them to prepare the radiation intensity profile (intensity versus distance) for a given fire situation. Such profiles are necessary as the basis for the development of on-site emergency plans for sites having storage/handling facilities for flammable substances. Two examples are given below: one for a pool fire at ground level and another for a tank fire.

Example 5.7

A 10-m-diameter 15-m-high storage tank containing benzene is located in an area surrounded by a dike. The area enclosed by the dike is 35 m×35 m, and the height of the dike is 1 m. The tank is 90% full when it develops a major leak at the bottom. The liquid coming out ignites to form a confined pool fire on the entire diked area. Determine the thermal radiation intensity versus distance profile on a target at ground level under two conditions: (a) wind speed is negligible and (b) wind speed is 2 m/s. In the second case, the target may be assumed to be downwind.

Data: mass burning rate for benzene=0.085 kg/s/m², density of benzene=870 kg/m³, calorific value of benzene=40,100 kJ/kg, radiation component of the liberated heat=30%, atmospheric temperature=27°C, and relative humidity=60%.

The cross-sectional area of the pool (neglecting the presence of the tank)=(35) (35)=1,225 m².

$$\text{Burning rate} = (0.085)\ (1225) = 104\ \text{kg/s}$$

$$\text{Total volume of benzene in the tank} = (0.9)\left(0.785\times10^2\right)(15) = 1{,}060\ \text{m}^3$$

$$\text{Mass} = (1{,}060)\,(870) = 922{,}450\,\text{kg}$$

$$\text{Total burning time} = 922{,}450/104 = 8{,}870\,\text{seconds, or } 2.46\,\text{hours.}$$

Based on Equation (5.1), the equivalent diameter of the pool = 35 m

a. Negligible Wind Speed
   The density of air at 27°C is 1.18 kg/m³. Substituting D = 35, m = 0.08 m, and
   ρ_a = 1.18 kg/m³ in Equation (5.4), we get L/D = 1.37.
   Hence, flame length, L = (1.37) (35) = 48.0 m.
   Rate of heat release as radiation = (104) (40,100) (0.30) = 1,251,000 W.
   The SEP of the flame in the absence of soot is:

$$1{,}251{,}000 / \left\{(\pi)(35)(48) + (2)(\pi/4)(35^2)\right\} = 173.7\,\text{kW/m}^2.$$

Assuming that 20% of the flame surface is covered by soot that has an emissive power of about 20 kW/m², the weighted average SEP of the flame is (0.8) (173.7) + (0.2) (20), or 143 kW/m².

For the calculation of view factor, H_r = 48/(35/2) = 2.86. We assume distances for the target from the axis of the flame from 30 to 120 m. In the absence of wind, the angle of inclination of the flame is zero. Values of the maximum view factor, F_{max}, calculated through Mudan's Equations (5.23a) to (5.23o), are given below.

From Equation (5.28), the vapor pressure of water at 27°C is 26.7 mmHg, whence partial pressure of water P_w at 60% relative humidity is 19 mmHg or 2,139 Pa. For assumed values of distance to the target, atmospheric transmissivity, τ, is calculated using Equation (5.27) and is given below.

Then, from Equation (5.21), with SEP = 140, radiation intensity values at various distances are calculated, and the results are as follows:

X, m	30	40	60	80	100	120
X/R	1.71	2.29	3.43	4.57	5.71	6.86
F_{max} (Mudan)	0.341	0.238	0.132	0.0810	0.0539	0.03814
T	0.746	0.727	0.701	0.683	0.669	0.658
I, kW/m²	35.6	24.2	13.0	7.74	5.05	3.51

For personnel safety, the maximum limit of intensity is usually about 4.5 kW/m². Hence, any distance less than 100 m from the flame's radius (about 80 m from the dike) would be unsafe. Using Equation (3.8a) from Chapter 3, we can show that – at this level of intensity – a person without appropriate protective clothing will have a 50% probability of first-degree burns at an exposure duration of 40 seconds. Unless active fire protection systems can be switched on in a matter of seconds, plant operating people and firefighting personnel would be at significant risk of injury or death.

We stress that the tank's presence within the flame has not been included in the modeling above. Unless the fire is extinguished promptly, the tank could rupture, increasing damage significantly.

b. Wind Speed = 2 m/s

Substituting the values of mass burning rate, flame diameter, air density, and wind speed in Equations (5.9), (5.9a), (5.9b), and (5.10), we get:

$$U_c = 2.91\,\text{m/s}, U_* = 0.687, \theta = 33°, L/D = 1.44 \text{ and } L = 50.4\,\text{m}.$$

Thus, the flame length is only slightly greater than that in Part (a), which would show a negligible effect of wind speed on flame length in this case.

View factors are calculated assuming downwind ($\theta = 33°$) and upwind ($\theta = -33°$) locations for the target, using Equations (5.23a) through (5.23o). Taking SEP equal to 140 kW/m^2 and atmospheric transmissivity to be the same as for Case (a), radiation intensity profiles are determined with the results given below:

X, m	30	40	60	80	100	120
X/R	1.71	2.29	3.43	4.57	5.71	6.86
T	0.746	0.727	0.701	0.683	0.669	0.658
F_{max} ($\theta = 33°$)	0.545	0.405	0.227	0.128	0.0776	0.0507
F_{max} ($\theta = -33°$)	0.197	0.136	0.0768	0.0494	0.0343	0.0252
I, kW/m^2, Downwind	56.9	41.2	22.3	12.2	7.274	4.67
I, kW/m^2, Upwind	20.6	13.8	7.54	4.72	3.21	2.32

These results show that view factors and, hence, radiation intensities are significantly lower at the same distance from the flame if the target is located upwind. Wind direction has a dramatic effect at even moderate wind velocities on radiation intensity. Therefore, it is a crucial consideration for plant operators and firefighters. Plant personnel are always advised to escape thermal radiation by running away from flames or cross-wind if the wind is blowing toward them from the flame.

Example 5.8

A manufacturing plant has several benzene storage tanks of 10 m in diameter and 15 m in height. A dike surrounds the storage area. The separation between tanks is 10 m. Overfilling in one of the tanks causes the tank top to rupture, leading to a fire. Estimate the radiation intensity (a) on the top of an adjacent tank and (b) at ground level. In both cases, the target positions are downwind, at a horizontal distance equal to the tank spacing. Assume wind speed to be 2 m/s. Flame length can be taken as 20 m. Atmospheric transmissivity of radiation, τ, is approximately 0.75.

The geometric configuration of the system is shown in Figure 5.6.

Data: mass burning rate for benzene = 0.085 kg/s/m^2, density of benzene = 870 kg/m^3, calorific value of benzene = 40,100 kJ/kg, radiation component of the liberated heat = 30%, atmospheric temperature = 27°C, and relative humidity = 60%.

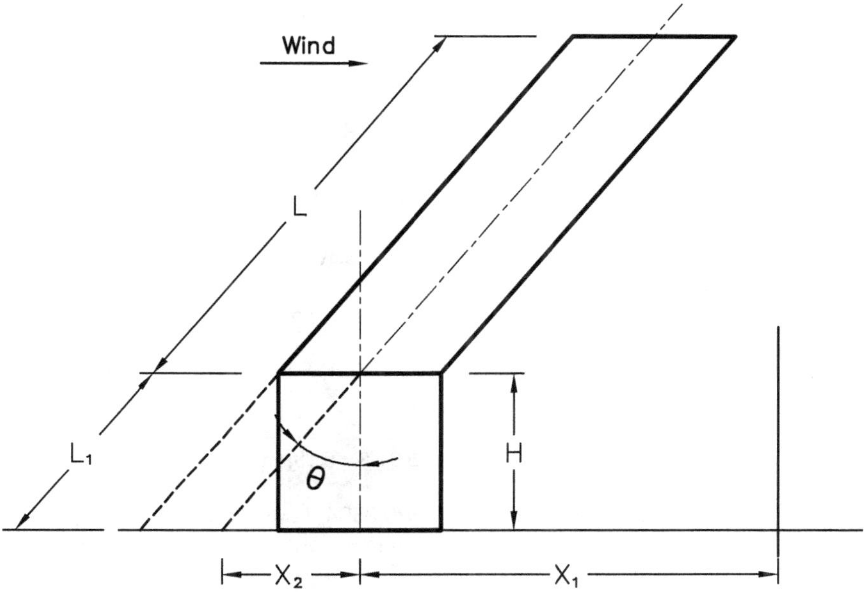

FIGURE 5.6 Flame/target configuration in case of a tank fire with wind.

From Equation (5.9b), $U_c = \{(9.81)(0.085)(10)/1.18\}0.333 = 1.92$ m/s
From Equation (5.9a), $U^* = U/U_c = 2/1.92 = 1.04$
From Equation (5.10), $\cos\theta = 0.7/U^*0.49 = 0.6860$, whence $\theta = 46.7°$
Flame length, $L = 20$ m. Tank height, $H = 15$ m.

SEP of the flame can be calculated in the same manner as in Example 5.7. In this case, tank cross-sectional area = (π/4) (D^2) = 78.54 and the burning rate = (78.54) (0.085) = 6.67 kg/s.

Taking the calorific value of benzene equal to 40,100 kJ/kg and fractional radiation equal to 0.3, the rate of heat release as radiation is: (6.67) (40,100) (0.3) = 80,240 kW.

Hence, the SEP for a clean flame is: 80,240/{(π) (10) (20) + (2)(π/4) (100)} = 102 kW/m^2.

Allowing 20% of the flame surface to be covered by soot with an SEP = 20 kW/m^2, the weighted average SEP of the flame is (102) (0.8) + (20) (0.2) = 85.6 kW/m^2.

Referring to Figure 5.6, X_1 = Tank spacing + Tank radius = 10 + 5 = 15 m.
With $\theta = 46.7°$, $X_2 = H \tan \theta = (15)\ (1.061) = 15.9$ m
$L_1 = H/\cos\theta = 15/0.6858 = 21.9$ m

a. Target on Top of Adjacent Tank

$$H_r = 20/5 = 4,\ X_r = 15/5 = 3, \text{ and } \theta = 46.7°$$

From Equations (5.23a) through (5.23o), $F_v = 0.136$, $F_h = 0.151$, $F_{max} = 0.204$

With $\tau = 0.75$, the maximum thermal radiation intensity on the tank is:

$$\mathrm{I} = (\mathrm{SEP})(F_{\max})(\mathrm{Transmissivity}) = (85.6)(0.204)(0.75) = 13.1\,\mathrm{kW/m^2}.$$

b. Target at Ground Level

View factor, F, for flame and tank combined, $L + L_1$:

$$L_1 = H/\cos\theta = 15/\cos 46.7° = 21.9\,\mathrm{m}, X_2 = H\tan\theta = 15\tan 46.7° = 15.9\,\mathrm{m}$$

$$H_r = (20 + 21.9)/5 = 8.38, X_r = (15 + 15.9)/5 = 6.18, \text{and } \theta = 46.7°$$

From Equations (5.23a) through (5.23o), $F_v = 0.0674$, $F_h = 0.0713$, $F_{max} = 0.0981$
View factor for tank alone:

$$H_r = 21.9/5 = 4.38, X_r = (15 + 15.9)/5 = 6.18, \text{and } \theta = 46.7°.$$

From Equations (5.23a) through (5.23o), $F_v = 0.0504$, $F_h = 0.0247$, $F_{max} = 0.0561$
Therefore, maximum view factor for the flame alone = 0.0981 − 0.0561 = 0.042
Hence, maximum radiation intensity = (85) (0.042) (0.75) = 2.68 kW/m^2

The estimated radiation intensity at ground level is below the upper limit of 4.5 kW/m^2 that usually is specified for personnel safety. The intensity on the top of an adjacent tank (13.1 kW/m^2) is below the 38 kW/m^2 maximum limit that is usually specified for the safety of adjacent tanks. However, as with all tank fires, unless the tank fire is extinguished promptly, prolonged heating at this level of intensity would be most undesirable: this could result in a high rate of vaporization and a significant escalation of the fire to the top of the second tank. Rapid cooling of surrounding tanks using water sprinklers is, therefore, necessary.

REFERENCES

1. Burgess, D. and Hertzberg, M.: *Radiation from Pool Flames*, in N. H. Afgan and J. M. Beer: *Heat Transfer in Flames, Chapter 27* (Scripta Book Company, Washington, DC, 1974).
2. Design Institute for Physical Property Research (DIPPR), *American Institute of Chemical Engineers* (2020)
3. Mannan, S.: *Lees' Loss Prevention in the Process Industries* (4th Ed., Butterworth-Heinemann, Oxford, 2012).
4. Thomas, P. H.: *The Size of Flames from Natural Fires, 9th Symposium on Combustion* (Academic Press, New York, pp. 844–859, 1963).
5. Heskestad, G.: Luminous Heights of Turbulent Diffusion Flames, *Fire Safety Journal*, 5, pp. 103–108 (1983).
6. Crocker, W. P. and Napier, D. H.: *Thermal Radiation Hazards of Liquid Pool Fires and Tank Fires* (Institution of Chemical Engineers Symposium No. 97, University of Manchester Institute for Science and Technology, England, 1986).
7. Welker, J. R. and Sliepcevich, C. M.: Bending of Wind-blown Flames from Liquid Pools, *Fire Technology*, 2, pp. 127–35 (1966).
8. Moorhouse, J. and Pritchard, M. J.: *Thermal Radiation Hazards from Large Pool Fires and Fireballs* (Institution of Chemical Engineers Symposium Series No. 71, Manchester, 1982).

9. Cline, D. and Koenig, L. N.: The Transient Growth of an Unconfined Pool Fire, *Fire Technology*, 19, pp. 149–162 (1983).
10. TNO: *Methods for the Calculation of Physical Effects (Yellow Book)* (3rd Ed., CPR 14E, Parts 1 and 2, TNO, The Hague, 1997).
11. Hottel, H. C. and Sarofim, A. F.: *Radiative Transfer* (McGraw-Hill, New York, 1967).
12. Modest, M.F.: *Radiative Heat Transfer* (McGraw-Hill, New York, 1993).
13. Rhine, J. M. and Tucker, R. J.: *Modelling of Gas-Fired Furnaces and Boilers* (British Gas, in association with McGraw-Hill, New York, 1991).
14. Stannard, J. H.: Thermal Radiation Hazards associated with Marine LNG Spills, *Fire Technology*, 13, pp. 35–41 (1977).
15. Mudan, K. S.: Geometric View Factors for Thermal Radiation Hazard Assessment, *Fire Safety Journal*, 12, pp. 89–96 (1987).
16. Houston Chronicle: *Spill From Tank Farm Fire Closes Houston Ship Channel* (March 25, 2019).
17. TNO: *Methods for the Calculation of Physical Effects (Yellow Book)* (2nd Ed., CPR 14E, TNO, The Hague, 1992).

6 Jet Fire

A jet fire results when a flammable gas (or gas-liquid mixture), under pressure, is released to the atmosphere and ignites. Common examples in the process industries are as follows:

- flames from gas burners in furnaces,
- flares,
- flames from the ignition of gases/vapors at the outlet of vents or relief valves.

Also, accidental releases or flashing of volatile hydrocarbon /inflammable products, generally at high velocity from holes in pipelines, pressure vessels, the lower sections of atmospheric storage tanks, and leaked gaskets, can get ignited and result in jet fires.

Jet fires caused by accidental releases can impinge on adjacent pipes and vessels, tanks, and other equipment. Impinging of jet fire on steel structure can result in buckling/failure of the supporting structure of other equipment, pipes, cables, etc., resulting in severe damage (Table 3.9). Such damages include loss of containment of inflammable or toxic materials and collapsing supporting structures; the consequences can be disastrous.

Sometimes, the release of a volatile hydrocarbon or flammable product, after flashing, does not find an ignition source. The highly inflammable vapor spreads quickly as a vapor cloud until it reaches an ignition source and ignites. Such a "vapor cloud fire" is the subject of Chapter 7.

The computation of the dispersed cloud geometry, the velocity, and concentration gradients of all components is done using dispersion modeling, discussed in detail in Chapter 11.

This chapter is concerned with the proper methodology for calculating the instantaneous release rate of a gas or gas-liquid mixture under pressure. Examples include the following:

- holes in pipelines or pressure vessels
- holes in the lower sections of storage tanks
- leaking gaskets, or
- releases from a relief valve on vessels open to the atmosphere.

This phenomenon has been the subject of numerous major publications from industry organizations such as the American Petroleum Institute (API) and textbooks by several distinguished authors.

We discuss the fundamental thermodynamic basis that should be used for describing these phenomena. This chapter seeks to provide a theoretically correct and consistent problem formulation based on the first law of thermodynamics as applied to free expansion phenomena, otherwise described as a Joule-Thompson (J-T) expansion.

DOI: 10.1201/9781003107873-6

J-T expansions are decidedly irreversible processes that can be modeled using the first law of thermodynamics without resorting to the concept of entropy.

We focus on the methods to estimate the release rate from a hole or leak in a pressurized container. We also describe procedures for the proper sizing of relief valves for given upstream and downstream conditions. In the literature, it is common to find formulas for such computations that assume that such a free expansion can be modeled using an isentropic expansion. While the numerical results from both methods may be close, we recommend using the procedures described below.

6.1 FLOW THROUGH A HOLE (FREE EXPANSION)

The reliable computation of flow through a leak (a hole in a pressure vessel, pipeline, or a blown-out gasket) is crucial for process safety calculations. An additional example would be releasing vessel contents into the atmosphere through a rupture disk or relief valve. The vessel or pipeline contents are then forcefully ejected into the atmosphere, forming a cloud with ambient air. A fire resulting from the uncontrolled release of combustible vapor/liquid mixtures can cause extensive damage to surrounding vessels and structures. Besides, the release of a toxic gas or vapor mixture, especially if it is heavier than air, poses grave dangers to people in the affected area.

Therefore, great attention must be paid to release rate calculations, atmospheric dispersion phenomena, proper distances between major process equipment, and sources of ignition.

While it is true that current safety and environmental regulations prohibit the release of flammable mixtures into the air, it is also true that – in many situations – safe releases to a flare system are not caused by improper facility design.

Releases are often caused when

a. sudden upsets occur in a steady running plant due to factors individually or collectively as (1) power failure, (2) process upsets, or (3) sabotage and
b. a plant is undergoing a start-up or shutdown.

For flammable components (or toxic components, whose effects are dealt with in Chapter 10), releases to the atmosphere cause an extremely hazardous situation. Such situations are deplorable, and the following discussion can quantify the magnitude of the problems caused by uncontrolled releases.

For ensuring that a relief valve is designed appropriately for release to a relief system header, it must be ensured that release rate computations are carried out in a theoretically sound and technically proficient manner. The thermodynamic and fluid flow phenomena for a relief valve are identical to those for flow through a hole in a pressure vessel or pipeline, as discussed below.

We first discuss the fundamental question of choosing the proper thermodynamic basis for such calculations. Our concern arose from studies of the relevant literature on this subject. One school of thought (which we think is theoretically correct) treats the flow of compressed gas, or gas-liquid mixture, through a rigidly held valve or hole in a pipeline as a Joule-Thompson expansion. Every J-T expansion is an irreversible adiabatic or isenthalpic process, i.e., a "free expansion". The alternative approach uses an isentropic (i.e., reversible adiabatic) assumption for these free expansion phenomena.

No work is done by a gas or gas/liquid mixture flowing out through a hole in a pressure vessel or a relief valve. The calculations for flow through a relief valve, which are similar in principle to flow through a hole in a vessel or pipeline, must also be considered applicable to free expansion.

This reasoning derives simply from the first law of thermodynamics as applied to a flow process, generally referred to as the steady-flow energy equation (SFEE). The following references illustrate the point quite unequivocally:

As early as 1947, Hougen and Watson[1] discussed "free expansion" of gases (p. 551) and noted as follows:

> *The unrestrained expansion of a gas is known as free expansion. Under conditions of no restraint no work is done, and under adiabatic conditions no heat is added. Free expansion under flow conditions is commonly known as throttling or as the Joule-Thompson effect referred to in Chapter XII.*
>
> *The Joule-Thompson effect is measured experimentally by expanding the gas slowly and in steady flow through a well-insulated porous plug; in this way, potential work is lost, and no heat is allowed to enter or leave the system through the walls. The fluid flows reversibly into and out of the process, but the expansion step is completely irreversible. Since no heat is added or lost, the process takes place under conditions of constant enthalpy, according to Equation (VII-11), page 207. However a change in internal energy results if any change in flow energy pV occurs.*

Similarly, in a discussion of valves and throttles, Elliott and Lira[2] observe as follows:

> *A throttling device is used to reduce the pressure of a flowing fluid without extracting any shaft work and with negligible fluid acceleration. Throttling is also known as Joule-Thompson expansion in honor of the scientists who originally studied the thermodynamics. An example of a throttle is the kitchen faucet. Industrial valves are modeled as throttles...Changes in kinetic and potential energy are small relative to changes in enthalpy as we just discussed. When in doubt, the impacts of changes in velocity can be evaluated as described in Example 2.9. The amount of heat transfer is negligible in a throttle. The boundaries are not expanding, and there is also no mechanical device for transfer of work, so the work terms vanish. Therefore, a throttle is isenthalpic.*

Finally, in a discussion of nozzles, Klein and Nellis[3] observe as follows:

> *...The potential energy terms of the inlet and outlet flows are not usually significant. Nozzles neither require nor produce power ($W_{in} = W_{out} = 0$) and are very nearly adiabatic. Note that the kinetic energy terms associated with the flows, particularly the outlet flow, cannot be neglected as the purpose of the nozzle is to increase the kinetic energy of the fluid. With these simplifications, an energy balance on the nozzle becomes:*

$$H_{in} + \frac{1}{2}V_{in}^2 = H_{out} + \frac{1}{2}V_{out}^2$$

In the light of these assertions by these noted authorities and others,[4,5] we describe the modeling of flow through a hole in a vessel that is open to the atmosphere, or a relief valve, as an isenthalpic (*irreversible* adiabatic) basis consistent with free expansion. An additional term corresponding to the change in kinetic energy is included. This term is necessary because, during a free expansion, a pipeline or vessel remains stationary, heat transfer is negligible, and absolutely no work is done by the gas flowing out through a valve or a hole open to the atmosphere.

When a relief valve is attached to a rigidly held pipe or vessel, no motion occurs, and no work is performed. For this reason, we emphasize that the flow through a hole in a pipe, or the opening of a relief valve, should be modeled as a stationary nozzle that does no work, consistent with a Joule-Thompson free expansion (isenthalpic). The same reasoning applies to flow through other devices such as venturis or orifices.

Estimation of gas properties at the inlet, throat, and outlet must be done correctly. The proper thermodynamic basis (isenthalpic or J-T expansion) can be used in all such computations. For an ideal gas, enthalpy is a function of temperature only. Therefore, free expansion of an ideal gas through a valve or a hole would not change its temperature significantly, except for the effect of a change in kinetic energy caused by a change in gas velocity.

For a real gas, however, enthalpy is, in general, a function of temperature, pressure, and composition. Therefore, a free expansion often results in a lower temperature (since the Joule-Thompson coefficient is negative for most gases). Note also that changes in kinetic energy, which generally are minor, may not always be negligible. This point is discussed in detail in the problem formulation in Section 6.1.1 below.

It is a well-known fact that all process simulators treat a valve in any conduit as an isenthalpic device. For ideal gases, this results in no temperature change. For high-pressure gases, especially under cryogenic conditions, a substantial temperature change may result even at constant enthalpy. This is because the enthalpy of a nonideal gas can be a strong function of both temperature and pressure. Such valves are also called J-T (for Joule-Thompson) valves in gas plants.

Conversely, for a given relief valve size, density calculations at the throat must be reliable to ensure that we predict the correct relief rate, even for an ideal gas. Errors in calculated gas or gas-liquid flow from a hole in a pipeline or a relief valve could lead to potentially serious process safety concerns.

For example, an isenthalpic expansion of methane gas from 50 bar and 26.85°C to 25 bar results in a temperature of 16.7°C. Similarly, we discuss free expansion of methane gas from 48.26 bar at −59.44°C to 6.89 bar. Using the Lee-Kesler equation of state (EOS) for isenthalpic expansion, the stream would remain 100% vapor, and the outlet temperature would be −102.9°C, neglecting changes in kinetic energy. Isentropic expansion, however, would implausibly predict that the outlet would be a two-phase mixture, with 12.4% as liquid, and at a temperature of −131.6°C, almost 30°C cooler. Calculated throat conditions such as temperature, density, and velocity must therefore be made reliably to ensure that relief valve sizing is adequate.

For these reasons, when calculating the release rate for gas at high pressure through a relief valve, the proper thermodynamic basis must be chosen for enthalpy and fluid properties at the throat. Reliable computations of density and other fluid thermodynamic and transport properties such as viscosity are mandatory. These matters are discussed later in this chapter.

The same considerations apply to gas releases to the atmosphere resulting from a rupture in a pipeline or vessel containing a gaseous mixture under pressure. Release rate calculations are performed in a manner that is identical to those for conditions at the throat of a relief valve.

To help clarify the proper methodology for a relief valve – or a hole in a pressurized vessel – the computations that should be performed are described below. We also show a few numerical examples.

Another important issue is that gas velocity in closed conduits or orifices can never exceed sonic velocity. In a pipeline, increasing upstream pressure does not increase flow beyond the limit imposed by the sonic velocity at the end of a pipe. (The extra energy of compression is dissipated as heat across a stationary shock wave at the end of the pipe). Sonic velocity may exist at the throat of the valve or the orifice in a ruptured vessel. This possibility is considered explicitly in the procedure described below. This development is based on the following assumptions:

1. The opening can be represented by a convergent nozzle of exit area equal to the area of the throat of the relief valve or hole in the case of a pipeline rupture
2. The compressed gas pipeline, or leaking container, is a large reservoir whose upstream pressure and temperature are maintained constant; therefore, we are dealing only with steady flow with no mass accumulation.
3. The compressed gas undergoes an irreversible, adiabatic process of expansion through the nozzle.

Regarding item 2 above, if the pressure vessel has no flow into it after the release has commenced, the vessel pressure would decrease over time. The gas temperature upstream of the hole or relief valve would also be lowered progressively as gas escapes (unsteady flow). However, in this situation, the instantaneous gas relief rate would also diminish with time until the vessel pressure becomes equal to the downstream pressure.

Time-dependent problems are not covered in the following discussion, however. They are solved using dynamic simulation. Formulating and solving (numerically) the appropriate algebraic, thermodynamic, and differential equations for unsteady flow can be arduous. Commercially available process simulation software is generally used for such problems,[6–9] as their scope is beyond what can be done conveniently with hand calculations or even spreadsheets.

6.1.1 Theoretical Basis

The theory and calculation methods for the estimation of release through an opening are given below for two cases:

- Compressed gas release
- Liquefied gas release

In both cases, the flow is assumed to be steady and continuous.

The flow through a hole in a punctured vessel or a relief valve from either (1) the inlet to the throat or (2) from the throat to the outlet follows the steady flow energy balance equation (Equation 9.21 in Himmelblau and Riggs[4]):

$$\Delta E = Q + W - \Delta(H + KE + PE) \tag{6.1}$$

where

ΔE = energy accumulation (zero at steady state), kcal/kg
Q = net heat transfer (zero both for a relief valve and an orifice), kcal/kg
W = net work done (zero for a relief valve), kcal/kg
ΔH = change in enthalpy, kcal/kg
ΔKE = change in kinetic energy, kcal/kg
ΔPE = change in potential energy (zero for a relief valve), kcal/kg

Clearly, for steady flow through an orifice, valve (or relief valve), ΔE, Q, W, and ΔPE are all zero.[1–4] Therefore, we are left with

$$\Delta(H + KE) = 0, \text{ or}$$

$$\Delta H = H_{out} - H_{in} = -\Delta KE, \text{ or} \tag{6.2}$$

$$H_{out} = H_{in} - \Delta KE$$

In other words, the outlet enthalpy is equal to the inlet enthalpy minus the change in kinetic energy in going through the throat or orifice. (It should be noted that this logic would apply for discharge through a hole in a pipe or vessel, a flow regulation valve, a relief valve, venturi, or an orifice plate).

If there is a negligible change in velocity, the outlet enthalpy will be the same as that of the inlet gas. As mentioned previously, this phenomenon is generally referred to as a Joule-Thompson (J-T) expansion. For an ideal gas, enthalpy is a function of temperature only; therefore, a J-T expansion would result in no change in temperature. However, for most real gases, a J-T expansion results in a decrease in temperature. However, there are a few notable exceptions, such as hydrogen or helium gas, for which the J-T coefficient has the opposite sign.[5] On the other hand, if the gas speeds up, its kinetic energy will increase, and therefore, its enthalpy will decrease according to Equation (6.2) above. This kinetic energy increase results in a decrease in outlet gas enthalpy, and therefore temperature, even for an ideal gas.

For ideal gases, the computation of enthalpy for ideal gases is straightforward, as it is a function of temperature only.[1] For real gases, however, enthalpy is also a function of pressure, temperature, and composition. The use of an EOS becomes necessary[2] to calculate the enthalpy departure from the ideal.

6.1.2 Compressibility Factor and Enthalpy for Real Gases

For real gases, enthalpy is found using a suitable EOS, such as the Soave-Redlich-Kwong (SRK), Peng-Robinson (PR), or Lee-Kesler (LK). This procedure may be summarized as follows:

At the given upstream pressure (P), temperature (T), and composition, solve the EOS for the compressibility factor (z). Typically, a cubic polynomial in z must be solved when using any cubic EOS, such as the SRK or PR. Either an explicit analytical (trigonometric) or numerical trial and error method can be used. The latter is generally more computationally efficient.

For the LK EOS, however, a trial-and-error procedure is always required. The LK EOS is not a simple cubic in z. It requires solving the 8-constant Benedict-Webb-Rubin[2] (BWR) equation twice: once for the "simple" fluid (methane) and again for the "reference" fluid (n-octane). All thermodynamic properties are then determined by interpolation based on the mixture's Pitzer's omega factor. With the appropriate thermodynamic expressions for the EOS, we use these values of z, for the simple and reference fluids, to calculate ΔH_d, the enthalpy departure from ideal.[2,10,11]

Calculate the ideal gas enthalpy, H_i (a function of temperature T_0 only). Find the real gas enthalpy using

$$H_r = H_i - \Delta H_d \tag{6.3}$$

where

H_r = real gas enthalpy, J/kg
H_i = ideal gas enthalpy, J/kg
ΔH_d = enthalpy departure from ideal, J/kg

This procedure must be followed for calculating enthalpy at both upstream and downstream (throat) conditions.

The gas velocity at the throat may be sonic or subsonic. Also, as the gas speeds up in passing through the valve throat or hole (since the cross-sectional area is lower than that of the pipe), its kinetic energy increases, and this results in a reduction of the enthalpy (Elliott and Lira,[2] Equation 2.53):

$$H_t = H_0 - \Delta KE \tag{6.4}$$

where

H_0 = upstream gas enthalpy, J/kg
H_t = exit gas enthalpy at the throat or hole, J/kg
ΔKE = change in kinetic energy, J/kg

ΔKE is found as follows (Elliott and Lira's[2] Equation 2.53):

$$\Delta KE = \left(u_t^2 - u_0^2\right)/2 \tag{6.5}$$

where

u_t = throat (or hole) velocity, m/s
u_0 = upstream velocity, m/s.

Clearly, the outlet and inlet enthalpies are equal only if the change in velocity is negligible.

6.1.3 Release Rate Calculation

The computation of the release rate from a pipe rupture should be done subject to the following principles and equations.

6.1.3.1 Bernoulli's Equation

This equation relates upstream and downstream conditions for an arbitrary flow geometry, as follows:

$$P_0 + \rho_0 u_0^2/2 + \rho_0 g\, z_0 = P_t + \rho_t u_t^2/2 + \rho_t g\, z_t + \Delta P_{\text{friction}} \tag{6.6}$$

Where the subscripts 0 and t refer to the inlet and exit planes, respectively,

P = pressure, Pa
ρ = density, kg/m^3
u = velocity, m/s
g = gravitational constant, m/s^2
z = elevation above datum, m
$\Delta P_{\text{friction}}$ = friction losses, Pa

The friction losses have been ignored for simplicity in this presentation.

Obviously, for a hole or throttle (or relief) valve, the elevations z_0 and z_t are the same.

6.1.3.2 Sonic Velocity

The sonic velocity in any medium[2] may be expressed as

$$u_s = \left(\gamma z RT/M\right)^{0.5} \tag{6.7}$$

where
u_s = sonic velocity, m/s
γ = gas C_p/C_v ratio,
z = gas compressibility factor,
R = gas constant = 8314.46, J/(kgmol K)
T = temperature, K
M = gas molecular weight, kg/kgmol

In Equation (6.7), properties at the throat must be used for choked flow calculations. For ideal gases, the compressibility factor z is equal to 1. However, for real gases, the compressibility factor z is, in general, not equal to 1 and can be found only using an appropriate EOS.

6.1.3.3 C_p, C_v, and $\gamma = C_p/C_v$ Ratio

Similarly, for ideal gases, the value of γ is found from the value of C_p, which is a function of temperature only, and the value of C_v is found from the following:

$$C_v = C_p - R \tag{6.8}$$

For real gases, however, the value of γ again requires the use of an EOS to calculate the ΔC_{pd} and ΔC_{vd} departures from ideality. The real gas C_p and C_v values are found using the ideal values and the calculated departures from ideality, by subtraction, as described above for enthalpy:

$$C_{p1} = C_{p0} - \Delta C_{pd} \tag{6.9}$$

where
C_{p1} = real gas C_p, J/(kg K)
C_{p0} = ideal gas C_p, J/(kg K)
ΔC_{pd} = departure in C_p from ideal gas value, J/(kg K)

Similarly,

$$C_{v1} = C_{v0} - \Delta C_{vd} \tag{6.10}$$

where
C_{v1} = real gas C_v, J/(kg K)
C_{v0} = ideal gas C_v, J/(kg K)
ΔC_{vd} = departure in C_v from ideal gas value, J/(kg K)

In a conduit, orifice, or valve, the actual gas velocity, u_e, can never exceed the sonic limit, u_s. In general, u_s is different at the upstream conditions compared to the value at throat conditions. The upstream and throat temperatures and pressures of the gas may differ, thanks to the kinetic energy effect, Equation (6.4).

At a sufficiently low upstream pipeline or vessel pressure, the hole velocity would be lower than the sonic velocity limit. Above some critical limit for the upstream pressure, however, the maximum discharge flow is limited by the sonic velocity criterion at the hole.

6.1.3.4 Density

Gas density is computed using

$$\rho = PM/(zRT) \tag{6.11}$$

where
ρ = density, kg/m^3
P = pressure, Pa
M = gas molecular weight, kg/kgmol
z = gas compressibility factor
R = gas constant = 8314.46, J/(kgmol K)
T = temperature, K

The gas compressibility factor is 1 for ideal gases; it is calculated from the EOS for real gases.

6.1.3.5 Velocity

The velocity in a pipe or hole (or valve throat) is found using

$$u = W/(\rho A) \quad (6.12)$$

where
u = velocity, m/s
W = mass flow rate, kg/s
ρ = gas density, kg/m^3
A = cross sectional area, m^2

For a circular pipe, the cross-sectional area is defined as

$$\mathrm{A} = \pi D^2/4 \quad (6.13)$$

where
D = inner diameter, m

For noncircular cross-sections, the equivalent diameter D_e is defined using the concept of the hydraulic radius, which is defined as

$$D_e = 4(\text{Cross-sectional-area}/\text{Wetted-perimeter}) \quad (6.14)$$

It can readily be seen that this definition holds for a circular pipe also:

$$4\pi D^2/4/(\pi D) = D$$

As noted earlier, for ideal gases, the compressibility factor z is equal to 1. For real gases, however, finding the gas density ρ requires first calculating z from the EOS. We do this using the procedure described above.

Note that conditions at the throat of a valve (or at the hole in a pipe) are unknown. Therefore, the gas flow rate computation requires solving a system of simultaneous nonlinear algebraic equations, using an iterative procedure. When using Microsoft Excel, this is best performed using the "Solver Add-in". Solver is a nonlinear optimization technique. It can simultaneously adjust the unknown quantities and solve the problem by enforcing the constraints imposed by the Bernoulli Equation (6.6) for throat velocity, the SFEE (6.1) for hole enthalpy, and, when appropriate, the sonic velocity limit, Equation (6.7).

Note that, since the calculated throat velocity can never exceed the sonic velocity limit, this constraint is added explicitly only when the upstream pressure exceeds the threshold value at which sonic velocity occurs. In general, for fixed inlet conditions and a specified downstream pressure, it is not known in advance whether the flow at the throat of a valve or orifice will be at the sonic velocity (choked flow) or not. The recommended procedure then is as follows:

1. Find the limiting upstream pressure required to make velocity at the throat equal to the sonic velocity, using specifications as shown above for Case (A), in Example 6.1 below.
2. If the actual upstream pressure is *equal to or greater than* this limiting value in step 1, the throat velocity will be sonic, and the flow *will be choked.*
3. If the actual upstream pressure is *less* than the limiting value in step 1, the throat velocity will be subsonic, and the flow *will not be choked.* Solve the problem as specified above for Case (B), also in Example 6.1 below.

When solving for an unknown gas flow rate, if the upstream pressure is sufficiently low, hole velocity could remain subsonic. However, if the upstream pressure is sufficiently high, both the pressure at the hole and the gas mass flow rate would vary in a manner that obeys the sonic velocity limit at the hole.

Example 6.1

Problem: Case (A): Methane is flowing through a 300 mm diameter, above-ground pipeline at an absolute pressure of 46 bar, and a temperature of 300 K. The pipeline develops a hole with an equivalent diameter of 50 mm. Calculate the mass rate of leakage to the atmosphere and the velocity of the gas jet emerging from the hole. The pressure inside the pipeline may be assumed to be constant owing to continuous supply from the source.

Case (B): Rework Case A assuming that the pipeline's pressure is 1.5 bar instead of 46 bar.

CASE (A) $P_O = 46$ BAR

Solve for release rate from a hole in a pipeline – sonic velocity at the throat.

This problem requires solving three simultaneous nonlinear algebraic equations for Case (A) (and only two simultaneous nonlinear algebraic equations for the subsonic Case (B), as discussed in Section 6.1.4.2 below). The solutions shown above for both cases were obtained using the Excel software package. Note that any other similar package can also be used. We verified this numerical solution using the MATHCAD and POLYMATH software packages also.

In Case (A), the upstream pressure is high enough to cause the hole velocity to reach the sonic limit. In Case (B), however, the upstream pressure was low enough for the throat velocity to be less than the sonic limit.

For comparison, for Case (A), if the gas were not assumed to be ideal, a rigorous solution using the Peng-Robinson EOS would yield a gas flow rate of 16.1 kg/s.

We show the following detailed solution for the ideal gas case only: The real gas case requires complex computer programs to iteratively solve the chosen EOS (e.g., Lee-Kesler or Peng-Robinson) for the required thermodynamic quantities. Since these routines are embedded in an outer loop for the unknowns (T and P at the throat and the mass flow rate), the computational burden is too great for hand calculations.

The Microsoft Excel® software (Excel 2016[12]) was used for solving this problem. All variables are defined in Table 6.1a and b. The problem statement for Case (A) for the ideal gas case is as follows:

 # Initial guesses:

$T_t(0) = 300$ # Initial guess for T_t, K
$W(0) = 20$ # Initial guess for W, Kg/s
$P_t(0) = 46$ # Initial guess for P_t, bar

 # Aly-Lee ideal gas enthalpy constants:

$A = 33298$ # Ideal gas enthalpy constant A
$B = 79933$ # Ideal gas enthalpy constant B
$C = 2086.9$ # Ideal gas enthalpy constant C
$D = 41602$ # Ideal gas enthalpy constant D
$E = 991.96$ # Ideal gas enthalpy constant E
$F = -120{,}280{,}000$ # Ideal gas enthalpy constant F

 # The following three functions must be made equal to zero by adjusting the three unknown variables W (kg/s), T_t (K), and P_t (bar):

$f(T_t) = h_t - h_{t2}$
$f(W) = u_t - u_{t2}$
$f(P_t) = u_t - u_s$

 # Initial guess ranges for nonlinear equations:

$W(\text{min}) = 1$
$W(\text{max}) = 50$
$T_t(\text{min}) = 200$
$T_t(\text{max}) = 400$
$P_t(\text{min}) = 10$
$P_t(\text{max}) = 46$

This problem requires solving the three simultaneous, nonlinear algebraic equations (defined above) by varying T_e (the hole temperature), W (the mass flow rate), and P_e (the hole pressure):

$f(T_t) = 0$
$f(W) = 0$
$f(P_t) = 0$

In this case, the upstream pressure is high enough to cause the hole velocity to reach the sonic limit. This problem requires solving three simultaneous nonlinear algebraic equations. The solution shown below for Example 6.1 (A) was obtained using the Microsoft Excel (2016) software. Note that any other similar package can also be used. We verified this numerical solution using the MATHCAD[13] and POLYMATH[14] software packages. Additionally, we verified these results using a custom Fortran program for the Lee-Kesler EOS. This procedure confirmed the correct handling of cases where the pure component, or gas mixture, is far from ideal (i.e., the compressibility factor is far from unity, and the enthalpy and C_p and C_p departures from ideality are all far from zero).

Note: The numerical solutions for Cases A and B are shown in Table 6.1a and b, following the problem statement for Case (B).

TABLE 6.1a
Sonic or Subsonic Flow through a Valve or Hole Variables and Equations

Variable	Description and Units	Explicit Equations
A	Ideal gas enthalpy constant A	$A = 33{,}298$
B	Ideal gas enthalpy constant B	$B = 79{,}933$
C	Ideal gas enthalpy constant C	$C = 2{,}086.9$
D	Ideal gas enthalpy constant D	$D = 41{,}602$
E	Ideal gas enthalpy constant E	$E = 991.96$
F	Ideal gas enthalpy constant F	$F = -120{,}280{,}000$
P_t	Throat pressure, bar	$P_t = 1.01325$
kcal_j	Conversion: 1 kcal = 4,184 J	kcal_j = 4,184
M	Gas molecular weight, kg/kgmol	$M = 16.0$
T_0	Inlet gas temperature, K	$T_0 = 300$
P_0	Inlet gas pressure, bar	$P_0 = 1.5$
z_0	Inlet gas compressibility factor	$z_0 = f(T_0, P_0, x)$
$(c_p/c_v)_0$	Inlet gas C_p/C_v ratio	$(c_p/c_v)_0 = 1.32$
R	Gas constant, J/(kgmol K)	$R = 8{,}314.46$
D_0	Diameter of pipe, m	$D_0 = 0.3$
D_t	Diameter of throat, m	$D_t = 0.05$
h_0	Inlet enthalpy, kcal/kg	$h_0 = (A\ T_0 + B\ C/\tanh(C/T_0) - D\ E/\tanh(E/T_0) + F)/M/\text{kcal\_j}$
ρ_0	Inlet density, kg/m^3	$\rho_0 = (P_0)(10^5)\ M/R/T_0$
A_0	Pipeline cross-sectional area, m^2	$A_0 = 3.14159/4\ (D_0^2)$
A_t	Throat flow area, m^2	$A_t = 3.14159/4\ (D_t^2)$
h_t	Throat enthalpy, kcal/kg	$h_t = (A\ T_t + B\ C/\tanh(C/T_t) - D\ E/\tanh(E/T_t) + F)/M/\text{kcal\_j}$
u_0	Upstream velocity, m/s	$u_0 = W/\rho_0/A_0$
ρ_t	Throat density, kg/m^3	$\rho_t = (P_t)(10^5)\ M/R/T_t$
u_t	Throat velocity, m/s	$u_t = W/\rho_t/A_t$
u_s	Sonic velocity at throat, m/s	$u_s = (z_0\ (c_p c_v)_0\ R\ T_t/M)^{0.5}$
u_{t2}	Throat velocity from Bernoulli Equation, m/s	$u_{t2} = (2/\rho_t\ (P_0)(10^5) + \rho_0\ u_0^2/2 - (P_t)(10^5))^{0.5}$
Δ_{KE}	Δ*Kinetic Energy* = $\Delta(u^2)/2$, kcal/kg	$dh_{KE} = (u_{t2}^2 - u_0^2)/2/\text{kcal\_j}$
h_{t2}	Calc. enthalpy at throat, kcal/kg = inlet enthalpy − $\Delta(u^2)$/2/kcal_kg	$h_{t2} = h_0 - \Delta_{KE}$
T_t	Temperature at throat, K	$f(T_t) = 0$
W	Mass flow rate, kg/s	$f(W) = 0$
P_t	Pressure at throat, bar	$f(P_t) = 0$

TABLE 6.1b
Sonic or Subsonic Flow through a Valve or Hole Case A and Case B Solutions (Using Excel®)

Variable	Case A Value (Sonic)	Case B value (Subsonic)
A	33,300.0	33,300.0
B	79,930.0	79,930.0
C	2,086.9	2,086.9
D	41,600.0	41,600.0
E	991.96	991.96
F	−120,300,000	−120,300,000
P_t	27.71686	1.01325
kcal_j	4184.	4184.
M	16.0	16.0
T_0	300.	300.
P_0	46.0	1.5
z_0	1.0	1.0
$(c_p/c_v)_0$	1.32	1.32
R	8,314.46	8,314.46
D_0	0.30	0.30
D_t	0.05	0.05
h_0	226.2098	226.2098
ρ_0	29.50683	0.96218
A_0	0.07069	0.07069
A_t	0.00196	0.00196
h_t	205.2641	210.3263
u_0	8.22527	7.71172
ρ_t	20.86568	0.73254
u_t	418.7383	364.6543
u_s	418.7383	427.2981
u_{t2}	418.7383	364.6543
Δ_{KE}	20.94576	15.88352
h_{t2}	205.2641	210.3263
T_t	255.6215	266.1833
W	17.15555	0.52433
P_t	27.71686	1.01325

CASE (B) P_O = 1.5 BAR

Solve for release rate from a hole in a pipeline. Assume that the throat velocity is subsonic (i.e., the outlet pressure is fixed at 1 atm = 1.01325 bar).

When the outlet pressure is fixed, we are left with two nonlinear equations. Also, for consistency with the methodology in (A) above, we use a very large upstream diameter to make the inlet velocity very small.

Note: If the calculated outlet velocity exceeds the sonic limit, we would revert to the problem formulation under (A) above. (This would add an equation requiring outlet velocity to equal sonic velocity and, thus, require solving three simultaneous nonlinear equations.)

As noted earlier, the real gas case requires the use of complex computer programs to solve the chosen EOS (Lee-Kesler) for the required thermodynamic quantities.

In this case, the upstream pressure is low enough so that the velocity at the orifice does not reach the sonic limit. Accordingly, this problem requires solving only two simultaneous nonlinear algebraic equations. The solution shown below was obtained using the Microsoft Excel software package. Note that any other similar package can also be used. As with Case (A), we verified this numerical solution using the MATHCAD® and POLYMATH® software packages.

The problem statement for Case (B) is as follows:

$$f(T_t) = h_t - h_{t2}$$

$$f(W) = u_t - u_{t2}$$

 # Initial guesses:

T_t (0) = 300 # Initial guess for T_t, K

W (0) = 20 # Initial guess for W, Kg/s

Using Microsoft Excel 2016, the solution for Case (A) shows that the gas release rate, W, for an ideal gas is 17.16 kg/s. As noted above, identical results were obtained using the MATHCAD and POLYMATH software packages.

CASE (A): SUMMARY OF RESULTS

This problem was also solved using the Lee-Kesler EOS to calculate nonideal (i.e., "real") gas properties, using a suitable computer program. This solution is provided below, with the ideal gas results quoted alongside for comparison:

	Case A	
	Ideal Gas	**Real Gas**
Inlet pressure, bar	46	46
Inlet temperature, K	300	300
Calculated mass flow, kg/s	17.28	18.25
Inlet velocity, m/s	8.27	8.09
Throat velocity, m/s	424.65	421.48
Sonic velocity at throat, m/s	424.65	421.48
Inlet enthalpy, kcal/kg	225.61	214.87
Delta kin. energy, kcal/kg	21.54	21.22
Outlet enthalpy, kcal/kg	204.07	193.64
Throat pressure, bar	27.32	25.73
Throat temperature, K	254.27	250.83

It should be noted that these results are quite similar for this case because the behavior of the gas (methane) is close to ideal at the conditions specified. However, these results would very likely differ significantly if the inlet temperature was considerably lower or the pressure much higher. The gas behavior would then be markedly nonideal.

CASE (B): SUMMARY OF RESULTS

Case (B) was also solved using the Lee-Kesler EOS to calculate nonideal (i.e., "real") gas properties. This solution is provided below, with the ideal gas results quoted alongside for comparison:

	Case B	
	Ideal GaS	**Real Gas**
Inlet pressure, bar	1.5	1.5
Inlet temperature, K	300	300
Calculated mass flow, kg/s	0.524	0.525
Inlet velocity, m/s	7.71	7.70
Throat velocity, m/s	364.7	364.2
Throat sonic velocity, m/s	435.06	435.3
Inlet enthalpy, kcal/kg	226.21	226.21
Delta kin. energy, kcal/kg	15.88	15.85
Outlet enthalpy, kcal/kg	210.33	210.36
Throat pressure, bar	1.01325	1.01325
Throat temperature, K	266.18	266.25

As might be expected, the gas mass flow rate is far lower for Case B compared to Case A. Also, for Case (B), it was found that the conditions at the throat are virtually identical for the ideal versus the nonideal gas.

6.1.4 Additional Examples

To illustrate the importance of the issues discussed above, we show a few additional examples of the differences in our numerical results when compared to others.

EXAMPLE 6.1 (C): GAS RELEASE FROM A HOLE IN A VESSEL

Problem: A 0.1 in hole forms in a tank containing nitrogen at 200 psig and 80°F. Determine the mass flow rate through this leak (Crowl,[15] Example 4.4, p. 143).

The solution for this problem using the Lee-Kesler EOS was 0.0426 (lb/s), compared to 0.0386 (lb/s) in the reference (gas velocity at the throat was sonic). This difference is over 10% and not negligible.

EXAMPLE 6.1 (D): FLOW THROUGH A NOZZLE

Problem: Air at $p_0 = 7.0e5$ (Pa) and $T_0 = 293$ (K) discharges to the atmosphere. Compute the discharge mass flux, G, the pressure, Mach number, and velocity at the exit.[16]

Using the Lee-Kesler EOS, the results are as follows:

Quantity	Lee-Kesler (Isenthalpic)	Ideal Gas (Isentropic)[16]	Difference
Mass flux kg/(m².s)	1,589	1,650	−3.68%
Exit temperature, K	238.5	244	−5.5 (K)
Exit pressure, Pa	212,041	370,000	−42.7%
Mach number	1.0	1.0	–
Sonic velocity, m/s	338	313	+8%
Velocity, m/s	338	313	+8%

Note that the nozzle is not predicted to be choked when the downstream pressure is below 2.12 bar, as opposed to 3.7 bar in the original reference (kinetic energy effects must be included). This difference is not negligible.

In those cases where a partial condensation of the gas could occur in passing through the throat, such computations are inordinately laborious to perform by hand or even in Excel or similar software. In that situation, the density of the two-phase mixture at the throat could be significantly different from that of the uncondensed gas, and rigorous flash calculations would be required at the throat to obtain the proper density of the vapor-liquid mixture. Only then would we have confidence in (1) the sizing of a relief valve in a pipe or (2) the estimation of the relief rate from a hole in a pressure vessel.

6.1.5 Flashing of Liquids

When a single-phase stream containing a single component or a multicomponent mixture suffers a pressure drop in flowing through a valve, it can separate into a two-phase vapor-liquid mixture. This phenomenon is known as flashing, and it is essential to predict the relative amounts of each phase, the phase compositions, and properties such as density and other transport properties. These properties are needed for calculating the flow rate through relief valves and pressure drop in downstream piping[17] correctly.

Even if there is no phase change, these properties must be known accurately, both for single-phase gas and liquid cases. These properties are required to enable proper equipment sizing or to determine the rate of relief for a system of fixed geometry. Accurate results are of paramount importance for equipment design and rating calculations that are made to guarantee meeting all requirements for process safety. These include personnel protection, equipment safety, corporate image, and governmental regulations.

The effect of flashing a liquid stream across a relief valve or any other type of valve is covered below.

The effect of pressure on liquid enthalpy is generally negligible, as liquids have very low compressibility. When liquids are flashed through a valve or nozzle, the change in enthalpy from the inlet to the outlet is usually minor because the change in the kinetic energy term ($\Delta v^2/2$) is small.

For example, if the velocity of a liquid is increased from $v_1 = 1$ m/s to $v_2 = 7$ m/s, the change in kinetic energy would be

$\Delta KE = \left(v_2^2 - v_1^2\right)/2 = \left(7^2 - 1^2\right)/2 = 24\,\text{J/kg} = 0.0057\,\text{kcal/kg}$, which would be negligible for practical purposes.

However, if a liquid is close to its bubble point and suffers a pressure drop across a valve, it could flash isenthalpically to form a vapor-liquid mixture. Flashing to a two-phase mixture would depend on the enthalpy of the inlet mixture. If it is lower than the enthalpy of the saturated vapor and higher than that of the saturated liquid after the flash, a two-phase vapor-liquid mixture will exist.

6.1.6 Flashing of Pure Components

For a pure component, this problem is solved readily from its enthalpy charts. Alternatively, a process simulator could also be used.

Example 6.2

Problem: Saturated liquid propane at 10 bar is flashed across a valve to atmospheric pressure. Find the percentage of liquid that will be vaporized.

OPTION 1 (PURE COMPONENT PROPERTY TABLES)

Note that this option is applicable only for pure components. For mixtures, such computations are quite complex algebraically and require the use of computerized software. Further, it is not feasible to use spreadsheet tools such as Excel for such work, unless extensive use is made of Visual Basic programming for the thermodynamic and phase equilibrium computations. The recommended option is to use general-purpose process simulators.

For this problem, we obtained the thermodynamic properties for pure propane from the NIST REFPROP[18] Program (Version 9.1), generally regarded as the most accurate source for pure component and mixture thermophysical and thermodynamic properties:

The saturation temperature of saturated pure propane liquid at 10 bar is 26.94°C, and its enthalpy is 270.41 kJ/kg, with an enthalpy datum of 0 kJ/kg at −88.61°C for saturated liquid propane.

At 1 atm, the saturation temperature of propane is −42.11°C, the saturated liquid enthalpy is 100.36 kJ/kg, and the saturated vapor enthalpy is 525.95 kJ/kg.

Since the feed enthalpy lies between the saturated vapor and liquid enthalpies after the flash, there will be a two-phase mixture at the outlet.

The material balance is

$$F = V + L \tag{6.15}$$

where,

F = feed rate, kg/h
V = flash vapor rate, kg/h
L = flash liquid rate, kg/h

The energy balance is

$$F \cdot H_F = V \cdot H_V + L \cdot H_L, \quad \text{or}$$

$$F \cdot H_F = V \cdot H_V + (F - V) \cdot H_L$$

where
H_F = feed enthalpy, kJ/kg
H_L = flash liquid enthalpy, kJ/kg
H_V = flash vapor enthalpy, kJ/kg

Rearranging, we get

$$H_F = (1 - V/F) \cdot H_L + V/F \cdot H_V, \quad \text{or}$$

$$V/F = (H_F - H_L)/(H_V - H_L) \tag{6.16}$$

$$H_F = 270.41,\ HV = 525.95\ \text{kJ/kg},\ H_L = 100.36\ \text{kJ/kg}$$

The flash vapor fraction, $V/F = (270.41 - 100.36)/(525.95 - 100.36) = 0.3996$ or 40.0%.

OPTION 2 (PROCESS SIMULATOR)

We also used the PD-PLUS process simulator for the same calculation, using the Peng-Robinson equation of state.

Note: The datum for the enthalpies used by the NIST REFPROP[18] program is different from those in the PD-PLUS process simulator.

$$H_F = 16.3\ \text{kcal/kg},\ H_L = -23.0\ \text{kcal/kg},\ H_V = 79.0\ \text{kcal/kg}.$$

Therefore, the flash vapor fraction, $V/F = (16.3 - (-23))/(79 - (-23)) = 0.385$ or 39%. This is quite close to the value from the NIST REFPROP program.

6.2 THERMODYNAMICS OF FLUID PHASE EQUILIBRIA

Before delving into the details of flash calculations for a liquid flowing through a valve, we first discuss the crucial issue of the best thermodynamic methods and procedures for calculating the *K*-values and enthalpies required in such calculations.

6.2.1 Phase Equilibria in Hydrocarbon Mixtures

For hydrocarbons, the *K*-values and enthalpies are found using an EOS, such as the Peng-Robinson. An EOS is an expression that predicts the pressure-volume-temperature behavior of both the vapor and the liquid phases for a pure hydrocarbon or a hydrocarbon mixture.

Dozens of such equations have been proposed over the decades. In a family of EOS expressions that require just two constants, the first was due to van der Waals.[19] Soave's modification of the Redlich-Kwong EOS (called the SRK) and the PR EOS are now prominent. All the two-constant equations of state result in a cubic polynomial for the compressibility factor, Z. The constants of these equations are themselves functions of

- Each component's critical temperature, critical pressure, and Pitzer's acentric factor[20]
- The mixing rules that were used to derive these constants for a mixture.

It is important to understand the physical significance of the acentric factor, a dimensionless quantity for any component, which is defined as follows:

$$\omega = -\log_{10}\left(P_{vp}\big|_{T_r=0.7} / P_c\right) - 1.0 \qquad (6.17)$$

where
ω = acentric factor
T_r = reduced temperature, defined as T/T_c, where T is the temperature, K
T_c = critical temperature of the pure component, K
$P_{vp}\big|_{T_r=0.7}$ = pure component vapor pressure, at $T_r = 0.7$ bar
P_c = critical pressure, of the pure component, bar

According to Pitzer, the acentric factor measures the deviation of intermolecular potential functions from that of simple spherical molecules. The closer the value of ω is to zero, the simpler the fluid is said to be. Large, highly aspherical molecules generally have ω values considerably higher than 0 (values of 0.5 or above).

It is also useful to discuss the critical point for pure compounds. While staying on the vapor pressure curve, the critical point is reached when any further increase in temperature results in the disappearance of two phases. The mixture then becomes a single homogeneous phase. This condition can be measured experimentally for many light hydrocarbons or chemicals that do not decompose thermally, polymerize, or otherwise degrade as the temperature increases. However, for heavy hydrocarbons and many naturally occurring substances, this condition cannot be reached, and the critical point becomes fictional. Nevertheless, for applying equations of state, the T_c, P_c, and ω parameters must be specified for all components. For hydrocarbons, well-established correlations are available to estimate these parameters when they cannot be determined experimentally.

In earlier work in the field of thermodynamics,[21] the third parameter was chosen to be the critical compressibility factor (Z_c), defined as $Z_c = P_c V_c/(R T_c)$. Unfortunately, measurements at the critical point cannot be made experimentally, not only for thermally unstable compounds, as mentioned previously, but also for compounds that could detonate and create experimental safety hazards.

In response to this conundrum, Pitzer[20] proposed that the acentric factor should be used instead, as it is far more readily measured for many pure components. Nevertheless, for thermally sensitive substances (e.g., very high-boiling hydrocarbons), it is still necessary to estimate the critical temperature and pressure using various correlations from the literature.[22] The acentric factor enables far more accurate predictions of the vapor pressure, a crucial requirement in phase equilibria computations.

Considerably more complex EOS expressions have also been proposed, chief among them the Benedict-Webb-Rubin (BWR),[23,24] Benedict-Webb-Rubin-Starling (BWRS),[25] Lee-Kesler (LK),[10] and Lee-Kesler-Plöcker (LKP).[26] For a discussion of their relative merits, see Elliott and Lira.[2]

The LK and LKP equations of state use the eight-constant BWR equation for defining two anchor points, one for a "simple" fluid (methane), and the other for a "reference" fluid (n-octane). Using optimized parameters, the eight-constant BWR EOS predicts the thermodynamic properties of these two compounds accurately. An interpolation based on the acentric factor is then used to find thermodynamic properties for other compounds; this is an application of the principle of corresponding states.[2] The LKP EOS, which is an extension of the LK EOS, allows user-defined binary interaction parameters and, therefore, is better suited to modeling phase equilibria than the LK EOS that has no provision for binary interaction parameters.

Wherever theoretically justifiable, there is a great advantage in using an EOS. First, we calculate the vapor and liquid compressibility factors (Z_V and Z_L). Then, using rigorous thermodynamic expressions, we can calculate all of the following properties *for each phase*, consistently:

- Vapor and liquid phase fugacity coefficients for each component
- Phase densities, kgmole/m^3, using Z_V and Z_L directly
- Phase enthalpy, J/kgmole, based on the EOS-predicted departure from the ideal gas value
- Phase entropy, J/(kgmole K), based on the EOS-predicted departure from the ideal gas value
- Phase heat capacity, J/(kgmole K), at constant pressure, based on the EOS-predicted departure from the ideal gas value
- Phase heat capacity, J/(kgmole K), at constant volume, based on the EOS-predicted departure from the ideal gas value
- Phase sonic velocity (m/s), required for choked flow conditions in relief valves and
- Other thermodynamic properties.

For the liquid phase density, some corrections may be needed with cubic equations of state. For example, Peneloux's "volume translation" method[27] has been proposed for the Peng-Robinson EOS.

The component fugacity coefficient is a correction factor for deviations from ideality (Dalton's law of partial pressures), applied for a given component in a mixture. For component i in the liquid phase, it is defined as follows:

$$\Phi_i^L = f_i^L / \left(P x_i \right) \tag{6.18}$$

Similarly, the component fugacity coefficient for component i in the vapor phase is

$$\Phi_i^V = f_i^V / \left(P y_i \right) \tag{6.19}$$

where
Φ_i^L = liquid phase fugacity coefficient
Φ_i^V = vapor phase fugacity coefficient
f_i^L = liquid phase fugacity, bar
f_i^V = vapor phase fugacity, bar
P = system pressure, bar
x_i = liquid phase mole fraction of component i
y_i = vapor phase mole fraction of component i

Since thermodynamic equilibrium requires that fugacity for any component be equal in all phases:

$$f_i^L = f_i^V \tag{6.20}$$

where
f_i^L = liquid phase fugacity, bar
f_i^V = vapor phase fugacity, bar

The K-values are then found as the ratios of the component fugacity coefficients:

$$K_i \equiv y_i / x_i = \Phi_i^L / \Phi_i^V \tag{6.21}$$

where
K_i = component ratio of vapor-to-liquid mole fractions
Φ_i^L = liquid phase fugacity coefficient
Φ_i^V = vapor phase fugacity coefficient
x_i = liquid phase mole fraction of component i
y_i = vapor phase mole fraction of component i

Improving the accuracy of K-value predictions for equations of state generally requires the use of an empirically determined interaction parameter for each binary mixture in the mixture.[2] Accurate phase equilibrium data must first be acquired for each binary mixture. These must be measured under closely controlled laboratory conditions. Then, a nonlinear regression is performed to match predicted K-values against experimentally measured ones. For the most common light hydrocarbon binary mixtures, such tuned interaction parameters are already incorporated in commercial process simulators.[7,8,17,18,28]

The number of binaries in any mixture is a combination of the number of components, n, taken two at a time, or ${}^nC_2$. For example, a six-component mixture contains a combination of six items taken two at a time: ${}^6C_2$ or (6) (5)/1/2 = 15 binary mixtures.

6.2.2 Phase Equilibria in Chemical Mixtures

For chemical mixtures that exhibit significant nonideality in the liquid phase, activity coefficient models such as the Wilson, Renon's NRTL, or Abrams' UNIQUAC are recommended.[22] All three models can provide activity coefficients for multicomponent mixtures based on the binary interaction parameters for each binary mixture in the system.

The NRTL and UNIQUAC activity coefficient models can be used for systems when two liquid phases are present (liquid phase splitting). Therefore, these can be used for vapor-liquid equilibria (VLE), liquid-liquid equilibria (LLE), and vapor-liquid-liquid equilibria (VLLE). The thermodynamic details can be found in Seader et al.[29] and Prausnitz et al.[22] In particular, the Wilson model is incapable mathematically of predicting the occurrence of two liquid phases and can only handle VLE problems.

The component activity coefficients can be regarded as corrections for deviations from ideality (Raoult's law for ideal solutions) in the liquid phase:

$$f_i^L = f_i^s x_i \gamma_i^L \tag{6.22}$$

where

f_i^L = fugacity of component i in the liquid phase, bar
f_i^s = fugacity of saturated vapor for component i (generally, close to the vapor pressure), bar
x_i = liquid mole fraction of i
γ_i^L = liquid phase activity coefficient for component i in the mixture

The fugacity of a pure component in a vapor phase at saturation is defined as

$$f_i^s = \Phi_i^s P_i^s \cdot \left(\int V_i^L/(RT)dP \right) \tag{6.23}$$

where

f_i^s = vapor fugacity at saturation for pure component i
Φ_i^s = vapor phase fugacity coefficient at saturation for pure component i
P_i^s = saturation vapor pressure, bar, of component i
V_i^L = liquid molar volume, m^3/kgmole, of component i
R = universal gas constant, J/(kgmole K)
T = absolute temperature, K
P = system pressure, bar

In Equation (6.23), the integral term in parentheses is referred to as the Poynting correction.[2] The integration is carried out between the lower limit of P_i^s and the upper limit of P.

For low-to-moderate pressures below about 5 atm, (e.g., for most complex chemical mixtures), the Poynting correction is generally ignored (the value is taken as 1).

The fugacity of component i in a vapor mixture is found using the expression previously shown for systems where an EOS was used, i.e.,

$$f_i^V = P y_i \Phi_i^V \tag{6.24}$$

where

f_i^V = fugacity of component i in the vapor phase, bar
P = system pressure, bar
y_i = vapor mole fraction of i
Φ_i^V = vapor phase fugacity coefficient for pure component i in mixture

Note the important distinction between Φ_i^s (for pure component i as a saturated vapor) and Φ_i^V (for component i in the mixed vapor).

Finally, the K-values (i.e., the mole fraction ratios) required for VLE, LLE, or VLLE calculations are found using the above expressions for f_i^V and f_i^L. The requirement is that fugacities for a given component be equal in each co-existing phase at thermodynamic equilibrium.

The vapor phase fugacity coefficient, Φ_i^V, is generally quite close to unity for chemical mixtures at low pressures. The major exceptions to this are mixtures containing organic aliphatic acids, such as acetic or propionic acids, because these compounds are known to exhibit a significant extent of dimerization and cross-dimerization.[2,30] The extent of this dimerization phenomenon depends on the equilibrium constants (that are themselves highly dependent on the temperature) and must be accounted for in phase equilibria computations. When properly calculated, values for Φ_i^V as low as 0.3–0.4 are found routinely in VLE regressions.[2,30] Commercially significant examples are the systems water-acetic acid and acetic acid-acrylic acid. When calculating mixture activity coefficients, care must be taken to include the partial vapor phase dimerization effects on fugacities at saturation, f_i^s in Equation (6.23).

Elliott and Lira[2] cite the following example to illustrate the extreme effect of dimerization on vapor phase nonideality even at very low pressures: at a pressure $P = 0.01$ bar, for acetic acid, the compressibility factor $Z = 0.625$ and the pure component saturation fugacity coefficient $\left(\Phi_i^s\right) = 0.4$. Accordingly, vapor densities calculated without including the partial dimerization effect will be too low by almost 60%. Also, pure component fugacities would be too low by 40%. These errors would be wholly unacceptable for engineering design applications.

Partial vapor phase dimerization and cross-dimerization phenomena also have a profound impact on vapor phase density. This aspect is crucial for the sizing of process equipment such as distillation columns, vessels, control valves, and relief valves. For safety-related process engineering work, such as relief rate calculations or valve sizing, this vital aspect must not be ignored.

As discussed previously, the liquid phase activity coefficients for all components in a liquid mixture are found using the Wilson, NRTL, or UNIQUAC equations. For illustration, the NRTL model, that was first described by Renon and Prausnitz[22] for liquid-phase activity coefficients in multicomponent mixtures, is described below:

Activity coefficient method: NRTL (nonrandom two-liquid)

$$\ln \gamma_i = \frac{\sum_j x_j \tau_{ji} G_{ji}}{\sum_k x_k G_{ki}} + \sum_j \left(\frac{x_j G_{ij}}{\sum_k x_k G_{kj}} \left(\tau_{ij} - \frac{\sum_m x_m \tau_{mj} G_{mj}}{\sum_k x_k G_{kj}} \right) \right)$$

where

$$G_{ij} = \exp(-\alpha_{ij}\tau_{ij}), \quad G_{ii} = 1$$

$$\tau_{ij} = a_{ij} + b_{ij}/T + e_{ij} \ln T + f_{ij}T, \quad \tau_{ii} = 0$$

$$\alpha_{ij} = c_{ij} + d_{ij}(T - 273.15)$$

Here, for each (*i*, *j*) pair, the dimensionless NRTL binary interaction parameters τ_{ij} and α_{ij} are functions of the absolute temperature *T* (K). Also, it should be noted that $\tau_{ij} \neq \tau_{ij}$. Also, in the expression for τ_{ij} above, the e_{ij} and f_{ij} terms are generally taken as zero. If insufficient data are available to express the values of τ_{ij} as a function of temperature, a_{ij} would also be set to zero. For many binary mixtures of industrial importance, several volumes of DECHEMA's Chemistry Data Series[31] are available in libraries. However, it should be noted that the DDBST mixture databank is far more extensive than the DECHEMA series, containing over 20 times as many datasets.

Owing to the algebraic complexity of the NRTL and similar activity coefficient models, it is tedious to compute them manually. Simple calculations can be made in a spreadsheet, but the formulas become unwieldy if cell references are used. In Microsoft Excel, it is more efficient and reliable to use the built-in Visual Basic for Applications (VBA) to program such equations. In any event, these models invariably are deeply embedded in a variety of other programs, for example, to calculate bubble/dew points or perform flash calculations.[29] These, in turn, might be part of a large-scale distillation or liquid extraction unit operation, and it is quite impractical to perform such computations reliably and efficiently except in a general-purpose process simulator.[9,32]

6.2.3 Flash Calculations for Mixtures

The equilibrium flash problem for mixtures is considerably more complex than for pure components, requiring extensive iterative calculations. Also, the vapor-liquid equilibria must be calculated, at each iteration, using rigorous thermodynamics:

i. EOSs for nonpolar mixtures, such as hydrocarbons or petroleum mixtures or
ii. liquid-phase activity coefficient models for highly nonideal chemical mixtures.[22,29]

In either case, the flash vapor fraction *V*/*F* is found using trial-and-error calculation methods. One such example is the Rachford-Rice[29] algorithm. Both hydrocarbon and chemical mixtures can be handled with this approach. The calculation methods for solving either the EOS or activity coefficient models for the *K*-values are too laborious for hand computation. Implementing such complex iterative procedures requires specialized software programs or process simulators, as discussed further below.

The Rachford-Rice flash equation is a function of the unknown vapor fraction and temperature. This function must equal zero at the proper solution for these two unknowns:

$$f(\beta,T) = -1 + \sum x_i = -1 + \sum z_i \big/ \left[1 + \beta(K_i - 1)\right] = 0 \qquad (6.25)$$

where

β = vapor fraction (= V/F)
T = flash temperature, K
z_i = mole fraction of component i in the feed
x_i = mole fraction of component i in the flash liquid
y_i = mole fraction of component i in the flash vapor
K_i = vapor-liquid equilibrium constant (K-value) of component i

For a flash calculation where the pressure (P) and heat input (Q) are specified ("P-Q flash"), the temperature (T) and flash vapor fraction (β) are both unknown. These must be solved iteratively using the Rachford-Rice flash equation, cited above, and a second equation that enables computation of the energy balance:

$$H_F = (1-\beta)\sum z_i H_i^L \big/ \left[1 + \beta(K_i - 1)\right] + \beta \sum z_i K_i H_i^V \big/ \left[1 + \beta(K_i - 1)\right] \qquad (6.26)$$

where

H_F = feed enthalpy, J/kg
H_i^V = vapor phase molar enthalpy of component i, J/kg
H_i^L = liquid phase molar enthalpy of component i, J/kg

[The other terms are defined immediately above for Equation (6.25)]

These two nonlinear equations, each a function of T and β, must be solved iteratively. We emphasize that the thermodynamic quantities above (K-values and enthalpies) for each phase are themselves functions of T, P, and phase composition.

Note: When using equations of state, real fluid enthalpies are determined by subtracting ΔH^{dep} (the enthalpy departure from the ideal) from the ideal gas mixture enthalpy, H^{id}. This is done for both phases after solving the EOS for their respective compressibility factors, Z_V and Z_L. The expression for H^F above is, therefore, modified accordingly.

Thus, it can be seen that solving the adiabatic flash problem requires extensive, nested trial and error computations that mandate the use of computerized software or process simulators. It is quite likely that, even for a three-component system, performing a single flash calculation *rigorously* by hand would take several weeks of intense effort. A process simulator, however, can solve this problem in a few milliseconds on a modern personal computer. Many highly efficient process simulators for solving a wide variety of chemical engineering problems are available commercially.[7,8,17,28,34] Based on the discussion above, we used a commercial process simulator to solve the free-expansion problem for a saturated hydrocarbon liquid mixture.

Example 6.3

Problem: For a mixture of ethane (50 mole%) and propane (50 mole%) at a pressure of 10 bar at its bubble point temperature, calculate the fractional vaporization when the mixture is flashed across a valve to atmospheric pressure.

Two commercially available process simulators[17,34] were used for this problem (using the Peng-Robinson EOS[19] for the thermodynamic properties of this nonpolar mixture). The results for the trial-and-error solutions from both simulators were virtually identical for this problem and are as follows:

***ADIABATIC* FLASH OF ETHANE-PROPANE MIXTURE**

Stream ID			**FEED**	**FLASHVAP**	**FLASHLIQ**
	MW	SpGr@60°F	kgmole/h	kgmole/h	kgmole/h
ETHANE	30.07	0.3554	500	280.881	219.119
PROPANE	44.1	0.5063	500	60.304	439.696
Total, kgmole/h			1000	341.185	658.815
			kg/h	kg/h	kg/h
ETHANE			15,035.0	8,446.1	6588.9
PROPANE			22,050.0	2,659.4	19,390.6
Total, kg/h			37,085.0	11,105.5	25,979.5
Temperature, °C			−11.3	−69.2	−69.2
Pressure, bar			10	1.013	1.013
Phase			Liquid	Vapor	Liquid
V/*F* ratio (molar)			0	1	0
Molecular weight			37.08	32.55	39.43
Heat capacity, kcal/(kg C)			0.6439	0.3332	0.502
Enthalpy, kcal/kg			25.339	101.715	−7.310
Energy, MMkcal/h			0.9397	1.1296	−0.1899
Density, kg/m^3			496.349	1.9993	587.0536

For this example, 34.1 mole% of the feed is flashed off as a vapor. These results also show that, after a significant pressure reduction under adiabatic conditions, the flash temperature can be much lower than the upstream value (the energy required to vaporize the mixture partially is drawn from its internal energy).

6.2.4 Laboratory Measurements Versus Estimation Methods in Phase Equilibria

The proper selection of the thermodynamic model (EOS or activity coefficients) in any process simulation is crucial to a successful simulation result. The selection of the wrong thermodynamic option in any process simulator is a fatal error, and *all* of the results from such a simulation are meaningless. Also, as mentioned earlier, both the EOS and activity coefficient models require binary interaction parameters obtained from the literature or private databanks.[33] This issue is discussed in greater detail below, along with some cautionary notes based on our experience.

Generally, binary interaction parameters should be obtained from regressions of accurate laboratory measurements, using all available datasets, for vapor-liquid equilibria (VLE), liquid-liquid equilibria (LLE), or vapor-liquid-liquid equilibria (VLLE).

For the given binary mixture, the available datasets should cover the full range of interest for both *P* and *T*. These data could be of the following types:

- *P-T-X-Y* for VLE
- P-T-X_1-X_2 for LLE
- P-T-X_1-X_2-Y for VLLE

where
Y = the vapor phase composition
X = the single-phase liquid composition
X_1 = component compositions in liquid phase 1
X_2 = component compositions in liquid phase 2

It should also be mentioned that experimental methods do exist for direct measurements of the activity coefficients at infinite dilution for binary chemical mixtures.[33] By solving two algebraic, nonlinear equations, the binary interaction parameters for any activity coefficient model can be obtained. Note that, for the NRTL equation, solving for the binary interaction parameters would first require fixing the value of the alpha (α) parameter arbitrarily. Generally, a value of 0.3 is used. See Walas[19] for further guidance.

Finally, for reliable simulation results, the full range of liquid phase compositions should also be covered. For guidance on the proper selection of thermodynamic methods, see Seader et al.,[29] Prausnitz et al.,[22] and Elliott et al.[2]

In the absence of measured phase equilibrium data, especially for complex chemical mixtures, it is possible to estimate the activity coefficients for a wide variety of mixtures using "group-contribution theory". This issue is discussed extensively by Elliott and Lira.[2]

Organic molecules are composed of molecular functional groups. Examples are as follows:

- OH – alcohols
- CHO – aldehydes
- COO – ketones
- COOC – esters
- COOH – organic acids and dozens of others

This theory postulates that the interactions between these molecular groups in different molecules can be used to predict the interactions between the molecules themselves in any mixture.

Group-contribution methods such as ASOG and UNIFAC[31,35,36] are well-known, and the UNIFAC method, in particular, has gained wide acceptance. UNIFAC group interaction parameters are already available in the published literature for many industrially significant types of mixtures. However, the interaction parameter table is not 100% populated, and significant gaps do exist wherever any of the following issues were encountered:

1. The researchers were unaware of published data available only in relatively obscure journals

2. The data available were of dubious quality
3. The main groups in question were not yet defined
4. No lab measurements for phase equilibria had yet been made
5. The number of measurements made was insufficient for reliable estimation
6. Inability to distinguish between isomeric compounds (e.g., xylenes or butanols)
7. Inability to handle proximity effects (nearby polar groups contain highly electronegative atoms such as oxygen or fluorine)
8. Chemical reactions among the main groups are known to occur.

Several inadequacies have been reported when applying the original (1975) version of UNIFAC. These are discussed at length in a series of papers published by Gmehling et al.[35] There are also available several modified versions of UNIFAC.[35,36] Overall, the best version, in our experience, is called the Dortmund UNIFAC method[33,35] or D-UNIFAC. D-UNIFAC incorporates a modification of the combinatorial term in the original UNIFAC method[35] but has been shown to have several highly significant and practical advantages. Chief among these is that the D-UNIFAC databank (a commercially licensed product) includes a much larger number of molecular functional groups and a far more extensive table of group interaction parameters, compared to the other methods. Also, these parameters are based on the *totality* of the world's published phase equilibrium data and also much privately measured data, verified independently. Major chemical companies and laboratories donated these data to DDBST, the company that licenses D-UNIFAC and also the databanks of measured phase equilibrium data. These features in D-UNIFAC enable applications for a wide range of mixtures of complex organic molecules.

The main advantage of the D-UNIFAC version over the original (public) UNIFAC method is its demonstrated ability to make more accurate predictions for the following properties, using a single table of group interaction parameters (some of which incorporate a quadratic temperature dependence):

1. VLE
2. LLE
3. VLLE
4. Solid-liquid equilibria (SLE)
5. Infinite-dilution activity coefficients (IDAC, γ_i^∞)
6. Excess heats of mixing (ΔH^E) (important for predicting the effect of temperature on binary infinite-dilution activity coefficients)
7. Excess heat capacities (ΔC_p^E) (important for predicting the temperature-dependence of excess heats of mixing)

For items 6 and 7 above, the term "excess" implies the deviation from the ideal mixture value.

However, *none* of the group contribution methods (including the D-UNIFAC) can distinguish among isomeric compounds[2,35,36]or handle cases where a molecule has more than 10 or so molecular functional groups. Much further work needs to be done in these areas.

It must always be remembered that estimates of activity coefficients are more reliable when based on interaction parameters derived from experimental measurements rather than predictions from group contribution methods. The latter should be used only when there are no data available, and the binary mixtures in question (especially for non-key components) do not affect the overall separation markedly.

An example would be the interaction of very low-boiling components with components having a much higher boiling point in a given mixture. Such cases frequently arise in mixtures with a large number of components. Typically, laboratory VLE, LLE, or VLLE measurements have been made only for those compounds that affect the key-component vapor-liquid or vapor-liquid-liquid split. Group contribution methods enable the estimation of the missing group interaction parameters for the non-key binaries.

Once the behavior of all the binary mixtures in a given multicomponent mixture has been characterized in this manner, the prediction of VLE, LLE, or VLLE for the complete mixture can be performed.[22] However, as noted by Prausnitz et al.[22]:

> ...when these equations are applied to ternary (or higher) systems, it is often not possible to predict multicomponent **liquid-liquid equilibria** using only experimental binary data" (emphasis added).

Further, they assert that "*Usually, only a few ternary LLE measurements are required to fit the "best" binary parameters*".

Accordingly, great care must be exercised when dealing with LLE or VLLE mixtures. We should always leave open the possibility that a few well-chosen LLE measurements may be required to ensure VLLE prediction accuracy for the relative amounts of the two liquid phases and their compositions.

6.2.5 Commercial Process Simulators

For ease of use, many commercial process simulators[7,8,12,17,28,34,37] provide built-in libraries of intermolecular interaction parameters for the primary EOS (e.g., SRK, Peng-Robinson) and activity coefficient models (Wilson, NRTL, or UNIQUAC). Unfortunately, the interaction parameters provided for the activity coefficient models, in particular, are rarely documented as to their source. Most often, we do not know whether these parameters were regressed from experimental phase equilibrium data or simply estimated using group contribution methods, such as the original (1975) version of UNIFAC,[9] sometimes referred to as the "public" version.

If the simulator's activity coefficient interaction parameters are based on the original, published UNIFAC parameters, the predicted activity coefficients could be significantly in error, sometimes by over 50%.[33,35] There are several reasons for this: (1) the database used for these group interactions was nowhere near as comprehensive as it is today, and (2) many group interactions were unknown and therefore assumed arbitrarily to be zero in the process simulator. In this situation, the simulator could, for example, predict no liquid phase split when it is known to happen. Conversely, it could falsely predict a phase split when there is none in reality. Using the D-UNIFAC model often improves this situation dramatically. However, no group contribution model can match the accuracy of activity coefficient interaction parameters based on experimental phase equilibria measurements for the key component binaries in any multicomponent mixture.

The user ultimately must accept responsibility for verifying *all* simulation results, especially when dealing with unusual polar mixtures that include a large number of components. Examples would be liquid-liquid extraction, or azeotropic and extractive distillation in the chemical industry. When using commercial process simulators,[7,8,12,17,34,37,38] the recommended approach to ensure proper selection of thermodynamic models and their interaction parameters is as follows:

1. Perform a literature search of laboratory phase equilibria measurements for VLE, LLE, and VLLE data, as appropriate, for the system at hand
2. Select the proper data sets
3. Select the proper thermodynamic model for the mixture at hand: this is a nontrivial decision; an incorrect choice could easily yield nonsensical results
4. Perform regressions based on measured phase equilibrium data to determine the activity coefficient interaction parameters for each binary
5. Check model predictions against the original lab measurements. If found inadequate, check whether the chosen thermodynamic model (Step 3) was the proper one. Alternatively, check the data used (Step 2) for their completeness and thermodynamic consistency[31]
6. Compare model predictions based on these data against the simulator's predictions for the same conditions
7. If these comparisons show that predictions based on the simulator's built-in parameters are unreliable, replace them with those obtained by data regression in step 4 and, finally,
8. Ensure that the pressure and temperature range of the lab data (Step 2) are within the range of process conditions where they will be applied.

For step 1 above, as stated earlier, DDBST's Dortmund Databank[33] (DDBSP) provides the largest, critically reviewed library of the world's published data. Also included are unpublished data donated by many chemical companies, and data measured in DDBST's own laboratories. The DDBSP software includes the properties of over 65,000 pure components and several tens of thousands of measured phase equilibria datasets (VLE, LLE, VLLE, IDAC, heats of mixing, etc.) for mixtures. Also provided are software packages to do model fitting by regression for most of the pure component properties and thermodynamic models of industrial interest in phase equilibria.

In the absence of experimental data, phase equilibria may be estimated based on group contribution methods. For doing this, we recommend the D-UNIFAC method.[33] Extensive tables of the D-UNIFAC group interaction parameters are provided by DDBST[33] to members of their industrial consortium. It should be re-emphasized that K-value estimates of phase equilibria obtained from the group contribution approach must be regarded as qualitative, at best. These are most useful only for estimating the separation of non-key components in a multicomponent mixture. When reliable predictions are essential, there is no substitute for laboratory measurements of phase equilibria.

When all predictive methods fail or are inappropriate for the problem at hand, the option to get the appropriate laboratory measurements made must always be kept in mind for phase equilibria and other thermodynamic properties. These precautions are especially vital before significant investments are made in new plant construction or revamps. There have been numerous reported instances where newly constructed plants failed to achieve their design performance because the underlying thermodynamic models (or their parameters) in the process simulations were flawed. Analogously, such errors can severely impact the adequacy of the process safety equipment installed at a plant. It is always the user's responsibility to pay the utmost attention to thermodynamic details of the type discussed above. When in doubt, one must always consult the appropriate domain experts.

A word of caution is in order: *plant data from commercial-scale units are never found to be sufficiently numerous, of sufficient accuracy, or covering a wide enough range of T, P, and composition to justify adjusting the binary interaction parameters for any system.* Few industrial mixtures contain just two components (no other components or impurities). We make this observation simply because we have witnessed numerous reckless attempts to "fix" a faulty simulation by adjusting interaction parameters. This practice is utterly indefensible from a fundamental theoretical viewpoint, as it relies on plant data that is subject to numerous measurement errors. This flawed approach also conceals the heavy distorting influence of unknown variables such as tray efficiency, trace impurities, etc. If a simulation does not match reality, we recommend that the practitioner must look elsewhere to diagnose the causes and use prudent, proven methods to resolve the problem.

6.2.6 Release of a Liquefied Gas: Two-Phase Flashing Flow

An important example of a two-phase flashing flow is encountered when the bottom liquid outlet from a vessel containing a liquefied gas at ambient temperature is ruptured.

Modeling of two-phase flow for vertical up-flow or down-flow, horizontal, or inclined flow configurations is a specialized discipline. Various models based on a wide range of assumptions are available in the literature. Many of these have been tested and incorporated in process simulators. However, experimental data in support of these models are limited. It is not the purpose of this book to go into a review of these models. For preliminary hazard assessment studies, however, approximate equations are often adequate. The AIChE/CCPS guideline[38] uses the following simple equation developed by Fauske and Epstein for two-phase choked flow:

$$m' = C_d A_o \left[\frac{F H_v}{\left(v_g - v_l\right)\sqrt{T_s C_{pl}}} \right] \tag{6.27}$$

where

m' = mass flow rate, kg/s

A_o = leakage area, m^2

C_d = discharge coefficient

F = frictional loss factor
H_v = heat of vaporization, J/kg
v_g = specific volume of vapor at storage P and T, m^3/kg
v_l = specific volume of liquid, m^3/kg
T_s = liquid temperature in the storage vessel, K
C_{pl} = specific heat of liquid, J/(kg K)

The frictional loss factor F, which depends upon the length-to-diameter (L/D) ratio of the exit tube between the source tank and the emission point, is as follows (Table 6.2):

An equation that reproduces the values in table 6.2 fairly accurately is as follows:

$$F = \left(p_0 + p_1 L/D\right)/\left(1 + q_1 L/D\right),$$

where

$p_0 = 1$, $p_1 = 2.492824 \times 10^{-3}$, and $q_1 = 6.573776 \times 10^{-3}$

Example 6.4

PROBLEM: A Horton sphere of diameter 15 m is used to store liquefied propane as a saturated liquid at 303 K and 10.6 atm absolute pressure. The diameter of the bottom outlet line is 400 mm, which starts leaking at the first isolation valve, which is at a distance of 12 m from the joint at the bottom of the sphere. Assume that the discharge coefficient $C_d = 1$. Data: Heat of vaporization of liquid propane is 348,000 J/kg, the density of liquid propane is 488.8 kg/m^3, propane vapor density is 23.47 kg/m3, the normal boiling point of propane is 231 K, and specific heat of liquid propane is 2,581 J/(kg.K). The leakage area is 10% of the cross-sectional area of the outlet pipe.

Specific volume of propane vapor at storage condition:

$$v_g = 1/23.47 = 0.0426 \text{ m}^3/\text{kg}$$

Specific volume of propane liquid $v_L = 1/488.8 = 0.002046$ m^3/kg

$L/D = 12/0.4 = 30$, whence $F = 0.8977$

Leakage area = (0.1) (π/4) (0.4)^2 = 0.01256 m^2

Substituting values in Equation (6.14), we get

$$m' = C_d A_0 \left[FH_v/\left(v_g - v_L\right)/\left(T_s C_{pL}\right)\right]; \text{ therefore}$$

TABLE 6.2
Frictional Loss Factor

L/D	0	50	100	200	400
F	1	0.85	0.75	0.65	0.55

$$m' = (1)\,(0.01256)\left[(0.8977)(348{,}000)/(0.0426 - 0.002046)/\left\{(303)\,(2{,}581)\right\}\right] = 109.5 \text{ kg/s}$$

On release to atmospheric pressure, part of the liquid will flash. T_0, the boiling point of propane at atmospheric pressure, is 231.1 K. Using tables for propane enthalpy:

$$\text{Saturated feed enthalpy, } H_f \text{ at 303 K} = -14.27 \text{ kJ/kg}$$

$$\text{Saturated liquid enthalpy, } H_L \text{ at 231.1 K} = -194.74 \text{ kJ/kg}$$

$$\text{Saturated vapor enthalpy, } H_v \text{ at 231.1 K} = 229.19 \text{ kJ/kg}$$

Fractional vaporization:

$$\left(H_f - H_L\right)/\left(H_v - H_L\right) = \left(-14.27 - (-194.74)\right)/\left(229.19 - (-194.74)\right) = 0.426$$

The unflashed liquid will be entrained in the gas jet as an aerosol.

6.2.7 Concluding Remarks for Release Rate Calculations

The release rate calculation from a relief valve (or a hole in a vessel) should be done using an isenthalpic (i.e., irreversible adiabatic) path. For such a Joule-Thompson free expansion for the flow process, the change in kinetic energy must be included. This procedure ensures a correct design when sizing relief valves or estimating the release rate from an existing valve or hole in a vessel.

It is also essential to use the proper thermodynamic option (e.g., Peng-Robinson or Lee-Kesler EOS) to estimate real gas properties reliably compared to the ideal gas assumption; this is especially important when gas conditions are far from ideal. The Bernoulli equation is also incorporated in this procedure. Finally, when the flow is choked at the throat, the sonic velocity limit must also be included for a complete definition of the problem. Suppose the upstream pressure is low enough and the throat is un-choked. In that case, the methodology previously described in Section 6.1 should again be followed to ensure a theoretically sound and consistent basis.

For complex non-ideal chemical mixtures, use of properly-tuned binary-interaction parameters for activity coefficient models (such as NRTL) is essential when computing phase equilibria. Such parameters should be based on regressions of accurately-measured laboratory data for VLE/LLE/VLLE. Group contribution methods, such as UNIFAC, should be used only for the non-key binaries. Iterative computations for nonideal cases (real gases) are generally too intensive for hand calculations and require computerized chemical engineering software (process simulators).

6.3 CALCULATIONS FOR JET FIRES

Jet fires result from the combustion of flammable gas as it is being released from a pressurized containment. A classic example is a Bunsen burner flame. Typical examples in the process industries are flames from gas burners in furnaces, flares, flames from the ignition of gases/vapors at the outlet of vents or relief valves, and ignition of accidental releases from pipeline or vessel ruptures.

We have dealt with methods for estimating the release rate through an opening in Sections 6.1 and 6.2. We now discuss methods to estimate the size and shape of flames and the intensity of thermal radiation as a function of distance from flames.

6.3.1 Size and Shape of Flames

This section discusses methods and procedures to calculate the size and shape of flames for several standard models used in industrial practice. In each case, the main formulas and terms are defined to enable efficient manual computations.

6.3.1.1 Hawthorn, Weddell, and Hottel Model[39]

This earliest model is based on the Sc.D. theses of Hawthorn and Weddell at the Massachusetts Institute of Technology. This model predicts the length of free turbulent flame jets in which buoyancy effects are small. Jets are vertical, implying that wind speed is negligible. The port diameters investigated ranged from 3 to 7.5 mm, and the predicted flame lengths varied from 40 to 290 nozzle diameters, with a maximum error of 20%. The equation is as follows:

$$L = \frac{5.3D_o}{C_t}\left[\frac{T_f}{\alpha_t T_a}\left\{C_t + (1 - C_t)\frac{M_a}{M_f}\right\}\right]^{\frac{1}{2}} \tag{6.28}$$

where

L = visible flame length, m

D_o = diameter of the nozzle, m

C_t = mole fraction of fuel in the unreacted stoichiometric mixture

α_t = ratio of moles of reactants/moles of products, for the stoichiometric mixture

T_f = adiabatic flame temperature, K

T_a = air temperature, K

M_a = molecular weight of air, kg/kgmole

M_f = molecular weight of fuel, kg/kgmole

It should be noted that the fuel flow rate does not appear in the above equation. However, it should be high enough to produce a turbulent jet. The reported experimental work gave the range of Reynolds number at the nozzle as 2,900 to 32,000. The term D_o in the above equation refers to the actual diameter for rounded nozzles and the "vena contracta" for sharp-edged orifices.

Example 6.5

Problem: Methane gas at 50 bar is escaping vertically upwards into the atmosphere from a pressurized pipeline through an opening of 50 mm. The gas ignites and burns as a turbulent diffusion flame. Ambient air is at 300 K, and adiabatic flame temperature for methane is 2,200 K. It is desired to calculate the flame length.

It can be shown that the mole fraction (C_t) of methane in the unreacted stoichiometric mixture of methane and air is 0.095. Therefore, C_t = 0.095. Also, the ratio

(α_t) of moles of reactants to moles of products is equal to 1.0. Substituting these values as well as values of hole diameter, flame temperature, air temperature, the molecular weight of air (29), and molecular weight of methane (16) in Equation (6.28), we obtain

$$L = 5.3 \times 0.05/0.095 \times \left(2,200/1/300 \times \left(0.095 + (1 - 0.095) \times 29/16\right)\right)^{0.5} = 9.95\,\text{m}$$

Neither the pipeline pressure nor the flow rate has been given in the problem. Hence, it is necessary to assume a range of flow rates to check if the flow is likely to be turbulent. The calculated value of flow rate, assuming a discharge coefficient of unity, is 17 kg/s at an upstream pressure of 50 bar, and 1 kg/s at an upstream pressure of 3 bar. At an approximate value of 0.01 cP for the viscosity of methane gas, the calculated values of Reynolds number are 4.4×10^7 and 2.6×10^6 for flow rates of 17 and 1 kg/s, respectively. These are much higher than the Reynolds number of about 2,500, at which transition from laminar to turbulent flow occurs.

6.3.1.2 API Model[40]

The ANSI/API Standard 521 model[6] is empirical and was developed originally for analysis of flares, but is now also applied for jet fires arising from accidental releases. The model assumes a plume-shaped flame whose length is determined by the rate of heat release. The relationship between the flame length and the heat release rate is shown graphically as a straight line on a log-log paper. For heat release rate between 10^8 and 10^{10} W, this relationship can be described by Equation (6.29):

$$L = 0.00326\,(m'\,Q_c)^{0.478} \tag{6.29}$$

where
L = flame length, m
m' = fuel release rate, kg/s
Q_c = heat of combustion of the fuel, J/kg

Equation (6.29) has been used by Cook, Bahrami, and Whitehouse[41] in the computer code SAFETI to calculate flame length based on the API standard. This standard does not specify any particular method for the estimation of the flame diameter. Based on the work of Baron,[42] Cook, Bahrami, and Whitehouse[41] quote the following equation for the diameter of the flame, D_s:

$$D_s = 0.58s\left[\log_{10}(L/s)\right]^{0.5} \tag{6.30}$$

where
D_s = flame diameter, m
L = visible flame length, m
s = distance along the centerline of the flame, m

The maximum diameter occurs at a center-line distance of about 60% of the flame length.

For a vertical release in the absence of wind, the flame is vertical. The wind tilts the flame in the direction the wind is blowing. The API has provided two methods for calculating the flame distortion caused by wind. The first method follows a simple approach, while the second is a more specific approach using Brzustowski and Sommer's methods.[11] Both the methods apply to subsonic flares, and API cautions that the flare manufacturer should be consulted for sonic flares.

In the simple model, distortion of the flame is defined in terms of the horizontal deviation $\sum\Delta x$ and vertical deviation $\sum\Delta y$ of the flame tip, both dependent on the ratio of the wind speed u_∞ and jet velocity at the exit plane u_j (see Figure 6.1).

The API model assumes that thermal radiation from the flame emerges from a point source at the mid-point of the straight line between the center of the exit plane and the flame tip. The mid-point displacement in the x-direction, defined as x_c, is assumed to be equal to $(\sum\Delta x)/2$. Similarly, the mid-point displacement in the y-direction, defined as y_c, is assumed to be equal to $(\sum\Delta y)/2$.

For the method based on Brzustowski and Sommer's approach, values of x_c (horizontal distance) and y_c (vertical distance) are presented below in equation form; these are based on nonlinear regressions of the original graphs in the API's API Standard 521, 6th Ed., 2014 for the horizontal distance x_c and the vertical distance y_c.

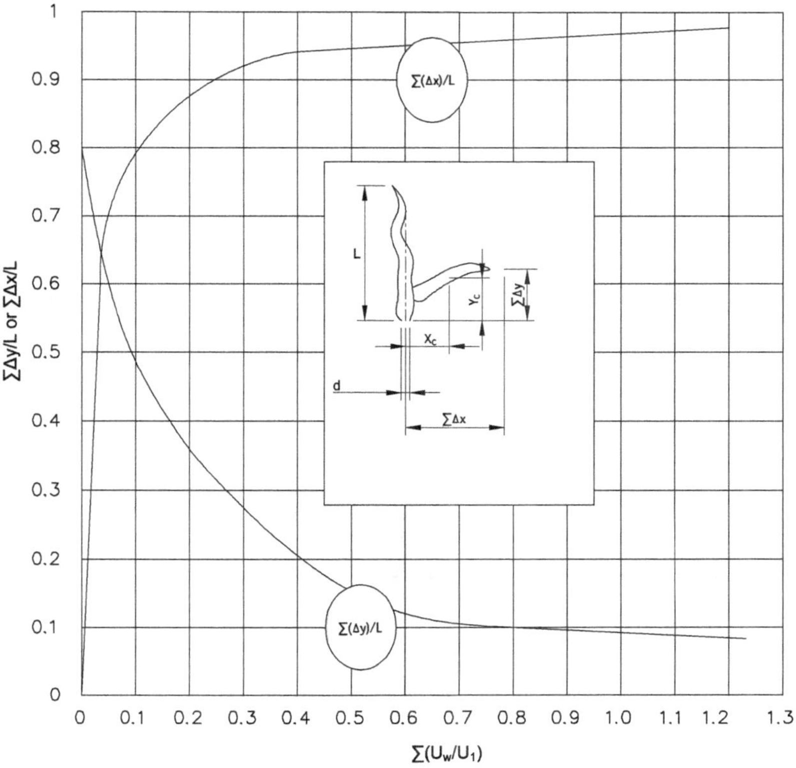

FIGURE 6.1 Distortion of a vertical flame due to wind. (With permission, API Standard 521, 6th Ed. (2014) – Approximate Flame Distortion Due to Lateral Wind on Jet Velocity from the Flare Stack.)

In both these equations, the x-coordinate is $\overline{C_L}$, defined by Equation (6.19), and the y-coordinate is either x_c (Equation 6.33 further below) or y_c (Equation 6.34 below), and (d_jR) is a parameter for jet thrust and wind thrust defined by Equation (6.32) below:

$$\overline{C_L} = C_L\left(\frac{u_j}{u_\infty}\right)\left(\frac{M_j}{29.0}\right) \tag{6.31}$$

$$d_jR = d_j\left(\frac{u_j}{u_\infty}\right)\left(\frac{T_\infty M_j}{T_j}\right)^{0.5} \tag{6.32}$$

where

$\overline{C_L}$ = x-coordinate defined by Equation (6.19)
C_L = lower explosive limit of the flammable gas in air (volume fraction)
d_jR is a parameter for jet thrust and wind thrust
u_j = jet velocity at the exit plane, m/s
u_∞ = wind speed, m/s
T_∞ = ambient air temperature, K, and
M_j = molecular weight of the flammable gas, kg/kgmol
(Molecular weight of air = 29)

6.3.1.2.1 *Equation for Flame Center for Flares and Ignited Vents –Horizontal Distance, x_c (SI Units)*

This equation is based on Figure 6.2 below, from API Standard 521, 6th Ed. (2014). The values in API's figure for Y may be approximated by the following expression, within the accuracy required for engineering work:

$$Y = 10^{\left[A+B\log_{10}(Z)+C\log_{10}(X)+D\{\log 10(X)\}^2\right]} \tag{6.33}$$

where

$Z = D_jR^a$ and
$A = -1.06296029$, $B = 0.99825131$, $C = -1.54403884$, $D = -0.181682433$

This expression is particularly useful when API's figure for x_c needs to be extrapolated, for either the X or the Z axes or both. Compared to the original API figure for x_c, the maximum and minimum percent deviations in Y in Equation (6.33) are generally within ±2%, respectively, and the standard deviation is around 1.3%.

6.3.1.2.2 *Equation for Flame Center for Flares and Ignited Vents – Vertical Distance, y_c (SI Units), Based on API Standard 521, (6th Ed., 2014)*

The values in Figure 6.3 below for y_c may be approximated by the following expression, within the accuracy required for engineering work (Note: this equation differs in form from Equation (6.70) because it contains five constants):

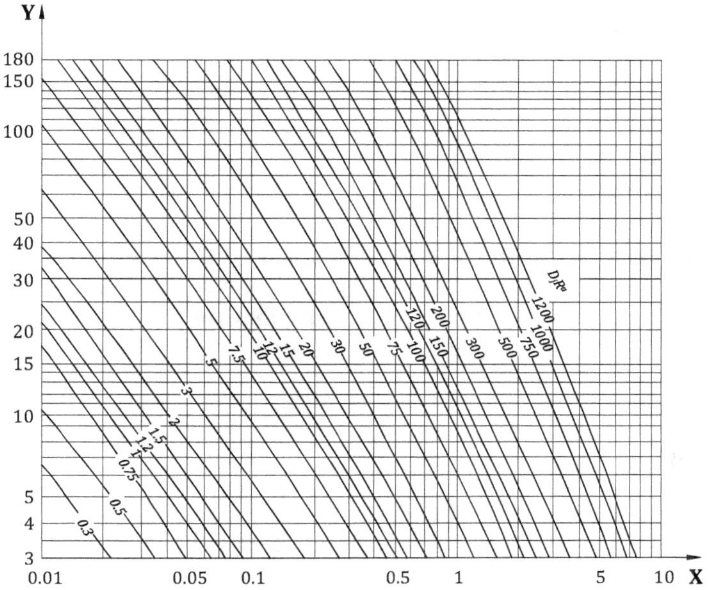

FIGURE 6.2 Flame center for flares and ignited vents – horizontal distance x_c – SI units. (With permission, API 521.)

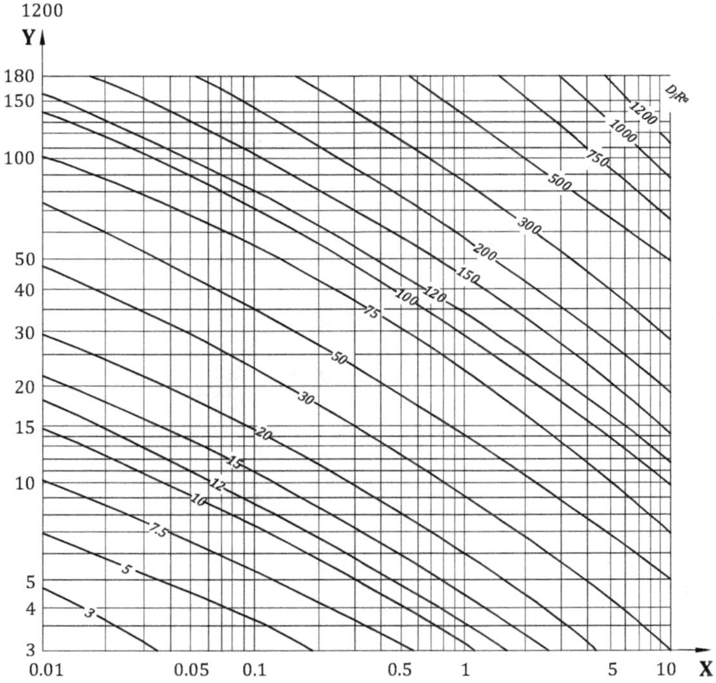

FIGURE 6.3 Flame center for flares and ignited vents – vertical distance y_c – SI units. (With permission, API 521.)

$$Y = 10^{\left[A+B\log_{10}(Z)\hat{}C+D\log_{10}(X)+E\{\log_{10}(X)\}\hat{}2\right]} \tag{6.34}$$

where

$Z = D_j R^a$

$A = -0.520398361,\ B = 1.00310326,\ C = 0.988559016,\ D = -0.428507076,$
$E = -0.044445017$

This expression is also particularly useful when API's figure for y_c needs to be extrapolated, for either the X or the Z axes, or both. For example, compared to the original API figure for y_c, the maximum and minimum percent deviations in Y in Equation (6.34) are generally accurate within ±4%, respectively, and the standard deviation is around 3.3%.

Example 6.6

Problem: A vertical 100 mm diameter overhead branch from a horizontal supply header undergoes full-bore rupture leading to the escape of methane gas to the atmosphere. The pressure and temperature in the supply header is 1.5 bar and 300 K, respectively. The gas, on emergence to the atmosphere, ignites, giving rise to a jet flame. For methane, the heating value is 50,000 kJ/kg, or 5×10^7 J/kg, γ is 1.32, and the lower explosive limit is 5% by volume. Calculate (1) the length of the flame, (2) the maximum diameter of the flame, and (3) displacements of the center of the flame from the axis of the gas jet at the exit plane, assuming wind speed to be 8 m/s. Atmospheric temperature is 298 K.

Using the methodology used for Example 6.1, Case (B), we obtain the following results:

$$\text{Exit temperature} = 266.2\,\text{K}$$

$$\text{Exit density} = 0.7325\,\text{kg/m}^3$$

$$\text{Mass flow rate} = 2.1\ \text{kg/s}$$

From Equation (6.29), at $m' = 2.1$ kg/s and $Q_c = 5\times10^7$ J/kg, the flame length L is calculated to be 22.2 m.

The maximum flame diameter occurs at a distance of 60% of the flame length measured from the center of the exit plane, i.e., at (22.21) (0.6) = 13.3 m. Hence, from Equation (6.30), at $s = 13.3$ m, the flame diameter D_s is calculated to be 3.64 m.

SIMPLE METHOD

$u_\infty/u_j = 8/340 = 0.024$, at which from Figure 6.2, $\sum\Delta x/L = 0.5$ and $\sum\Delta y/L = 0.65$.

Hence, $\sum\Delta x = (0.5)(21.4) = 10.7$ m, and $\sum\Delta y = (0.65)(21.4) = 13.9$ m

$x_c = 0.5\ \sum\Delta x = (0.5)(10.7) = 5.35$ m

$y_c = 0.5\ \sum\Delta y = (0.5)(13.9) = 6.95$ m

Method using Brzustowski and Sommer's approach:

Using Equation (6.31), $\overline{C_L} = 0.05(340/8)(16/29) = 1.17$

Also, from Equation (6.32), $d_jR = 0.1(340/8)(298)(16/268.17)(0.5) = 17.9$

At $\overline{C_L} = 1.17$ and $d_jR = 17.9$, the value of the flame diameter, y_c, from Equation (6.34) above is 5.05 m, and the value of x_c from Equation (6.33) above is 1.21 m, which happens to be outside the range of API's original graphs. Therefore, x_c is easily found analytically using Equation (6.33). Similarly, using Equation (6.34) above, the value for y_c again matches the value from API's figure quite closely. Thus, these examples show that the numerical approximations developed for x_c and y_c, which are certainly more convenient to use than the original API figures, also have acceptable accuracy.

However, for this example, the values calculated by the simple and the Brzustowski and Sommer methods seem to differ considerably. These results indicate that predictions of flame length and diameter should generally be considered only to be approximate.

6.3.1.3 Shell Model

Shell Research developed two important jet fire models at their Thornton laboratory in the UK. One of these models, described by Chamberlain,[43] predicts thermal radiation from vertical and inclined flares. The other model, described by Johnson, Brightwell, and Carsley,[44] predicts thermal radiation from horizontally released jet fires. In either case, the flame is modeled as the truncated frustum of a cone, as shown in Figures 6.4 and 6.5.

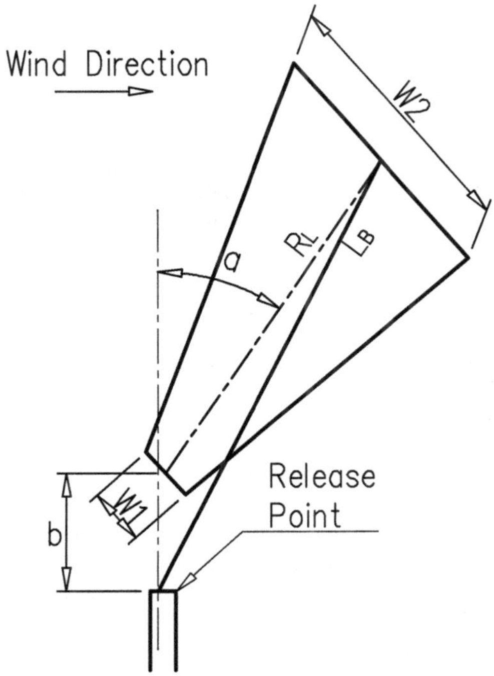

FIGURE 6.4 Shell model for vertically released flame shape with wind.

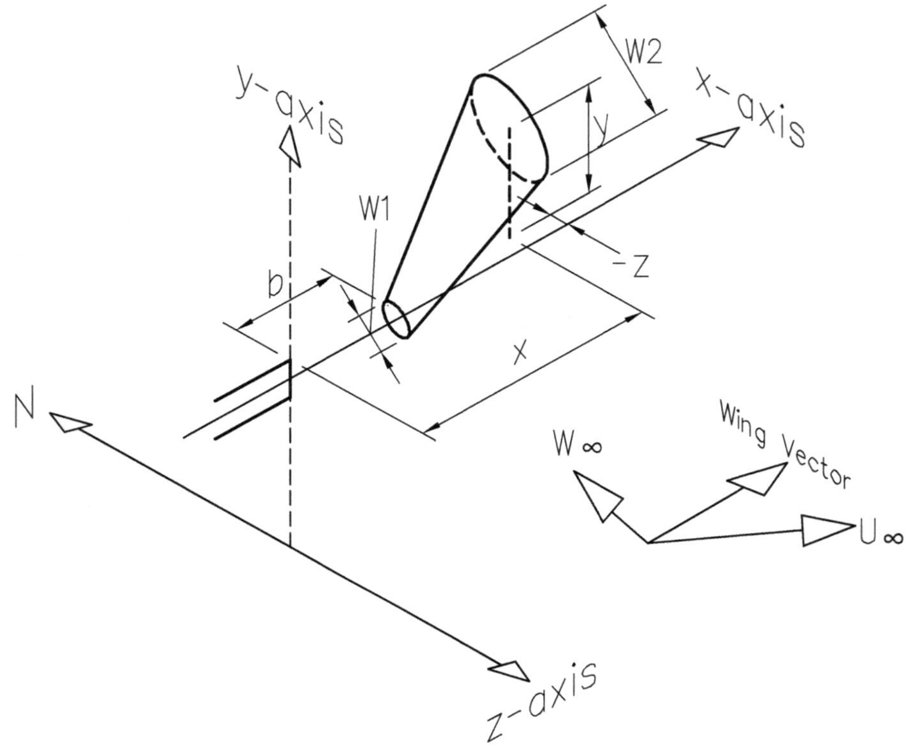

FIGURE 6.5 Shell model for horizontally-released flame shape with wind.

The flame size is calculated in terms of frustum length, the width of the frustum base, and the frustum tip width. The flame's orientation is expressed in terms of the angle of tilt of the flame and the lift-off distance of the flame.

6.3.1.3.1 Vertically Released Jet Flames

This section covers the methodology for jet flames where the axis of the hole is vertical. For those cases where the hole axis is tilted at an angle from vertical, interested readers are referred to the paper by Chamberlain.[43]

The methodology starts with the calculation of two characteristic lengths, namely L_B and D_s. L_B is the length measured from the flame tip to the center of the gas exit plane, and D_s is an effective source diameter. The physical significance of D_s is that it is the throat diameter of an imaginary nozzle from which air of density ρ_a at ambient temperature issues with the same mass flow and momentum as the fuel gas. The length scale, L_{Bo}, is used as the flame length of vertical flames in still air.

The effective source diameter D_s is calculated as follows:

$$D_s = d_j \left[\frac{\rho_j}{\rho_a} \right]^{1/2} \tag{6.35}$$

where
D_s = effective source diameter, m
ρ_j = jet density after expansion to atmospheric pressure, kg/m^3
ρ_a = density of ambient air, kg/m^3
d_j = expanded jet diameter, m

For subsonic flow, d_j can be assumed to be equal to the hole diameter, while in the case of choked flow, d_j is the diameter of a virtual source where the jet has expanded to atmospheric pressure.

If the mass flow rate is known, then the jet diameter d_j for choked flow can be calculated as follows:

$$d_j = \left(\frac{4\ m'}{\pi\ \rho_j u_j} \right)^{1/2} \tag{6.36}$$

where
d_j = expanded jet diameter, m
m' = the mass flow rate, kg/s
ρ_j = the expanded jet density, kg/m^3
u_j = the expanded jet velocity, m/s

The mass fraction, W, of fuel in a stoichiometric mixture with air and for a paraffin of molecular weight W_g is given by

$$W = W_g / \left(15.816\ W_g + 39.5\right) \tag{6.37}$$

where
W = the mass fraction of fuel in a stoichiometric mixture with air
W_g = paraffin molecular weight, kg/kgmole

The length L_{Bo} is correlated in terms of two dimensionless terms, as follows:

$$\psi = \left(\frac{2.85\ D_s}{L_{Bo} W} \right)^{2/3} \tag{6.38}$$

$$\xi(L_{Bo}) = \left(\frac{g}{D_s^2 u_j^2} \right)^{1/3} L_{Bo} \tag{6.39}$$

$$\psi = 0.2 + 0.024\ \xi(L_{Bo}) \tag{6.40}$$

where, in Equations (6.38–6.40)
ψ = a parameter
D_s = flame diameter, m
W = the mass fraction of fuel in a stoichiometric mixture with air
L_{Bo} = length of the flame in still air, including the lift-off distance, m

$\xi(L_{Bo})$ = the Richardson number based on L_{Bo}
g = gravitational acceleration = 9.81 m/s
u_j = jet velocity at the exit plane, m/s

$\xi(L_{Bo})$ is known as the Richardson number based on L_{Bo}. The value of L_{Bo} is to be found from a trial-and-error solution of Equations (6.38), (6.39), and (6.40).

L_{Bo} represents the length of the flame in still air (including the lift-off distance). With a vertical axis for the hole and in the presence of wind, the flame length L_B, measured from the tip of the flame to the center of the gas exit plane, is given by

$$L_B = L_{Bo}\left(0.51e^{-0.4v} + 0.49\right) \tag{6.41}$$

where
L_B = length measured from the tip of the flame to the center of the gas exit plane
L_{Bo} = length of the flame in still air (including the lift-off distance), m
v = wind speed, m/s.

Equation (6.41) is useful for calculating the reduction in flame length caused by the wind.

Having determined the value of L_B and $\xi(L_{Bo})$, angle (α) between the hole axis and the flame axis, frustum lift-off distance (b), frustum length (R_L), the width at frustum base (W_1), and width at frustum tip (W_2) can be calculated.

6.3.1.3.1.1 Angle, α

$$\alpha\xi(L_{Bo}) = 8{,}000R, \quad R \le 0.05 \tag{6.42}$$

$$\alpha\xi(L_{Bo}) = 1726(R - 0.026)^{0.5} + 134, \quad R > 0.05 \tag{6.43}$$

where
α = angle of flame from the vertical, radians
R = ratio of wind speed (v)/jet velocity (u_j)
$\xi(L_{Bo})$ = the Richardson number based on L_{Bo}

6.3.1.3.1.2 Frustum Lift-Off Distance, b

$$b = L_B\frac{\sin(K\alpha)}{\sin(\alpha)} \tag{6.44}$$

$$K = 0.185e^{-20R} + 0.015, \quad 0.005 \le R \le 3 \tag{6.45}$$

where:
b = frustum lift-off distance, m

L_B = the length measured from the tip of the flame to the center of the gas exit plane, m
α = angle of flame from the vertical, radians
K = a parameter
R = wind speed (v)/jet velocity (u_j)

6.3.1.3.1.3 Frustum Length, R_L

$$R_L = \sqrt{L_B{}^2 - b^2 sin^2(\alpha)} - b\cos(\alpha) \tag{6.46}$$

where
R_L = frustum length, m
b = frustum lift-off distance, m
L_B = the length measured from the tip of the flame to the center of the gas exit plane, m
α = angle of flame from the vertical, radians

6.3.1.3.1.4 The Width of Frustum Base, W_1

$$W_1 = D_s\left(13.5e^{-6R} + 1.5\right)\, F_1 \tag{6.47}$$

$$F_1 = 1 - \left(1 - \frac{1}{15}\left(\frac{\rho_a}{\rho_j}\right)^{1/2}\right) e^{-70\ \xi(D_s)\ (C)(R)} \tag{6.48}$$

$$C = 1{,}000e^{-100R} + 0.8 \tag{6.49}$$

where
W_1 = width of frustum base, m
D_s = effective source diameter, m
F_1 = a local parameter
ξ (D_s) = Richardson number based on D_s

$$\xi(D_s) = \left(\frac{g}{D_s{}^2 u_j{}^2}\right)^{1/3} D_s \tag{6.50}$$

where
D_s = effective source diameter, m
ξ (D_s) = Richardson number based on D_s
g = gravitational acceleration, m/s^2
u_j = jet velocity at the exit plane, m/s

6.3.1.3.1.5 Width at Frustum Tip, W_2

$$W_2 = L_B\left(0.18e^{-1.5R} + 0.31\right)\left(1 - 0.47e^{-25R}\right) \tag{6.51}$$

where

W_2 = width at frustum tip, m

L_B = length measured from the tip of the flame to the center of the gas exit plane, m

R = ratio of wind speed (v)/jet velocity (u_j)

Example 6.7

Problem: Methane gas (MW=16) is flowing through a horizontal pipeline at a pressure of 46 bar and a temperature of 300 K. A vertical branch from this pipeline develops a major leak that ignites. The leakage area can be approximated by a circular hole of 50 mm diameter. The distance between the leakage area and the main pipeline is sufficiently small, so that pressure drop in the vertical branch up to the point of leakage can be neglected. Assume discharge coefficient $C_d = 1.0$. Wind velocity is 5 m/s, and atmospheric temperature is 300 K. Calculate the size and shape of the flame.

Temperature and pressure inside the pipeline, leakage diameter, and the discharge coefficient are identical to Example 1, Case (A). Hence, the mass flow rate is equal to 18.23 kg/s. Also, as before, conditions after expansion to atmospheric pressure are as follows:

The pressure of the jet = 1 atm

The temperature of the jet, T_j = 234.3 K

Gas density of the expanded jet = 0.8332 kg/m^3

Mach number of the jet = 1

Velocity of the jet, u_j = 381.6 m/s

Mass flow rate of methane from jet = 17.16 kg/s

The next step is to calculate L_B and D_s.

Ambient air density, $\rho_a = (P)(MW)/R/T$ = (1) (29)/0.08205/291.6 = 1.178 kg/m^3

From Equation (6.36), expanded jet diameter, d_j = [(4) (16)/((3.1416) (0.8332) (381.6))]$^{0.5}$ = 0.2533 m

From Equation (6.35), effective source diameter, D_s = 0.2533 (0.8332/1.178)$^{0.5}$ = 0.213 m

From Equation (6.37), mass fraction fuel in stoichiometric mixture, W = 16.0/((15.816) (16.0)+39.5) = 0.0547

After substituting values of D_s and W in Equation (6.38) and also D_s and u_j in Equation (6.39), and solving Equations (6.38), (6.27), and (6.40) by trial and error, we get L_{Bo} = 54.06 m, and ξ (L_{Bo}) = 6.167 m. Hence, at a wind speed of 5 m/s, from Equation (6.41), we get L_B = 30.22 m.

Angle α between the hole axis (which is vertical in this case) and the flame axis

$$R = \text{wind speed/jet velocity} = 5/381.6 = 0.0131$$

Since R is less than 0.05, from Equation (6.43), α = (8,000) (0.0131)/6.167 = 17.00° or 0.2966 radians (in the downwind direction).

Frustum left-off distance:

At R = 0.01174, from Equation (6.45), K = 0.1574

From Equation (6.44), lift-off distance, b = 30.22 sin ((0.1574)(0.2966))/sin (0.2966) = 4.824 m

Frustum length R_L:

From Equation (6.46), R_L = (30.22 – 4.824 sin²(0.2966))$^{0.5}$ – 4.824 cos(0.2966) = 30.11 m

Width of Frustum base, W_1:

From Equation (6.49), at R = 0.0131, C = 270.6

From Equation (6.50), at D_s = 0.2130,

ξ (D_s) = [9.81/((0.2130) (381.6))]$^{(1/3)}$ (0.2130) = 0.0243 m

Substituting values of ρ_a, ρ_j, C, R, and ξ (D_s) in Equation (6.36), we get F_1 = 0.9978

Substituting values of D_s, R, and F_1 in Equation (6.47), we get W_1 = 2.971 m

Width of Frustum tip, W_2:

From Equation (6.51), at L_B = 30.22 m, and R =0.0131, we get W_2 = 9.722 m.

6.3.1.3.2 Horizontally Released Jet Fires[44]

In this case, the flame shape is specified in (X, Y, Z) coordinates, with the X-axis pointing in the direction of the release, the Y-axis pointing vertically upwards, and the Z-axis pointing horizontally, perpendicular to the release direction (see Figure 6.5). The variables defining the flame shape are as follows:

- the coordinates (X, Y) of the center of the end of the frustum
- the maximum flame width, W_2
- the minimum flame width, W_1 and
- the flame lift-off, b.

The length scales used are the same as for vertical flames in still air, namely L_{Bo} and D_s. These are calculated in the same manner as in the Chamberlain model for vertical/tilted flares. In the horizontal release model, two additional parameters are used:

$$\Omega_x = \left(\frac{\pi}{4}\frac{\rho_a}{G}\right)^{1/2} L_{Bo}u_a \tag{6.52}$$

$$\Omega_z = \left(\frac{\pi}{4}\frac{\rho_a}{G}\right)^{1/2} L_{Bo}u_w \tag{6.53}$$

where

Ω_x = parameter defined by Equation (6.52)

Ω_z = parameter defined by Equation (6.53)

ρ_a = density of ambient air, kg/m³

L_{Bo} = length of the flame in still air, including the lift-off distance, m

u_a = the wind speed component in the release direction, m/s

u_w = the wind speed component in the direction perpendicular to the release direction, m/s

G = the expanded jet momentum flux, N.

G is found using:

$$G = \frac{\pi \rho_j u_j^2 d_j^2}{4} \tag{6.54}$$

where

G = the expanded jet momentum flux, N.
u_j = jet velocity at the exit plane, m/s
ρ_j = the expanded jet density, kg/m^3
d_j = expanded jet diameter, m

The coordinates (X, Y) of the center of the end of the frustum are calculated as follows:

$$X/L_{Bo} = f(\xi)\left(1 + r(\xi)\,\Omega_x\right) \tag{6.55}$$

where

X = distance in the X-direction, m
L_{Bo} = length of the flame in still air, including the lift-off distance, m
$f(\xi)$ = function of ξ defined in Equation (6.56)
$r(\xi)$ = function of ξ defined in Equation (6.57)
Ω_x = parameter defined by Equation (6.52)

$$f(\xi) = 0.55 + 0.45\exp(-0.168\xi), \quad \xi \le 5.11$$

$$= 0.55 + 0.45\exp\left(-0.168\xi - 0.3(\xi - 5.11)^2\right), \quad \xi > 5.11 \tag{6.56}$$

and

$$r(\xi) = 0, \quad \xi \le 3.3$$

$$= 0.082\left(1 - \exp\left(-0.5(\xi - 3.3)\right)\right), \quad \xi > 3.3 \tag{6.57}$$

$$Y/L_{Bo} = h(\xi)\left(1 - c(\xi)\,\Omega_x\right) \tag{6.58}$$

where

$\xi = \xi(L_{Bo})$ defined in Equation (6.39)
Y = distance in the Y-direction, m
L_{Bo} = length of the flame in still air, including the lift-off distance, m
$h(\xi)$ = function of ξ defined in Equation (6.59)
$c(\xi)$ = function of ξ defined in Equation (6.60)
Ω_x = parameter defined by Equation (6.52)

$$h(\xi) = 1/\left(1 + 1/\xi\right)^{8.78} \tag{6.59}$$

and

$$c(\xi) = 0.02\ \xi \tag{6.60}$$

$$L_{Bxy} = (X^2 + Y^2)^{1/2} \tag{6.61}$$

$$W_2/L_{Bxy} = -0.004 + 0.0396\ \xi - \Omega_x\left(0.0094 + 9.5 \times 10^{-7} \xi^5\right) \tag{6.62}$$

$$b = 0.141\ \left(G\rho_a\right)^{1/2} \tag{6.63}$$

$$W_1/b = -0.18 + 0.081\ \xi \tag{6.64}$$

where

L_{Bxy} = parameter defined in Equation (6.59), m
W_2 = width at frustum tip, m
$\xi = \xi\ (L_{Bo})$ defined in Equation (6.39)
b = parameter defined in Equation (6.63)
G = the expanded jet momentum flux, N.
ρ_a = density of ambient air, kg/m^3
W_1 = width of frustum base, m

Note: ξ in the above equations represents $\xi\ (L_{Bo})$ defined in Equation (6.39). The model specifies that W_2 must be greater than or equal to the minimum flame width W_1 and less than L_{Bxy} for a physically realistic flame shape prediction. Also, the minimum value of the parameter W_1/b has been set at 0.12.

The deflection of the flame by crosswind is correlated as follows:

$$\tan(\alpha) = Z/(X - b) = 0.178\ \Omega_z \tag{6.65}$$

where

α = angle that the projection of the frustum axis onto the horizontal plane makes with the release axis, radians
Z = z-position of the end of the flame, m
X = distance in the X-direction, m
b = parameter defined in Equation (6.63)
Ω_z = parameter defined by Equation (6.53)

Example 6.8

Problem: Rework Example 6.7, assuming that the release occurs horizontally. The wind speed is 5 m/s in the release direction and 3 m/s in a direction perpendicular to the release.

Since the conditions in the problem are the same as in Example 6.7 except for the release direction and the presence of wind vector in a direction perpendicular to the direction of release, the following values can be taken from Example 6.7:

$L_{Bo} = 54.06$ m, $\rho_a = 1.178$ kg/m^3, $\rho_j = 0.8332$ kg/m^3, $u_j = 381.6$ m/s, $d_j = 0.2533$, m and $D_s = 0.2130$ m. Also, $\xi = (L_{Bo}) = 6.167$.

From Equation (6.54), jet momentum flux, $G = \pi\ \rho_j\ u_j^2\ d_j^2/4 = \pi\ (0.8332)$ $(381.6^2)\ (0.2533^2)/4 = 6114$ N. (It might be noted that in Reference [9], the range of G studied was 383–6,828 N).

Using Equation (6.42), $\Omega_x = \left(\dfrac{(\pi)(1.178)}{((4)(6114))}\right)^{1/2} (54.06)(5) = 3.3252$

Using Equation (6.43), $\Omega_z = \left(\dfrac{(\pi)(1.178)}{((4)(6114))}\right)^{1/2} (54.06)(3) = 1.9951$

From Equation (6.55) $\left(\text{at } \xi = 5.814, \text{ which is } > 5.11\right)$,

$$f(\xi) = 0.55 + 0.45\ exp\left((-0.168)(6.167) - 0.3(6.167 - 5.11)^2\right) = 0.6642$$

From Equation (6.57), at $\xi = 6.167,\ r(\xi) = 0.082\left(1 - exp\left(-0.5(6.167 - 3.3)\right)\right) = 0.06245$

Substituting values of $f(\xi)$, $r(\xi)$, and Ω_x in Equation (6.55), we get $X/L_{Bo} = 0.6642$ $(1 + (0.06245)\ (3.3252)) = 0.8021$,

whence $X = (0.8021)\ (54.06) = 43.36$ m

At $\xi = 6.167$, Equation (6.59) gives $h\ (\xi) = 0.2673$, and Equation (6.60) gives $c\ (\xi) = 0.1233$.

Substituting these values in Equation (6.58), we get

$$\frac{Y}{L_{Bo}} = 0.2673\ \left(1 - (0.1233)\ (3.252)\right) = 0.1577, \text{ whence } Y = (0.1577)\ (54.06) = 8.5230 \text{ m}$$

From Equation (6.61), $L_{Bxy} = \left(43.36^2 + 8.523^2\right)^{0.5} = 44.19$ m

At $\xi = 6.167$ and $\Omega_x = 3.3252$, Equation (6.62) gives $W_2/L_{Bxy} = 0.1808$,

whence $W_2 = (0.1808)\ (44.19) = 7.9886$ m

From Equation (6.63), flame lift-off, $b = 0.141\ ((6114)\ (1.178))^{0.5} = 11.97$ m

At $\xi = 6.167$, from Equation (6.64), $W_1/b = -0.018 + (0.081)\ (6.167) = 0.3195$,

whence $W_1 = (0.3195)\ (11.97) = 3.8236$ m

From Equation (6.65), at $\Omega_z = 1.9951$, and the angle of deflection of the flame due to crosswind is

$$\alpha = \tan^{-1}\left((0.178)\ (1.9951)\right) = 0.3412 \text{ radians.}$$

We find the Z-position of the end of the flame, from Equation (6.65):

$$Z = (X - b)tan(\alpha) = (43.36 - 11.97)tan(0.3412) = 10.71 \text{ m.}$$

6.4 ESTIMATION OF RADIATION INTENSITY

Models for the estimation of radiation intensity differ on two basic concepts: the point source model, such as the API model, and the surface emitter model, such as the shell model. In the API model, the radiation is assumed to emanate from a flame epicenter that is taken as the mid-point of the flame along its length. In the Shell model, the radiation is assumed to emanate from the surface of the flame. Therefore, the methodology for radiation intensity is similar to that in a pool fire.

TABLE 6.3
Fractional Radiation from Gaseous Diffusion Flames[40]

Gas	Burner Diameter, cm	F_R
Hydrogen	0.51	0.095
	8.4	0.156
	40.6	0.169
Butane	0.51	0.215
	8.4	0.291
	40.6	0.299
Natural gas (95% CH_4)	20.3	0.192
	40.6	0.232

6.4.1 Fractional Radiation

Both API and shell models require data on fractional radiation F_R, which is the fraction of the radiation heat to the total heat liberation rate in the combustion process. The API standard[40] gives the following data for F_R from gaseous diffusion flames (Table 6.3):

Based on data from field trials, Chamberlain[43] gives the following correlation for F_R, which is used in the shell model:

$$F_R = 0.21\ e^{-0.00323\ u_j} + 0.11 \tag{6.66}$$

where
F_R = fractional radiation absorbed by object
u_j = jet velocity, m/s.

This correlation gives values of F_R equal to 0.26 and 0.15 at u_j values of 100 and 500 m/s, respectively. Based on these data, a value of F_R between 0.25 and 0.3 should be acceptable for hazard assessment.

6.4.2 Radiation Intensity by the API method

In this method, the radiation intensity on a receiver is calculated as follows:

$$I = \frac{MQ_c F_R \tau}{4\ \pi\ X^2} \tag{6.67}$$

where
I = intensity on the receiver, kW/m^2
M = mass flow rate of combustible gas, kg/s
Q_c = lower heating value of the combustible gas, kJ/kg
F_R = fractional radiation

τ = atmospheric transmissivity
X = distance between the flame epicenter and the receiver, m

Since the API method assumes the radiation to emanate from a point source, the results are susceptible to large errors, particularly when the receiver is located close to the flame.

6.4.3 Radiation Intensity by the Shell Method

In the shell method, the intensity of radiation I_x, kW/m^2, is calculated in the same manner as in the case of pool fire, as follows:

$$I_x = \left(\text{SEP}\right)\left(\text{View Factor}\right)(\tau) \tag{6.68}$$

where
I_x = intensity of radiation, kW/m^2
SEP = the surface emissive power of the flame, kW/m^2
τ = the transmissivity.

For calculating the SEP, A_f (the surface area of the frustum, m^2, including the ends) needs to be calculated first.

$$A_f = \frac{\pi}{4}\left(W_1^2 + W_2^2\right) + \frac{\pi}{2}(W_1 + W_2)\left(R_L^2 + \left(\frac{W_2 - W_1}{2}\right)^2\right)^{0.5} \tag{6.69}$$

$$\text{SEP} = \frac{MQ_cF_R}{A_f} \tag{6.70}$$

where
A_f = the surface area of the frustum, including the ends, m^2
W_1 = width of frustum base, m
W_2 = width at frustum tip, m
R_L = frustum length (considered as a cylinder), m
SEP = the surface emissive power of the flame, kW/m^2
M = mass flow rate of combustible gas, kg/s
Q_c = lower heating value of the combustible gas, kJ/kg
F_R = fractional radiation

The view factor is calculated in the shell model by numerical integration. For an approximate estimate, the frustum could be considered as a cylinder of length = R_L and uniform diameter = $(W_1 + W_2)/2$. For a vertically released jet in the presence of wind, the flame would be tilted through an angle α from vertical. For such a tilted flame, the view factor can be estimated by following the procedure for pool fires given in Chapter 5.

Example 6.9

Problem: A vertical flare, 10 m high, releases methane gas that ignites, giving rise to a jet flame. Conditions of release and dimensions of the flame are the same as in Example 6.7. Calculate the thermal radiation intensity on an object located at ground level at a downwind distance of 15 m from the flare axis (see Figure 6.6). The heat of combustion of the gas is 50,000 kJ/kg. Assume atmospheric transmissivity is equal to 0.7.

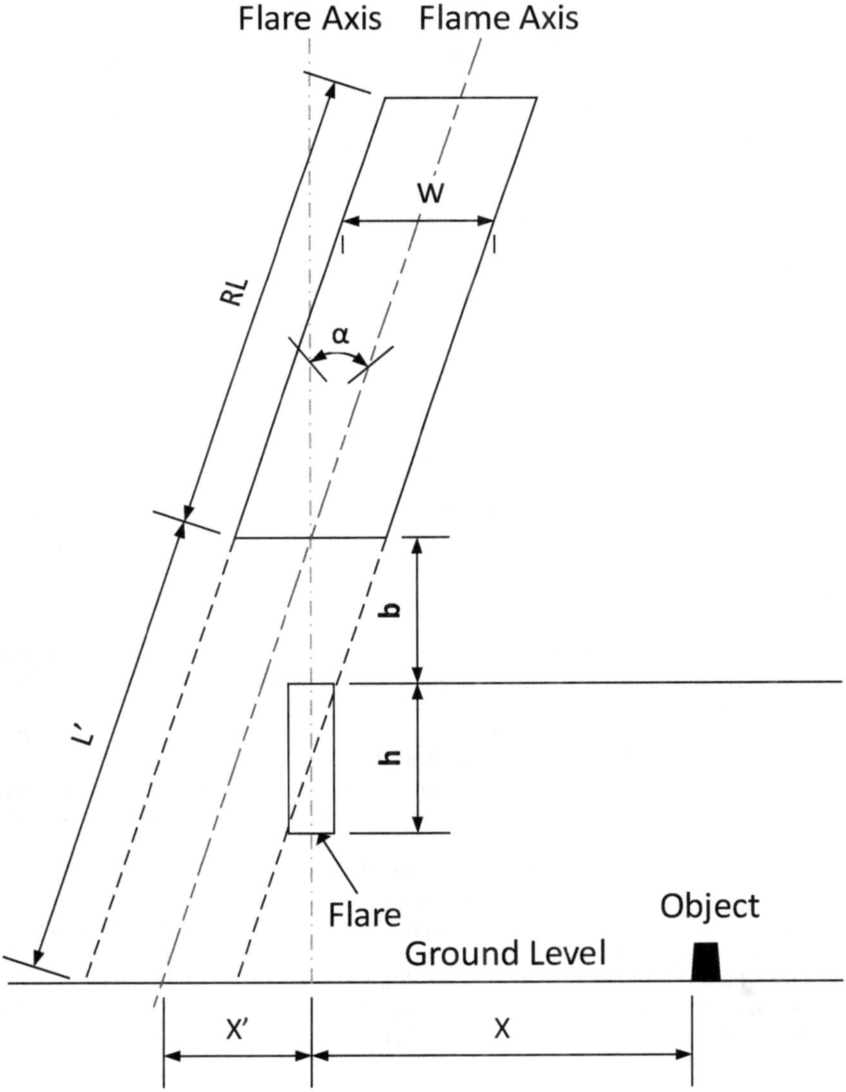

FIGURE 6.6 Approximation of jet flame model for view factor calculations.

Please see Figures (6.4–6.6) for an explanation of the geometrical terms used.

Referring to Example 6.7, $W_1 = 3.8236\,\text{m}$, $W_2 = 7.9886\,\text{m}$, $R_L = 30.11\,\text{m}$, $b = 4.824\,\text{m}$, and $\alpha = 17.00°$.

Average diameter of the cylindrical flame W is $(2.971 + 9.722)/2 = 6.346\,\text{m}$ and The radius of the cylindrical flame = 3.173 m.

Using Equation (6.66), at a gas velocity of 381.6 m/s, we get the fractional radiation, $F_R = 0.1712$

Substituting values of W_1, W_2, and R_L in Equation (6.69),

A_{fi} the surface area of the flame, is calculated to be 685.2 m^2.

Hence, from Equation (6.70), SEP = (16) (50,000) (0.1712)/685.2 = 200.1 kW/m^2

Referring to Figure 6.6, $L' = (4.824 + 10)/\cos(17.00°) = 15.50\,\text{m}$, and

$X' = (4.824 + 10)\tan(17.00°) = 4.531\,\text{m}$.

Following the procedure due to Mudan,[45] as described in Chapter 5, the view factor between the object and a tilted flame of length equal to (15.50 + 30.11) = 45.61 m is calculated to be 0.1610, and the view factor between the object and a tilted flame of length 15.53 m is 0.1339. Hence, the view factor of the object and the actual flame is

$$(0.1610 - 0.1339), \text{ or } 0.0271.$$

Finally, radiation intensity on the object = (SEP) (view factor) (transmissivity) = (195) (0.0271) (0.7) = 3.799 kW/m^2.

REFERENCES

1. Hougen, O. A. and Watson, K. M.: *Chemical Process Principles* (1st Ed., Part II, p. 551, John Wiley & Sons, New York, and 2nd Ed. with Ragatz, R. A. Vol. II, p. 653, Chapman and Hall, London, 1947, and 1959).
2. Elliott, J. R. and Lira, C. T.: *Introductory Chemical Engineering Thermodynamics* (2nd Ed., pp. 68–69, Pearson Education, Inc. publishing as Prentice Hall, Old Tappan, NJ, 2012).
3. Klein, S. and Nellis, G.: *Thermodynamics* (1st Ed., p. 168, Cambridge University Press, Cambridge, 2011).
4. Himmelblau, D. M. and Riggs, J. B.: *Basic Principles and Calculations in Chemical Engineering* (8th Ed., Fig. 9.18, Item 3, p. 525, Pearson Education, Upper Saddle River, NJ, 2012).
5. Lewis, G.N., and M. Randall: *Thermodynamics* (2nd Ed., Revised by K. E. Pitzer and L. Brewer, p. 48, McGraw-Hill, New York, 1961).
6. Cutlip, M. B. and Shacham, M.: *Problem Solving in Chemical and Biochemical Engineering with POLYMATH, Excel, and MATLAB* (2nd Ed., Prentice-Hall, Upper Saddle River, NJ, 2007).
7. *Aspen Plus Process Simulator* (Aspen Technology, Cambridge, MA).
8. *PRO II Process Simulator* (Aveva, Cambridge, UK).
9. *VMGSim Process Simulator* (Virtual Materials Group, a Schlumberger Company, Calgary).
10. Lee, B. I. and Kesler, M. G.: A Generalized Thermodynamic Correlation Based on Three-Parameter Corresponding State, *AIChE Journal*, 21, p. 510 (1975).
11. Brzustowski, T. A. and Sommer, E. C. Jr.: *Predicting Radiant Heating from Flares* (Proceedings – Division of Refining, Vol. 53, pp. 865–893, American Petroleum Institute, Washington, DC, 1973).
12. *Excel® Spreadsheet Program* (Microsoft Corp., 2019).

13. Mathcad 2001 Professional: *Mathcad User's Guide with Reference Manual* (Mathsoft Engineering and Education, Inc., Cambridge, 2001).
14. Freund, S.M., M. Jones, and J.L. Starks: *Microsoft Excel 2013: Complete* (1st Ed., Shelly Cashman Series, Course Technology, Boston, MA, 2013).
15. Crowl, D. A. and Louvar, J. F.: *Chemical Process Safety* (3rd Ed., pp. 130–131, Pearson Education, Englewood Cliffs, NY, 2011).
16. Green, D. W. and Perry, R. H.: *Perry's Chemical Engineers' Handbook* (8th Ed., Section 6, pp. 23–24, McGraw-Hill, New York, 2008).
17. *Symmetry Process Simulator* (Virtual Materials Group, a Schlumberger Company, Calgary).
18. National Institute of Standards and Technology (NIST): *REFPROP – Computer Program for Thermodynamic and Transport Properties* (Release 10, 2018).
19. Walas, S.: *Phase Equilibria in Chemical Engineering* (Butterworth-Heinemann, Boston, MA, 1985).
20. Pitzer, K. S.: *Origin of the Acentric Factor*, in T. S. Storvick and S. I. Sandler: *Phase Equilibria and Fluid Properties in the Chemical Industry* (ACS Symposium Series 60, pp. 1–10, American Chemical Society, Washington, DC, 1977).
21. Hougen, O. A., Watson, K. M., and Ragatz, R. A.: *Chemical Process Principles, Part 2, Thermodynamics* (2nd Ed., Wiley, New York, 1954).
22. Prausnitz, J. M., Lichtenthaler, R. N., and Gomes de Azevedo, E.: *Molecular Thermodynamics of Fluid Phase Equilibria* (3rd Ed., Prentice-Hall, Upper Saddle River, NJ, 1999).
23. Benedict, M., Webb, G. B., and Rubin, L. C.: An Empirical Equation for Thermodynamic Properties of Light Hydrocarbons and their Mixtures I. Methane, Ethane, Propane and n-Butane. *The Journal of Chemical Physics*, 8, p. 334 (1940).
24. Benedict, M., Webb, G. B., and Rubin, L. C.: An Empirical Equation for Thermodynamic Properties of Light Hydrocarbons and their Mixtures II. Methane, Ethane, Propane and n-Butane. *The Journal of Chemical Physics*, 10, p. 747 (1942).
25. Starling, K. E.: *Fluid Thermodynamic Properties of Light Petroleum Systems* (Gulf Publishing, Houston, TX, 1973).
26. Plöcker, U., Knapp, H., and Prausnitz, J. M.: Calculation of High-Pressure Vapor-Liquid Equilibria from a Corresponding-State Correlation with Emphasis on Asymmetric Mixtures, *Industrial & Engineering Chemistry Process Design and Development*, 17, p. 324 (1978).
27. Peneloux, A., Rauzy, E., and Freze, R.: A Consistent Correction for Redlich-Kwong-Soave Volumes, *Fluid Phase Equilibria*, 8, p. 7 (1982).
28. *Hysys Process Simulator* (Aspen Technology, Cambridge, MA).
29. Seader, J. D., Henley, E. J., and Roper, D. K.: *Separation Process Principles, Chapter 2* (3rd Ed., Wiley, New York, 2011).
30. Prausnitz, J. M., Anderson, T. F., Grens, E. A., Eckert, C. A., Hsieh, R., and O'Connell, J. P.: *Computer Calculations for Multicomponent Vapor-Liquid and Liquid-Liquid Equilibria* (Prentice-Hall, Englewood Cliffs, NJ, 1980).
31. Fredenslund, A., Gmehling, J., and Rasmussen, P.: *Vapor Liquid Equilibria Using UNIFAC* (Elsevier, Amsterdam, 1977).
32. Russell, R. A.: *PD-PLUS Process Simulator* (Deerhaven Technical Software, Burlington, MA).
33. Dortmund Databank and Software Technologies (DDBST), Dortmund, Germany (2018).
34. Russell, R. A.: *PD-PLUS Process Simulator* (Deerhaven Technical Software, Burlington, MA, 2019).
35. Gmehling, J., Li, J., and Schiller, M. A.: A Modified UNIFAC model. 2. Present Parameter Matrix and Results for Different Thermodynamic Properties, *Industrial & Engineering Chemistry Research*, 32, pp. 178–193 (1993).

36. Larsen, B. L., Rasmussen, P., and Fredenslund, A.: A Modified UNIFAC Group-Contribution Model for the Prediction of Phase Equilibria and Heats of Mixing, *Industrial & Engineering Chemistry Research*, 26, pp. 2274–2286 (1987).
37. *PROSIM Process Simulator* (PROSIM, Labege).
38. Fredenslund, A., Jones, R. L., and Prausnitz, J. M.: Group Contribution Estimation of Activity Coefficients in Nonideal Liquid Mixtures, *AIChE Journal*, 21, pp. 1086–1098 (1975).
39. Hawthorne, W. R., Weddell, D. S., and Hottel, H. C.: *Mixing and Combustion in Turbulent Gas Jets* (Third Symposium on Combustion, Flame and Explosion Phenomena, Williams and Wilkins, Baltimore, MD, 1949).
40. ANSI/API Standard 521: *Pressure-Relieving and Depressuring Systems, 5th Ed.* (2014).
41. Cook, J., Bahrami, Z., and Whitehouse, R. J.: A Comprehensive Program for Calculation of Flame Radiation Levels, *Journal of Loss Prevention in Process Industries (UK)*, 3, pp. 150–155 (1990)
42. Baron, T.: The Turbulent Diffusion Flame, *Chemical Engineering Progress*, 50(2), pp. 73–76, (AIChE, New York, 1954).
43. Chamberlain. G. A.: Developments in Design Methods for Predicting Thermal Radiation from Flares, *Chemical Engineering, Research and Design*, 65, pp. 299–309 (1987).
44. Johnson, A. D., Brightwell, H. M. and Carsley, A. J.: A Model for Predicting the Thermal Radiation Hazards from Large-Scale Horizontally Released Natural Gas Jet Fires, *Process Safety and Environmental Protection*, 72, pp. 157–166 (1994).
45. Mudan, K.S.: Geometric View Factors for Thermal radiation Hazard Assessment, *Fire Safety Journal*, 12, pp. 89-96 (1987)

7 Vapor Cloud Fire

A vapor cloud fire, also called "flash fire", refers to the combustion of a flammable vapor cloud in the air without generating any significant overpressure. A flammable gas or vapor may be released to the atmosphere according to three cases, as follows:

a. Release of liquid from a vessel followed by evaporation from the resulting liquid pool
b. Release of gas from a pressurized containment (either continuously through a leak, or instantaneously from rupture of a pressure vessel), and
c. Release of a mixture of vapor and liquid droplets instantaneously from the rupture of a pressurized containment of liquefied gas.

If an immediate ignition occurs, the result will be a pool fire, case (a), and a jet fire in the continuous release, case (b). For case (c) and the instantaneous release case, the result will be a fireball. The consequences of pool fires have been covered in Chapter 5. Those of jet fires and fireballs are covered in Chapters 6 and 8, respectively.

In the absence of immediate ignition, the released gas or vapor will travel mainly downwind and crosswind. As the gas travels, its concentration in the air will fall to values determined by conditions affecting dispersion. Methods of estimating concentration contours for the flammable cloud have been covered in Chapter 11.

In case ignition occurs at some point away from the source, the result may be a flash fire or a vapor cloud explosion. It is usual to assume that the cloud encounters the ignition source while traveling outwards, and once ignited, the flame travels backward through the cloud towards the source of release. In a calm, turbulence-free environment, no overpressure is created, and the damage is in the form of flame engulfment and heat radiation from the flame. If the release continues even after the flash fire has reached the source, the fire will continue, either as a pool fire or a jet fire, depending on the nature of the release. In case there are obstacles or other confinements in the path of the flash fire, the result will likely be a rapid escalation to a vapor cloud explosion with significant overpressure effects.

The characteristics of flash fires and their effects have been considered in this chapter. Vapor cloud explosions have been covered in Chapter 9.

7.1 FLASH FIRE ACCIDENTS AND EXPERIMENTS

A flash fire causes much less damage than a vapor cloud explosion or a boiling liquid, expanding vapor explosion (BLEVE). For this reason, accident reports are usually much less focused on flash fires. Instances have been cited[1] where the transition phenomenon from flash fires to vapor cloud explosions has not received adequate attention.

DOI: 10.1201/9781003107873-7

TABLE 7.1
Summary of Tests on Vapor Cloud Fires

Test Program	Fuel	Release Rate (kg/s)/ Quantity (kg)	No. of Tests	Primary Objectives of Tests
Maplin Sands, Shell, 1980	LNG	20–40 (continuous), 3,500–5,000 (instantaneous)	3 2	Flame propagation, thermal radiation, and overpressure
	Liquefied propane	20–55 (continuous), 4,500 (instantaneous)	3 1	
Coyote, China Lake, LLNL, 1980	LNG	100–120 (continuous), 12,000 (instantaneous)	4	Flame propagation and thermal radiation
	Liquefied methane	100 (continuous), maximum11,000 (instantaneous)	1	
Mussel banks, Terneuzen, TNO, 1983	Liquefied propane	1,000–4,000 (dispersed cloud inventory)	7	Flame propagation and overpressures with and without obstacles
China Lake, NWC, 1978	LNG	25–35 (continuous), maximum 2,500 (instantaneous)	6	Flame propagation and thermal radiation
China Lake, US DOE, 1977	LPG	30–40 (continuous), maximum 2,500 (instantaneous)	3	Flame propagation and thermal radiation

Appendix 1 of Lees[2] contains a list of accidents between 1911 and 1995. These include 149 accidents, out of which 120 (80%) were vapor cloud explosions, and the balance 29 were flash fires. Of the 29 flash fire accidents, no death or injury has been reported in 14 accidents. Nine of the accidents resulted in 1–5 deaths. One was a colossal accident (Mexico City, which escalated into a BLEVE) with 650 deaths and some 6,400 injured. Three were LPG tanker road accidents that incurred 68–216 deaths and 100–220 injuries. These statistics do not include the Feyzin accident (listed under BLEVE), which is known to have started with a flash fire that culminated in a BLEVE. The Feyzin ignition was caused by an automobile traversing a nearby public road when it drove through the LPG vapor cloud that had spread beyond the plant boundary.

In the "Review of Flash Fire Modelling" by Rew et al.,[3] a summary of various test conditions has been presented, as described below. All these tests involved spillage onto land or water, which produced dense low-lying vapor clouds (Table 7.1).

These tests have enabled data collection on flame speeds, flame sizes, and radiative heat fluxes from flames, which have been considered in the following sections.

7.2 FLAME SPEED[2,3]

The combustion of the vapor cloud generally starts with the burning of the premixed part of the cloud. It is followed by the diffusive burning of the fuel-rich part to the source of fuel release. The premixed burning stage is characterized by a bluish flame that contains insignificant quantities of soot. The non-premixed or diffusive burning stage is characterized by a relatively high radiation level caused by incandescent soot particles within the flame. The flame velocity for a premixed flame is determined by the burning velocity, a fundamental property of the reaction mixture. For a nonpremixed flame, the flame velocity is limited by the rate of air entrainment by the flame.

7.2.1 Premixed Flame

In the premixed region for a turbulent flame, the flame speed U_f relative to the unburnt gas mixture ahead of the flame is defined as follows:

$$U_f = EU_t \tag{7.1}$$

where

U_f = flame speed relative to the unburnt gas mixture ahead of the flame, m/s
U_t = turbulent burning velocity, m/s
E = expansion ratio.

The turbulent burning velocity is defined as the speed at which a turbulent flame front or reaction zone moves relative to the unburned gas mixture ahead of it. It is a characteristic of the gas mixture. The expansion ratio, E, is a factor that allows for the push given to the flame front by the expansion of the burnt gases behind it and is defined as follows:

$$E = \frac{T_f}{T_i}\frac{N_p}{N_r} \tag{7.2}$$

where

E = expansion ratio
T_f = flame temperature to which the burned products are raised, K
T_i = initial temperature of fuel-air mixture, K
N_p/N_r = molar ratio of combustion products to reactants

For a stoichiometric mixture of propane in air at an initial temperature of 18°C, the adiabatic flame temperature is 1,925°C; from Equation (7.2), E is 7.86.

For determining the flame speed U_g relative to the ground, we also need to account for the wind speed, in addition to the turbulent flame velocity and the expansion ratio. Assuming that the flame is propagating in the direction opposite to that of the wind, the flame speed U_g is given by

$$U_g = U_f - U_w \tag{7.3}$$

where

U_g = flame speed relative to the ground, m/s
U_w = wind speed, m/s.

For a laminar flame, Equations (7.1) through (7.3) are based on the laminar burning velocity, U_L, of the flame, instead of the turbulent burning velocity. U_L is the relative speed of a laminar flame front to the unburned mixture ahead of it.

A considerable amount of work has been done to obtain a relationship between the turbulent burning velocity and the laminar burning velocity. This aspect has been discussed by Rew et al.[3] and by Lees.[2] A correlation presented graphically by Lees[2] – based on Bradley, Lau, and Lawes – shows that the ratio of turbulent flame-to-laminar flame velocities ranges from about 2 to 18.

The laminar burning velocity of a fuel-air mixture is usually determined using a Bunsen burner. For a premixed laminar flame, the burning velocity is determined by dividing the volumetric gas flow rate by the flame front cone area. For paraffinic hydrocarbons at atmospheric pressure and temperature, laminar burning velocities range from a few centimeters per second near the flammability limits to a maximum of about 45 cm/s for stoichiometric mixtures. Experimental maximum values of laminar burning velocity quoted by Lees[2] are 36–45 cm/s for alkanes, 69 cm/s for ethylene, and 173 cm/s for acetylene.

Experimental values of flame speeds obtained in various trials have been quoted by Lees and by CCPS. These are given in Table 7.2 below:

Thus, an average value between 3 and 15 m/s is considered appropriate for flammable cloud flame speeds in the premixed region. These are well below a value of about 150 m/s often quoted as necessary to generate any significant overpressure.

7.2.2 Nonpremixed Flame

Rew et al.[3] have presented a correlation due to Raj and Emmons[1] for flame propagation velocities through the rich section of the gas cloud on land. According to this correlation, which is based on experimental data on dense gas clouds, the propagation velocity relative to the cloud is directly proportional to the wind speed and is given by

$$S = 2.3U_w \tag{7.4}$$

where

S = flame speed, m/s
U_w = mean wind speed, m/s

TABLE 7.2
Experimental Data on Flame Speed in Vapor Cloud Fires

Lees[2]	CCPS[4]
Maplin Sands Trials (on sea): *LNG*: Average 4 m/s with a maximum of 10 m/s *Propane*: Average 12 m/s with a maximum of 20 m/s *Coyote Trials* (on water): *LNG*: 11.9–18.9 m/s for wind speeds of 4.6–10.0 m/s (6.9–12.9 m/s for flame speed minus the wind speed)	*HSE Experiments on land (1983)*: *LPG* (propane): 3.2–11.3 m/s after adding/subtracting wind speed of 2–7 m/s, depending on upwind or downwind locations

Data on flame speed for calculation of flame height using the Raj and Emmons model[1] are based on Equation (7.4).

7.3 FLAME DIMENSIONS[1,2,4]

The flame size for flash fires is usually estimated based on the model by Raj and Emmons.[1] It applies to nonpremixed flames where air diffusion rate determines the rate of burning into the flame. The model assumes a two-dimensional, turbulent flame front with a thickness W at the base of the flame. A flame of height H propagates at a constant velocity S into the unburned air-fuel cloud of depth d, as shown schematically in Figure 7.1.

According to this model, the flame height is given by

$$H = 20d\left[\frac{S^2}{gd}\left(\frac{\rho_o}{\rho_a}\right)^2\frac{wr^2}{(1-w)^3}\right]^{\frac{1}{3}} \tag{7.5}$$

where

H = height of the visible flame, m

d = depth of the unburned gas cloud, m

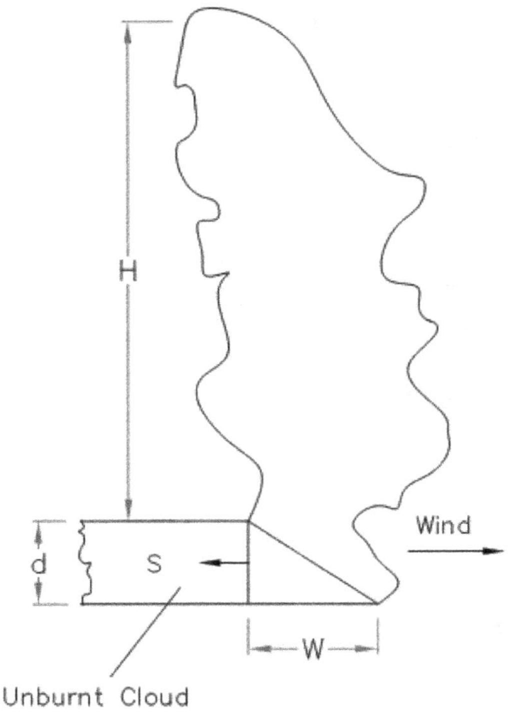

FIGURE 7.1 Schematic representation of a flash fire.

g = gravitational acceleration = 9.81, m/s^2
ρ_o = density of the fuel-air mixture, kg/m^3
ρ_a = density of air, kg/m^3
w = a parameter (defined below)
r = stoichiometric air-fuel mass ratio
S = rate of propagation of the flame through the cloud, m/s, given by Equation (7.4) relative to the gas cloud.

The CCPS defines the parameter w as follows[4,5]:

$$w = 0, \quad \phi \le \phi_{st} \tag{7.6a}$$

$$w = \frac{\phi - \phi_{st}}{\alpha\left(1 - \phi_{st}\right)}, \quad \phi > \phi_{st} \tag{7.6b}$$

where
α = constant pressure expansion ratio for stoichiometric combustion
φ = mole fraction of the fuel in the air-fuel mixture
φ_{st} = mole fraction of the stoichiometric mixture of the fuel in air.

For hydrocarbons, the value of α is typically about 8.
The stoichiometric air-fuel mass ratio r is given by

$$r = \frac{\left(1 - \varphi_{st}\right) M_{\text{air}}}{\varphi_{st}\ M_{\text{fuel}}} \tag{7.7}$$

where
r = stoichiometric air-fuel mass ratio
φ_{st} = mole fraction of the stoichiometric mixture of the fuel in air.
M_{air} = molecular weight of air, kg/kgmol
M_{fuel} = molecular weight of fuel, kg/kgmol

The density ratio is given by

$$\frac{\rho_o}{\rho_a} = \frac{\left(1 - \varphi\right) M_{\text{air}} + \varphi\ M_{\text{fuel}}}{M_{\text{air}}} \tag{7.8}$$

where
φ = mole fraction of the fuel in the air-fuel mixture
M_{air} = molecular weight of air, kg/kgmol
M_{fuel} = molecular weight of fuel, kg/kgmol
ρ_o = density of the fuel-air mixture, kg/m^3
ρ_a = density of ambient air, kg/m^3

7.4 EFFECT OF FLAME EXPOSURE

It is usual to assume 100% fatality for people who are engulfed by a flame. Outside the flammable cloud, an injury will be caused by thermal radiation from the flame. People exposed to the gas cloud upstream of the propagating flame will be subjected to the combined effect of thermal radiation and asphyxiation.

For determining the number of people who might be engulfed by the flame, it is necessary to carry out a dispersion study to estimate the LFL contour of the released fuel at ground level. Thermal radiation intensity on a body outside the flame can be determined by following the procedure given in Chapter 5 for pool fires.

Example 7.1

Input data of this example are given below:

Rate of release of propane vapor: 10 kg/s (point source)

Atmospheric stability: Category F, wind speed: 2 m/s

NOTE: Data are identical to those for Example 11.5. This is because equations from Chapter 11 are used to determine the shape of the LFL contour, in the x-y plane, resulting from a continuous release of propane vapor at ground level. These results are reproduced below as these are required for demonstrating the flash fire calculation procedure.

LFL (0.038 kg/m^3 propane in air) contour:

Downwind distance, x, m	100	125	140	150	200	235
Crosswind distance, $\pm y$, m	6.8	7.2	7.2	7.1	5.5	0

Downwind 150 m from the source, and at the axis of the cloud ($y = 0$), the fuel concentration at ground level is 0.077 kg/m^3. At that location, the estimated concentration drops to the LFL at a height z of about 4 m.

For estimating fatalities from a flash fire, the area enclosed by the LFL contour at ground level is estimated using Equation (11.19). In Example 11.5, this is estimated as 5870 m^2. The flammable mass in the cloud, estimated by Equation (11.18), is 437 kg, or approximately 440 kg. This mass is required when the consequence of a vapor cloud explosion needs to be estimated. The number of fatalities resulting from a flash fire can now be estimated by multiplying the area and the population density enclosed within the LFL contour.

As shown in Figure 7.2, the release occurs at the point where the x and y coordinates are both zero. Ignition is assumed to occur at the downwind tip of the LFL contour, where the value of x is 235 m (ignition could occur at other locations as well). Following ignition, the flame travels upwind at a velocity of $4.6 - 2 = 2.6$ m/s relative to the ground. Therefore, the approximate total time of travel of the flame following ignition is $235/2.6 = 90$ seconds.

A person is assumed to stand stationary at a target location 50 m downstream of the ignition point, as shown in the diagram. He would receive thermal radiation at time-varying rates as the two-dimensional flame travels upwind. At an intermediate location 150 m downstream of the release point, the plan view of the

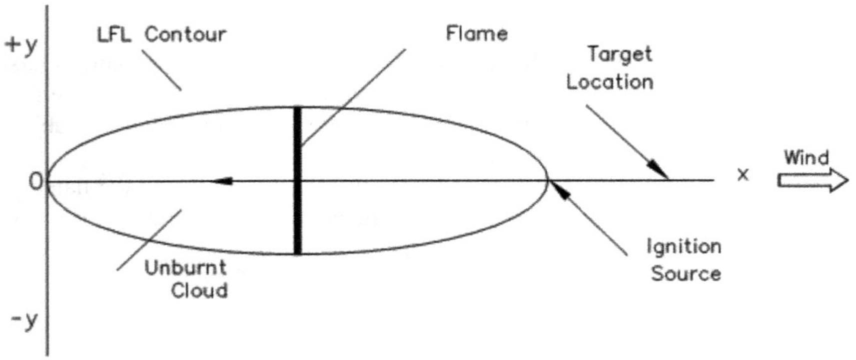

FIGURE 7.2 Plan view of the area affected by propane release.

flame is shown by a thick line. The distance between this flame and the person is $235 - 150 + 50 = 135$ m. At this location, the flame width would be the total cross-wind width of the LFL contour, i.e., approximately 14 m. The approximate height of the unburned cloud, d, at this location would correspond to the value of z of the LFL contour at $x = 150$ m and $y = 0$, i.e., approximately 4 m.

Therefore, with a propane release rate of 10 kg/s and a wind speed of 2 m/s, the average concentration of propane in the air is 10/[(14) (4) (2)] = 0.089 kg/m^3. At an ambient temperature of 298 K, this concentration corresponds to a mole fraction in the air of (0.089/44) (22.4) (298/273) = 0.05 approximately. Hence, for use in Equation (7.6), $\varphi = 0.05$. It can be shown that for a stoichiometric mixture of propane in air, the mole fraction φ_{st} is 0.04.

The flame height can now be calculated as follows:

From Equation (7.4), $S = 4.6$ m/s
From Equation (7.6b), parameter $w = 0.0013$
From Equation (7.7), $r = 15.8$
From Equation (7.8), $\rho_o/\rho_a = 1.026$
Unburned cloud height, $d = 4$ m

Substituting these values in Equation (7.5), the flame height $H = 45$ m.

For calculating the radiation intensity of the flame on the person, the view factor is calculated using the equivalent radiator model, Equations (5.24a–d), in Chapter 5. The calculated value is 0.01. The transmissivity of radiation can be calculated using Equations (5.27) and (5.28); for this problem, a value of 0.7 has been assumed.

The surface emissive power (SEP) is often quoted to be 170–230 kW/m^2. At an average SEP value of 200, the intensity on the person is estimated to be (200) (0.01) (0.7) = 1.4 kW/m^2, well within the recommended upper limit of 4.5 kW/m^2.

This example shows a simple method for calculating the consequence of flash fires. The methodology is subject to the simplifying assumptions involved in modeling flame height and in dispersion modeling. The estimated consequence should, therefore, be regarded as an approximate order-of-magnitude estimate, although it is good enough for risk analysis and development of site emergency plans.

REFERENCES

1. Raj, P.P.K. and Hammons, H. W.: *On the Burning of a Large Flammable Vapor Cloud* (Paper presented at the joint technical meeting of the Western and Central State Sections of the Combustion Institute, San Antonio, TX, April, 1975).
2. Mannan, S.: *Lees' Loss Prevention in the Process Industries* (4th Ed., Butterworth-Heinemann, Oxford, 2012).
3. Rew, P. J., Deaves, D. M., Hockey, S. M., and Lines, I. G.: *Review of Flash Fire, HSE Contract Research Report No. 94, Modelling* (1996).
4. AIChE/CCPS: *Guidelines for Vapor Cloud Explosion, Pressure Vessel Burst, BLEVE and Flash Fires* (July, 2010).
5. AIChE/CCPS: *Guidelines for Use of Vapor Cloud Dispersion Models* (1996).

8 Fireball

The term "fireball" refers to the atmospheric burning of a fuel-air cloud in which the shape of the flame is roughly spherical. A typical example is a rising fireball following ignition of the vapor cloud released to the atmosphere from a "boiling liquid, expanding vapor explosion" (BLEVE) in a vessel containing a liquefied gas under pressure. Two examples of such fires (Feyzin and Mexico City) were covered in Chapter 2.

Although a BLEVE is the predominant source of fireball accidents, other situations could give rise to a fireball. For example, if a flammable gas-containing pressure vessel bursts suddenly and releases a significant quantity of gas that ignites immediately, the result could be a fireball. A momentary flash associated with the explosion of a propellant or high explosive could also be considered a type of fireball.

However, the scope of treatment in this chapter has been restricted to BLEVE fireballs. The storage and transportation of liquefied flammable gases is an essential part of chemical plant operations. Understanding the measures necessary to avoid a potential BLEVE and the associated fireball is crucial.

This chapter deals with the BLEVE mechanism and methods for estimating a fireball's size and duration. The method for calculating thermal radiation intensity as a function of distance from the flame has also been covered. In the literature, the level of attention devoted to this case has been somewhat limited compared to pool fires or jet fires. While this treatment is simplified and empirical, its accuracy is considered adequate for hazard assessment.

8.1 BLEVE

A BLEVE occurs when there is a sudden loss of containment in a pressure vessel containing a superheated liquid or a liquefied gas. The primary cause of such an event is usually an external flame (from a pool fire or a jet fire) that impinges on the vessel's shell above the liquid level, thereby weakening the vessel and leading to its sudden rupture. The rupture may occur well before the pressure has reached the design pressure of the relief valve provided on the vessel.

As a precaution, a fire emanating from a relief valve should be treated as a critical warning sign of an imminent BLEVE. For example, a blocked-in pump in LPG service can result in a BLEVE by this mechanism.

Before the vessel ruptures, the liquid in it is in equilibrium with the saturated vapor above it. Upon rupture, the vapor is vented, and the pressure in the liquid drops sharply. Then, the liquid flashes at the liquid-vapor interface, the liquid-vessel-wall interface, and, depending on the temperature, throughout the liquid volume. If the liquid temperature is higher than its superheat limit temperature, a large fraction of the released liquid can vaporize within milliseconds. The limit to which a liquid may be heated (before spontaneous nucleation throughout the liquid gives rise to spontaneous vaporization) is called the superheat limit temperature.

DOI: 10.1201/9781003107873-8

Reid's[1] Equation (8.1) below is usually used to estimate the superheat limit temperature:

$$T_{sl} = 0.895\ T_c \tag{8.1}$$

where

T_{sl} = superheat limit temperature, K
T_c = critical temperature of the liquid, K

For example, the critical temperature for propane is 369.8 K. Therefore, its estimated superheat limit temperature from Equation (8.1) is 331 K (58°C). A measured superheat limit temperature for propane has been quoted as 53°C.[2]

The elapsed time from the start of an engulfing fire to BLEVE is a complex function of several factors such as the following:

a. The area of engulfment or impingement
b. The temperature of the flame
c. The extent of filling of the vessel.

For the Feyzin sphere in 1966, the time between the ignition of the leakage and vessel rupture was about 90 minutes. For other storage vessels, and road tankers in transport accidents, the time has been much shorter, from 5 to 30 minutes.[2] Therefore, water sprinklers or other cooling systems for the vessels should become operational within about 5 minutes following the detection of an external fire close to the vessels.

If the material released to the atmosphere upon vessel rupture is flammable, instantaneous ignition occurs, producing a fireball. Although there is some blast effect from the tank rupture, the main hazards are the thermal effects of the fireball and the mechanical damage from flying vessel fragments.

The mass of fuel in a fireball consists of the mass of vapor released because of the BLEVE, plus about an equal amount of entrained liquid. For example, if a vessel containing liquid propane at 30°C undergoes a BLEVE, the theoretical flash fraction would be about 39%. Hence, the estimated mass of vapor and entrained liquid in the fireball would be 78% of the tank's liquid. The entrained liquid vaporizes almost immediately. Therefore, the rule of thumb is that the mass of fuel in a BLEVE fireball equals the entire vessel inventory.

8.2 DIAMETER AND DURATION OF FIREBALL

Initially, as the flammable fluid is released, its momentum causes turbulence and in-mixing of air, resulting in rapid evaporation and combustion. As combustion continues, the fireball's increasing buoyancy causes it to rise vertically, and buoyancy-induced in-mixing of air becomes more significant. Further expansion of the fireball occurs, followed by shrinkage as combustion is completed.

There are many correlations for the maximum diameter of the fireball, but the one most widely used is as follows[3]:

$$D_{max} = 5.8M^{1/3} \tag{8.2}$$

where
D_{max} = maximum diameter of the fireball, m
M = mass of fuel in the fireball, kg

The vertical rise of the center of fireballs is known to vary approximately between 75% and 100% of the maximum fireball diameter.[4,5]

Fireball duration is usually correlated as follows.[3]

$$t_d = 0.45M^{1/3}, \quad M \le 30{,}000 \tag{8.3a}$$

$$t_d = 2.6M^{1/6}, \quad M > 30{,}000 \tag{8.3b}$$

where
t_d = fireball duration, seconds
M = mass of fuel in the fireball, kg

8.3 INTENSITY OF THERMAL RADIATION

8.3.1 Fractional Radiation

Based on the work of Roberts,[6] TNO[7] recommends the following equation for calculation of the fraction F_R of the combustion energy released through radiation:

$$F_R = 0.00325(P_{sv})^{0.32} \tag{8.4}$$

where
F_R = the combustion energy released through radiation, J
P_{sv} = saturated vapor pressure of the liquid before release, Pa

For example, if a vessel containing liquid propane is at 45°C (vapor pressure = 15 bar) just before it undergoes a BLEVE rupture, then the fractional radiation from the resulting fireball would be about 0.31.

CCPS,[4] quoting Hymes,[8] suggested the following values of F_R for fireballs from vessels:

0.3, when bursting below the relief valve set pressure and
0.4, when bursting above the relief valve set pressure.

8.3.2 Surface Emissive Power

The surface emissive power of a fireball used in radiation intensity calculations is usually based on the maximum radius of the fireball, the initial mass of fuel in the fireball, and the fireball duration:

$$\text{SEP} = \frac{M\Delta H_c F_R}{\pi D_{\max}{}^2 t_d} \tag{8.5}$$

where

SEP = surface emissive power, kW/m^2
M = initial mass of fuel in the fireball, kg
ΔH_c = net heating value of the fuel, kJ/kg
F_R = the combustion energy released through radiation, J
$D_{\max}$ = diameter of the fireball at the maximum surface area, m
t_d = fireball duration, seconds

8.3.3 View Factor

The geometric configuration of the fireball relative to an object at ground level is shown in Figure 8.1. *X* is the horizontal distance of the object at ground level from the vertical centerline of the fireball. H is the vertical height of the center of the fireball from ground level. For view factor calculations, TNO[7] follows the Bagster model,[9] which takes *H* equal to twice the maximum radius of the fireball.

For the configuration shown in Figure 8.1, and with *H* equal to twice the maximum radius of the fireball, the maximum view factor is given by[7]

$$F_v = \frac{R_{fb}{}^2}{X_r{}^2} = \frac{R_{fb}{}^2}{4R_{fb}{}^2 + X^2} \tag{8.6}$$

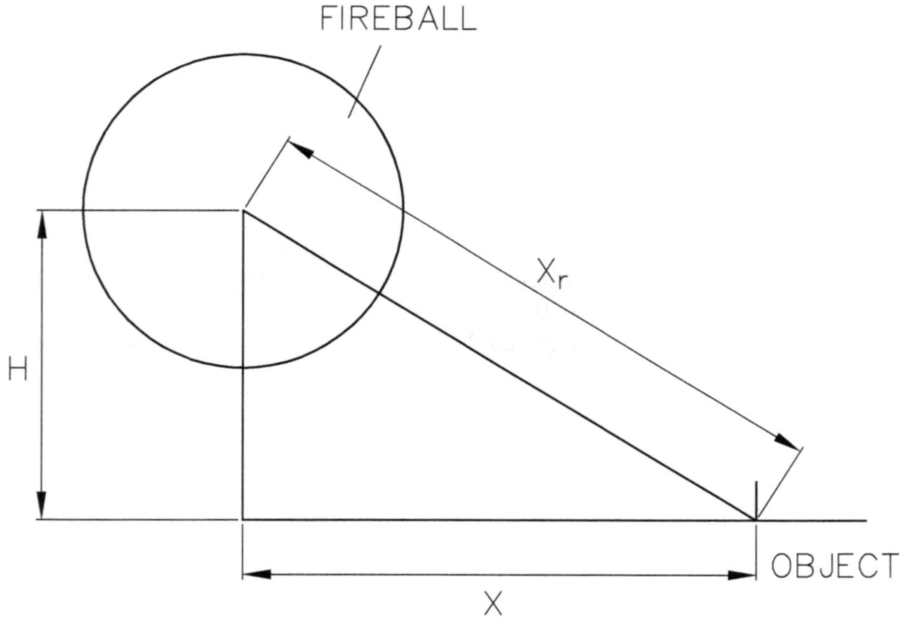

FIGURE 8.1 Configuration of a fireball relative to an object.

where

F_v = maximum view factor

R_{fb} = maximum radius of the fireball, m

X = horizontal distance of the object from the vertical centerline of the fireball, m

X_r = distance of the object from the centerline of the fireball, m

8.3.4 Atmospheric Transmissivity

The atmospheric transmissivity, τ, of thermal radiation from the fireball to an object can be calculated using Equation (5.27) in Chapter 5, the distance of the path length being equal to $X_r - R_{fb}$.

Example 8.1

A vessel containing 100 tons of liquefied propane at ambient temperature undergoes a BLEVE. The temperature of liquid propane just before vessel rupture is 45°C (propane vapor pressure, P_{sv} = 15.1 bar or 1.51×10^6 Pa).

The atmospheric temperature is 30°C at which the vapor pressure of water is 0.042 bar. The relative humidity is 70%. The heating value of propane is 46,000 kJ/kg.

Estimate:

a. the intensity of thermal radiation on a person standing at ground level at a horizontal distance of 300 m from the vertical centerline of the fireball, and
b. the effect of this radiation on the person.

PART (A)

Using Equation (8.2), maximum fireball diameter = $5.8(100{,}000)^{0.333}$ = 268 m

Using Equation (8.3b), duration of the fireball = $2.6(100{,}000)^{1/6}$ = 17.7 seconds

Using Equation (8.4), fractional radiation = $0.00325[(1.5 \times 10^6)^{0.32}]$ = 0.31

Using Equation (8.5), SEP = $(100{,}000)(46{,}000)(0.31)/[(3.14)(268^2)(17.7)]$ = 357 kW/m^2.

Fireball radius R_{fb} = 134 m

Horizontal distance X of the person from the vertical centerline of the fireball = 300 m

Therefore, from Equation (8.6), view factor = $134^2/[(4)(134^2) + 300^2]$ = 0.11

Distance between fireball center and the person $X_r = (268^2 + 300^2)^{0.5}$ = 402 m

Radiation path length = 402 – 134 = 268 m

Partial pressure of water vapor = (vapor pressure) (fractional humidity) = (0.042) (0.7) = 0.0294 bar, or 2,940 Pa.

From Equation (5.27), Transmissivity $\tau = 2.02\ [(2{,}940)(268)]^{-0.09}$ = 0.595

Hence, intensity of thermal radiation on the person = (357) (0.11) (0.595) = 23.4 kW/m^2.

PART (B)

Thermal load on the person (for an intensity of 23.4 kW/m^2 and exposure duration of 17.7 s) is $(23.4^{1.33})(17.7)$ = 1,185

For first-degree burns, from Equation (3.8a):

Probit = −12.03 + 3.0186 ln (1,185) = 9.33 (probability > 99%)

For a fatality, from Equation (3.8c), probit = −12.8 + 2.56 ln (1,185) = 5.32 (62% probability)

Hence, the estimated probability of fatality is about 60%, while that of first-degree burns exceeds 99%.

8.4 MEASURES TO PREVENT BLEVE

For preventing a BLEVE from a flammable liquefied gas under pressure, the vessel's shell above the liquid level must not be heated by an external fire to an unsafe temperature. The following methods may be used.

8.4.1 Cooling the Vessel by Water Deluge or Spray

The cooling water rate should be about 10 L/min/m^2 of the surface area of the vessel. The cooling system should become operational as quickly as possible, preferably within 5 minutes following fire detection. Once the sprays are operational, emergency personnel should retreat to a safe distance.

8.4.2 Insulation of the Vessel

Providing fire-resistant insulation, such as vermiculite concrete, on the vessel's outer surface and its supports will serve as an immediate barrier to heat input. The insulation material is required to withstand fire for a minimum of 90 minutes.

8.4.3 Providing an Earth Mound around the Vessel

Nowadays, earth-mounded vessels are used by many companies in preference to vessels with conventional insulation. Figure 8.2 shows a typical diagram for such a vessel. The vessel is completely covered with clean sand to avoid any corrosion from any impurities in the sand. Portions of the covering are removed whenever the outside of the vessel is inspected.

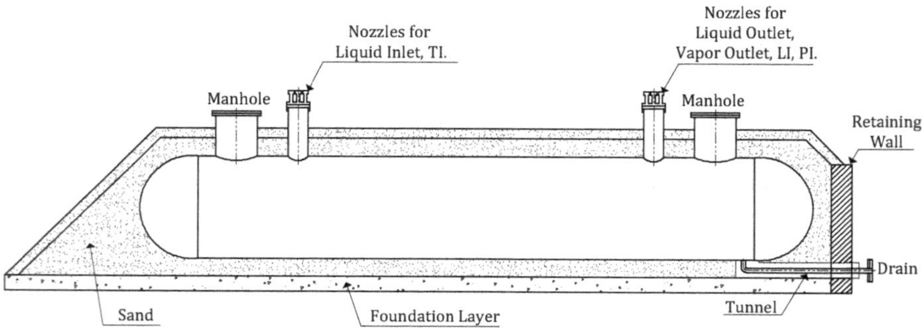

FIGURE 8.2 Schematic of an earth-mounded storage vessel.

8.5 MEASURES IN CASE OF IMMINENT BLEVE

On detection of any sign of an imminent BLEVE from a pressure vessel, such as a flame emanating from a relief valve vent, all personnel (including firefighters and emergency responders) must withdraw immediately to a safe distance. All efforts should be made to maintain a remote-controlled continuous water spray on the vessel surface.

REFERENCES

1. Reid, R. C.: Superheated Liquids, *American Scientist*, 64, pp. 146–156 (1976).
2. Mannan, S.: *Lees' Loss Prevention in the Process Industries* (4th Ed., Butterworth-Heinemann, Oxford, 2012).
3. Marshall, V. C.: *Major Chemical Hazards* (Ellis Horwood, Chichester, 1987).
4. AIChE/CCPS: *Guidelines for Chemical Process Quantitative Risk Analysis* (2nd Ed., AIChE, New York, 2000).
5. Loss Prevention Bulletin (No. 068): *The Effect of Explosions in the Process Industries by Overpressure Working Party, Major Hazards Assessment Panel* (The Institution of Chemical Engineers, April, 1986).
6. Roberts, A. F.: Thermal Radiation Hazards from Release of LPG from Pressurized Storage, *Fire Safety Journal*, 4, pp. 197–212 (1982).
7. TNO: *Methods for the Calculation of Physical Effects (Yellow Book)* (3rd Ed., CPR 14E, Parts 1 and 2, TNO, The Hague, 1997).
8. Hymes, I.: *SRD R275, The Physiological and Pathological Effects of Thermal Radiation* (U.K. Atomic Energy Authority, Culcheth, 1983).
9. Bagster, D. F. and Pitblado, R. M.: Thermal Hazards in the Process Industry, *Chemical Engineering Progress*, 85, pp. 69–75 (1989).

9 Explosion

In engineering terms, "explosion" refers to a sudden release of energy of moderate to violent intensity. The intensity of the explosion depends both on the quantum of energy released and the rate of release. TNO[6] defines explosion as a sudden release of energy that causes a rapidly propagating pressure or shockwave in the atmosphere with high pressure, high density, and high particle velocity.

9.1 KINDS AND TYPES OF EXPLOSIONS

There are several kinds of energy which may be released in an explosion:

a. Chemical energy, e.g., the energy released to the atmosphere in a vapor cloud explosion (VCE) caused by loss of flammable containment from a pressurized container, an uncontrolled chemical reaction in a reactor vessel, or detonating explosives for civil and military use
b. Physical energy released upon the bursting of a vessel under high internal pressure (e.g., when the vessel is suddenly connected to a pressure vessel operating at a higher pressure than its maximum pressure rating). Such malfunctions can occur for several reasons:
    i. Operator error
    ii. During a pneumatic pressure test when vessel pressure relief is inadequate
    iii. Excessive vessel corrosion or fatigue
    iv. Failure of the emergency shutdown system, or
    v. After a flawed maintenance shutdown, when critical vessel components were replaced improperly or welds not stress relieved. Failures to observe the restrictions imposed by applicable pressure vessel codes are distressingly common and have contributed to several major industrial accidents.
c. *Nuclear energy*: Nuclear explosions have not been considered in this book

Explosions in the process industries are of the following types:

1. Vapor cloud
2. Condensed phase
3. BLEVE
4. Chemical reactor
5. Dust
6. Physical

This chapter first introduces the types of VCE mechanisms and available mathematical models to describe their effects. VCE accidents are a significant cause of concern in the oil, gas, refining, and petrochemical process industries. Estimation methods

DOI: 10.1201/9781003107873-9

for blast damage, with examples and ways for prevention and mitigation, have also been included. We then describe condensed phase explosions, which are generally of interest in the explosives manufacturing industry. Finally, we introduce the other types of explosions listed above, except for boiling liquid, expanding vapor explosions (BLEVE), a phenomenon of great concern in industries handling highly flammable gases, that has already been discussed in Chapter 8.

9.2 EXPLOSION MECHANISMS

9.2.1 Deflagration

A deflagration occurs when, in an explosion of a mixture of fuel and air, the velocity of flame propagation is subsonic (i.e., less than sonic velocity). Deflagrations occur both in gas and VCEs. Figure 9.1a shows a typical pressure wave build-up for a deflagration process.[1] A vapor cloud can be created by either gas releases or by the evaporation of a liquid released from containment. After ignition, the ignition source generates a flame front that causes the temperature to rise quickly. Because of the expansion of the gases in a fast-moving flame front, a pressure build-up occurs.

TNO[6] defines "deflagration" as a propagating chemical reaction of substances in which the propagation of the reaction front is determined by conduction and molecular diffusion.

During deflagration, the flame front advances into the unreacted material at subsonic velocities (generally below 350 m/s in ambient air). In an unconfined, uncongested environment, flame speeds rarely exceed 20–25 m/s. Deflagration in open environments results in the rapid burning of the vapor cloud for a prolonged period, causing little or no over-pressure. This phenomenon is generally referred to as a "vapor cloud fire" (see Chapter 7), or more commonly as a "flash fire" that can cause burn injuries and fire damage. It generally does not cause any structural damage by "overpressure" and is far less destructive.

Vapor cloud deflagrations in a confined and highly congested environment, or high-pressure jet releases that result in turbulent mixing into an open environment, are particular concerns. Turbulent flame speeds can cause considerable peak over-pressure and, thereby, significant damage. Flame speeds over about 100 m/s are necessary to develop significant over-pressure effects in deflagration.

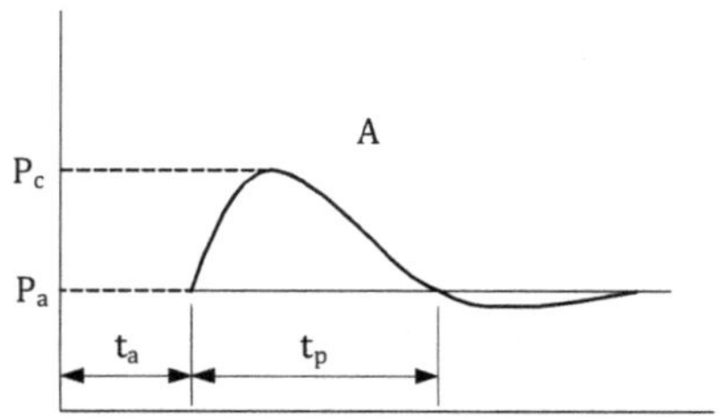

FIGURE 9.1a Typical shape of a pressure wave: deflagration.

Figure 9.1a shows the pressure profile of a deflagration starting after a time t_a, with pressure gradually rising from ambient pressure P_a to a peak value P_c. After that, it decays during the "positive phase duration", t_p.

9.2.2 Detonation

A detonation occurs mostly with condensed (high) explosives and, occasionally, with VCEs that undergo a "deflagration-to-detonation" (DDT) transition. DDTs occur under unusual circumstances and have been witnessed recently in some devastating industrial accidents.[2,3] Propagating flame speeds in detonation are not only supersonic but can be up to several times sonic velocity.

TNO[6] defines "detonation" as a propagating chemical reaction of a substance in which the propagation of the reaction front is determined by compression beyond the auto-ignition temperature.

Figure 9.1b shows a shock wave, where the pressure build-up, P_c, is virtually instantaneous but decays over the positive phase lasting for time t_p.

The shock wave results from a static detonation followed closely by a combustion wave that releases the energy required to sustain the shock wave. The velocity of the detonation flame front can reach six to eight times sonic velocity. A typical detonation may result in a peak pressure of the order of 20 bar in a closed vessel compared to a peak pressure of around 8 bar that might result from the DDT of a hydrocarbon–air mixture.

There are two possible scenarios for a detonation to take place:

a. A detonation wave being initiated by explosives and
b. A DDT transition in a VCE.

A direct detonation initiation requires extremely high ignition energy in the premixed explosive mixture. Ignition energies of about 10^6 J would lead to instantaneous detonations. For deflagrations, however, the ignition energy is of the order of 10^4 J.[6]

The detonation (for condensed explosives) is initiated directly and instantaneously. Its propagation is self-sustained with a very high detonation velocity. Direct or static detonations are typical with condensed explosives, as discussed in detail in Section 9.4.

For a shock wave, the positive phase duration is followed by a negative pressure phase of a smaller magnitude. As the shock front expands, pressure decays back to

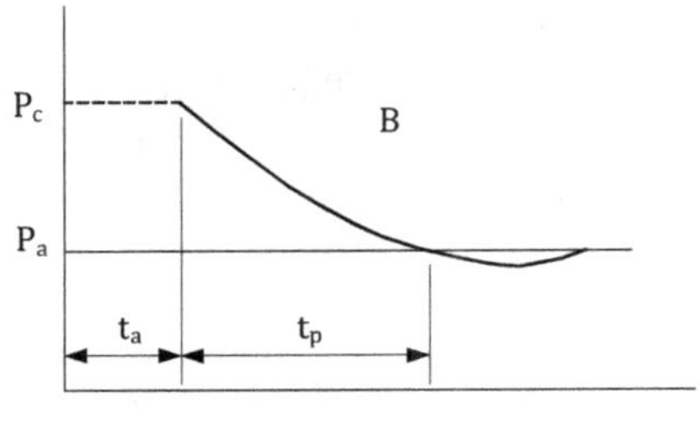

FIGURE 9.1b Typical shape of a pressure wave: detonation.

ambient pressure, and a negative pressure phase occurs that is usually longer in duration than the positive phase. For design work, the negative phase is usually taken as being less important than the positive phase. However, this approach is not always appropriate. There have been instances where walls designed to handle the detonation's positive pressure phase collapsed inward because the designer ignored the negative phase.

9.2.3 DDT

A "DDT" transition is caused by an ignition source of low ignition energy. A laminar flame is formed during the initial stages. If the flame front encounters confinement and congestion, or a flammable mixture forms because of a high-pressure jet release, the flame front accelerates quickly. As it moves through the fuel mixture, it turns into a turbulent combustion flame. Recent studies and research[2,3] have established that the catastrophic VCE explosions in Buncefield, U.K. (2005),[3,4] Jaipur, India (2009),[5] and Puerto Rico, USA (2009)[2] were all caused by the DDT phenomenon.

For the types of accidents mentioned above, a realistic prediction of DDT blast effects is crucial in the process industries. Detonations caused by very high energy VCEs have massive destructive power. This problem is a complex, challenging scientific area because of the nonlinear interactions among the underlying physical processes such as turbulence, shock interactions, and energy release.

In the first decade of the 21st century, detailed investigations, field experiments, and accompanying research have found that the catastrophic consequences of major VCEs have been caused by DDTs. It is now accepted that factors such as the following can all contribute to this most intense and dangerous detonation phenomenon[2,3]:

i. Congestion,
ii. Confinement
iii. Atmospheric conditions
iv. Jet releases causing turbulent mixing and
v. Rich fuel-air mixtures (even with low-energy ignition initiation)

9.3 VCE

A VCE can occur after an accidental release of flammable gas or vapor from a vessel or containment followed by ignition.

After the accidental release of pressurized vapors/gases from a leaky joint, or the rupture of a pipeline, a VCE may occur. The released vapor disperses into the air and forms a flammable mixture that reaches a flammable range soon after the release. If this mixture is ignited quickly, the result would either be a jet fire emanating from the release point or a pool fire close to it. However, if the release does not soon encounter an ignition source, as is common in process plants, it would form a vapor cloud that keeps spreading if the leak continues.[25] Upon meeting an ignition source, this large cloud would catch fire at its edges. The flame would then propagate backward through the vapor cloud, igniting the flammable mixture on its way and result in a VCE deflagration or detonation.

To summarize, a VCE occurs if the following conditions are satisfied.

- The flammable mixture in the vapor cloud is within the flammability limit
- The cloud encounters confinement in a highly congested environment
- Even in areas with minimal or no congestion, provided the flammable mixture is highly reactive, jet releases of unusually long duration that release large quantities of flammable material can create huge vapor clouds. Flames propagating through these clouds generate a very high degree of turbulence.

Several free and commercial mathematical models are available for estimating the damage potential of a VCE. They are all empirical or semi-empirical, and all of them are based on the explosion energy of a flammable vapor cloud. Models commonly used to determine VCE effects (with worked examples) are listed below and explained in detail with examples worked out. More complex models in use commercially are based on computational fluid dynamics (CFD), and these are also mentioned. Details of such modeling are, however, beyond the scope of this book.

9.3.1 TNT Equivalent Model

The TNT equivalent model is the simplest of all models, easy to use, and useful for quick estimations, despite some of the limitations mentioned in Section 9.3.7. The method requires calculating an "equivalent mass" of TNT that could generate a shock wave equivalent to that from a fraction of the energy released during an explosion of the cloud. This fraction is called a "yield factor", η. It is also called an equivalent factor, expressed as a percentage.

The equivalent mass of TNT is obtained from Equation (9.1) below:

$$W_{Eq} = \frac{\eta M_f E_f}{E_{\text{TNT}}} \tag{9.1}$$

where

W_{Eq} = the "equivalent mass" of TNT, kg
M_f = he mass of flammable substance in the cloud, kg
E_f = he energy of explosion of the flammable material, kJ/kg
E_{TNT} = he energy of explosion of TNT (approximately 4,600 kJ/kg)
η = he yield factor

The yield factor (η) depends mainly on the fraction of the total flammable mass that lies between the flammable limits and the fraction of the total cloud volume in the congested region. Based on studies of several actual explosions, it has been suggested that a value of 0.04 for the yield factor should be used for hydrocarbons and other common flammable substances. For highly reactive gases such as ethylene oxide, higher values around 0.10 are recommended.[8]

The next step is to calculate the scaled distance, a parameter for estimating the over-pressure at a given distance from the cloud center. The scaled distance (m/kg$^{1/3}$) is determined from Equation (9.2) below:

$$Z = \frac{X}{W_{Eq}^{\;1/3}} \tag{9.2}$$

where

X = distance of the target from the center of the explosion, m
Z = scaled distance, m/kg$^{1/3}$
W_{Eq} = he equivalent mass of TNT, kg

As for condensed phase explosives, the next step is to estimate the overpressure, using Equation (9.3) below:

$$P^o = 1{,}128\ Z^{-2.019} \quad Z < 5$$

$$P^o = 427Z^{-1.411} \quad Z \geq 5 \tag{9.3}$$

where

Z = scaled distance, m/kg$^{1/3}$
P^o = overpressure, kPa

For convenience, the scaled peak overpressure, P^o, is plotted against scaled distance, Z, in Figure 9.2 that is based on Equation (9.3).

Figure 9.2 is based on a publication by the Major Hazards Assessment Panel, U.K.[8] The Y-axis is the peak overpressure P^o, kPa. The X-axis is the scaled distance Z, m/kg$^{1/3}$, in Equation (9.2). The numerator X is the distance, m, from the center of the explosion, and the denominator W_{Eq} is the quantity, kg, of TNT raised to the one-third power. For example, for an explosion involving 1,000 kg of TNT, the scale factor at a distance of 150 m is 150/1000$^{1/3}$ = 15 m/kg$^{1/3}$.

Example 9.1

A road tanker carrying 30 tons of liquefied propane at 298 K (T_p) meets with an accident, spilling its contents. The resulting vapor cloud explodes. Determine the overpressure, and estimate the damage at a distance of 200 m from the explosion center.

Data: Boiling point of liquid propane at atmospheric pressure = 231 K (T_{amb}), latent heat of vaporization, L_p of propane = 426 kJ/kg, specific heat of liquid propane, C_p = 2.45 kJ/(kg.K), heat of combustion of propane, $E_{f\,=}$ 46,000 kJ/kg.

$$\text{Fractional flash} = 1 - e^{\left[-C_p\left(T_p - T_{amb}\right)/L_p\right]} = 1 - e^{\left[-(2.45)(298-231)/426\right]} = 1 - 0.68 = 0.32$$

The flash vapor will entrain with it a large fraction of liquid droplets in the form of a mist or spray. The mass of entrained liquid is usually taken as twice the flash

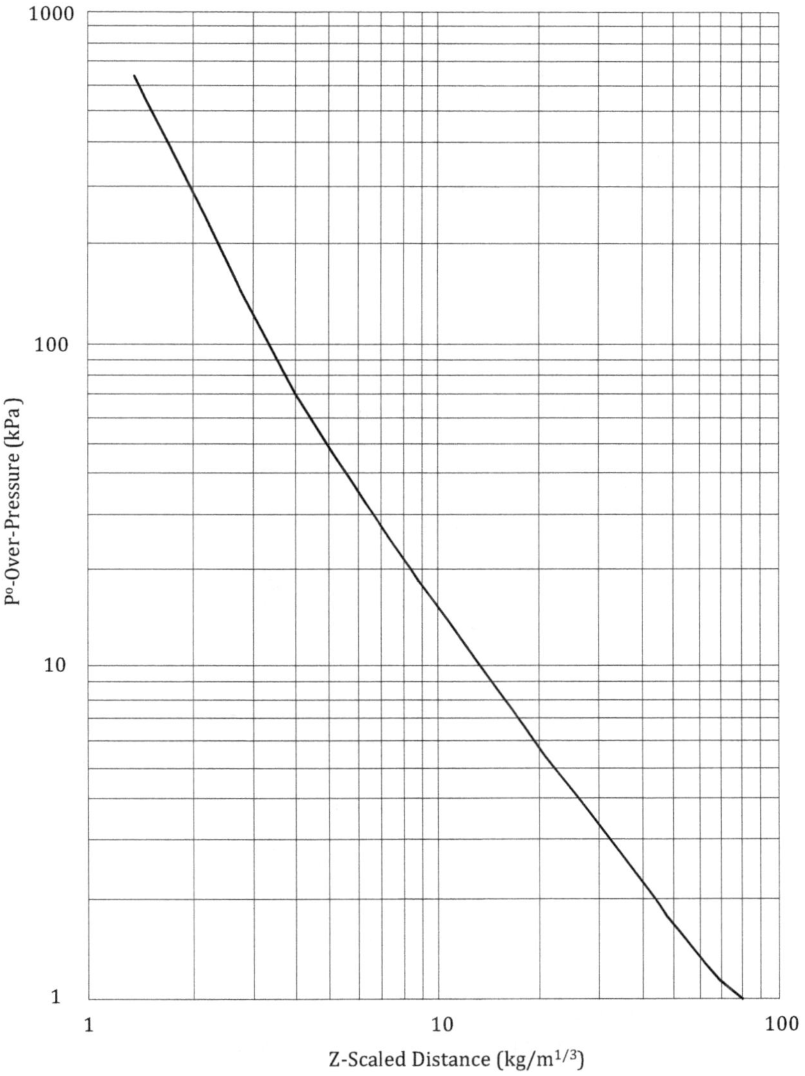

FIGURE 9.2 Peak overpressure, P^o, vs. scaled distance "Z" for TNT explosion.

vapor generated. Therefore, the estimated mass of propane in the vapor cloud would be estimated as

$$Q_{\text{ext}} = (30{,}000)\,(2)\,(0.32) = 19{,}200 \text{ kg}$$

From Equation (9.1), $W_{Eq} = (0.04)(19{,}200)(46{,}000)/4{,}600 = 7{,}680$ kg

From Equation (9.2), scaling distance, $Z = 200/7{,}680^{1/3} = 10.2$ m/kg$^{1/3}$

From Equation (9.3), at Z = 10.2, the value of over-pressure is calculated to be {427} {$10.2^{(-1.411)}$}, or 16.1 kPa. Referring to Figure 9.10 and Table 9.9, the damage would be significant. A typical U.S. house (item 14 in Table 9.8) would suffer severe damage (ref. Figure .9.10), and so will a typical British house (ref. Table 9.9). Houses will become uninhabitable, with a partial or total collapse of the roof and severe damage to external load-bearing walls.

The calculation method for the mass of flammable material in the above example is for a batch release. If the release to the atmosphere occurs continuously, for example, from evaporation from a liquid pool, the flammable mass in the cloud must be calculated by dispersion modeling. This approach has been discussed in Chapter 11.

The principal limitations of the TNT equivalent model are as follows:

i. arbitrariness in the choice of the value for yield factor, or the TNT equivalence factor (η), and
ii. unrealistic prediction of overpressure in the near field.[8]

9.3.2 TNO Correlation Model

The TNO correlation model[10] is an empirical model for estimating the damage circle radius for four defined categories of damage:

$$R(s) = C(s)\,(\eta\; E)^{1/3} \tag{9.4}$$

where

$R(s)$ = maximum radius for a defined category of damage, m
$C(s)$ = constant characteristic of the damage, m/J$^{1/3}$
η = yield factor (= 0.1)
E = energy of combustion of the total quantity of fuel in the cloud, J

The TNO method gives four categories of damage and the corresponding values of $C(s)$ as follows:

Damage	C(s)
Heavy damage to buildings and processing equipment	0.03
Repairable damage to buildings and façade damage to dwellings	0.06
Glass damage, causing injury	0.15
Glass damage (10% of panes)	0.40

The method is applicable for the energy of combustion, E, between 5×10^9 J (approximately 100 kg of a hydrocarbon) and 5×10^{12} J (approximately 100 te of a hydrocarbon). In this case, the hydrocarbon is propane.

Example 9.2

In earlier Example 9.1, calculate the maximum radius of the damage circle for each of the categories of damage defined in the TNO correlation.

The estimated mass of propane in the vapor cloud Q_{ext} = 19,200 kg (Example 9.1)

Combustion energy of propane $E_f = 46{,}000$ kJ/kg (Example 9.1)

$E = Q_{ext} E_f = (19{,}200)(4.6)(10^7) = 8.8 \times 10^{11}$ J, which is within the limits of applicability of the TNO correlation model. Calculated radii are as follows:

Damage	R(s)
Heavy damage to buildings and processing equipment	$(0.03)(0.1)(8.8 \times 0^{11})^{1/3} = 132$ m
Repairable damage to buildings and façade damage to dwellings	$(0.06)(0.1)(8.8 \times 10^{11})^{1/3} = 265$ m
Glass damage causing injury	$(0.15)(0.1)(8.8 \times 10^{11})^{1/3} = 662$ m
Glass damage (10% of panes)	$(0.40)(0.1)(8.8 \times 10^{11})^{1/3} = 1{,}770$ m

The TNT correlation model can be considered an extension of the equivalence model, where an equivalence factor of 10% (or yield as 0.1) is considered. Both are based on a TNT blast.

A TNT blast is inadequate for modeling VCEs. A TNT explosion produces a shockwave of a very high amplitude but a very short duration, whereas a VCE produces a blast wave or shock wave of lower amplitude and longer duration.

9.3.3 TNO Multienergy Model

The TNO multi-energy model (MEM) was first developed by van den Berg.[11] The method assumes VCE can occur only when flammable vapor is partially confined; only the combustion energy of that part of the flammable cloud that is confined or obstructed contributes to pressure generation in a VCE. Combustion energy and the reactivity of the fuel-air mixture both contribute to blast strength. The remaining parts of the vapor cloud simply burn as a flash fire without causing any significant over-pressure.

The amount of energy released during a VCE is limited either by

- The volume of the partially confined portion of the flammable vapor cloud (if the flammable vapor cloud is larger than the partially confined region) or
- The volume of the vapor cloud (if the vapor cloud is smaller than the volume of the portion of the partially confined region)

This model assumes that a blast of steady flame speed is produced by a hemispherical cloud of stoichiometric hydrocarbon-in-air composition.

A vapor cloud may have several blast regions that would cause a series of sub-explosions corresponding to various confined and unconfined regions. This phenomenon can occur when a vapor cloud is present at a site with many buildings and plants, each in a confined or congested region.

The blast from each source must be modeled as a blast from an equivalent hemispherical fuel-air charge of volume $V_{c\,=} E/E_v$ ($E_v = 3.5$ MJ/m^3, Equation 9.6). The regions can be very close or be separated by considerable distances. If separated, blast energies need to be considered as additional blast sources for each of the obstructed regions. The center of this blast source would be calculated as an "energy-weighted average location" after working out each of the separate blast force centers (r_0), using Equation (9.9).

The key aspect is the blast or charge strength that ranges from 1 (insignificant) to10 (detonation strength). Each curve represents a blast or charge strength representing the extent of confinement and congestion. MEM blast curves consist of scaled blast wave properties (peak pressure and positive phase duration versus scaled distance). These are presented in families of ten curves with source strength (over-pressure) as the parameter.

These pressure parameter plots are shown below in Figures 9.3 and 9.4 and the impulse parameter plot in Figure 9.5. After selecting one of the ten blast curves, the blast parameters, at any scaled distance R_o from the ignition source, can be obtained using these three figures, based on the scaled distance R_s'.

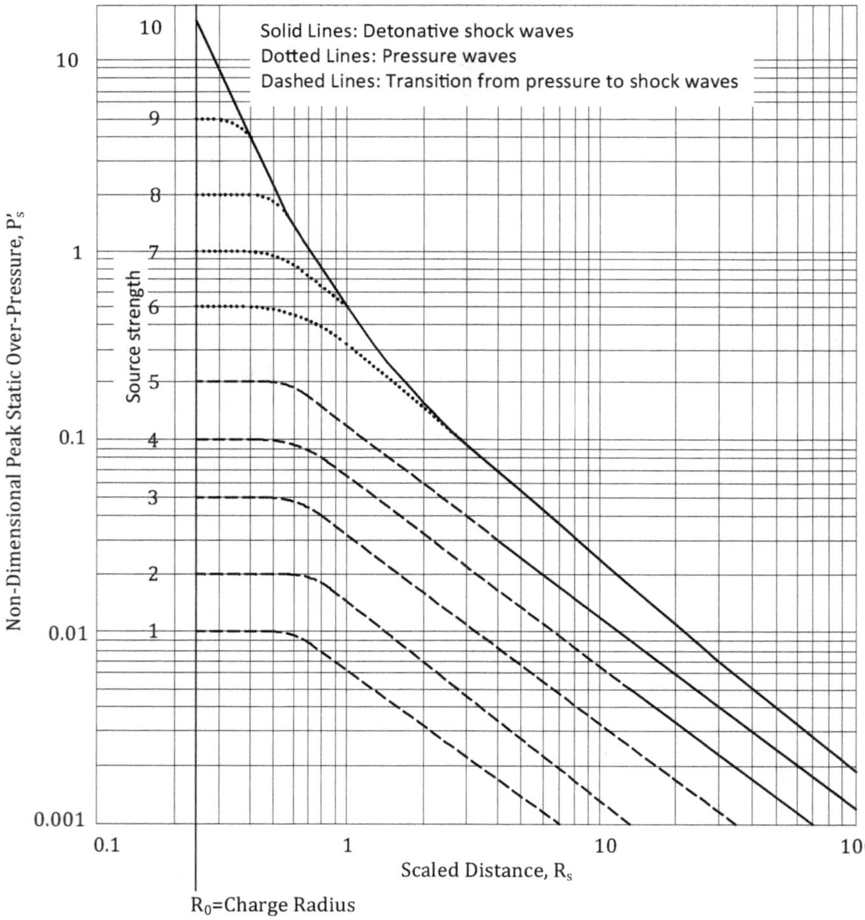

FIGURE 9.3 MEM blast chart: peak static overpressure vs. scaled distance.

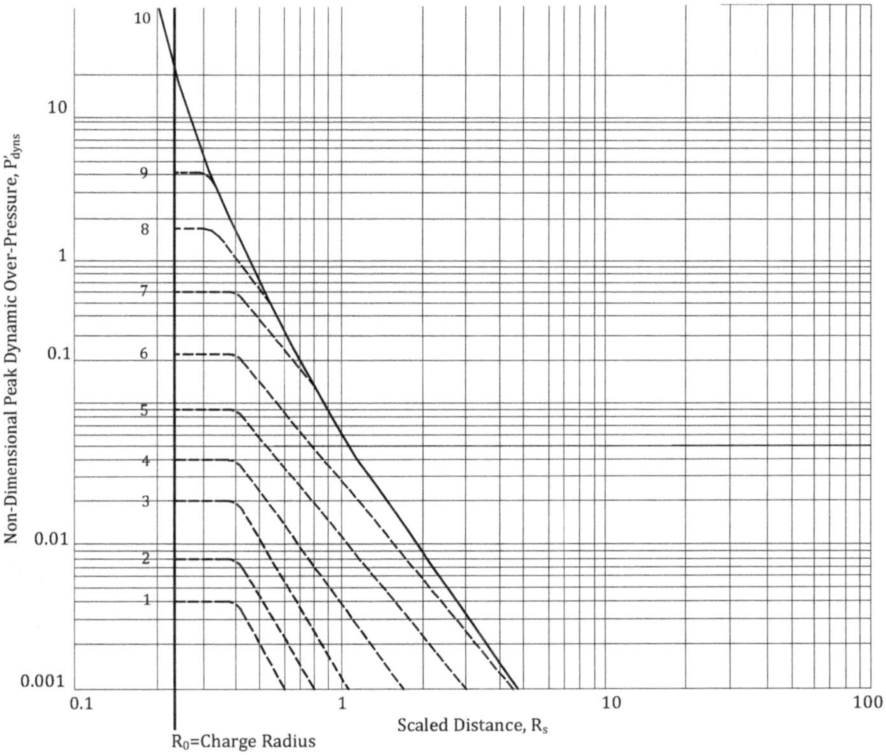

FIGURE 9.4 MEM blast chart: peak dynamic pressure vs. scaled distance.

Blast charts are provided[10] for VCEs with a combustion energy E_v of 3.5 MJ/m^3, valid for most hydrocarbon mixtures in stoichiometric concentration with air.

In Figures 9.3–9.5, high-strength shock wave blasts are represented by solid lines. Pressure waves of low initial strength are indicated by dotted lines that may steepen to shock waves in the far-field. In between, there is a transition from pressure waves to shock waves, indicated by dashed lines.

The exercises involved in an MEM are as follows:

a. Determine the location of the vapor cloud by a dispersion study (Chapter 11)
b. Identify factors for flame acceleration, confinement (1, 2, or 3-dimensional), depending on the plant or site conditions, congestion or obstruction (plant equipment, buildings, car parks) and high-pressure jet release
c. Cloud size (volume) to be determined for each zone:
   1. Assign portions of the cloud to different confined and unconfined areas
   2. Determine fuel volume V_c (m^3) of fuel-air mixture present in each zone as follows:

$$V_c = Q_{\text{ext}} / (\rho / c_s) \tag{9.5}$$

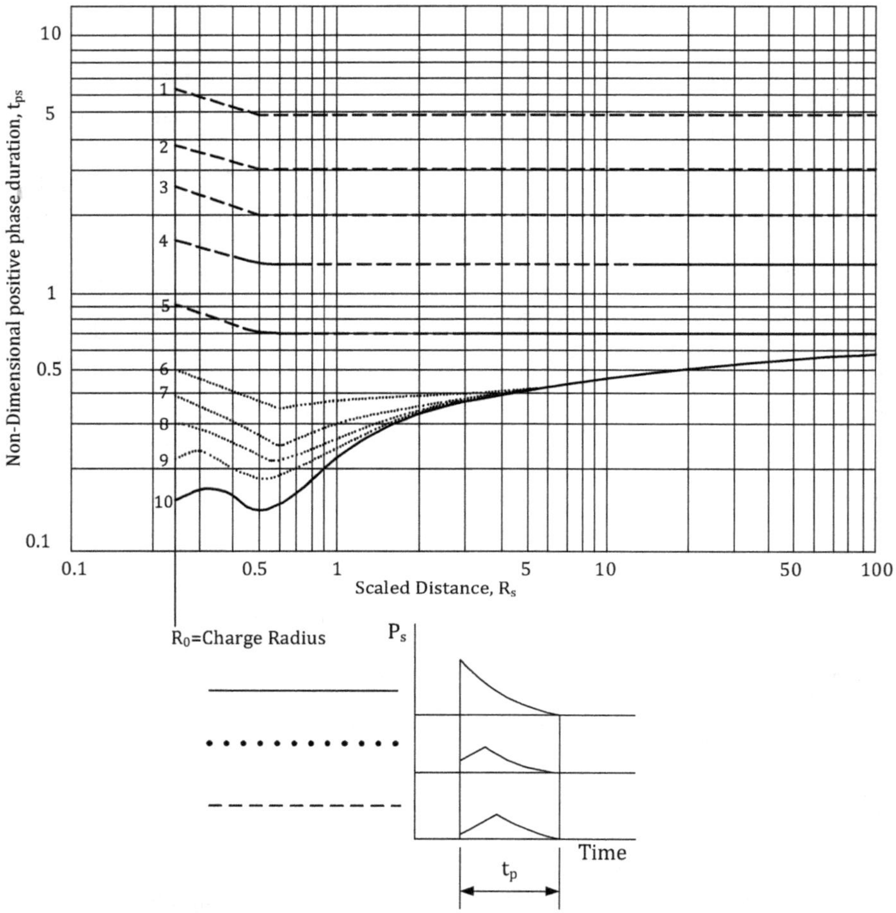

FIGURE 9.5 MEM blast chart: positive phase duration and blast-wave shape.

where:

V_c = cloud volume in each zone, m^3

Q_{ext} = quantity of Inflammable gas/vapor, kg

ρ = density of gas/vapor, kg/m^3

c_s = he stoichiometric concentration, %

d. Determine the "cloud energy" or "source strength" of each region:

$$E = V_c E_v \tag{9.6}$$

where

E = he energy of the explosion, J

V_c = cloud volume in each zone, m^3

E_v = combustion fuel energy (3.5 MJ/m^3 at stoichiometric concentration for most hydrocarbons)

e. Assign blast or charge strength to each of the zones based on their degree of confinement – Refer to Table 9.3 or Table 9.4.
f. Calculate "scale distance" (R_s) for blast charge at each region:

$$R_s = R_o \Big/ \left(E/P_a\right)^{1/3} \quad (9.7)$$

where
- R_s = scale distance for blast charge at each region
- R_o = arget distance from the center of the explosion, m
- P_a = atmospheric pressure, Pa

g. Refer to Figure 9.3 (plot of P_s' vs. R_s) and Figure 9.4 (plot of P_{dyn}' P_a vs. R_s) and determine the following:
   1. Scaled peak static overpressure P_s' for each zone and
   2. Scaled peak dynamic overpressure P_{dyn}' for each zone
h. Calculate the peak static and dynamic pressures as follows:

$$\text{Peak static overpressure } P_s = P_s' P_a \quad (9.8a)$$

$$\text{Peak dynamic overpressure } P_{\text{dyn}} = P_{\text{dyn}}' P_a \quad (9.8b)$$

i. If the blast zones are close to each other and ignitions are almost simultaneous, a VCE can be considered to be simultaneous, and over-pressure is superimposed at the target distance. If a distance separates the zones, the effect of each is considered as explained earlier.
j. Refer to Figure 9.5 (plot of t_{ps} vs. R_s), and determine the scaled positive phase duration for each zone.
k. Calculate t_p, the positive phase duration period, as follows:

$$t_p = t_{ps}\, C_0 \Big/ \left(E/P_a\right)^{1/3} \quad (9.8c)$$

where:
- t_p = he positive phase duration of blast-wave, seconds
- t_{ps} = he scaled positive phase duration of blast-wave, seconds
- C_0 = he velocity of sound in ambient air, m/s
- E = he energy of the explosion, J
- P_a = he atmospheric pressure, Pa

l. Determine the blast cloud radius as

$$R_o = \left[3/2E/\left(E_v p\right)\right]^{1/3} \quad (9.9)$$

where
- R_o = distance of the receptor from the center of the explosion, m
- E = he energy of the explosion, J
- P_a = he atmospheric pressure, Pa
- t_p = he positive phase duration of blast-wave, seconds
- t_{ps} = he scaled positive phase duration of blast-wave, seconds
- C_0 = he velocity of sound in ambient air, m/s

For assessing blast damage, in addition to the value of the over-pressure, the magnitude of the impulse is required. The impulse, I_s, is taken to be approximately equal to half of the product of peak overpressure and positive phase duration:

$$I_s = \tfrac{1}{2} P_s t_p \tag{9.10}$$

where

I_s = Impulse, Pa.s
P_s = Peak overpressure, Pa
t_p = Positive phase duration, seconds

Note that the positive phase duration period becomes nearly independent of its initial strength at a certain distance and at higher strengths (between blast curves 6 and 7).

The graphs in Figure 9.3 and 9.5 have been converted to equations to facilitate calculations using a computer or a calculator, as shown below.

The nondimensional scaled peak static overpressure versus scaled distance for selected source strength values can be calculated using the equations in Table 9.1. Also, values for nondimensional positive phase duration versus scaled distance, for selected source strength values, are given in Table 9.2.

The multienergy model, or MEM, is useful as a quick exercise for estimating peak overpressure and use as a guide for safe distances between equipment, facilities, plants, and "drag load" on equipment. By itself, it lacks clear guidance on the selection of the charge strength curve of required severity.[11] It merely states the factors to be considered:

TABLE 9.1
Equations for Nondimensional Peak Overpressure in MEM

Source Strength	Range for R_s	Equation for P_s'
10	$R_s \geq 3$	$P_s' = 0.3403\ R_s^{-1.162}$
	$R_s < 3$	$P_s' = 0.558\ R_s^{-1.8017}$
7	$R_s \geq 3$	$P_s' = 0.3403\ R_s^{-1.162}$
	$1 < R_s < 3$	$P_s' = 1/(4.457\ R_s^{-2.373})$
	$0.5 < R_s < 1$	$P_s' = 0.48\ R_s^{-1.045}$
	$R_s < 0.5$	$P_s' = 1$
6	$R_s \geq 3$	$P_s' = 0.3403\ R_s^{-1.162}$
	$0.7 < R_s < 3$	$P_s' = 0.32\ R_s^{-1.1}$
	$R_s < 0.7$	$P_s' = 0.5$
5	$R_s \geq 4$	$P_s' = 0.1247\ R_s^{-1.032}$
	$0.6 < R_s < 4$	$P_s' = 0.12/R_s$
	$R_s < 0.6$	$P_s = 0.2$
3	$R_s \geq 1$	$P_s' = 0.0324\ R_s^{-0.983}$
	$0.6 < R_s < 1$	$P_s = 1/10^{(1.0176+0.487 R_s)}$
	$R_s < 0.6$	$P_s' = 0.05$

TABLE 9.2
Equations for Nondimensional Positive Phase Duration in MEM

Source Strength	Range for R_s	Equation for t_{ps}'
3	>0.6	$t_{ps}' = 2$
4	>0.6	$t_{ps}' = 1.4$
5	>0.6	$t_{ps}' = 0.7$
7	0.6–6	$t_{ps}' = 0.30\, R_s^{0.212}$
	6–100	$t_{ps}' = 0.364\, R_s^{0.106}$
10	0.6–6	$t_{ps}' = 0.22\, R_s^{0.387}$
	6–100	$t_{ps}' = 0.364\, R_s^{0.106}$

- Ignition strength
- Type of confinement and
- The extent of obstruction

Further work by reputed researchers such as Kinsella[12] in 1993, and Roberts and Crowley[13] in 2004, provides guidance linking the obstacle factors with blast class/charge strength curves in tabular form, Tables 9.3 and 9.4, respectively.

The use of Tables 9.3 and 9.4 may sometimes result in gross underestimation. A conservative approach in selecting the blast/charge strength is, therefore, prudent.

For source strength, the guidelines in Tables 9.3 and 9.4 provide categorical classes and not particular values. Further, the decision on obstacle density will generally be subjective.

TABLE 9.3
Guidelines for Selecting Charge Strength (Kinsella[12])

Ignition Energy		Confinement			Confinement/Obstacle Density		
Low	High	No	Existing	No	High	Low	Strength
	X		X		X		7–10
	X			X	X		7–10
X			X		X		5–7
	X		X			X	5–7
	X			X		X	4–6
	X	X	X				4–6
X				X	X		4–5
	X	X		X			4–5
X			X			X	3–5
X				X		X	2–3
X		X	X				1–2
X		X		X			1

TABLE 9.4
Guidelines for Selecting Charge Strength (Roberts and Crowley[13])

		Charge Strength		
		Confinement/Obstacle Density		
Types of Flame Expansion	Mixture Reactivity	High	Medium	Low
1-D	High	10	10	10
	Medium	9–10	9	7–8
	Low	9–10	7–8	6
2-D	High	9	7–8	4–5
	Medium	7–8	6–7	2–3
	Low	6	5–6	1–2
3-D	High	6	3	1
	Medium	3–4	2	1
	Low	3	2	1

To address the subjective element, TNO subsequently sponsored two projects to determine the parameters for estimating the source strength class more precisely. The first project was titled "Guidance for Application of the Multi-Energy Method" GAME, followed by a further second phase, GAME2. This report[26] would help choose the source strength and a range of class numbers more objectively for better results.

Example 9.3

Rework Example 9.2 using the MEM at two levels of source strength, namely 7 and 5. The velocity of sound in the air can be taken as 350 m/s.

Energy of explosion, $E = (19{,}200)\ (46{,}000)\ (1{,}000) = 8.8 \times 10^{11}$ J
Receptor distance, $R = 200$ m
Atmospheric pressure, $P_a = 1.013 \times 10^5$ Pa
Sound velocity, $C_0 = 350$ m/s
From Equation (9.6), scaled distance is

$$R_s = 200 \Big/ \left[8.8 \times 10^{11} \Big/ \left(1.013 \times 10^5\right) \right]^{0.333} = 0.978\ (\text{nearly}\,1)$$

At this value of R_s, values of the nondimensional peak static overpressure (P_s) and nondimensional positive phase duration (t_{ps}) are calculated at source strengths of 7 and 5 using the equations given in Tables 9.1 and 9.2, respectively, or from Figures 9.3 and 9.5, respectively. The results are as follows:

Source strength	7	5
Scaled distance, R_s	1.0	1.0
Nondimensional peak static overpressure, P_s	0.47	0.12
Nondimensional positive phase duration, t_{ps}	0.30	0.70

The peak overpressure, P^o, is calculated by multiplying the nondimensional peak static overpressure with atmospheric pressure. The values of positive phase duration t_p are calculated from the nondimensional positive pressure duration, using Equation (9.6b). In each case, the impulse is also calculated, as this is needed to estimate the damage. The results are as follows:

Source strength	7	5
Peak overpressure, P^o, Pa	47,452	12,443
Positive phase duration, t_p, seconds	0.175	0.409
Positive phase impulse, i_p, Pa.s	4,100	2,500

9.3.4 Baker-Strehlow-Tang (BST) Method[18,20,21]

The basic principle of the BST method of VCE explosion modeling is similar to the TNO MEM. The BST method selects source strength based on flame speed, whereas the MEM method uses over-pressure. Only the energy in the fuel (in the vapor cloud) present in the confined or congested zone is considered as explosion energy in both methods.

The BST blast curves show plots for scaled blast wave properties (peak pressure and positive phase duration versus scaled distance). These are presented in families of nine curves, each with a flame Mach number as the parameter. The flame Mach number is the apparent flame speed divided by the ambient sonic velocity.

The BST method also uses numerically determined curves with dimensionless parameters (pressure and impulse values) against distance plotted for each of the nine Mach numbers from 0.2 to 5.2. The BST-2 method uses a set of curves, plotted as a function of Eulerian Mach numbers, M_f. The pressure, impulse, and distance are nondimensionalized using Sach's scaling law.

As stated for earlier models, most large sites have multiple confinement or congestion zones, termed as potential explosion sites (PESs) in the BST Method. If they adjoin or are very close to each other, they can be treated as a composite zone or PES with a common explosion scenario. On the other hand, if the PESs are separated by sufficient uncongested and unconfined spaces (4.6 m in case of almost empty gaps or 9.2 m if the separation distance includes obstacles), the blast effect of each PES needs to be considered separately.

Flame propagation speed (expressed as a Mach number) is the most important parameter in selecting the explosion blast intensity from the blast curves in the BST method. The steps are as follows:

i. Determine the extent of confinement:
   In the BST method, confinement is generally rated in three categories: 2D, 2.5-D, and 3-D. 2-D and 3-D denotes two- and three-dimensional confinement, respectively, whereas 2.5-D is the simple average of 2-D and 3-D confinement values for the same congestion and fuel reactivity.
ii. Estimate the degree of congestion or obstacles:

Congestion occurs because of obstacles (plant, equipment, buildings, etc.) that impede the flame and restrict the flame front path, creating turbulence. Congestion increases the flame speed (depending on obstacle density) while allowing expansion of the flame front. The obstacle density is classified into three categories: low, medium, and high, depending on the area blockage ratio (ABR), pitch, and the number of layers of obstacles. Table 9.5 provides guidelines for congestion categories.

iii. Determine the stoichiometric concentration (η):

$$\eta = \text{ratio of moles fuel/moles fuel plus air}$$

iv. Determine n, the moles of gas in combustion volume:

$$n = P_0 V/(RT) \tag{9.11}$$

where

n = moles of gas in combustion volume
P_0 = atmospheric pressure, Pa
V = volume of congestion for fuel-air mixture, m^3
R = universal gas constant, kJ/(kgmol.K)
T = emperature, K

TABLE 9.5
Guidance on Congestion based on ABR, Pitch, and Number of Layers

Type	Obstacle Blockage Ratio/Plane	Pitch for Obstacle Layers	Geometry
Low	Less than 10%	One or two layers of obstacles	
Medium	Between 10% and 40%	Two to three layers of obstacles	
High	Greater than 40%	Three or more fairly closely spaced obstacles	

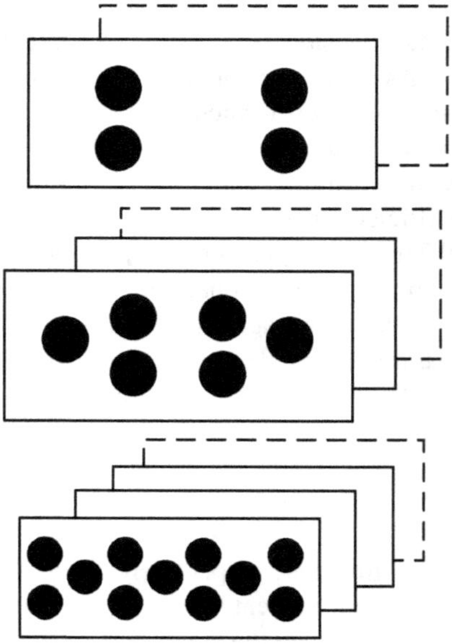

v. Combustion energy of the flammable cloud:

$$E = 2n\eta MW\ H_c \tag{9.12}$$

where
"2" is a "ground reflection factor" considered in BST
n = moles of gas in combustion volume
η = ratio of moles fuel/moles fuel plus air
MW= the fuel molecular weight, kg/kgmol
H_c = he heat of combustion, J

vi. Select flame speed from Table 9.5, based on the extent of confinement and congestion as determined earlier
vii. Calculate the scaled target distance:
Use a similar equation as for MEM:

$$R_s = R_0/\left(E/P_0\right)^{1/3} \tag{9.13}$$

where
R_s = scaled target distance, m
R_0 = distance to the center of ignition, m
E = explosion energy, J
P_0 = atmospheric pressure, Pa

viii. Using the two blast curves (Figures 9.6 and 9.7), read the scaled positive overpressure, $p' = (P - P_0)/P_0$, and scaled positive specific impulse

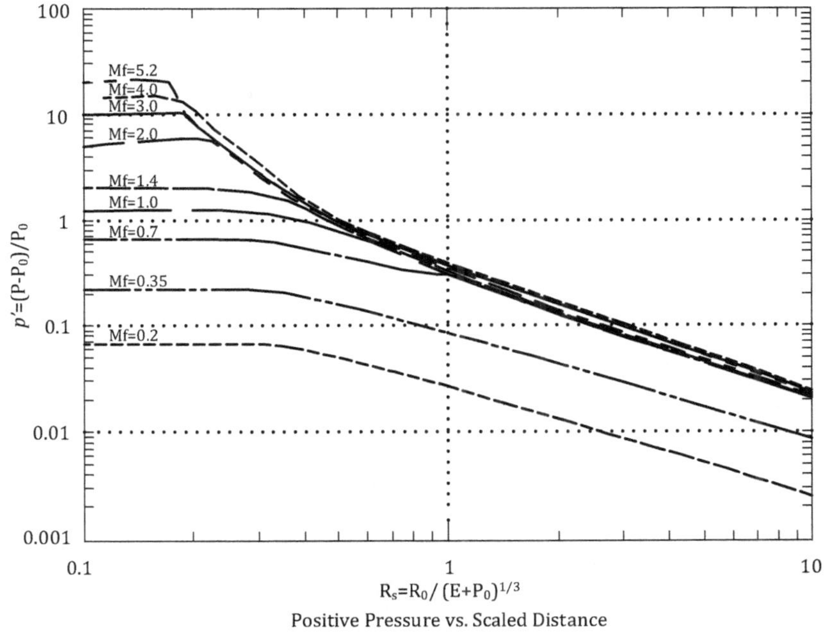

FIGURE 9.6 Positive scaled overpressure vs. distance for various flame speeds.

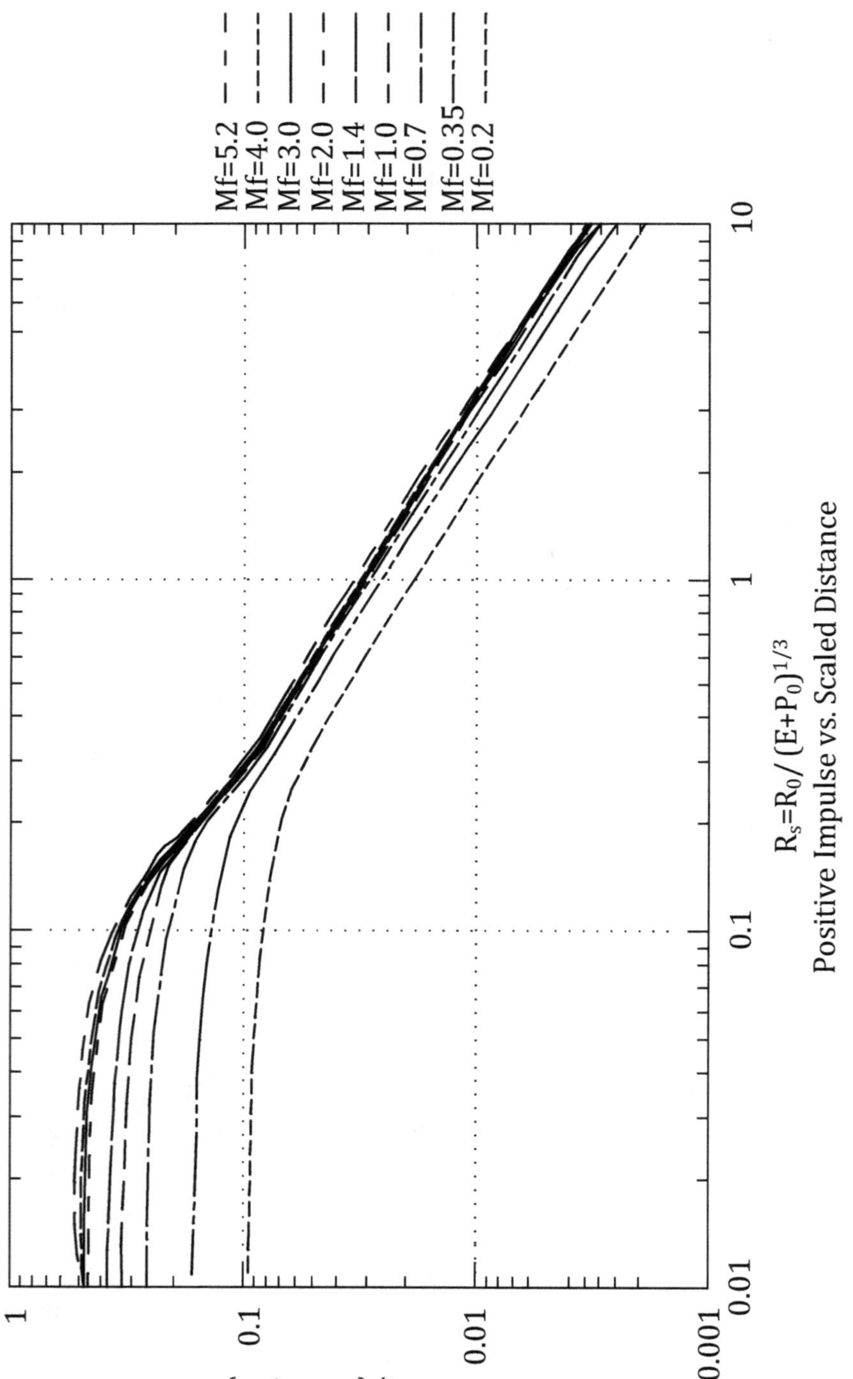

FIGURE 9.7 Positive scaled impulse vs. distance for various flame speeds.

ix. Calculate actual positive blast absolute pressure P from the equation below:

$$P = (p')(P_0) \tag{9.14}$$

where

P = positive blast absolute pressure, Pa
p' = scaled positive overpressure (dimensionless)
P_0 = atmospheric pressure, Pa

x. Calculate actual positive impulse I from the scaled positive impulse i':

$$I = i' \left(E^{1/3}\right)\left(P_a^{2/3}\right)/a^o \tag{9.15}$$

where

I = he positive impulse, Pa.s
i' = he scaled positive impulse, Pa.s
E = combustion energy of flammable liquid, J
a^o = he velocity of sound at atmospheric conditions (= 350 m/s)

As explained here, the BST method provides guidance for the determination of maximum flame speed, based on empirical data for selecting the applicable blast curves for the appropriate Mach number M_f for the flame velocity.

After 2004, several VCE test experiments were carried out. The flame speed correlation table was developed (Table 9.6 below), taking into account confinement, congestion, and reactivity. This table also displays the possibilities of the DDT phenomenon for VCEs.

TABLE 9.6
BST Correlation[21] for Flame Speed (Mach No. = M_f)

		Congestion		
Confinement	**Reactivity**	**Low**	**Medium**	**High**
2-D	High	0.59	DDT	DDT
	Medium	0.47	0.66	1.6
	Low	0.079	0.47	0.66
2.5-D	High	0.47	DDT	DDT
	Medium	0.29	0.55	1.00
	Low	0.053	0.35	0.50
3-D	High	0.36	DDT	DDT
	Medium	0.11	0.44	0.50
	Low	0.026	0.23	0.34

Example 9.4

A spillage of propane occurs from a pipeline leak in a processing unit covering an open processing area 24 m long, 18 m wide, and 8 m high. The congestion level is high. The leak causes the entire area to be filled with a stoichiometric mixture of propane and air.

$$\text{Congested volume } V = (24)(18)(8) = 3{,}456 \text{ m}^3$$

In an open facility, there is no confinement. Hence, the confinement level is 3D. The entire congested volume is considered to be filled with a stoichiometric mixture of propane and air.

FUEL AND OTHER PROPERTIES

The heat of combustion of propane, $H_c = 4.6 \times 10^7$ J
Atmospheric pressure, $P_o = 1.013 \times 10^5$ Pa
Universal gas constant, $R = 8.3$ kJ/(kgmol.K)
Speed of sound, $a^o = 350$ m/s
Fuel molecular weight, $MW = 44$ kg/kgmol
Ambient temperature, $T = 300$ K
Congested volume, $V = 3{,}456$ m^3
Stoichiometric Concentration of Fuel:
For complete combustion of propane,

$$C_3H_8 + 5(O_2 + 3.76N_2) = 3CO_2 + 4H_2O + 18.8N_2$$

Stoichiometric concentration of propane:

$$\eta = \text{moles of fuel/mole of fuel and air} = 1/\left[1 + (5)(4.76)\right] = 0.04$$

Now, $n = P_0V/R/T = \left(1.013\times10^5\right)\left(3{,}456/8.206\times10^3/300\right) = 142.25 \text{ kgmol}$

Combustion energy, $E = 2n\eta MWH_c$

$$\text{Or, } E = (2)(142.25)(0.04)(44)\left(4.6\times10^7\right) = 2.30\times10^{10} \text{ J}$$

Ground reflection doubles the energy.
From Table 9.6, $M_f = 0.50$
Scaled standoff distance, R_s, is determined using

$$R_s = R_0/\left(E/P_0\right)^{1/3} \tag{9.13}$$

where
R_0 = he distance to the center of ignition (taken as 50 m)
E = combustion energy, J
P_0 = atmospheric pressure, Pa
$R_s = (50)/(4.38 \times 10^{10}/1.013 \times 10^5)^{1/3} = 0.661$

From the blast curve (Figure 9.6) for $M_f = 0.50$, and $R_s = 0.661$, scaled peak overpressure $p' = 0.24$.

Therefore, peak over-pressure (Equation 9.6), $P = p'P_0 = (0.24)(1.013\times10^5) =$ 24 kPa or 0.24 barg.

From the impulse curve (Figure 9.7), for $M_f = 0.50$, for $R_s = 0.69$, scaled positive impulse, $i' = 0.05$.

$$\text{Positive impulse, } i = i'\left(E^{1/3}P_0^{2/3}\right)/a^o$$

Therefore, $i = (0.05)\left(4.38\times10^{10}\right)^{1/3}\left(1.013\times10^5\right)^{2/3}/350 = (0.05)(3{,}525)(2{,}173)/350$,

$$i = 1{,}094\,\text{Pa.s}$$

9.3.5 Congestion Assessment Method[18,19]

The congestion assessment method (CAM), originated by Cates[14,15] and updated by Puttock[16,17], is a simple method to derive blast overpressure from plant layout and flammable gas or vapor properties.[14] It also estimates the overpressure P_{max} of the vapor cloud and the blast pulse as a function of the distance from the source. The source strength of an explosion was derived from a decision tree, and the decay of the blast wave was obtained from a simple formula. Since this initial development, the CAM model was modified to include estimating pulse duration and shape by Puttock[16,17] and can make more detailed predictions of obstructed explosions.

In CAM, the source overpressure (the pressure generated in a congested section of the plant area) is estimated based on the extent of congestion and the fuel characteristics. For congestion assessment, a series of rows of obstructions in two horizontal and vertical directions are identified. For each of these, the area blockage (A.B.) is estimated. Overpressure beyond the congested area is predicted as a function of the following:

a. The source overpressure
b. The congested volume and
c. Distance from the congestion.

The effect of the reactivity of the fuel on the blast loads is determined as follows. First, the fuel factor (F) is determined. F is characterized by two properties of the fuel: (1) laminar burning velocity and (2) the ratio of the density of unburnt fuel to the burnt fuel. The following equation is used for F

$$F = \left(U_0(E-1)\right)^{2.71}\Big|_{\text{Fuel}} \Big/ \left(U_0(E-1)\right)^{2.71}\Big|_{\text{Propane}} \tag{9.16}$$

where

F = fuel factor
U_0 = laminar burning velocity, m/s
E = he ratio of the density of unburnt fuel to the burnt fuel

The fuel factor, expansion ratio, and laminar burning velocity for a few common fuels are given in Table 9.7 below:

TABLE 9.7
Fuel Factor *F* and Expansion Ratio *E* for Common fuels

Fuel	Fuel Factor, *F*	Expansion Ratio, *E*	Laminar Burning Velocity U_0 (m/s)
Methane	0.6	7.75	0.448
Propane	1.0	8.23	0.464
Pentane	1.0	8.34	
Methanol	1.0	8.22	
Ethylene (ethene)	3.0	8.33	0.735

The congested volume V_o, m^3, needs to be defined. For this, the nature of confinement and the obstacles in the gas-filled area of congestion are modeled.

In the CAM method, the peak overpressure is thought to be generated in a symmetrical, unconfined, congested region consisting of a regular array of cylinders. The symmetrical assumption implies that the ignition is assumed to occur in the center of the congested region. In this arrangement, the flame will encounter the same congestion or blockage level in progressing towards the east, west, north, south, and upwards (assuming no roof).

The area is assumed to be filled with obstacles placed in the two horizontal directions, *x* and *y*, and the vertical *z*-direction.

A cylinder (obstacle) is thus considered to be of dimension

i. length along *x*-direction: $2L_x$
ii. width along *y*-direction: $2L_y$
iii. height along z-direction: L_z

The numbers of rows of obstacles along the three directions (*x*, *y*, and *z*) are as follows: $2\ n_x$, $2\ n_y$, and n_z, respectively.

The blockages caused by obstacles along the three directions are represented as the factors b_x, b_y, and b_z, respectively.

The source volume recommended for use in the CAM model is defined by the length ($2L_x$), width ($2L_y$), and height (L_z) of the congested volume plus an additional 2 m beyond the last obstacle in each direction, i.e.,

$$\text{Source volume} = (2L_x + 4)(2L_y + 4)(L_z + 2)\ldots \tag{9.17}$$

Sharp-edged obstacles, e.g., beams, have a more significant effect than pipes, and their blockages have an increased effect by a factor of 1.6.

An additional "complex factor", f_c, is considered because the obstacles in realistic situations will be of quite different shape and form. Accordingly, the flame would have to propagate through quite geometrically complex surfaces relative to simple cylinders. A substantially more complex flame shape and surface will result than what might be caused by rows of uniform cylinders contemplated in the CAM Method. The result would be faster flame acceleration and, therefore, higher overpressures. For these reasons, the complex factor f_c *is* taken as 4.

In VCE, when one hydrocarbon fuel is replaced by a more reactive one, the more active fuel is considered to give rise to much higher overpressures. If the less reactive one results in an overpressure of 40 mbar, the overpressure may double to 80 mbar with the higher reactivity fuel. However, this holds only at the low range of overpressure. A hydrocarbon fuel reaches a high overpressure of around 8 bar when burning at constant volume. There will be no flow under such conditions and, therefore, no turbulence to create a high overpressure. At the high range of overpressure, the proportionate increase and decrease of overpressure are much lower than at the low range of overpressure.

The Severity Index (*S*) accounts for this reduction of expansion of the volume and flame acceleration mechanism at the higher range of overpressures. It ensures that a realistic source overpressure is maintained when extrapolating experimental data to real-world conditions.

The *S* is a function of congestion parameters and is calculated using the following equations:

For the *x*, *y*, and *z* directions, when there is no roof in the facility:

$$S_i = 0.89 \times 10^{-3} f_c F\, L_i^{0.55} n_i^{1.99} e^{[(6.44)(bi)]} \tag{9.18}$$

For the *x*, *z* directions, when there is a roof:

$$S_i = 1.1 \times 10^{-3} f_c F\; L^{0.55} n_i^{1.66} e^{[(7.24)(bi)]} \tag{9.19}$$

where

S_i = Severity Index
f_c = complex factor (= 4)
F = fuel factor (Table 9.7)
i denotes the dimension (x, y, or z)
L_i = length, m
n_i = number of obstacles through dimension i
bi = blockage factor along dimension i

Calculate the *S* for all directions using the above equations and then the average *S* for all directions as follows:

$$\text{Without roof: } S = \left(S_x + S_y + S_z\right)/3 \tag{9.20}$$

$$\text{With roof: } S = \left(S_x + S_y\right)/2 \tag{9.21}$$

The source overpressure, P_{max}, is determined as a function of the *S*. The relationship between the *S S* and the source overpressures was determined experimentally in 1999 by Puttock[17] for a wide range of conditions:

$$S = P_{max} \exp\left[0.4 P_{max} / \left(E^{1.08} - 1 - P_{max}\right)\right] \tag{9.22}$$

where
S = Severity Index, bar
E = expansion ratio relative to propane
P_{max} = peak or source overpressure
[exp denotes exponentiation]

P_{max} can be determined from the S from Equation (9.22) by trial and error.

Alternatively, Equation (9.23) and Figure 9.8 can be used for obtaining P_{max} more simply, as recommended by the authors of the CCPS book.[18]

a. Determine X, a factor based on the expansion ratio E from Table 9.7:

$$X = E^{1.08} - 1 \tag{9.23}$$

where
E = expansion ratio for the fuel gas (Table 9.7)
Calculate the S (severity)/X ratio.

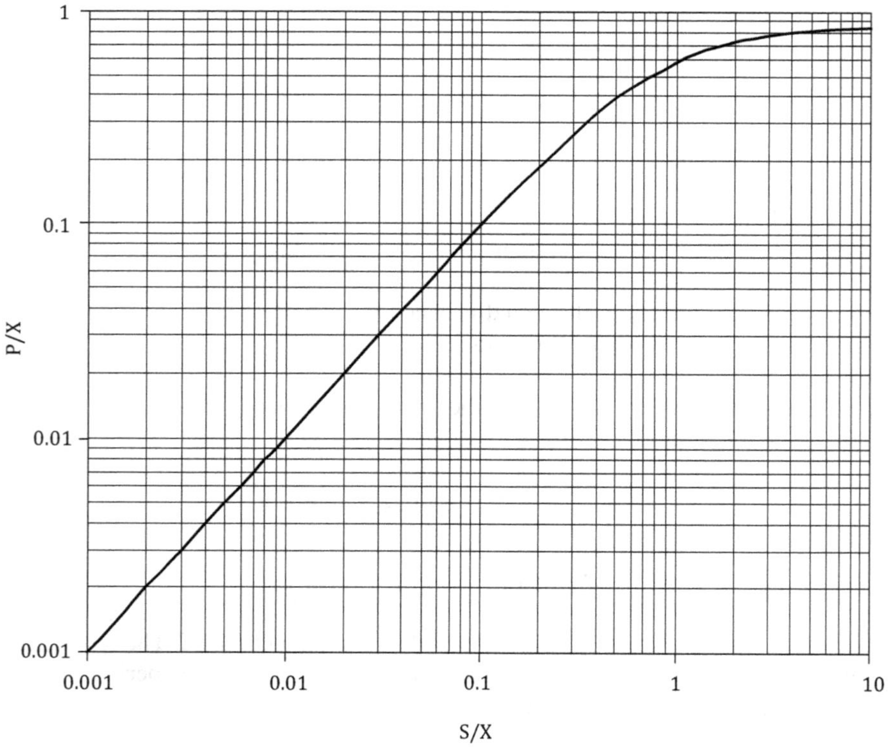

FIGURE 9.8 Scaled source overpressure as a function of scaled severity index.

b. Refer to Figure 9.8 above and, for a given *S*/*X* ratio along the *X*-axis, read the corresponding *P*/*X* value (*Y*-axis).
c. Multiply the *P*/*X* ratio by *X* to determine the estimated source overpressure, P_{max}:

$$P_{max} = (P/X)X \tag{9.24}$$

where
P_{max} = source overpressure, Pa
P/X = ordinate value from Figure 9.8
X = factor based on expansion ratio, E (Equation 9.23)
For ease of use, the following equation may be used to approximate the curve in Figure 9.9 with reasonable accuracy:

$$Y = \left(P_0 + P_1 \cdot X + P_2 \cdot X^2 + P_3 \cdot X^3\right) / \left(1 + Q_1 \cdot X + Q_2 \cdot X^2 + Q_3 \cdot X^3 + Q_4 \cdot X^4\right)$$

where
$Y = \log_{10}(P/X)$, $X = \log_{10}(S/X)$, and
$P_0 = -0.233751472$
$P_1 = 0.471956126$
$P_2 = -0.488324231$
$P_3 = 1.33821\text{E-}01$
$Q_1 = -0.015676453$
$Q_2 = 0.34946128$
$Q_3 = 0.04950662$
$Q_4 = 4.46697E\text{-}03$

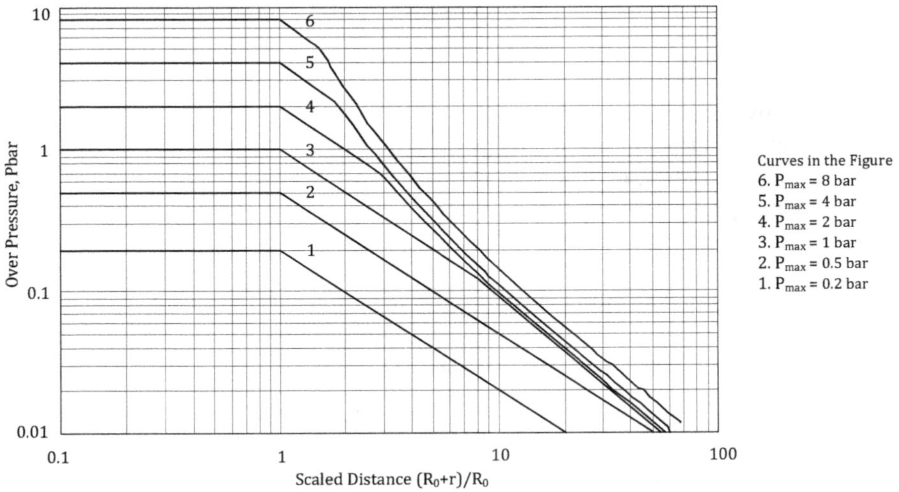

FIGURE 9.9 CAMS pressure decay as a function of scaled distance.

d. The effective radius (R_0) can be calculated from the total congested volume:

$$R_0 = \left[3V_0/(2\pi)\right]^{1/3} \tag{9.25}$$

where
R_0 = effective radius, m
V_0 = volume of the blockage, m^3

e. Calculate the free-field pressure at any distance
The predicted overpressure P_r (bar) at a given distance r (m), from the edge of the congested area, can be read in Figure 9.10. This figure shows a plot of overpressure P against the scaled distance $(R_0+r)/R_0$, choosing one of the six pressure decay curves each for peak overpressure P_{max} value of 0.2, 0.5, 1, 2, 4, and 8 bar.

Estimation of pulse duration and shape: The duration of the pulse at this distance is found by defining a dimensionless distance parameter, d_f, as follows:

$$d_f = \left(r/R_0\right)\left(P_{max}/P_0\right)^2 \tag{9.26}$$

where
d_f = dimensionless distance parameter
r = distance from the edge of the cloud, m
R_0 = effective radius, m
P_{max} = source overpressure, Pa
P_0 = atmospheric pressure, bar

The positive phase duration t_d is estimated by

$$t_d = C\,R_0/\left(P_0'/\rho_{air}\right)^{1/2} \tag{9.27}$$

where
t_d = positive phase duration, seconds
C = coefficient (based on the table below)
R_0 = effective radius, m
P_0' = atmospheric pressure, Pa
ρ_{air} = the density of air (about 1.2 kg/m^3)

Values of C are found as follows:

Range	*C*
<5°C	0.65
5°C–20°C	$0.65(d_f+10)/15$
>20°C	15

For calculating impulse (I), a triangular linear decay from the instantaneous pressure rise to atmospheric pressure is considered using

$$\mathrm{I} = 1/2\, P_r\, t_d \tag{9.28}$$

where
I = impulse, Pa.s
P_r = the overpressure from the edge of the cloud at a distance r, m
t_d = the positive phase duration obtained from Equation (9.27), seconds

Example 9.5

Let us consider an example when a spillage of 30 tons of propane occurred in a facility from a pipeline leak in a processing unit. The facility has a roof.

Congestion height = 8 m (= L_z)
Congestion width = 18 m (= $2L_y$)
Congestion length = 24 m (= $2L_x$)
No. of obstacles through length = 8 (= $2n_x$)
No. of obstacles through width = 8 (= $2n_y$)
No. of obstacles through height = 4 (n_z)
Area blockage ratio: $b_x = b_y = b_z = 0.7$

Table 9.7 shows that the fuel factor for propane is 1, and the expansion ratio E is 8.23.

Hence, $L_x = 12$, $L_y = 9$, $L_z = 8$, and $n_x = n_y = n_z = 4$
Volume of the blockage, $V_0 = (2L_{x+}4)\ (2L_{y+}4)\ (L_z+2) = (28)\ (22)\ (10) = 6{,}160\ \mathrm{m}^3$.
Thus, factor $X = E^{1.08} - 1 = 8.23^{1.08} - 1 = 8.74$.

CALCULATION OF S

The facility is taken to have a roof.
The complexity factor $f_c = 4$, as explained earlier.
Now, $S_i = 1.1\times10^{-3} f_c F\ L^{0.55} n_i^{1.66} e^{[(7.24)(bi)]}$ (9.19)

$$S_x = \left(1.1\times10^{-3}\right)(4.0)(1)\left(12^{0.55}\right)\left(4^{1.66}\right)\left(e^{[(7.24)(0.7)]}\right) = 27.38$$

Similarly, $S_y = \left(1.1\times10^{-3}\right)(4.0)(1)\left(9^{0.55}\right)\left(4^{1.66}\right)\left\{\exp\left[(7.24)(0.7)\right]\right\} = 23.37$, and

$$S = \left(S_x + S_y\right)/2 = 25.38 \tag{9.21}$$

OVERPRESSURE

Calculate $S/X = 25.38/8.74 = 2.90$.
For $S/X = 2.90$, from Figure 9.9 (read along y-axis), $P/X = 0.79$,
Maximum overpressure, $P_{max} = (P/X)X = (0.79)\ (8.74) = 6.90$

INITIAL CLOUD RADIUS

$$R_0 = \left[3V_0/(2\pi)\right]^{1/3} \tag{9.25}$$

Hence, $R_0 = \left[3V_0/(2\pi)\right]^{1/3} = \left[(3)(6{,}160)/(2)(3.1416)\right]^{1/3} = (2{,}943)^{1/3} = 14.3$ m.

CALCULATION OF PRESSURE AT A CERTAIN DISTANCE FROM THE EDGE OF THE CLOUD

Consider a distance (r) from the edge of cloud =25 m,

Calculated scaled distance for r is $(R_0+r)/R_0$ = (14+25)/14 = 2.79

From Figure 9.9, read P_r (at a scale distance of 2.79 m) as 1.2 bar.

PULSE DURATION AND IMPULSE AT A DISTANCE *R* = 25 M

Calculate distance parameter (dimensionless) d_f using Equation (9.26):

$$d_f = \left(r/R_0\right)\left(P_{\max}/P_0\right)^2$$

$$\text{Or, } d_f = \left(25/14\right)\left(6.90/1.013\right)^2 = (1.786)(6.881)^2 = 84.56$$

PULSE DURATION (t_d)

Use Equation (9.27)

$$t_d = C\,R_0/\left(P_0'/\rho_{\text{air}}\right)^{1/2}$$

Now, $C = 0.65$, $P_0 = P_{\max}$ = 6.90 bar or 690,000 Pa, ρ_{air} = 1.2 kg/m³

Therefore, t_d = (0.65) (14)/(758) = 0.012 seconds, or 12 ms

IMPULSE

For calculating impulse, a triangular linear decay from instantaneous pressure rise to atmospheric is considered following Equation (9.28):

$$I = \tfrac{1}{2}(P_r)(t_d)$$

P_r is the overpressure at a distance r (25 m in this example) from the edge of the cloud, and it is calculated here as 1.20 bar; the duration, t_d, is 12 m.s.

$$\text{Impulse } I = 0.5(1.20)(0.012) = 0.0072 \text{ bar.s or } 7.2 \text{ bar.ms}$$

9.3.6 CFD Models

CFD has developed as a most useful tool for analyzing and solving fluid flow problems in complex geometries, using numerical methods and data structures.

In oil, gas, and petrochemical industries (onshore or offshore), it is essential to consider the possible consequences of a large-scale accidental release of flammable gas or vapor. The risk of a large VCE represents an imminent hazard. This hazard must be assessed appropriately, as an integral part of the design process for any new installation, and for assessing the effectiveness of layers of protection (LOPA) or barrier analysis for existing facilities.

The underlying physical and chemical processes involved are quite complex. There is also a lack of reliable scaling laws to link small-scale experimental tests to full-scale explosion behavior. For these reasons, accurate prediction of the effects of an accidental VCE in large industrial facilities by simple computations is not considered acceptable. Accordingly, in the earlier examples in Sections 9.3.3–9.3.5, all three blast models would not provide results anywhere near as accurate as those from a competent CFD model. As mentioned previously, those models are qualitative and are useful only as rough guides for screening broad alternatives in the design of new facilities.

CFD makes it possible to carry out numerical simulations of accidental explosions by portraying the large-scale geometry of a facility much more accurately than was possible with earlier methods. Automated inputs of design-optimized initial conditions such as land topography, wind direction, and ignition location are feasible. CFD makes it possible to arrive at a highly realistic prediction of VCE consequences in a typical industrial facility, quickly and far more economically than full-scale testing. Such models run on standard desktop computers and make it possible to analyze many scenarios and plant designs quickly and at a modest cost.

For detailed design work, consequence analysis, etc., for offshore and onshore installations – which are complex facilities – CFD models are employed routinely. These models divide the flow field into hundreds of thousands of cells (using finite element methods). In each cell, the basic equations of fluid dynamics (Navier-Stokes) and other equations for mass and energy transfer are solved simultaneously. Real-life field data and physical sub-models of confinement and congestion are used to determine turbulence and combustion.

CFD models can be classified as "simple" and "advanced". The simple CFD models commonly in use are FLACS, EXSIM, and AutoReaGas. The advanced models are CFX-4, COBRA, NEWT, and McNEWT.

Three simple CFD models, FLACS, EXSIM, and AutoReaGas, are now used quite commonly for offshore and large or critical onshore facilities. They are mentioned below. It is beyond this book's scope to discuss CFD modeling principles in any detail, however.

9.3.6.1 FLACS (FLame ACceleration Simulator)

FLACS is a proprietary product of GEXCON AS (Norway-based fire and explosion safety consultants) to simulate gas explosions in offshore oil and gas platforms. It solves the compressible Navier-Stokes equations on a 3-D Cartesian grid using the finite volume method. The FLACS code was initially developed by the Christian Michelsen Research (CMR) Institute in Norway to simulate gas explosions in offshore modules. The FLACS code has been enhanced significantly over the last decade. FLACS is continuously being developed by CMR-GEXCON/FLACS and is used by many oil and gas companies for offshore and onshore facilities.

FLACS is a powerful tool that can reconcile the predicted overpressures and dynamics of an explosion with the observed near- and far-field blast damage. Based on the ignition source location, FLACS can be used to evaluate the release and migration of the gas or vapor, the strength of an explosion, the local flame accelerations, the resulting overpressures, and the blast waves.

9.3.6.2 EXSIM™ (EXplosion SIMulator)

EXSIM is a proprietary product of DNV-GL (Norway-based independent experts in risk management and quality assurance). EXSIM was developed at the Telemark Technology R&D Centre (Tel-Tek) in Norway. The CFD code EXSIM has been developed in collaboration with Shell Oil and is used for advanced industrial 3-D gas explosion analyses since 1989. At the end of 2014, the rights to the EXSIM™ source code were transferred to ComputIT for further development and integration with their KFX™ simulation technology.

EXSIM uses a finite volume code based on a structured Cartesian grid to solve the time-averaged conservation equations of fluid flow and chemical reactions. It determines the pressure rise time, pressure pulse duration, flame speed, and dynamic pressure for various geometries, scales, degrees of confinement, and congestion. EXSIM can be used to simulate effects such as ignition point location, vent arrangements, different geometries, scaling effects, and gas reactivity. Finally, EXSIM has been validated extensively and has been reported to be Shell's primary gas explosion simulation tool for more than 20 years.

9.3.6.3 AutoReaGas Model

TNO and Century Dynamics jointly developed the AutoReaGas software program. This code integrates solvers for the ReaGas and BLAST codes developed by the TNO Prins Maurits Laboratory (PML) in the Netherlands. The gas explosion solver in the ReaGas code is used to analyze gas explosions, including flame propagation, turbulence, and the effects of obstacles in the flow. The blast solver used in the BLAST code is used for accurate and efficient capture of shock phenomena and blast waves.

This code has also been applied to case studies in offshore platforms and process plants.

9.3.7 Comparison of Various Models

TNT equivalence or correlation models are simple methods and are easy to use. However, these methods consider the cloud's total energy for detonation and may yield unrealistically conservative results. At best, these provide a general estimate of the maximum damage potential during the early stages of a project.

The MEM, BST, and CAM models furnish more realistic results over short-to-medium distances.

The MEM and BST models are vapor blast curve methods and consider only the energy of the fuel present in the blast-generating area in a VCE. The MEM blast curves are defined by source or explosion strength, whereas the BST curves use flame strength. Both approaches consider confinement, congestion, and resulting flame acceleration to be the factors responsible for VCE. At a given location, the source strength is determined considering its distance from the confined cloud center.

The CAM method uses pressure decay curves. The source strength at any location depends on the distance from the edge of the confined cloud. The CAM method uses a single curve of severity factor against scaled maximum overpressure, followed by a plot of overpressure versus distance (from the edge of the cloud) with six pressure decay curves for maximum overpressure from 0.2 to 8 bar. CAM also considers confinement, geometry, and fuel reactivity to determine the peak pressures.

All these methods (MEM, BST, and CAM) improve the accuracy of prediction for blast effects substantially, compared to TNT-based methods.

However, their results correlate reasonably with experimental results and VCE accident findings over small-to-medium distances (again varying with the flammable vapor reactivity). These methods are based on one-dimensional inviscid fluid flow equations and assume symmetrical (hemispherical or spherical) blast effects. For blast effects over large distances, their results are not likely to be reliable.

In real-life scenarios, nonuniform confinement and congestion, non-homogenous mixtures, directional effects, and ignition location all play a significant part in accelerating turbulent combustion flames. These factors need especially to be considered for realistic prediction of VCE effects over large distances. For consequence analysis and project siting decisions, CFD models use sophisticated problem definitions based on the fundamental chemical engineering principles of fluid flow and heat and mass transfer with chemical reactions. Today's powerful computers enable rapid and reliable numerical solutions to such complex simulation problems. Therefore, CFD-based approaches consider all the pertinent factors realistically for accurate modeling of the flame acceleration mechanism in VCE explosions. Their use is considered mandatory for the final design stages of large-scale projects.

9.3.8 Precautionary Measures to Prevent and Minimize Damage in VCEs

The large-scale oil, gas, and petrochemical industries and medium to small-scale chemical industries incorporate many units that process flammable gases and chemicals. These face the risk of serious VCEs resulting from uncontrolled releases. Their process safety performances vary widely from "generally good" in the larger facilities to "reasonable but needing improvement" in the medium and smaller-scale industrial sectors. The safety measures recommended for all industries handling flammable highly volatile may be summarized as follows.

- Site layout must follow safety distances between manufacturing buildings, offsites, and non-plant buildings, and be based on risk assessments that ensure meeting facility ALARP guidelines (See Section 13.4.5). Minimum separation distances between different processing areas of a plant must adhere to the applicable codes/standards/guidelines and statutes
- Provide suitable buffer zones around all large onshore oil/gas/petrochemical installations to prevent and mitigate the catastrophic offsite consequences of accidents affecting residential areas and public places
- Control rooms must be located away from potential blast areas in locations that enable easy access to the emergency exits, bypassing processing areas
- Fire stations and fire water storage tanks must be located away from potential blast areas with easy access from main entrances
- Strict compliance with applicable codes/standards/industry guidelines (following the most conservative recommendations) for limiting the storage of hazardous chemicals, whether they are raw materials, intermediate, or

finished products. The number of full-time employees physically present in manufacturing and storage buildings/locations must be minimized consistent with safety requirements

- Mandatory provision of gas detectors and flame detectors in sensitive areas
- Mandatory provision of appropriate flame-proof and dust-proof electrical fittings in manufacturing buildings and loading/unloading areas, based on "area classifications" that follow country or international guidelines/standards
- Loading/unloading of highly flammable liquid and gases should be confined to the daylight hours (except for ship loading/unloading) and follow industry standards and guidelines, as applicable (NFPA, EN, BS, API, PESO, etc.)
- Mandatory use of Personal Protective Equipment (such as safety shoes, Nomex clothing, safety belts, goggles, gas masks, etc.) by all personnel. The use of non-sparking tools must be preferred with attention to cleanliness to avoid potential sources of friction, impacts, or sparks. Plant personnel must avoid clothing made from synthetic fibers
- Vehicles for the transportation of flammable liquefied gases and liquids must conform strictly to applicable standards and be labeled clearly as per the U.N.'s GHS (Globally Harmonized System of Classification and Labelling of Chemicals standard)
- Only diesel vehicles with flame arresters must be allowed in plant battery limits and also storage and utility areas
- Ensure strict adherence to work permit/vessel entry/electrical and other permit systems, management-of-change systems (process and hardware), together with regular internal and external audits, inspections, mock drills (onsite and offsite emergency plans), and safety review meetings with participation of senior staff
- Mandatory HAZOP studies before approval of any new plants or major modifications
- Annual third-party safety audits

9.3.9 Damage Caused by VCE

The overpressure and shockwave generated by a VCE causes structural damage, and also results in fatalities and injuries to people surrounding the explosion site. These damages depend on the following parameters that are determined by one of the three blast wave methods given in earlier paras 9.3.3, 9.3.4 and 9.3.5.

a. Source overpressure
b. Positive phase duration
c. Impulse

9.3.9.1 Damage to Structures – TNO[26]

The TNO Green Book (2005)[26] "Methods of Determining Possible Damages" shows extensive work on classifying damages to structures in three categories: (1) light, (2) moderate, and (2) heavy. The TNO book also provides empirical damage data for thirty-seven (37) types of structures listed in Table 9.8:

TABLE 9.8
Description of Structure Whose Explosion Damage Category is Shown in Figures 9.10 and 9.11

Type	Description of Structure
1.	City
2.	Refinery
3.	Glass windows, large and small
4.	Corrugated asbestos siding
5.	Corrugated steel or aluminum paneling
6.	Wood siding panels, standard house construction
7.	Concrete or cinder-block wall panels, 8″ or 12″ thick (not reinforced)
8.	Brick wall panel, 8″ or 12″ thick (not reinforced)
9.	Blast-resistant reinforced concrete windowless building
10.	Multi-storey wall-bearing building, brick apartment house type, up to three storeys
11.	Wood frame house
12.	Multi-storey reinforced concrete building with concrete walls, small window area, three to eight storeys
13.	Multi-storey wall-bearing building, monumental type, up to four storeys
14.	Typical American-style house
15.	Typical brick-built English house
16.	45 kg LPG tank
17.	68,000 L LPG bulk gas plant
18.	Floating or conical roof tanks, empty
19.	Light steel frame industrial building, single storey with up to 5ton crane capacity, low strength walls which fail quickly
20.	Heavy steel frame industrial building, single storey with 25–50ton crane capacity, lightweight low strength walls which fail quickly
21.	Heavy steel frame industrial building, single storey with 60–100ton crane capacity, lightweight low strength walls which fail quickly
22.	Multistorey steel frame officetype building, 3–10 storeys. lightweight low strength walls which fail quickly, earthquakeresistant design
23.	Multistorey steel frame officetype building, 3–10 storeys. lightweight low strength walls which fail quickly, nonearthquakeresistant design
24.	Multistorey reinforced concreteframe officetype building, 3–10 storeys Lightweight low strength walls which fail quickly, earthquakeresistant construction
25.	Multistorey reinforced concreteframe officetype building, 3–10 storeys. lightweight low strength walls which fail quickly, nonearthquakeresistant construction
26.	Railroad girder bridges, single track deck or through, open floors, span 200 ft (60 m)
27.	Railroad girder bridges, single track deck or through, open floors, span 75 ft (23 m)
28.	Pipe bridge
29.	Fractioning column
30.	Truckmounted engineering equipment (unprotected)
31.	Earthmoving engineering equipment (unprotected)

(Continued)

TABLE 9.8 (*Continued*)
Description of Structure Whose Explosion Damage Category is Shown in Figures 9.10 and 9.11

Type	Description of Structure
32.	Transportation vehicles
33.	Unloaded railroad cars
34.	Loaded boxcars, flatcars, full tanks cars and gondola cars (sideon orientation)
35.	Merchant shipping
36.	Telephones lines (transverse)
37.	Average deciduous forest stand

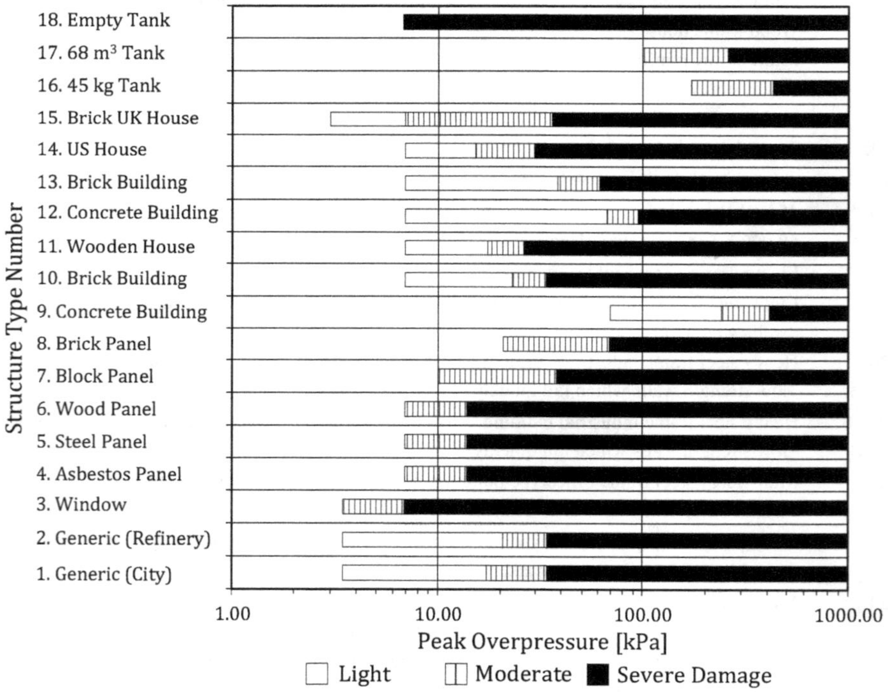

FIGURE 9.10 Damage categories of 18 types of structure (Table 9.8) against peak overpressure.[27]

Figure 9.10 above shows the damage categories of eighteen (18) structures against peak overpressure:

Figure 9.11 below shows the damage categories of the next nineteen (19) structures against peak overpressure:

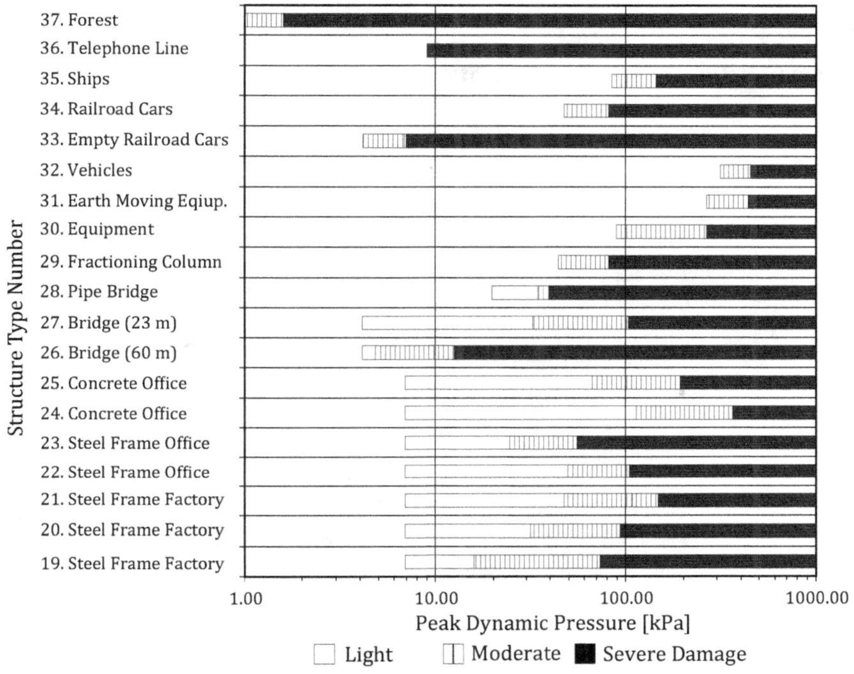

FIGURE 9.11 Damage categories of 19 types of structure (Table 9.8) against peak overpressure.[27]

9.3.9.2 Damage to Structures – Major Hazard Assessment Panel (IChemE, U.K.)[8]

The Overpressure Working Party of the Major Hazards Assessment Panel (U.K.)[8] compiled damage versus overpressure data for various structures. These are based on the work of Brassie and Simpson[9] and are reproduced in Table 9.9.

9.3.9.3 Damage to Storage Tanks – TNO[27]

The TNO Green Book provides useful data on damage to storage tanks under external pressure resulting from a VCE in a facility.

Figure 9.12 below shows a plot of peak overpressure on the y-axis against the tank height/diameter (*H*/*D*) ratio. The plot shows six curves covering tank diameters of 15, 22, and 30 m, with different fill ratios from empty to 90% full. The peak overpressure for deformation or collapse of the tank is read against the tank's *H*/*D* ratio on the selected curve, based on the fill ratio.

9.3.9.4 Effect on People – Major Hazard Assessment Panel (IChemE U.K.)[8]

Injuries caused by blast waves to people arise either from rupture of organs because of overpressure, being struck by missiles, or structural collapse. An indication of casualty probability as a function of overpressure is given in Table 9.10.[8] In these data, casualties include fatalities and injuries.

TABLE 9.9
Damage Versus Overpressure for Structures[8]

Structure	Failure	Peak Overpressure (kPa)
Windowpane	5% Broken	0.7–1
	50% Broken	1.4–3
	90% Broken	3–6
House	Tiles displaced	3–5
	Doors and window frames broke	6–9
	Inhabitable after repair – some damage to ceilings, windows, and tiling	1.4–3
	Major structural damage, partitions, and joinery wrenched from fixings	3–6
	Uninhabitable; partial or total roof collapse, partial demolition of one or two external walls, severe damage to load-bearing partitions	14–28
	50%–75% External brickwork destroyed or rendered unsafe	35–80
	Almost complete demolition	80–260
Telegraph poles	Destroyed	70–170
Large trees	Destroyed	170–380
Missiles	Unlikely at distances corresponding to overpressures less than	0.7–1.4
Rail freight wagons	Derailment unlikely at distances corresponding to overpressures less than	80–190

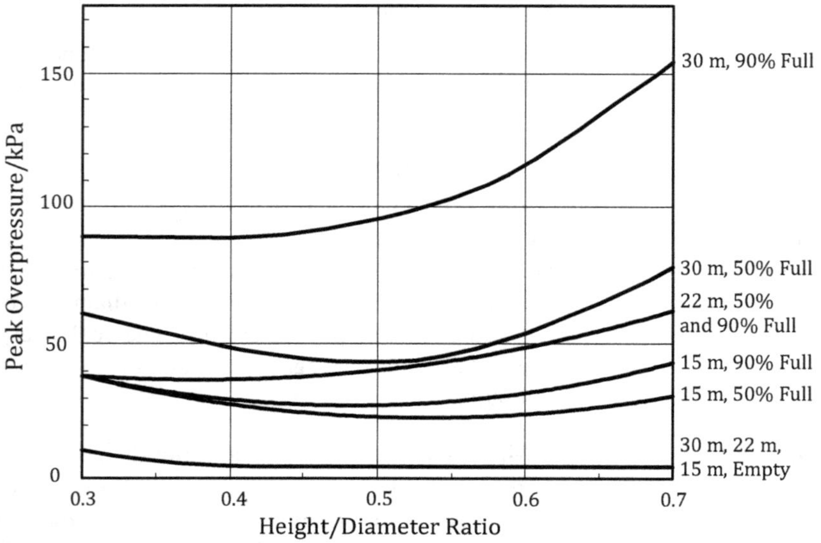

FIGURE 9.12 Overpressure of atmospheric tanks against *H/D* ratio.

TABLE 9.10
Overpressure vs. Casualty Probability

Overpressure (kPa)	Casualty Probability
<7	0
7–21	0.1
21–34	0.25
34–48	0.7
>48	0.95

9.4 CONDENSED PHASE EXPLOSION

An explosion of a high explosive is referred to as a condensed phase explosion. Trinitrotoluene (TNT), gelatine dynamites, based mainly on nitroglycerine and ammonium nitrate (AN), dry blasting agents such as an AN/fuel oil mixture (ANFO), pentaerythritol tetranitrate (PETN), and cyclonite (RDX) are examples of condensed phase explosives. Such explosives are used in mining operations, military operations, and demolition work to construct roads, tunnels, etc. Discussion on the manufacture or use of such explosives is not within the scope of this book. However, certain basic principles and empirical approaches for safety considerations in their manufacture, storage, and handling are relevant in the process industries. A brief description of high explosives has, therefore, been included in this section.

Until the early 1970s, commercial explosives for blasting operations were based on nitroglycerine and AN, and these were manufactured as cartridges in diameters ranging from 25 to 175 mm. These have since been replaced by non-nitroglycerine slurries and formulations containing mainly aqueous AN solutions (and other nitrates), oils, emulsifiers, and sensitizers. Some of the formulations also contain metallic powders, such as aluminum, for increased power. In small diameter applications (25–32 mm), diameter cartridges continue to be used. In larger diameters (75–250 mm), the explosive is often site-mixed and loaded in bulk by pumping directly into boreholes.

High explosives for mining or demolition work are used in conjunction with detonators or detonating fuses to initiate the high explosions to detonation. Ordinary detonators, about 6 mm in diameter and 35 mm long, are made of copper or aluminum and contain a charge of approximately 25 mg PETN, which is fired by either an electric spark or a safety fuse. The safety fuse consists of a core of black powder (a mixture of potassium nitrate, charcoal, and sulfur) tightly wrapped and enclosed in an envelope of textiles and waterproofing materials. The detonating fuse is also about 6 mm in diameter and consists of a core of PETN within a reinforced waterproof covering.

The burn speed of black powder in a safety fuse is of the order of 1 cm/s, and the velocity of detonation of the PETN core in a detonating fuse is of the order of 7,000 m/s. Relatively less sensitive high explosives such as ANFO or slurry blasting agents require a high explosive booster and a detonating fuse as the primary initiator. A commonly used booster is cast Pentolite, a 50:50 mixture of PETN and TNT by weight.

TABLE 9.11
Explosive Power of Materials[8]

Material	Power (TNT = 100)
AN	56
Nitroglycerine	169
PETN	177
RDX	157
Sodium chlorate	15

The power of an explosive is usually expressed relative to TNT's power, which is taken as 100. Data for some of the common explosives are given in Table 9.11:

The calculation of the damage resulting from the explosion of a high explosive is shown in Example 9.6.

Example 9.6

A van carrying 5,000 kg of gelatine dynamites suffers an explosion. Estimate the damage 200 m from the center of the explosion. The dynamite may be assumed to contain 1,300 kg nitroglycerine and 2,600 kg of AN, the balance being other nonexplosive materials.

Using data from Table 9.11, 1,300 kg of nitroglycerine is equivalent to (1,300) (1.69) or 2,200 kg of TNT, and 2,600 kg of AN is equivalent to (2,600) (0.56) or 1,456 kg of TNT. Therefore, the total quantity of explosive ingredients in the explosion is equivalent to 3656 kg of TNT.

Scaled Distance at a distance of 100 m is $200/3{,}656^{1/3} = 13.0\ \mathrm{m/kg^{1/3}}$

From Equation (9.3), Overpressure = $(427)\ (13.0^{(-1.411)}) = 11.4$ kPa. Referring to both Figure 9.10 and Table 9.9, this value of overpressure will cause moderate to near severe damage to buildings. Further, from Table 9.10, the probability of casualties would be about 10%.

9.4.1 Precautionary Measures to Minimize Damage in Condensed Phase Explosion

The explosives industry historically has followed high standards of safety. These generally could serve as examples of good site-wide safety in the process industries. The measures followed are primarily empirical, and a few of these are mentioned below:

- Site layout with specified safety distances between manufacturing buildings and also between buildings and site boundaries
- Provision of mounds around manufacturing and storage buildings to direct the blast wave upwards and minimize damage to nearby structures. Also, depending on the site location, provision of blow-out panels on one wall of the buildings to direct the blast in a direction in which there would be minimal damage or none at all

- Strict compliance with specified and licensed explosives storage limits and workforce limits in manufacturing and storage buildings
- Provision of appropriate flame-proof and dust-proof electrical fittings in manufacturing buildings
- Working during daylight hours only in explosives storage magazines, to avoid the need to provide any electrical fittings therein
- Use of safety shoes and safety uniforms by all personnel; use of non-sparking tools; use of antistatic/conducting floors and maintaining their cleanliness to avoid potential sources of friction, impacts, or sparks
- Use of specially designed and licensed vehicles for transportation of explosives or for mixing and bulk delivery into boreholes
- Strict adherence to permit-to-work system and modification procedures (MOC), together with regular audits, inspections, and review meetings with senior-level participation
- Mandatory HAZOP studies before approval of any new plants or major modifications
- For AN, large manufacturers, as well as environment and safety agencies, provide "safe handling and storage" guidelines that must be followed

9.4.2 Formation of Explosive Mixture – Ammonium Nitrate (AN)

Ammonium nitrate is prominent in the explosives and fertilizer industries. It forms an explosive mixture when it comes into contact with fuel oil or dirt, grease, and other contaminants. Massive AN explosions occur because of the shock-to-detonation transition and have caused massive devastation in several cases, such as Kriewald, Morgan, Oppau, Tessenderlo, and Traskwood. The accident at Grand Paroisse, Toulouse is relatively recent and is discussed in detail in Chapter 2, along with recommended prevention and mitigation measures.

Even more disturbingly, on August 4, 2020, a mass of 4.35 tons of AN that had been stored for over 6 years in a warehouse at the Port of Beirut in Lebanon exploded and destroyed not only the Port but also a significant portion of the city itself. There were many civilian injuries and considerable loss of life. The economic devastation that will ensue from this disaster will be felt for many years to come. This accident points both to the incredible devastation large amounts of AN can cause, and the laxity of some governments in regulating hazardous substance storage in general. There is no earthly reason why such a vast amount of high explosives should remain unsupervised for years, despite the reported complaints made repeatedly by the local Port management to senior government officials.

9.4.3 Effect of Mechanical or Electrical Shock

Explosive chemical reactions also occur when compounds undergo a rapid, violent release of energy. An explosive compound may nominally be stable, but a triggering event – such as a mechanical or electrical shock that breaks chemical bonds in the molecules – can cause sudden instability. When this happens, some molecules release energy, setting off a chain reaction in neighboring molecules. Chain reactions occur at a very high speed, consuming the explosive substance in a few thousandths of a second, releasing massive amounts of energy, and resulting in a shock wave.

Sodium chlorate and other oxidizers can have similar effects. Explosions are reported with sodium chlorate and manganese dioxide when contaminated with organics. Acetone and peroxide mixtures are other common examples. Also, fine metal powders should be handled with utmost care following their safety data sheets.

9.5 EXPLOSIONS IN A CHEMICAL REACTOR

In a chemical reactor, reactions occur between chemicals, usually in the liquid or gas phases. Many chemical reactions are highly exothermic, i.e., they release large amounts of heat as reactions proceed. Typical reactor types are batch, continuous, continuous stirred tank, tubular, and fluidized bed. These reactors usually operate within specified limits, and problems arise when deviations exceed safe operating limits.

For example, a batch reactor designed for operation under atmospheric pressure could implode if subjected to external pressure or explode if the reactor pressure exceeds a specified limit. First, a Hazop or What-If study needs to be carried out in a manner described in Chapter 12. If these studies establish significant hazards that pose a risk of severe consequences, a LOPA or barrier analysis must be carried as described in Chapter 13 before plant start-up to provide and ensure adequate independent protection layers (IPL), or barriers, to minimize the risk of untoward incidents. An explosion is far more likely in reactors where there is a possibility of an exothermic reaction runaway.

It is usual practice to protect chemical reactors against overpressure by providing pressure relief valves. Bursting discs are also provided occasionally as additional protective measures. Vessels can be designed for full containment of the explosion pressure, although this is not common practice. Depending on foreseeable venting scenarios, vents are designed to relieve the overpressure safely. Vent sizing of reactors is technically complicated and comes under the purview of specialists. A good overview has been provided by Lees.[22]

Explosions in nitration reactors (for explosives such as nitroglycerine, PETN, or TNT) result in severe damage. Such plants are usually located in areas that provide adequate safety distances and use mounds to safely redirect a blast while confining the debris or fragments to a small radius. The design of such plants is the domain of experts.

9.6 DUST EXPLOSION

A dust explosion is a significant hazard in the process industries. Explosions of dust suspensions in the air cause a rapid increase in pressure in the containing structure. If the equipment is not strong enough to withstand the explosion or is not adequately protected, extensive damage and injuries can occur. The IChemE (U.K.) provides useful guidance on dust explosion prevention and protection.[7,8]

The following conditions must all be satisfied for a dust explosion to occur:

- *Combustible* dust
- Dust in suspension in an *atmosphere* capable of supporting combustion (air or oxygen)
- Dust *particle size distribution* that will propagate flame
- Dust *concentration* in the suspension within the explosible range
- Dust suspension in contact with an *ignition source* of sufficient energy.

The explosibility of dust can be determined by tests outlined in the IChemE Guide.[23,24] Some everyday products that are handled or processed as dusts and powders are as follows:

- Wood
- Coal
- Food (e.g., starch, flour, cocoa)
- Chemicals (e.g., drugs, dyestuffs)
- Metals (e.g., aluminum, magnesium)
- Fine sugar.

Some of the industrial operations where dusts are generated or handled are as follows:

- Size reduction (crushing/grinding mills)
- Conveying (bucket elevators, screw/belt conveyors, pneumatic conveyors)
- Dust separators (e.g., cyclones, dust filters)
- Dryers (e.g., fluidized bed dryers, spray dryers)
- Storage silos.

Generally, the explosibility of combustible dust is enhanced if the particle size is reduced. A decrease in particle size reduces the minimum ignition energy and increases the pressure rise rate. Also, fine particles stay in suspension more readily than coarse particles, increasing the probability of an explosible concentration. Particles greater than 500 μ are generally considered unlikely to cause dust explosions, although the possibility of coarser materials producing fines by attrition during handling should be borne in mind.

Minimum explosible concentrations of dust in the air are typically from 10 to 500 g/m^3. Such concentrations should be expected to occur very close to a dust source or within an enclosed space where the dust cloud cannot spread.

The primary ignition sources against which precautions must be taken are the following:

- Flames
- Hot surfaces
- Burning material
- Spontaneous heating

- Welding or cutting operations
- Friction heating
- Impact/electric/electrostatic discharge sparks.

Dusts are usually classified according to the K_{st} value, which is a characteristic constant related to the rate of pressure rise in a vessel of volume V as follows:

$$K_{st} = \left(\frac{dP}{dt}\right)_{\max} V^{1/3} \tag{9.29}$$

where

P = the absolute pressure, Pa
t = time, seconds
V = volume of the vessel, m^3

The subscript "st" derives from the German word "staub" for dust. Based on measurements in a 1 m^3 test apparatus and a strong ignition source of 10 kJ, dusts are usually classified as follows[7] (Table 9.12):

The IChemE Guide[24] provides nomographs for determining the vent areas of vessels. Information required for the use of these nomographs includes the following:

i. the vessel volume;
ii. the St classification of the dust;
iii. P_{stat}, the vent opening pressure (barg), and
iv. P_{red}, the reduced pressure (barg).

P_{red} is the maximum pressure in the vented vessel. Readers interested in using these nomographs are referred to the IChemE Guide[24] or Lees.[22]

Inert gases such as nitrogen or carbon dioxide are commonly used to create an inert atmosphere in dust handling equipment. The maximum permissible oxygen concentration for preventing an explosion for many dusts ranges from 8% to 15% with carbon dioxide and 6% to 13% with nitrogen.[8]

For precautions against ignition sources, our recommendations in Chapters 4 and 5 should generally be applicable. An additional concern for dusts is the likelihood of spontaneous heating when powdery materials (coal, sawdust, reactive metals, or

TABLE 9.12
Dust Explosion Class

Dust Explosion Class	K_{st} (bar m/s)	Characteristics
St 0	0	No explosion
St 1	0–200	Weak explosion
St 2	200–300	Strong explosion
St 3	>300	Very strong explosion

dusts impregnated with vegetable oils) are held in bulk storage. Due caution should be exercised while designing such storage facilities.

A major dust explosion in a U.S. sugar company in early 2008 is covered in detail in Chapter 2, with recommended prevention and mitigation measures.

9.7 PHYSICAL EXPLOSION

Physical explosions refer to mechanical failures of pressurized systems that are not related to chemical reactions. Typical examples are the following:

a. The implosion of a vessel caused by the creation of a vacuum inside the vessel
b. Failure of a pressure vessel being exposed accidentally to overpressure. Examples include connection to another vessel that is at a pressure higher than the vessel's pressure rating
c. Rupture during a pneumatic or inert gas pressure test that exceeds the maximum allowable test pressure
d. Overheating caused by radiant heat from flame exposure/fire[25] engulfment (e.g., BLEVE)
e. Cooling to extremely low temperatures – such as brittle-failure of vessels subjected to temperatures below the lower metallurgical limit, etc.

REFERENCES

1. TNO: *Methods for the Determination of Possible Damage (Green Book)* (1st Ed., CPR 16E, TNO, The Hague, 1992).
2. Chamberlain G. A., Oran, E., and Pekalski, A.: *Detonation in Industrial vapor cloud explosion* (12th Industrial Symposium on Hazards, Prevention and Mitigation of Industrial Explosions, ISPHMIE, Kansas City, MD, 2018).
3. Oran, E. S., Chamberlain, G., and Pekalski, A.: Mechanisms and occurrence of detonation in vapor cloud explosion, *Progress in Energy and Combustion Science*, 77, p. 100804 (2019).
4. Buncefield Major Investigation Board: *The Buncefield Incident, 11th December, 2005* (The Final Report of Major Investigation Board, Vol. 2a, 2008).
5. Johnson D. M.: Vapor cloud Explosion in IOC Oil Terminal in Jaipur, *Loss Prevention Bulletin*, 229, pp. 11–18 (2013).
6. TNO: *Methods for the Calculation of Physical Effects (Yellow Book)* (3rd Ed., 2nd Revised Print, CPR 14E, TNO, The Hague, 2005).
7. Marshall, V.C.: *Major Chemical Hazards* (Ellis Horwood, Chichester, 1987).
8. Loss Prevention Bulletin (No. 068): *The Effect of Explosions in the Process Industries by Overpressure Working Party, Major Hazards Assessment Panel* (The Institution of Chemical Engineers, Rugby, 1986).
9. Brasie, W. C. and Simpson, D. W.: *Guidelines for Estimating Damage Explosion* (2nd Symposium on Loss Prevention, AIChE, New York, Vol. 1, 1968).
10. TNO: *Methods for the Calculation of Physical Effects (Yellow Book)* (1st Ed., CPR 14E, TNO, The Hague, 1992).
11. van den Berg, A. C.: The Multi-Energy Method: A Framework for Vapor Cloud Explosion Blast Prediction, *Journal of Hazardous Materials*, 12, pp. 1–10 (1985).

12. Kinsella, K. G.: *A Rapid Assessment Methodology for the Prediction of Vapor Cloud Explosion Overpressure* (Proceedings of the International Conference and Exhibition on Safety, Health and Loss Prevention in the Oil, Chemical and Process Industries, Singapore, 1993).
13. Roberts, M. R. and Crowley, W. K.: *Evaluation of Flammability Hazards in Nonnuclear Safety Analysis* (EFCOG Safety Analysis Workshop, San Francisco, CA, 2004).
14. Cates A. T.: *Fuel Gas Explosion Guidelines* (Conference on Fire and Explosion Hazards, Fire Risk Management Institute, Moreton-in-Marsh, U.K., April, 1991).
15. Cates, A. T., and Samuels, B.: A Simple Assessment Methodology for Vented Explosions, *Journal of Loss Prevention in the Process Industries*, 4, pp. 287–296 (1991).
16. Puttock, J. S.: *Fuel Gas Explosion Guidelines – The Congestion Assessment Method* (2nd European Conference on Major Hazards Onshore and Offshore, IChemE, Manchester, U.K., 1995).
17. Puttock, J. S.: *Improvements in Guidelines for Prediction of Vapor Cloud Explosions* (International Conference and Workshop on Modelling and Consequences of Accidental Releases of Hazardous Materials, San Francisco, CA, September–October. Symposium Series No. 151, #2006 Shell Global Solutions International B.V. 14, The Netherlands, 1999).
18. CCPS (Center for Chemical Process Safety, AIChE): *Guidelines for Vapor Cloud Explosion, Pressure Vessel Burst, BLEVE, and Flash Fire Hazards* (AIChE, Hoboken, NJ, 2011).
19. Baker, Q. A., Tang, M. J., Scheier, E. A., and Silva, G. J.: Vapor Cloud Explosion Analysis, *Process Safety Progress*, 15(2), pp. 106–109 (1989).
20. Tang, N. G. and Baker, Q. A.: A New Set of Blast Curves for Vapor Cloud Explosion, *Process Safety Congress*, 18(3), pp. 235–240 (1999).
21. Pierozarieo, J. A., Thomas, J. K., Baker, Q. A., and Ketchum, G. E.: An Update to Baker-Strehlow-Tang Vapor Cloud Explosion Prediction Methodology, Flame Speed Table, *Process Safety*, 24(1), pp. 59–65 (2005).
22. Mannan, S.: *Lees' Loss Prevention in the Process Industries* (4th ed., Butterworth-Heinemann, Oxford, 2012).
23. Schofield, C.: *Guide to Dust Explosion Prevention and Protection, Part 1 – Venting* (The Institution of Chemical Engineers, Rugby, 1984).
24. Schofield, C. and Abbott, J. A.: *Guide to Dust Explosion Prevention and Protection, Part 2 – Ignition Prevention, Containment, Inerting, Suppression and Isolation* (The Institution of Chemical Engineers, Rugby, 1988).
25. Loss Prevention Bulletin (No. 082): *Calculation of the Intensity of Thermal Radiation from Large Fires, First Report by the Thermal Radiation Work Group of the Major Hazards Assessment Panel* (The Institution of Chemical Engineers, Rugby, 1988).
26. TNO Report, PML 1998-C53: *Application of Correlations to Quantify the Source Strength of Vapor Cloud Explosions in Realistic Situations. Final Report for the Project: 'GAMES'.*
27. TNO: *Methods for the Determination of Possible Damage (Green Book)* (2nd Ed., CPR 16E, TNO, The Hague, 2005).

10 Toxic Releases

Methods for assessing thermal radiation effects from fires were discussed in Chapter 3 and those of explosions in Chapter 9. The third significant hazard in chemical plants is the release of toxic substances, which is considered in this chapter.

Toxic substances, as commonly understood, can destroy life or injure health when introduced into or absorbed by a living organism. We do not attempt any discussion in this chapter on toxicology in the medical sense. However, we consider the effects of toxic releases on people who work in or live around chemical plants in broad terms.

Toxic chemicals can enter the human body in three ways: (1) inhalation, (2) ingestion (through eating or drinking), and (3) external contact through skin or the mucous membranes (eyes, mouth, or throat). Generally, gases, vapors, fumes, and dusts are inhaled, and liquids and solids are ingested. For workers in factories and those residing in the neighborhood, inhalation of gases and vapors is by far the most significant route of entry. Therefore, this has been considered in greater detail in this chapter.

Data on toxicity are estimates based on human experimentation, evidence from wars, major accidental releases, and animal experimentation, the last category being the most significant. Data collected through experiments on animals vary widely depending upon experimental conditions, e.g., specimens, species, and caging conditions,[1] sometimes by many orders of magnitude.[2] The toxic effect on people varies considerably depending upon their age, sex, and state of health. Published toxicity data from various organizations and from manufacturers of chemicals through Safety Data Sheets (SDSs) are therefore approximate.

Toxic effects are classified as either acute or chronic. Acute effects result from a single exposure to a high concentration of the chemical for a short duration. Chronic effects result from exposure to low concentrations over prolonged periods.

For a sizeable accidental release, the acute effect on the exposed population is preeminent. For workers in factories, the chronic effect of daily exposure over their working lives is relevant (occupational safety). Also, where people residing around chemical factories are exposed to continuous environmental pollution from stacks, the chronic effects may sometimes be quite severe and, therefore, are relevant.

10.1 PROCESS SAFETY CONCERNS – ACUTE EFFECTS/ EMERGENCY EXPOSURE LIMITS

The sudden release of toxic chemicals in the form of gases or high vapor pressure liquids creates emergencies. People in surrounding areas are exposed to a gas cloud of high concentrations. The level of harm needs to be assessed based on the chemicals' acute toxicity.

DOI: 10.1201/9781003107873-10

10.1.1 Emergency Response Planning Guidelines

Emergency Response Planning Guidelines (ERPGs) are prepared by an industry task force in the USA and are published by the American Industrial Hygiene Association. Three concentration ranges are provided as a consequence of exposure to a specific substance:

- The ERPG-1 is the maximum airborne concentration below which it is believed that nearly all individuals could be exposed for up to 1 hour (1) without experiencing any symptoms, other than mild transient adverse health effects, or (2) without perceiving a clearly defined objectionable odor.
- The ERPG-2 is the maximum airborne concentration below which it is believed that nearly all individuals could be exposed for up to 1 hour without experiencing or developing irreversible or other serious health effects or symptoms that could impair their abilities to take protective action.
- The ERPG-3 is the maximum airborne concentration below which it is believed nearly all individuals could be exposed for up to 1 hour without experiencing or developing life-threatening health effects.

ERPG data for some selected chemicals are given below in Table 10.1 in ppm by volume and in mg/m³ at 20°C:

10.1.2 Toxic Endpoints

The toxic endpoint values promulgated by the U.S. Environmental Agency are valuable for air dispersion modeling of toxic releases as part of the risk management plan. 40 CFR Part 68 (Appendix A) gives values for 77 chemicals, some of which are quoted in Table 10.2.

TABLE 10.1
ERPGs[5]

	ERPG-1		ERPG-2		ERPG-3	
Chemical	**ppm**	**mg/m³**	**ppm**	**mg/m³**	**ppm**	**mg/m³**
Acetaldehyde	10	18	200	366	1,000	1,831
Ammonia	25	18	200	142	1,000	708
Benzene	50	162	150	487	1,000	3,247
Carbon disulfide	1	3	50	158	500	1,583
Chlorine	1	3	3	8.7	20	59
Ethylene oxide	NA	4.6	50	92	500	916
Hydrogen chloride	3	0.14	20	30	100	152
Hydrogen cyanide	NA	266	10	11	25	28
Hydrogen sulfide	0.1	0.06	30	43	100	142
Methanol	200	192	1000	1332	5,000	6,660
Methyl isocyanate	0.025		0.5	1.2	5	11.9
Toluene	50		300	1149	1,000	3,830

TABLE 10.2
Toxic Endpoints for Selected Chemicals[3]

	Toxic Endpoint (mg/m^3)
Ammonia	140
Carbon disulfide	160
Chlorine	8.7
Hydrogen sulfide	42
Methyl isocyanate	1.2
Phosgene	0.81

It may be observed from a comparison of Tables 10.3 and 10.6 that toxic endpoints are the same as the ERPG-2 values.

10.1.3 Acute Exposure Guideline Levels

Acute exposure guideline level (AEGL) values represent threshold levels for the general public. It is stated by US EPA that, in 2001, the National Academies published procedural guidance or "standard operating procedures" to make the development of AEGLs systematic, consistent, documented, and transparent to the public.

AEGLs are used by emergency planners and responders worldwide as guidance in dealing with rare, usually accidental releases of chemicals into the air. AEGLs are expressed as specific concentrations of airborne chemicals at which health effects may occur. They are designed to protect the elderly, children, and other individuals who may be susceptible.

AEGLs are calculated for five relatively short exposure periods – 10 and 30 minutes and 1, 4, and 8 hours – as differentiated from air standards based on longer or repeated exposures. AEGL "levels" are dictated by the severity of the exposure's toxic effects, with Level 1 being the least and Level 3 being the most severe.

10.1.3.1 Level 1

Notable discomfort, irritation, or specific asymptomatic, nonsensory effects. However, these effects are not disabling and are transient and reversible upon cessation of exposure.

10.1.3.2 Level 2

Irreversible or other serious, long-lasting adverse health effects or an impaired ability to escape.

10.1.3.3 Level 3

Life-threatening health effects or death.

For controlling chronic and acute effects of toxic chemicals, the commonly used measures and criteria are covered in this chapter.

10.2 OCCUPATIONAL SAFETY CONCERNS – TOXICITY MEASURES AND ASSESSMENT

Some of the measures commonly used for expressing toxicity are defined below.

10.2.1 Median Lethal Dose (LD_{50})

The term median lethal dose, LD_{50}, is commonly used for expressing the toxicity of a chemical that enters the body orally. It is defined as the single dose that would kill 50% of a large population of test animals, such as rats, mice, or guinea pigs. It is a measure of acute toxicity, mg/kg body weight. When the substance is administered through the skin, the term is called LD_{50} (dermal).

10.2.2 Median Lethal Concentration (LC_{50})

When the chemical enters the body through inhalation, the toxicity measure is called median lethal concentration, LC_{50}. It is defined as the concentration of a chemical in the air that kills 50% of the test animals in a given time. The exposure duration is usually taken as 30 minutes for LC_{50} data for emergency planning against major accidental releases of toxic gases or vapors.

10.2.2.1 Toxic Load

When a population is exposed to toxic gases or vapors in air, the proportion of the people suffering a defined degree of injury is often expressed in the form of a probit function:

$$\mathrm{Pr} = a + b\ln\left(C^n t\right) \tag{10.1}$$

where

Pr = probit function value,
C = the concentration of the toxic chemical in the air, mg/m^3
t = exposure duration, minutes
(a, b, and n are constants)

The product (C^n t) is known as the toxic load. U.S. values of a, b, and n, based on the TNO Green Book, are shown in Table 10.3.[4] Derived mainly from animal experimentation data, these are intended to be used for the calculation of percentage fatalities in the human population when exposed to a given concentration of the chemical in the air.

The values of LC_{50} for various chemicals listed in Table 10.3 have been calculated using these probit constants and assuming a 30-minute exposure duration. The calculated values in Table 10.4 below and the values from the Green Book show excellent agreement. This comparison illustrates the usefulness of the probit function.

TABLE 10.3
Probit Function Constants for Lethal Toxicity[4]

Chemical	a	B	N
Acrylonitrile	−4.1	1	1
Ammonia	−15.8	1	2
Carbon monoxide	−7.4	1	1
Chlorine	−14.3	1	2.3
Ethylene oxide	−6.8	1	1
Hydrogen cyanide	−9.8	1	2.4
Hydrogen sulfide	−11.5	1	1.9
Methyl isocyanate	−1.2	1	0.7
Nitrogen dioxide	−19.6	1	3.7
Phosgene	−0.8	1	0.9

TABLE 10.4
Calculated values of LC_{50} using the Probit Function

	LC50 (mg/m³)	
	Calculated	Green Book
Acrylonitrile	2,554	2,533
Ammonia	6,000	6,164
Carbon monoxide	8,093	
Chlorine	1,005	1,017
Ethylene oxide	4,442	4,443
Hydrogen cyanide	116	114
Hydrogen sulfide	986	987
Methyl isocyanate	55	57
Nitrogen dioxide	235	235
Phosgene	14	14

10.2.3 Immediately Dangerous to Life and Health

CCPS[5] quotes the IDLH definition given by the American National Institute for Occupational Safety and Health (NIOSH). Immediately dangerous to life and health (IDLH) corresponds to a condition "that poses a threat of exposure to airborne contaminants when that exposure is likely to cause death or immediate or delayed permanent adverse health effects or prevent escape from such an environment". If the IDLH concentration limits are exceeded, all unprotected workers must leave the area immediately. For flammable vapors, the IDLH is defined as 10% of the LFL concentration.

Because IDLH values were developed to protect healthy worker populations, they must be stricter for sensitive populations such as older, disabled, or chronically ill (Table 10.5).

TABLE 10.5
IDLH Values for a Few Common Chemicals

	IDLH	
	ppm by volume[6]	mg/m^3
Acetone	2,500	6,036
Acrylonitrile	500	1,103
Ammonia	300	212
Benzene	500	1,624
Carbon disulfide	500	1,583
Carbon monoxide	1,200	1,397
Chlorine	10	29
Ethylene oxide	800	1,465
Hexane	1,100	3,940
Hydrogen chloride	50	76
Hydrogen cyanide	50	56
Hydrogen sulfide	100	142
Nitrogen dioxide	20	38
Phosgene	2	8
Toluene	500	1,915
Xylene	900	3,972

10.3 REGULATORY CONTROLS

Besides the LD_{50} and LC_{50} concepts considered under Section 10.1, other measures are used for specifying regulatory standards and guidelines on the use of toxic chemicals. These come under two categories: occupational exposure standards and emergency exposure limits. It should be mentioned that although occupational exposure standards are mentioned below, these usually are not considered in process safety.

10.3.1 Occupational Exposure Standards

Occupational exposure standards (also called occupational hygiene standards) are intended to control chronic effects on exposure. These apply to workers in factories. The regulatory authorities are the OSHA in the USA, the HSE in the U.K., and the Directorate of Factories in India.

The American Conference of Governmental Industrial Hygienists (ACGIH) recommends threshold limit values on airborne concentrations of chemical contaminants for use in the practice of industrial hygiene and by other qualified professionals to protect worker health. These limits are updated routinely. Permissible exposure limits (PELs) laid down by OSHA in the USA, workplace exposure limits (WELs) by HSE in the U.K., and Permissible Levels of Chemical Substances in Work Environment in India under the Factories Act are, by and large, based on the ACGIH recommendations.

ACGIH limit values are defined as follows[5]:

- Threshold Limit Value, Time Weighted Average (TLV-TWA). It is the time-weighted average concentration for a normal 8-hour workday or 40-hour workweek to which nearly all workers may be repeatedly exposed, day after day, without adverse effect.
- The threshold limit value, short-term exposure limit (TLV-STEL) is the maximum concentration to which workers can be exposed for 15 minutes continuously without suffering from (1) intolerable irritation, (2) chronic or irreversible tissue change, or (iii) significant narcosis that increases proneness to accidents, impairs self-rescue, or materially reduces worker efficiency, provided, however, that no more than four excursions per day are permitted, with at least 60 minutes between exposure periods and also provided that the daily TLV-TWA is not exceeded.
- The threshold limit value, ceiling (TEL-C) is the concentration that should not be exceeded even instantaneously. Substances that are given a TLV-C are those which are predominantly fast acting and require a limit related to this aspect.

PEL and WEL values applicable in the USA, the U.K., and India are given in Table 10.6.

EH40 document containing WEL values for compliance in the U.K. clarifies that these values relate to exposure control at the workplace and should not be readily adapted to evaluate or control nonoccupational exposure, e.g., levels of contamination in the neighborhood close to an industrial plant. For the latter category, there are additional pollution control requirements that must also be observed. For example, in India, regulatory emission standards cover maximum pollutant concentrations in the stack gases and at ground level within the plant boundary.

TABLE 10.6
Permissible Exposure Limits (ppm by volume) for Airborne Chemicals

	USA[7]		U.K.[8]		India[9]	
Substance	**PEL**	**STEL**	**WEL**	**STEL**	**TWA**	**STEL**
Acetaldehyde	25		20	50	100	150
Acetone	500	750	500	1500	750	1,000
Acrolein	0.1		0.1	0.3	0.1	9.3
Ammonia	25	35	25	35	25	35
Aniline	2		1		2	
Benzene	1	5	1		10	30
n-Butyl alcohol	50				50	
Carbon disulfide	1	12	5		10	
Carbon monoxide	25		30	200	50	400
Chlorine	0.5	1	0.5		1	3
Ethyl acetate	400		200	400	400	

(Continued)

TABLE 10.6 (*Continued*)
Permissible Exposure Limits (ppm by volume) for Airborne Chemicals

	USA[7]		U.K.[8]		India[9]	
Substance	**PEL**	**STEL**	**WEL**	**STEL**	**TWA**	**STEL**
Ethyl alcohol	1,000		1,000		1,000	
Ethylene oxide	1	5	5		1	2
Ethyl ether	400	500				
Formaldehyde	0.75	2				
Gasoline	300	500			300	500
Hydrogen sulfide	5	15	5	10	10	15
Isopropyl alcohol	400	500				
Methyl alcohol	200	250	200	250	200	250
Methyl ethyl ketone	200	300				
Methyl isobutyl ketone	50	75			50	75
Methyl isocyanate	0.02			0.02	0.02	
Nitric acid	2	4			2	5
Nitrobenzene	1				1	
Phenol			2	4	5	
Phosgene	0.1		0.02	0.06	0.1	
Sulfur dioxide	2	5			2	5
Toluene	10	150			100	150
Vinyl chloride	1		3		5	
Xylene	100	150			100	150

10.4 EMERGENCY PLANNING

The release of toxic gases or vapors into the atmosphere can lead to an emergency, depending upon the release size and nature. The health impact of a toxic release may be many times more significant than a fire or an explosion, as evident from a review of past case studies in Chapter 2.

The main accidental scenarios that need to be considered are as follows:

i. Release in the form of a gas or a liquid from a pressurized containment and
ii. Release of a refrigerated liquid close to atmospheric pressure.

Releases may be continuous or instantaneous. For instantaneous liquid releases, vaporization proceeds at a finite rate. Vapors are released continuously until the entire liquid spill has evaporated. The released gas or vapor spreads, getting diluted as it travels mainly in downwind and crosswind directions. The movement of the gas, air entrainment, mixing, and gradual dilution is known as dispersion.

For a given size of the release, emergency planning requires an assessment of the maximum distance in any direction from the point of release to reach an endpoint concentration. Within the area bounded by the contour for such an endpoint concentration, the facility should work with the local community to develop, communicate, and practice emergency response plans in the event of a release. While one might think that "everyone within the contour boundary of endpoint concentration would be equally threatened," this is not entirely correct. Firstly, the release will not diffuse equally in all directions; it will potentially threaten only some who live within that boundary. Secondly, careful planning will enable the community to protect themselves, avoiding injuries and fatalities.

The procedure for estimation of the concentration versus distance contours has been covered in detail in Chapter 11.

REFERENCES

1. Withers, J.: *Major Industrial Hazards* (Gower Technical Press, Aldershot, 1988).
2. Milnes, M. H.: Formation of 2,3,7,8-Tetrachloro-Dibenzo-Dioxin by Thermal Decomposition of Sodium 2,4,5-Trichlorophenate, *Nature*, 232, pp. 392–396 (1971).
3. U.S. Environment Protection Agency: *40 CFR Part 68, Appendix A.*
4. TNO: *Methods for the Determination of Possible Damage (Green Book)* (16th Ed., CPR, TNO, The Hague, 1992).
5. CCPS (AIChE): *Guidelines for Chemical Process Quantitative Risk Analysis* (2nd Ed., AIChE, New York, 2000).
6. American National Institute for Occupational Safety and Health (NIOSH, USA): *IDLH Values* (2004).
7. Department of Industrial Relations, California (USA): *Permissible Exposure Limits for Chemical Contaminants* (2012).
8. Health and Safety Executive (UK): *EH40/2005 Workplace Exposure Limits* (2nd Ed., Health and Safety Executive, Bootle, 2011).
9. Government of India: *The Factories Act 1948, Including the 1987 Amendment.*

11 Dispersion of Gases and Vapors

Whenever a hazardous gas or vapor is released into the atmosphere, it spreads and is transported downwind. During spreading and transportation, the air is entrained. The entrained air mixes with the released vapor (or gas) and dilutes it. Dispersion refers to the combined process of transport, entrainment, mixing, and the resultant dilution of the vapor.

11.1 PURPOSE OF DISPERSION STUDIES

Whenever a hazardous substance is released into the atmosphere as a gas or vapor, a hazardous zone is created around the source. Dispersion studies estimate the extent of this hazardous zone so that appropriate precautionary measures can be taken. The zone boundary is determined by the concentration, often called the Endpoint Concentration, above which the potential to cause harm is significant.

For flammable vapors, the endpoint is usually the lower flammability limit (LFL). Sometimes, to allow for an additional safety margin, a value of half of the LFL is used as the endpoint.

Various measures and criteria are used for the endpoint for toxic gases, depending on the study purpose. The measures commonly followed for emergency planning are as follows:

- Immediately dangerous to life and health (IDLH),
- Emergency response planning guideline (ERPG), and
- Toxic endpoint limits.

For occupational exposure planning, the measures used are the following:

- Occupational or permissible exposure limit (OEL or PEL),
- Time weighted average (TWA), and
- Short time exposure limit (STEL).

These terms have also been discussed in Chapter 10.

11.2 EMISSION SOURCE MODELS

A release from containment may be in the form of a liquid, gas, or vapor, or a vapor mixed with fine droplets or a two-phase mist; it may be instantaneous or continuous. Instantaneous releases, also called "puff" releases, result from the sudden

DOI: 10.1201/9781003107873-11

vessel rupture; they are short in duration, of the order of a few seconds. Continuous releases occur for long durations, and the rate of release may be steady or time-varying. The term "source strength", often used in dispersion modeling, denotes either the total quantity released instantaneously or the release rate for continuous releases.

Two principal types of sources are used in dispersion modeling: the point source and the area source. An escape through a hole in a vessel or from a ruptured pipe is treated as a point source, while a liquid pool generating vapor is treated as an area source.

11.2.1 Liquid Releases

For the flow of a single-phase liquid that has a vapor pressure substantially below the static pressure at any point in the flow path, the rate of discharge through a hole from the bottom of a tank to the atmosphere is found using

$$Q_L = C_d A_h \rho_L \left[2\left(\frac{P - P_a}{\rho_L} \right) + 2gH_L \right]^{1/2} \quad (11.1)$$

where

Q_L = mass rate of flow, kg/s
C_d = discharge coefficient (usually 0.6 to 0.8)
A_h = cross-sectional area of the hole, m^2
g = gravitational acceleration, 9.81 m/s^2
H_L = liquid height above the level of discharge, m
P = pressure in the tank above the top of the liquid, Pa
P_a = atmospheric pressure, Pa
ρ_L = liquid density, kg/m^3

11.2.2 Gas Jet Releases[1]

As explained in Chapter 6, the procedure for calculating the flow rate of compressed gas through an opening depends on whether the flow is sonic or subsonic. As the source pressure increases, the release rate reaches a maximum when gas velocity equals the sonic velocity. The flow is then said to be "choked". The proper computational procedures for release rate calculations are explained in Chapter 6.

11.2.3 Two-Phase Releases

For releases of pressurized liquids, e.g., liquefied gases from storage tanks, the release occurs as a two-phase jet, consisting of a mixture of gas and liquid. The emission rate is somewhere between the rate for a gas and that for a liquid.

For releases of pressurized saturated liquids to the atmosphere, the fraction φ of the vapor in the jet is calculated by using a simplified heat balance for adiabatic flashing:

$$\varphi = \frac{C_{pl}(T_s - T_b)}{\lambda_v} \tag{11.2}$$

where

φ = vapor fraction of the jet
C_{pl} = specific heat of liquid, J/(kg.K)
T_s = storage temperature, K
T_b = normal boiling temperature of the liquid, K
λ_v = latent heat of vaporization of the liquid, J/kg

An alternative expression that purports to take into account the differential nature of the vaporization is[2]

$$\varphi = 1 - \exp\left[-\frac{C_{pl}(T_s - T_b)}{\lambda_v}\right] \tag{11.3}$$

where

φ = vapor fraction of the jet
C_{pl} = specific heat of liquid, J/(kg.K)
T_s = storage temperature, K
T_b = normal boiling temperature of the liquid, K
λ_v = latent heat of vaporization of the liquid, J/kg

For our purposes, we recommend Equation (11.2) over Equation (11.3).

There is enormous literature on the modeling of two-phase flow in pipelines. We have adopted an equation based on Fauske and Epstein, as quoted by CCPS.[1]

$$Q = \frac{A_h F \lambda_v}{(v_g - v_l)\sqrt{T_s C_{pl}}} \tag{11.4}$$

where

Q = mass flow rate of the two-phase mixture, kg/s
A_h = flow area, m^2
F = a frictional loss factor
λ_v = latent heat of vaporization of the liquid, J/kg
v_g = specific volume of vapor at storage temperature and pressure, m^3/kg
v_l = specific volume of liquid, m^3/kg
T_s = storage temperature of the liquid, K
C_{pl} = specific heat of liquid, J/(kg.K)

The frictional loss factor, F, accounts for the frictional dissipation based on the length-to-diameter (*L*/*D*) ratio of the exit pipe. Based on the work of Fauske and Epstein, the following values of F are often used:

L/*D*	0	50	100	200	400
F	1	0.85	0.75	0.65	0.55

Values of the frictional loss factor, F, may be interpolated within 1% of the values shown in the table above using the following approximation:

$$F = \left[A + B\left(L/D + 500\right)\right]^{(-1/C)} \tag{11.5}$$

where

$A = -6.3084664$
$B = 0.014606555$
$C = 3.1951816$

Example 11.1

Liquid propane stored in a spherical vessel at ambient temperature starts leaking from a pipeline connection at the vessel bottom. It is desired to calculate the rate of leakage. Data:

Storage temperature: 298 K	Pipeline diameter: 50 mm
Storage pressure: 9.4 atm	Leakage diameter: Full-bore rupture
Molecular weight: 44	Leakage area: 0.001963 m^2
Normal boiling point: 231 K	L/D ratio of exit pipe: 150
Liquid density: 494 kg/m^3	Frictional loss factor, F: 0.696
Latent heat of vaporization: 427,000 J/kg	
C_p for liquid: 2,713 J/(kg.K)	
C_p for vapor: 1,868 J/(kg.K)	

Specific volume of liquid = 1/0.494 = 0.0020 m^3/kg
Specific volume of vapor at 298 K and 9.4 atm = 0.0485 m^3/kg
Substituting values in Equation (11.4), leakage rate, as vapor and mist = 11.68 kg/s
From Equation (11.2), fractional flash = 2713(298 − 231)/427,000 = 0.4262
From Equation (11.3), fractional flash = 0.34

For reference, a rigorous flash calculation yields a fractional flash of 0.3932, closer to the value from Equation (11.2).

Taking a value of 0.40 for fractional flash, release rate of vapor = (11.68)(0.4) = 4.67 kg/s.

11.2.4 Evaporation from Liquid Pools

Evaporation from liquid pools is an essential item while considering the dispersion of vapors from area sources. Such pools can be formed on land or water when large quantities of liquid are spilled, usually from the failure of storage vessels or transfer pipelines. Two categories of liquid pools cover most of the scenarios that need to be considered in risk analysis, as follows:

- The liquid in the pool is LNG, ethane, or propane (i.e., normal boiling temperature well below the ambient temperature). The pool's heat source is mainly via conduction from the land or water below.

- The liquid in the pool is a nonboiling liquid, such as hexane, heptane, methanol, or gasoline, where the normal boiling temperature is well above the ambient temperature. The evaporation rate is controlled mainly by the rate of vapor diffusion through the liquid-air interface. A heat balance is used to determine the temperature at the surface of the evaporating liquid. The inputs for this calculation are: (1) convection from the air, (2) solar radiation, and (iii) for a continuous spillage of liquid into the pool, heat in the incoming liquid. The output is the heat utilized for vaporization. Heat conducted from the ground or water underneath the pool is neglected in this case.

In both cases, the pool may be a confined pool enclosed by a dike, or it may be an unconfined spreading pool where there is no dike. The methodology for estimating evaporation rates for these scenarios has been considered below for spillage of liquid on land. For cases involving spillage on water, interested readers are referred to Lees[2] or Shaw and Briscoe.[3]

11.2.4.1 Evaporation of Cryogenic Liquids

The equation often used for estimating the heat flux to the liquid pool from concrete or soil below the pool is given by[3,4]:

$$q = \frac{Ck_s(T_s - T_b)}{(\pi \alpha t)^{1/2}} \tag{11.6}$$

where

q = heat flux, J/(s.m²)
k_s = thermal conductivity of the soil, W/(m.K)
α = thermal diffusivity of the soil, m²/s
T_s = temperature of the soil, K
T_b = normal boiling point of the liquid, K
t = time from the start of the liquid spillage on the pool, seconds
C = surface roughness factor

C is often 1 for impenetrable surfaces and 3 for other surfaces, e.g., soils.[4] The thermal properties of soils and concrete are shown in Table 11.1.

TABLE 11.1
Thermal Properties of Concrete and Soils[4]

Material	Density (kg/m³)	C_p [J/(kg.K)]	k_s [W/(m².K)]	A (m²/s)
Concrete	2,300	961	0.92	4.16×10^{-7}
Average soil	2,500	836	0.96	4.59×10^{-7}
Dry, sandy soil	1,650	794	0.26	1.98×10^{-7}
Moist, sandy soil	1,750	1,003	0.59	3.36×10^{-7}

Example 11.2

Liquid propane at 231 K is spilled instantaneously on a concrete floor with a dike, 20 m diameter × 1 m high. The temperature of concrete is 293 K. The thermal properties of concrete are given in Table 11.1. Assuming that the spilled liquid covers the entire floor instantaneously, calculate the heat flux to the liquid pool and the rate of vaporization as a function of time in the range of 10–300 seconds. The latent heat of vaporization of propane is 427,000 J/kg. The value of surface roughness factor C can be taken as 3.

$T_s = 293$ K, $T_b = 231$ K, $\lambda_v = 427{,}000$ J/kg

$k_s = 0.92$ W/(m.K), $\alpha = 4.16 \times 10^{-7}$ m²/s

Surface area of the pool = (0.785) (20²) = 314 m²

Substituting values in Equation (11.4), we get the following:

Time, seconds	10	50	100	200	300
Heat flux, J/(s.m²)	47,347	21,174	14,972	10,587	8,644
Mass flux, kg/(s. m²)	0.1109	0.0496	0.0351	0.0248	0.0202
Vapor rate, kg/s	34.82	15.57	11.01	7.79	6.36

An average value of the vapor release rate in this example, over 5 minutes from the start of the liquid spillage, is around 15 kg/s.

If spillage of liquid into the pool is continuous and the spillage rate exceeds the evaporation rate, the dike's liquid level will continue to rise, and the liquid may spill over. Therefore, an important safety measure is to ensure that the volume of liquid in a storage vessel is always less than the dike's containment volume.

11.2.4.2 Evaporation of High Boiling Liquids[1]

In this case, the evaporation rate is determined by the mass transfer rate from the pool to ambient air. At steady state, this rate is given by

$$Q_v = \frac{k_g A_p P_s M_f}{R T_f} \tag{11.7}$$

where

Q_v = evaporation rate, kg/s

k_g = mass transfer coefficient of the interface film, m/s

A_p = surface area of the liquid pool, m²

P_s = vapor pressure of the evaporating liquid at the liquid temperature, atm

M = molecular weight of the vapor, kg/kgmole

R = universal gas constant = 0.08206 m³.atm/(kgmole K)

T_f = mean temperature of the interface, K

Ideally, the liquid temperature should be calculated from heat balance by taking into account the heat inputs such as (1) convection from air to the pool and (2) radiation from the sun. However, evaporation rates in these cases are relatively low. For risk analysis, it is appropriate to assume that the liquid temperature is equal to the ambient air temperature.

The mass transfer coefficient can be evaluated using standard correlations for the Sherwood number for forced convection flow over a flat plate[1]:

$$k_g = N_{\text{Sh}} D_v / L \tag{11.8}$$

$$N_{\text{Sh}} = 0.664\ N_{\text{Sc}}^{1/3} N_{\text{Re}}^{0.5}, \quad N_{\text{Re}} < 320{,}000 \tag{11.9}$$

$$N_{\text{Sh}} = 0.037 \quad N_{\text{Sc}}^{1/3}\left[N_{\text{Re}}^{0.8} - 15{,}200\right], \quad N_{\text{Re}} < 320{,}000 \tag{11.10}$$

$$N_{\text{Re}} = L u_w \rho_g / \mu_g \text{, and } N_{\text{Sc}} = \mu_g / (\rho_g D_v) \tag{11.11}$$

where

D_v is the diffusivity of the vapor in the air, m²/s
L = a characteristic length equal to the pool diameter for a circular pool and the downwind length for a rectangular pool, m
N_{Sh} = Sherwood number
N_{Re} = Reynolds number
N_{Sc} = Schmidt number
u_w = wind speed, m/s
μ_g = viscosity, kg/(m.s)
ρ_g = density of the air-vapor mixture at the mean film temperature, kg/m³

When the evaporation rate is low, the viscosity and density values can be taken equal to those of air.

Example 11.3

Acrylonitrile (normal boiling point = 352 K) at a temperature of 300 K is spilled instantaneously on a concrete floor enclosed by a dike 20 m diameter and 1 m height. The ambient temperature is 300 K, and wind speed is 5 m/s. Assuming that the spilled liquid covers the entire floor instantaneously, calculate the evaporation rate from the pool. The vapor pressure of acrylonitrile at 300 K is 126 mmHg or 0.165 atm.

The diffusivity of acrylonitrile vapor in air at 300 K is 1.15×10^{-5} m²/s. Air has a kinematic viscosity of 1.8×10^{-5} m²/s and a density of 1.18 kg/m³. The molecular weight of acrylonitrile is 53.

Substituting values in Equations (11.6) through (11.9), we get

$$N_{\text{Re}} = (20)(5)(1.18)/\left(1.8\times10^{-5}\right) = 6.55\times10^{6}$$

$$N_{\text{sc}} = \left(1.8\times10^{-5}\right)/\left[(1.18)\left(1.15\times10^{-5}\right)\right] = 1.33$$

$$N_{\text{sh}} = (0.037)\left(1.33^{0.333}\right)\left[\left(6.5510^{6}\right)^{0.8} - 15{,}200\right] = 10{,}900$$

$$\text{Mass transfer coefficient, } k_g = (10{,}900)\left(1.15\times10^{-5}\right)/20 = 6.27\times10^{-3}\ \text{m/s}$$

Substituting values in Equation (11.5),

$$\text{Evaporation rate, } Q_v = 6.27\times10^{-3}\,(0.785)\left(20^2\right)(0.165)(53)/\left[(0.082)(303)\right] = 0.69\ \text{kg/s.}$$

11.3 DISPERSION MODELS

Models are used to estimate the gas concentration in air as a function of distance from the source. These fall into three broad categories:

- Neutral or passive dispersion
- Dense gas dispersion
- Jet dispersion

11.3.1 Passive Dispersion

Passive dispersion (also called neutral or Gaussian dispersion) occurs when air movement is the main driving force for dispersion. Here, the density of the gas released is close to the density of ambient air. For dense gases, the gas cloud density is higher than that of air in the vicinity of the release source. However, it comes down progressively with downwind distance as mixing with air occurs. Therefore, at some downwind distance sufficiently away from the release source, a transition to passive dispersion occurs.

11.3.1.1 Factors Affecting Passive Dispersion

For passive dispersion, the crucial parameter that affects dispersion is atmospheric stability. This stability is a measure of the degree of turbulent mixing in the atmosphere. Unstable atmospheric conditions lead to better mixing and dispersion; stable conditions are the least effective for mixing.

The atmospheric conditions are generally classified according to six Pasquill stability classes (*A* through *F*). "A" represents the least stable conditions, whereas "F" represents the most stable. The stability classes are related to wind speed and sunlight intensity, as shown in Table 11.2.

Table 11.2 shows that increased wind speeds result in higher atmospheric stability during the day, while the reverse is true at night. This phenomenon could be explained by the fact that higher wind speeds yield a higher heat transfer coefficient at the interface between the ground and the air above. Therefore, on a sunny day when the ground is warmer than the ambient air, there is a higher air heating rate. The reverse is true at night when the ground is cooler than the ambient air.

It is common in dispersion calculations to consider two cases: (1) stability class D, together with an average wind speed of 3–5 m/s, representing neutral conditions, and (2) stability class *F*, together with a low wind speed of 2 m/s, representing stable conditions. *D* represents the worst situation during the day, while *F* represents the worst situation at night.

TABLE 11.2
Pasquill stability classes[2]

	Daytime Sunlight Intensity			Night Conditions	
Wind Speed at 10 m Height (m/s)	Strong	Moderate	Slight	Thinly Overcast or ≥4/8 Cloud	≤3/8 Cloud
<2	*A*	*A*–B	*B*	*F*	–
2–3	*A–B*	*B*	*C*	*E*	*F*
3–4	*B*	*B–C*	*C*	*D*	*E*
4–6	*C*	*C–D*	*D*	*D*	*D*
>6	*C*	*D*	*D*	*D*	*D*

A, extremely unstable; *B*, moderately unstable; *C*, slightly unstable; *D*, neutral; *E*, slightly stable; *F*, Stable.

The Pasquill stability classification is based on the weather conditions in England. Therefore, the effect of local conditions should be considered as appropriate. In particular, for coastal sites, land and sea breezes can become predominant.

The ground roughness is another parameter that affects passive dispersion modeling. Guidance from AIChE[5] on surface roughness lengths under different terrain conditions is summarized in Table 11.3.

For continuous releases, models are used to estimate the concentration profile as a function of downwind distance from the source. For instantaneous releases, a concentration versus time profile is used at specified locations downwind. The term "plume model" is used for continuous releases and "puff model" for instantaneous releases.

Figure 11.1 shows a plume originating from the top of a stack. It is directed along the mean wind direction. The *x* coordinate is measured downwind from the stack's vertical axis, *y* is the coordinate in the crosswind direction measured

TABLE 11.3
Data on Surface Roughness[5]

	Example	Roughness Length, z_o (m)
Flatland	Reclaimed land, level plain land with grass and few trees	0.1
Farmland	Airfields, agricultural land, reclaimed land with many trees	0.1
Cultivated land	Open area with much overgrowth, scattered house	0.3
Residential area	Ares with densely located but low buildings, wooded area, industrial areas with obstacles that are not too high	1.0
Urban area	A city with high buildings, industrial areas with high obstacles	3.0–6.0
Placid water/desert	Large lakes, water bodies, flay desert land	0.001
Sea/snowy plains	Calm oceans, snow-covered flats	0.001
Modern oil refinery	Tall columns, large tanks, etc.	1.0

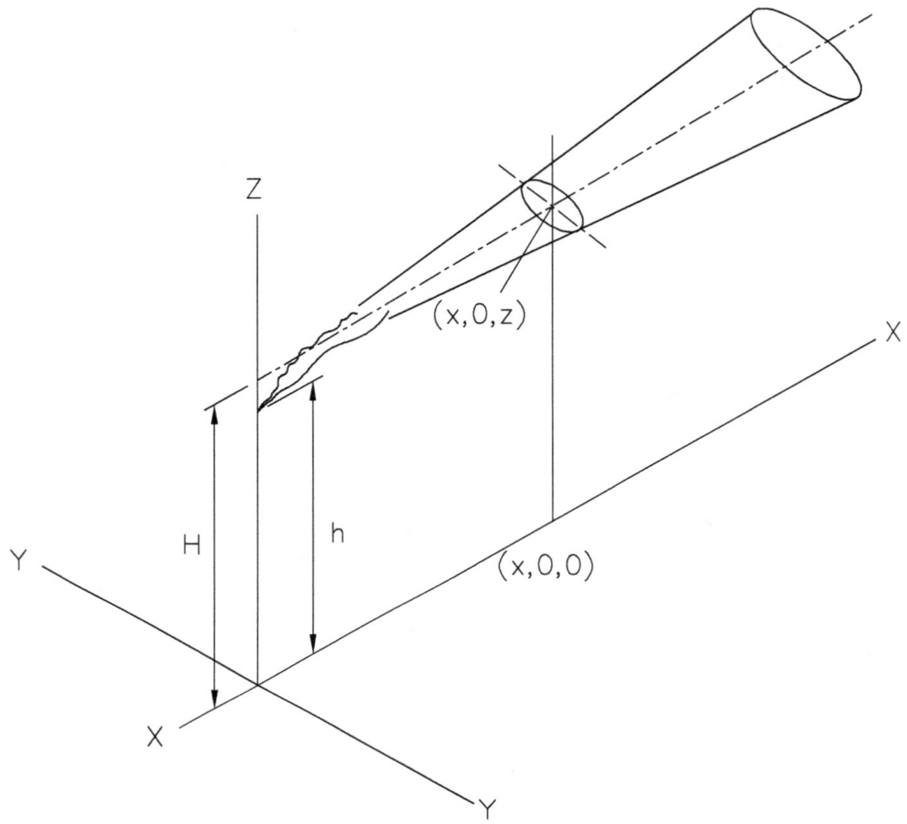

FIGURE 11.1 Coordinate system for a typical plume dispersion.

from the plume's axis, and z is the coordinate in the vertical direction measured from the ground.

As the plume disperses downwind, it entrains ambient air and is thus progressively diluted. At any downwind distance, the maximum concentration occurs at the axis of the cloud. The usual procedure is to define dispersion coefficients σ_x, σ_y, and σ_z as the standard deviations of the concentrations in the downwind, crosswind, and vertical (x, y, z) directions, respectively. The dispersion coefficients are a function of the atmospheric conditions and the distance downwind from the point of release.

One of the objectives of dispersion calculations is to determine the cloud boundary at ground level for a chosen concentration. The boundaries obtained by joining points of equal concentration are called isopleths or contours. Figure 11.2 shows such typical isopleths.

The boundary concentrations for isopleths are chosen based on safe permissible values. It is usual for a flammable gas to take 50% of the LFL. For a toxic gas, the ERPG or some other convenient measures could be used.

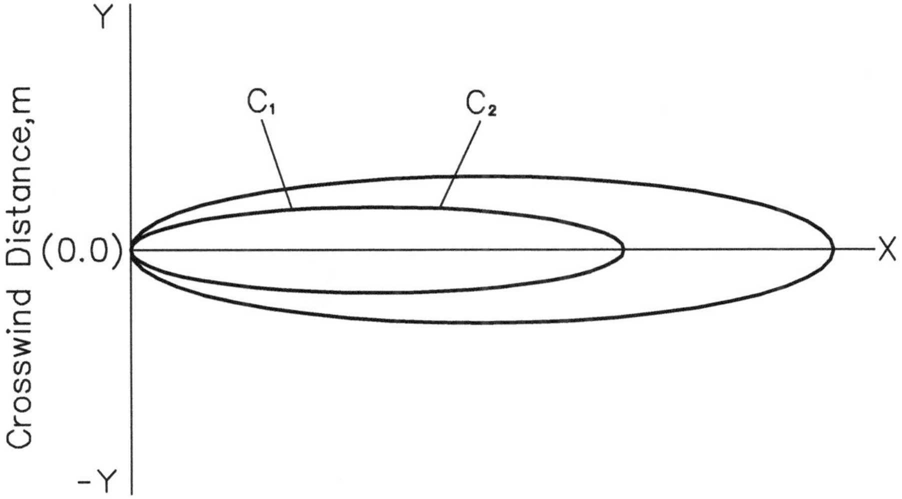

FIGURE 11.2 Typical isopleths (contours) at ground level for continuous release at ground level ($C_1 > C_2$) from a point source.

11.3.1.2 Dispersion Calculations

The calculation procedure is given below for the following scenarios:

1. Continuous release from a point source
2. Instantaneous release from a point source
3. Continuous release from an area source
4. Instantaneous release from an area source

11.3.1.2.1 Release from a Point Source

Several models, e.g., the Roberts, Sutton, Pasquill, and Pasquill-Gifford models, are available for passive dispersion. The Pasquill-Gifford model is more widely used and is described below. In this model, empirical correlations for dispersion coefficients are as given below:

$$\sigma_y = ax^b \tag{11.12}$$

$$\sigma_z = cx^d \tag{11.13}$$

where

σ_x and σ_y are dispersion coefficients, m
x is the downwind distance, measured from the point of release, m
a, b, c, and d are constants given in Table 11.4 for continuous releases.

For instantaneous releases, available data are quite limited, and only approximate values are given in Table 11.5.

TABLE 11.4
Values of Constants for Approximate Calculation of Dispersion Coefficients in Case of Continuous Release

Pasquill Stability Category		*a*	*B*	*c*	*d*
Very unstable	*A*	0.527	0.865	0.28	0.90
Unstable	*B*	0.371	0.866	0.23	0.85
Slightly unstable	*C*	0.209	0.897	0.22	0.80
Neutral	*D*	0.128	0.905	0.20	0.76
Stable	*E*	0.098	0.902	0.15	0.73
Very stable	*F*	0.065	0.902	0.12	0.67

TABLE 11.5
Values of Constants for Approximate Calculation of Dispersion Coefficients in Case of Instantaneous Release

Pasquill Stability Category		*a*	*b*	*c*	*d*
Unstable	*A, B*	0.14	0.92	0.53	0.73
Neutral	*C, D*	0.06	0.92	0.15	0.70
Stable	*E, F*	0.024	0.89	0.05	0.61

Values of constants given in Tables 11.4 and 11.5 are valid for downwind distances 100 m or more. For a distance smaller than 100 m, a value of σ (m) obtained by linear interpolation for distances between 0 and the value at 100 may be used.

It should also be noted that the values of the constants given above are valid for a flat terrain ($z_o = 0.1$ m). For continuous release scenarios, TNO[5] suggests incorporating a correction term C_{zo} for σ_z to account for variation in z_o. This correction is as follows:

$$C_{zo} = (10z_o)^{0.53x^{-0.22}} \tag{11.14}$$

$$(\sigma_z)_{\text{corrected}} = C_{zo}\,(\sigma_z)_{\text{at } z_o=0.1} \tag{11.15}$$

For a continuous release from an elevated point source, the equation for the concentration at various downwind locations is given by

$$C(x,y,z) = \frac{Q_c}{2\pi\sigma_y\sigma_z u}\exp\left(-\frac{y^2}{2\sigma_z^2}\right)\left[\exp\left(-\frac{(z-h)^2}{2\sigma_z^2}\right)+\exp\left(-\frac{(z+h)^2}{2\sigma_z^2}\right)\right] \tag{11.16}$$

where

$C(x, y, z)$ = concentration at a location whose coordinates are x, y, and z (measured from the projection of the release point on ground level), mg/m^3

Q_c = continuous rate of release, kg/s

σ_x, σ_y, and σ_y are dispersion coefficients, m
h = height of the point of release above ground level, m
u = wind speed at a 10 m height, m/s.

For an instantaneous point source at ground level, the equation for the concentration at various downwind locations at time t (counted from the time of release) is given by

$$C(x,y,z,t) = \frac{2Q_i}{(2\pi)^{3/2}\sigma_x\sigma_y\sigma_z}\exp\left(-\frac{1}{2}\left[\frac{(x-ut)^2}{\sigma_x{}^2}+\frac{y^2}{\sigma_y{}^2}+\frac{z^2}{\sigma_z{}^2}\right]\right) \quad (11.17)$$

where
C = concentration, mg/m^3
Q_i = quantity released, mg
σ_x, σ_y, and σ_y are dispersion coefficients, m
t = time from the instant of release, seconds
u = wind speed, m/s.

Here, σ_x is taken to be equal to σ_y.

For the continuous release of flammable gas from a point source at ground level, the mass of gas within the flammable region and the area at ground level enclosed by a contour of specified concentration is given by

$$\frac{Q_f}{Q_c} = \left(\frac{b+d}{b+d+1}\right)\left(\frac{x_L - x_U}{u}\right) \quad (11.18)$$

$$\text{Area} = \frac{a\sqrt{2\pi}}{b+d}\left(\frac{Q_c}{\pi u C_{\text{contour}}\, ac}\right)^{\frac{b+1}{b+d}}\left(\frac{b+d}{b+1}\right)^{3/2} \quad (11.19)$$

where
Q_f = mass of gas in the flammable region, kg
Q_c = continuous rate of release, kg/s
x_L and x_U = maximum distances to the lower and upper flammability limits, respectively, m
u = wind speed, m/s
[a, b, c, and d are dispersion coefficients from Tables 11.4 and 11.5]

In Equation (11.16), the area refers to the area (m^2) enclosed by a concentration contour C_{contour}, kg/m^3, Q_c is the continuous release rate, kg/s, and u is the wind speed, m/s.

These calculation methods are now illustrated with examples.

Example 11.4

Propane (molecular weight = 44) is released as a gas at ground level at 10 kg/s. Calculate the downwind distance at ground level where the maximum concentration in air is equal to half the LFL under two atmospheric conditions: (i) Pasquill stability category D at a wind speed of 5 m/s and (ii) Pasquill stability category F at a wind speed of 2 m/s. In each case, also calculate the flammable mass in the cloud. The upper and lower flammability limits for propane in air are 9.5% and 2.1% by volume, respectively. The ambient temperature is 298 K. Surface roughness z_o can be taken as 0.1 m.

$Q_c = 10$ kg/s

UFL = 9.5% by volume = (9.5/100) (44/22.4) (273/298) = 0.17 kg/m^3

LFL = 2.1% by volume = (2.1/100) (44/22.4) (273/298) = 0.038 kg/m^3

Since the release is at ground level, $h = 0$, the concentration at a downwind distance will be the maximum if y is zero and z is zero.

i. Stability category = D. Choose, $x = 100$ m. Then, from Equations (11.10) and (11.11) and values of a, b, c, and d from Table 11.4

$$\sigma_y = (0.128)100^{0.905} = 8.26\,\text{m}$$

$$\sigma_z = (0.20)\left(100^{0.76}\right) = 6.62\,\text{m}$$

Substituting values for Q_c, σ_y, σ_z and $u = 5$ m/s in Equation (11.14), we get $C = 0.012$ kg/m^3, which is much less than ½-LFL (half the LFL).

By trial and error, we find $x = 78$ m. Therefore,

$$\sigma_y = \left(78/100\right)8.26 = 6.44\,\text{m}$$

$$\sigma_z = \left(78/100\right)6.62 = 5.16\,\text{m}$$

Substituting the values for Q_c, u, and the new values for σ_y and σ_z, in Equation (11.14), we get $C = 0.019$ kg/m^3, which is equal to ½-LFL. Hence, the estimated distance to ½-LFL is 78 m.

Proceeding in the same manner, estimated distances to LFL and UFL are as follows:

$$x_{\text{LFL}} = 55\,\text{m, and } x_{\text{UFL}} = 26\,\text{m}$$

Substituting these values in Equation (11.15),

$$Q_f/Q_c = \left[(0.905+0.76)/(0.905+0.76+1)\right]\left[(55-26)/5\right] = 3.62, \text{ whence}$$

$$Q_f = 36\,\text{kg}$$

ii. Stability category = F. Again, by trial and error, we find $x = 370$ m. Then from Equations (11.10) and (11.11), and the values of a, b, c, and d from Table 11.4

$$\sigma_y = (0.065)\left(370^{0.902}\right) = 13.47\,\text{m}$$

$$\sigma_z = (0.12)\left(370^{0.67}\right) = 6.31\,\text{m}$$

Substituting for Q_c, σ_y, σ_z and u = 2 m/s in Equation (11.14), we get C = 0.019 kg/m^3, which is equal to ½-LFL.

Proceeding as in the case (i), we get the distance to ½-LFL = 370 m, x_{LFL} = 235 m, x_{UFL} = 92 m, and Q_f = 437 kg.

Example 11.5

For Example 11.5 (ii), draw the contour for a ground-level concentration corresponding to ½-LFL = 0.019 kg/m^3, and estimate the area enclosed by this contour.

At an assumed distance downwind, the methodology involves estimating the crosswind distance where the concentration becomes equal to 0.019 kg/m^3. For example, at a downwind distance x = 200 m, using values of a, b, c, and d for stability category F, and wind speed of 2 m/s, we get

$$\sigma_y = (0.065)\left(200^{0.902}\right) = 7.73\,\text{m, and}$$

$$\sigma_z = (0.12)\left(200^{0.67}\right) = 4.18\,\text{m}$$

Then, substituting values of Q_c = 10 kg/s, σ_y = 7.73 m, σ_z = 4.18 m, u = 2 m/s, h = 0, x = 200 m, and a trial value of 10.7 m for y in Equation (11.14), we get C = 0.019 kg/m^3.

The calculations are repeated for other values of the downwind distance. The results are summarized below.

Downwind distance x, m	100	150	200	250	300	370
Crosswind distance y, m	8.4	10.0	10.7	10.2	8.5	0
Concentration, kg/m^3	0.019	0.019	0.019	0.019	0.019	0.019

The contour diagram is shown in Figure 11.3. The area enclosed is estimated, by graphical integration, to be 5,840 m^2. The area can also be calculated based on Equation (11.16). Using values of a, b, c, and d for stability category F, a wind speed of 2 m/s, Q_c = 10 kg/s, and $C_{contour}$ = 0.019 kg/m^3, the calculated area is 5,870 m^2.

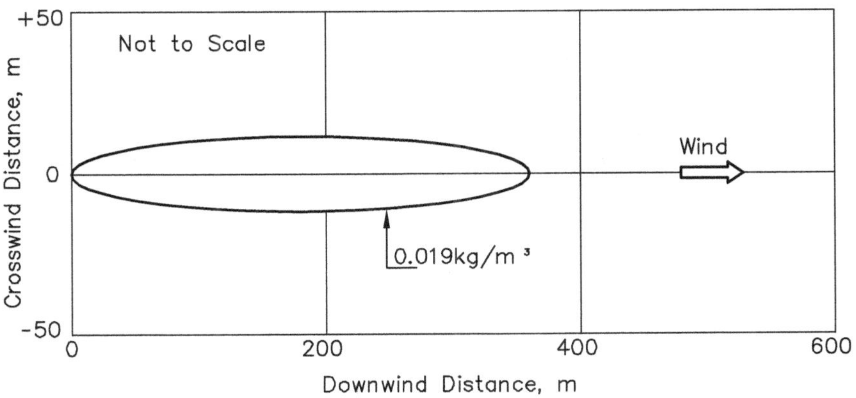

FIGURE 11.3 Contour diagram for Example 11.5.

Example 11.6

1,000 kg of chlorine gas (molecular weight = 71) is released instantaneously as a ground-level point source. Draw a contour diagram for a concentration of 35 ppm or 102 mg/m^3 in the air at ground level. Atmospheric condition: Pasquill stability category F at a wind speed of 2 m/s. The ambient temperature is 298 K. Surface roughness z_o can be taken as 0.1 m.

The steps involved are as follows:

- Select a time t (seconds) counted from the time of release
- Calculate the downwind distance (m) by multiplying the wind speed (m/s) with time
- Compute the dispersion coefficients σ_y and σ_z using values of a, b, c, and d from Table 11.5 for the "stable" stability category for the calculated value of x. Assume $\sigma_x = \sigma_y$.
- Based on Equation (11.13) and assuming $z = 0$ and $h = 0$, compute the value of y at which the concentration becomes equal to 102 mg/m^3. (A few trials will be required).
- Assume different times and compute the corresponding values of x and y in the same manner.
- Plot $+y$ and $-y$ versus x, as shown in Figure 11.4.
- The calculated values of x and y at different values of time are as follows:

t (seconds)	1,800	3,600	5,400	7,200	9,000	12,000	14,100
x (m)	3,600	7,200	10,800	14,400	18,000	24,000	28,200
y (m)	110	166	200	216	215	166	0

For example, at t = 9000 seconds, the cloud width is (2) (215) = 430 m. At a wind speed of 2 m/s, the cloud pass-over time is 215 seconds.

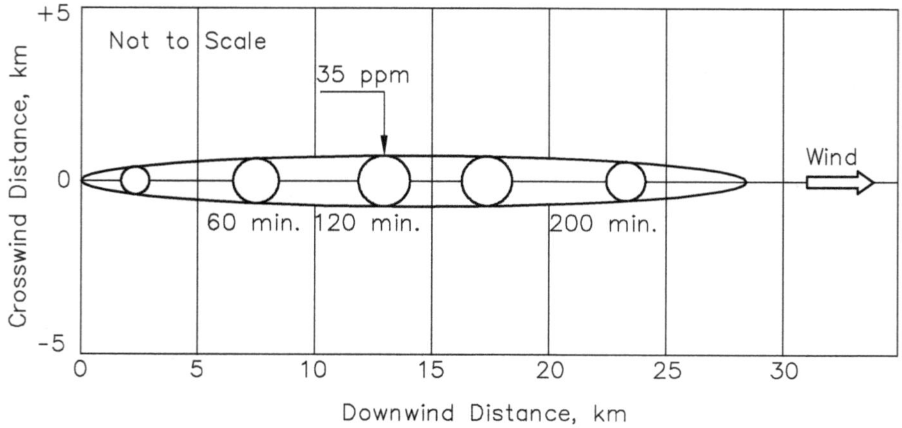

FIGURE 11.4 Contour diagram for Example 11.6.

11.3.1.2.2 Release from an Area Source

The models considered above for release from a point source need to be modified for release from an area source of finite dimensions. This situation arises when (1) the release is in the form of vapors rising from an evaporating liquid pool or (2) there is instantaneous flashing off of vapor at the source to form a relatively large cloud. The modified model involves the use of a "virtual source" concept.

In the virtual source model, a point A is located at an upwind distance x_v from the center B of a real area source (see Figure 11.5) at ground level, such that if a point source were present in A, it would produce a plume which, at B, would have the same crosswind width $2L_y$ of the real source. For the virtual source model, it is necessary to know the dimensions of the real source's actual release area. Details of this method are given in the TNO Yellow Book.[5]

In the virtual source model, the coordinates x, y, and z are referred to the virtual point source A, and the distance x_v is estimated based on the actual value L_y of the source in the y-direction. The model assumes that the L_y equals 2.15 times the value of σ_y at point B. This assumption implies that the concentration at the source's crosswind limit has dropped to 10% of the maximum concentration at point B. Hence,

$$L_y = 2.15\ \sigma_y = 2.15\ \left(a x_{vy}{}^{b}\right)$$

$$\text{or,} \quad x_{vy} = \left(\frac{L_y}{2.15\ a}\right)^{1/b} \tag{11.20}$$

Similarly,

$$x_{vz} = \left(\frac{L_z}{2.15c}\right)^{1/d} \tag{11.21}$$

where

L_y is the source dimension in the y-direction.
L_z is the source dimension in the z-direction.

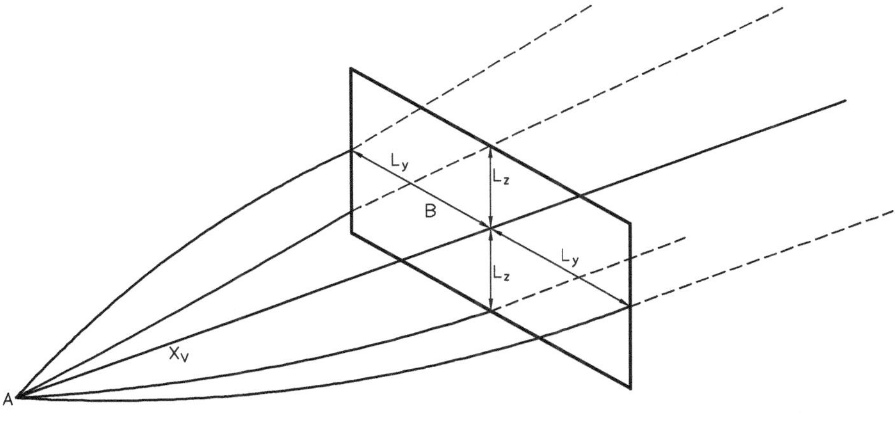

FIGURE 11.5 Schematic representation of the use of a virtual point source.

The downwind and crosswind concentrations can be estimated in the same manner as for point source releases using the values of x_{vy} and x_{vz}; these are found by substituting $(x_{vy} + x)$ and $(x_{vz} + x)$ in place of x while computing σ_y and σ_z, respectively.

Example 11.7

Liquefied propane at 231 K is released instantaneously onto an open concrete floor at ground level where it is fully contained within a dike of diameter = 30 m. The liquid evaporates continuously from the pool at a diminishing rate. One minute after the release commences, the estimated vaporization rate is 17 kg/s. Calculate the downwind distance at ground level where the maximum concentration of propane in the air is equal to ½-LFL (0.019 kg/m³). Atmospheric conditions: Pasquill stability category is *F* at a wind speed of 2 m/s. Surface roughness, z_o, can be taken as 0.1 m.

$Q_c = 17$ kg/s and $u = 2$ m/s

Since the release is at ground level, $h = 0$. The downwind concentration will be the maximum if *y* and *z* are zero.

For stability category *F*, values of (*a*, *b*, *c*, and *d*) are, from Table 11.4:

$$a = 0.065, b = 0.902, c = 0.12, d = 0.67$$

Source dimension L_y in the *y*-direction = 15 m. Source dimension L_z in the *z*-direction = 0. Substituting these values as well as those of *a*, *b*, *c*, and *d* in Equations (11.17) and (11.18), we get

$$x_{vy} = 178\,\text{m}$$

$$x_{vz} = 0$$

The next step involves assuming a value of downwind distance *x* from point *B*, calculating σ_y and σ_z, and substituting values in Equation (11.15) to calculate the concentration. The procedure is repeated for different assumed values of *x* until the calculated value of the composition is equal to the desired value; in this case, a value equal to ½ the LFL.

The value of *x* obtained by trial and error, as above, is 420 m, for which the results are shown below.

$$\sigma_y = a\left(x + x_{vy}\right)^b = 0.065(420+178)^{0.902} = 20.8\,\text{m}$$

$$\sigma_z = c\left(x + x_{vz}\right)^d = 0.12(420+0)^{0.67} = 6.9\,\text{m}$$

$$C = (17)(2)/[(2)(3.14)(20.8)(6.9)(2)] = 0.019\,\text{kg/m}^3$$

11.3.2 Dense Gas Dispersion

A dense gas is defined simply as one whose density exceeds the density of the air in which it is dispersed. The higher density usually results from a higher molecular weight of the gas and sometimes from a lower release temperature than the ambient air.

The mechanisms of dense gas dispersion differ markedly from those of passive dispersion. The notable characteristics of dense gas dispersion are as follows:

- In the immediate neighborhood of a release, the cloud slumps, spreading in the upwind, crosswind, and downwind directions (with passive dispersion, there is no upwind movement)
- As the cloud slumps, the air is entrained at the cloud front and also at the top of the cloud surface
- The entrained air dilutes the cloud and also changes its temperature. The relative density, therefore, drops progressively as the cloud moves downwind
- A transition point is reached when the density of the cloud and the air become equal. Thereafter, the passive dispersion model can be used

A dense gas dispersion model would represent the actual situation more closely than the passive model near the release point. Therefore, in the near field, a dense gas model would better estimate hazard distances than the passive model. In this region, the passive dispersion model would likely underestimate the lateral spread and overestimate the cloud height. The upwind hazard distance can be estimated only by using a dense gas model.

A large number of models are available in the literature for dense gas dispersion. Some of these are empirical, and some are semi-empirical. Some of them are theoretically elegant, but few provide the background data used for experimental validation.

Britter and McQuaid[6] developed a simple correlation for dense gas dispersion modeling by starting with fundamental equations, using dimensional analysis. They correlated their experimental data in terms of a set of dimensionless groups, and the correlations are shown graphically in Figures 11.6 and 11.7 for continuous and instantaneous releases, respectively.

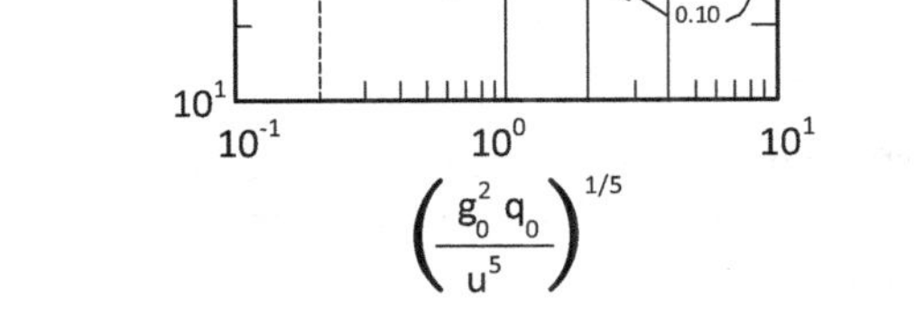

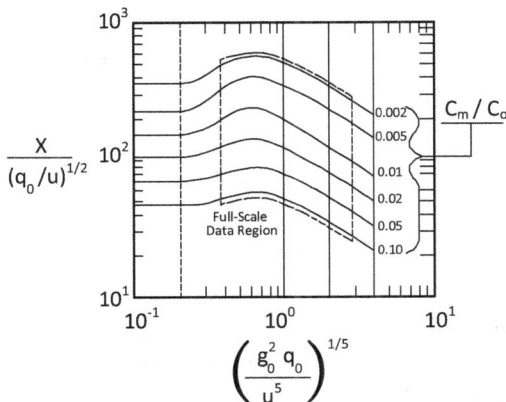

FIGURE 11.6 Britter and McQuaid correlation for dense gas dispersion – continuous release model.

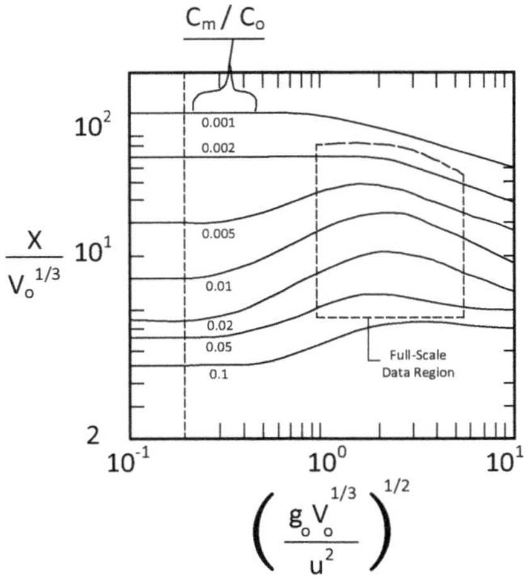

FIGURE 11.7 Britter and McQuaid correlation for dense gas dispersion – instantaneous release model.

The first step is to determine if the dense gas model is applicable. For this purpose, several variables are defined:

Reduced gravity of the initial release, g_o:

$$g_o = g\frac{\rho_o - \rho_a}{\rho_a} \tag{11.22}$$

where

g_o = reduced gravity of the initial release
g = gravitational acceleration = 9.81 m/s^2
ρ_o = density of the released gas or vapor, kg/m^3
ρ_a = density of the ambient air, kg/m^3

Characteristic source dimension: D_c (m) for continuous release and D_i (m) for instantaneous releases:

$$D_c = \left(\frac{q_o}{u}\right)^{1/2} \tag{11.23}$$

$$D_i = V_o^{1/3} \tag{11.24}$$

where

D_c = characteristic source dimension for a continuous release, m
D_i = characteristic source dimension for an instantaneous release, m
q_o = volumetric rate of the continuous release, m^3/s
V_o = volume of gas or vapor released instantaneously, m^3
u = wind speed at a 10 m height above ground, m/s.

The criteria for the gas cloud to be sufficiently dense for justifying the use of the dense gas model are as follows:

$$\text{Continuous releases: } \left(\frac{g_o q_o}{u^3 D_c}\right)^{1/3} \geq 0.15 \tag{11.25}$$

$$\text{Instantaneous releases: } \left(\frac{g_o V_o^{1/3}}{u D_i}\right)^{1/2} \geq 0.20 \tag{11.26}$$

The criteria for determining whether the release is continuous or instantaneous are determined by the value of a dimensionless group, as follows:

$$\text{Continuous: } \frac{u T_d}{x} \geq 2.5 \tag{11.27}$$

$$\text{Instantaneous: } \frac{u T_d}{x} \leq 0.6 \tag{11.28}$$

where
u = wind speed at a 10 m height above ground, m/s
T_d = duration of discharge, seconds
x = distance downwind, m.

If this group's calculated value lies between 0.6 and 2.5, calculations are done using both the continuous and instantaneous models, and the lower concentration result is selected.

The various symbols used in Figures 11.6 and 11.7 are summarized below:

q_o = release rate for continuous releases, m^3/s
V_o = quantity released as puff, m^3,
C_o = initial concentration of the gas released, vol%
C_m = concentration of the gas at a downwind distance x, vol%
ρ_o = initial density of the gas released, kg/m^3
ρ_a = density of ambient air, kg/m^3
g_o = reduced gravity of the initial release, $g\,(\rho_o - \rho_a)/\rho_a$, m/s^2
u = wind speed 10 m above ground level, m/s
x = downwind distance from the source, m

In Figures 11.6 and 11.7, the horizontal axis is a source Richardson number, and the vertical axis is a dimensionless distance. To facilitate computer calculations, CCPS[5] has provided equations corresponding to various curves in these figures, as shown in Tables 11.6 and 11.7 below:

TABLE 11.6
Equations for Graphical Correlations in Figure 11.6

$\varphi = \left(g_o^2\, q_o/u^5\right)^{0.2}$, $y = x/\left(q_o/u\right)^{1/2}$, $\alpha = \log(\varphi)$, $\beta = \log(y)$

Concentration Ratio, C_m/C_o	Validity Range for α	Equation for β
0.10	$\alpha \le -0.55$	$\beta = 1.75$
0.10	$-0.55 < \alpha \le -0.14$	$\beta = 0.24\,\alpha + 1.88$
0.10	$-0.14 < \alpha \le 1$	$\beta = 0.50\,\alpha + 1.78$
0.05	$\alpha \le -0.68$	$\beta = 1.92$
0.05	$-0.68 < \alpha \le -0.29$	$\beta = 0.36\,\alpha + 2.16$
0.05	$-0.29 < \alpha \le -0.18$	$\beta = 2.06$
0.05	$-0.18 < \alpha \le 1$	$\beta = -0.56\,\alpha + 1.96$
0.02	$\alpha \le -0.69$	$\beta = 2.08$
0.02	$-0.69 < \alpha \le -0.31$	$\beta = 0.45\,\alpha + 2.39$
0.02	$-0.31 < \alpha \le -0.16$	$\beta = 2.25$
0.02	$-0.16 < \alpha \le 1$	$\beta = -0.54\,\alpha + 2.16$
0.01	$\alpha \le -0.70$	$\beta = 2.25$
0.01	$-0.70 < \alpha \le -0.29$	$\beta = 0.49\,\alpha + 2.59$
0.01	$-0.29 < \alpha \le -0.20$	$\beta = 2.45$
0.01	$-0.20 < \alpha \le 1$	$\beta = -0.52\,\alpha + 2.35$
0.005	$\alpha \le -0.67$	$\beta = 2.40$
0.005	$-0.67 < \alpha \le -0.28$	$\beta = 0.59\,\alpha + 2.80$
0.005	$-0.28 < \alpha \le -0.15$	$\beta = 2.63$
0.005	$-0.15 < \alpha \le 1$	$\beta = -0.49\,\alpha + 2.56$
0.002	$\alpha \le -0.69$	$\beta = 2.60$
0.002	$-0.69 < \alpha \le -0.25$	$\beta = 0.39\,\alpha + 2.87$
0.002	$-0.25 < \alpha \le -0.13$	$\beta = 2.77$
0.002	$-0.13 < \alpha \le 1$	$\beta = -0.50\,\alpha + 2.71$

The effect of nonisothermal mixing is an additional factor that needs to be considered while using the Britter and McQuaid models when the release temperature is significantly different from the ambient temperature. The concentration C_o is taken as unity, assuming that the released gas comprises a single component. For a nonisothermal release, the concentration C_m in the above correlations is given by

$$C_m = \frac{C_i}{C_i + (1 - C_i)\left(T_a/T_o\right)} \tag{11.29}$$

where

C_i = concentration (volume fraction) under isothermal mixing
T_a = ambient temperature, K
T_o = temperature of the gas/vapor on release, K.

TABLE 11.7
Equations for Graphical Correlations in Figure 11.7

$\psi = \left(g_0 V_o^{1/3}/u^2\right)^{1/2}$, $y = x/V_0^{1/3}$, $\alpha = log\ (\psi)$, $\beta = log\ (y)$

Concentration Ratio, C_m/C_o	Validity Range for α	Equation for β
0.10	$\alpha \leq -0.44$	$\beta = 0.70$
0.10	$-0.44 < \alpha \leq 0.43$	$\beta = 0.26\ \alpha + 0.81$
0.10	$0.43 < \alpha \leq 1$	$\beta = 0.93$
0.05	$\alpha \leq -0.56$	$\beta = 0.85$
0.05	$-0.56 < \alpha \leq 0.31$	$\beta = 0.26\ \alpha + 1.00$
0.05	$0.31 < \alpha \leq 1$	$\beta = -0.12\ \alpha + 1.12$
0.02	$\alpha \leq -0.66$	$\beta = 0.95$
0.02	$-0.66 < \alpha \leq 0.32$	$\beta = 0.36\ \alpha + 1.19$
0.02	$0.32 < \alpha \leq 1$	$\beta = -0.26\ \alpha + 1.38$
0.01	$\alpha \leq -0.71$	$\beta = 1.15$
0.01	$-0.71 < \alpha \leq 0.37$	$\beta = 0.34\ \alpha + 1.39$
0.01	$0.37 < \alpha \leq 1$	$\beta = -0.\ 38\ \alpha + 1.66$
0.005	$\alpha \leq -0.52$	$\beta = 1.48$
0.005	$-0.52 < \alpha \leq 0.24$	$\beta = 0.26\ \alpha + 1.62$
0.005	$0.24 < \alpha \leq 1$	$\beta = -0.30\ \alpha + 1.75$
0.002	$\alpha \leq 0.27$	$\beta = 1.83$
0.002	$0.27 < \alpha \leq 1$	$\beta = -0.32\ \alpha + 1.92$
0.001	$\alpha \leq -0.10$	$\beta = 2.075$
0.001	$-0.10 < \alpha \leq 1$	$\beta = -0.27\ \alpha + 2.05$

Example 11.8

17 kg/s of liquefied propane vapor at 231 K is generated from a pool at ground level. The duration of release is sufficiently long for the process to be considered as continuous. Calculate the downwind distance at which the concentration of propane in the air drops to ½ LFL ($C_i = 0.0105$). Ambient air temperature is 298 K. Wind speed is 2 m/s, 10 m above the ground.

The first step is to check if the dense gas model is applicable.

$$\rho_o = (44/22.4)(273/231) = 2.321\,\text{kg/m}^3$$

$$\rho_a = (29/22.4)(273/298) = 1.186\,\text{kg/m}^3$$

$$g_o = 9.81(2.321 - 1.186)/1.186 = 9.39\,\text{m/s}^2$$

$$\text{Vapor generation rate, } q_o = 17/2.321 = 7.324\ \text{m}^3/\text{s}$$

Characteristic source dimension, from Equation (11.25):

$$D_c = (7.324/2)^{0.5} = 1.91\,\text{m}$$

The value of the criterion for the dense gas model, Equation (11.27), is $\left(g_o q_o/(u_3 D_c)\right)0.333 = \left[(9.39)(7.324)/\{(23)(1.91)\}\right]0.333 = 1.65$. This value being more than 0.15, the dense gas model is applicable.

Referring to Table 11.6, source Richardson number

$$\varphi = \left(g_o^2 q_o/u^5\right)^{0.2} = \left[\left(9.39^2\right)(7.324)/2^5\right]^{0.2} = 1.824$$

$$\alpha = \log(1.824) = 0.26$$

From Equation (11.31):

$$C_m = 0.0105/\left[0.0105 + (1 - 0.0105)\left(298/231\right)\right] = 0.0082\,(\cong 0.01)$$

From Table 11.6, at $\alpha = 0.62$ and $C_m/C_o = 0.01$,

$$\beta = (-0.52)(0.26) + 2.35 = 2.215, \text{ whence}$$

$$\text{Downwind distance, } x = 102.215\left(7.324/2\right)0.5 = 314 \text{ m}$$

Example 11.9

Liquefied propane stored at 298 K is released at ground level from a ruptured bottom outlet. The initial quantity of flash vapor is 40,000 kg. The duration of the release is sufficiently small for the process to be considered instantaneous. Calculate the downwind distance at which the propane concentration in the air drops to ½ LFL ($C_i = 0.0105$). The ambient air temperature is 298 K, and the temperature of propane after flashing is 231 K. Wind speed is 2 m/s, 10 m above the ground.

The first step is to check if the dense gas model is applicable.

$$\rho_o = \left(44/22.4\right)\left(273/231\right) = 2.321 \text{ kg/m}^3$$

$$\rho_a = \left(29/22.4\right)\left(273/298\right) = 1.186 \text{ kg/m}^3$$

$$g_o = 9.81(2.321 - 1.186)/1.186 = 9.39 \text{ m/s}^2$$

$$\text{Initial volume of vapor released, } V_o = 40{,}000/2.321 = 17{,}200 \text{ m}^3$$

Characteristic source dimension, from Equation (11.26):

$$D_i = 17{,}200^{0.333} = 25.7 \text{ m}$$

Criterion for dense gas model, Equation (11.28): $\left(g_o V_o\right)^{0.5}/(u D_i) = \left[(9.39)(17{,}200)\right]^{0.5}/\left[(2)(25.7)\right] = 7.82$. This value being more than 0.20, the dense gas model is applicable.

Referring to Table 11.7, source Richardson number

$$\psi = \left(g_o V_o^{0.333}/u^2\right)^{0.5} = \left[(9.39)\left(17,200^{0.333}/2^2\right)\right]^{0.5} = 7.77$$

$$\alpha = \log(7.77) = 0.89$$

From Equation (11.31):

$$C_m = 0.0105/\left[0.0105 + (1 - 0.0105)\left(298/231\right)\right] = 0.0082 \quad (\cong 0.01)$$

From Table 11.7, at $\alpha = 0.89$ and $C_m/C_o = 0.01$:

$$\beta = (-0.38)(0.89) + 1.66 = 1.322, \text{ whence}$$

$$\text{Downwind distance, } x = \left(10^{1.322}\right)\left(17,200^{0.333}\right) = 540 \text{ m}$$

11.3.3 Jet Dispersion

It is common practice in the process industries to discharge materials into the atmosphere. Intended discharges occur through chimneys, vents, and relief devices. Accidental releases from pressurized systems could be at high velocities, which cause entrainment of ambient air by momentum transfer and thereby dilute the material in the plume path. For upwardly pointed discharges, the plume rises to some height because of the momentum of the jet.

Two cases have been considered in this section – (1) dense gas jets and (2) positively buoyant jets.

11.3.3.1 Dense Gas Jet Dispersion

The model considered in this section is due to Hoot, Meroney, and Peterka (commonly known as the HMP model), which is frequently used and has been described by Lees[2] and CCPS.[5] The model assumes an upward-pointing dense plume from a vertical stack. The ambient atmosphere is defined by density ρ_a and the wind speed u at an altitude equal to the stack height. The model provides equations to calculate

a. the height at which the dense plume stops rising and starts to bend along the wind direction toward the ground
b. the touchdown distance and
c. the concentration on the axis of the plume at the point of maximum rise as well as when the plume strikes the ground

For the initial plume rise:

$$\frac{\Delta h}{2R_o} = 1.32\left(\frac{w_o}{u}\right)^{\frac{1}{3}}\left(\frac{\rho_o}{\rho_a}\right)^{\frac{1}{3}}\left[\frac{w_o^2 \rho_o}{2R_o g(\rho_o - \rho_a)}\right]^{1/3} \tag{11.30}$$

For the touchdown distance, i.e., the downwind distance at which the centerline of the plume strikes the ground:

$$\frac{x_t}{2R_o}=\frac{w_o u \rho_o}{2R_o g(\rho_o-\rho_a)}+0.56\left\{\left(\frac{\Delta h}{2R_o}\right)^3\cdot\left[\left(2+\frac{h_s}{\Delta h}\right)^3-1\right]\cdot\frac{u^3\rho_a}{2R_o g w(\rho_o-\rho_a)}\right\}^{1/2} \tag{11.31}$$

The ratio of the maximum concentration C to the initial concentration C_o at the point of the maximum rise of the plume:

$$\frac{C}{C_o}=1.688\left(\frac{w_o}{u}\right)\left(\frac{\Delta h}{2R_o}\right)^{-1.85} \tag{11.32}$$

The ratio of the maximum concentration C to the initial concentration C_o at the point x_t where the centerline of the plume strikes the ground:

$$\frac{C}{C_o}=2.43\left(\frac{w_o}{u}\right)\left(\frac{h_s+2\Delta h}{2R_o}\right)^{-1.95} \tag{11.33}$$

In Equations (11.30–11.33):

C = concentration at a specified downwind position, vol%.
C_o = initial concentration at stack exit, vol%
R_o = radius of the stack, m
h_s = height of the stack, m
Δh = maximum plume rise, m
u = wind speed, m/s
w_o = velocity of the released gas at stack exit, m/s
x = distance downwind, m
x_t = touchdown distance at ground level from the base of the stack, m
ρ_o = density of the released gas at stack exit, kg/m^3
ρ_a = density of the ambient air, kg/m^3

Example 11.10

Butane is being released continuously from the top of a vertical stack of 0.75 m internal diameter and 20 m height. The temperature of butane and the velocity of release at stack exit are 278 K and 12.7 m/s, respectively. The ambient air temperature is 298 K, and the wind speed is 4 m/s. Calculate the maximum plume rise, the touchdown distance at ground level, the concentration ratio at maximum plume rise, and at the touchdown distance.

$$\rho_o=(58/22.4)(273/278)=2.54\ \text{kg/m}^3$$

$$\rho_a=(29/22.4)(273/298)=1.19\ \text{kg/m}^3$$

$$R_o=0.375\ \text{m}$$

$$h_s = 20 \text{ m}$$

$$w_o = 12.7 \text{ m/s, and}$$

$$u = 4 \text{ m/s}$$

Substituting these values in Equation (11.32), $\Delta h/(2R_o) = 8.62$, whence

$$\Delta h = (8.62)(2)(0.375) = 6.5 \text{ m}$$

Substituting the values of w_o, u, ρ_o, ρ_a, R_o, h_s, and Δh in Equations (11.28), (11.29), and (11.30), we get the touchdown distance = 104 m.

$$C/C_o \text{ at max plume rise} = 0.10, \text{ and } C/C_o \text{ at touchdown} = 0.005.$$

11.3.3.2 Positively Buoyant Jet Dispersion

The TNO Yellow Book[5] refers to Brigg's method and the Dutch national model to calculate the plume rise for positively buoyant jets. The equation for the plume trajectory, neglecting any rise due to momentum, is given by

$$\Delta h_B(x) = 1.60 \, F_o^{1/3} \, x^{2/3}/u \tag{11.34}$$

where

x = distance downwind, m
$\Delta h_B(x)$ = plume rise, measured from stack height at x, m
u = wind speed at a minimum 10 m release height, m/s, and
F_o = buoyancy flux factor, defined as

$$F_o = g b_o^2 w_o \left(1 - \frac{\rho_o}{\rho_a}\right) \tag{11.35}$$

where

g = gravitational acceleration= 9.81, m/s^2
b_o = radius of the stack, m
w_o = velocity of the gas released at stack exit, m/s
ρ_o = density of the gas at stack exit point, kg/m^3
ρ_a = density of the ambient air, kg/m^3

For neutral or unstable atmospheres (Pasquill category A, B, C, and D), the final plume rise is obtained by substituting $x = x_r$ in Equation (11.36), where x_r is given by

$$\begin{aligned} x_r &= 49 \quad F_o^{5/8} \quad \text{if} \quad F_o \le 55 \\ x_r &= 119 \; F_o^{2/5} \quad \text{if} \quad F_o > 55 \end{aligned} \tag{11.36}$$

where

F_o = buoyancy flux factor (Equation 11.35)

As a rule of thumb, $x_{r\,max}$, the maximum value of x_r is specified as

$$x_{r\,max} = 87u \tag{11.37}$$

Hence, if the value of x_r calculated from Equation (11.36) exceeds $x_{r\,max}$ from Equation (11.37), $x_{r\,max}$ should be used.

The TNO Yellow Book also quotes Brigg's expressions for plume rise when momentum forces dominate. These expressions are

$$\Delta h_m = 6\ b_o w_o / u$$
$$\text{and} \quad x_r = 18\ b_o w_o / u \tag{11.38}$$

where

Δh_m is the plume rise, under momentum-dominated conditions, m

In cases where both momentum and buoyancy contribute to the final plume rise, it is calculated as follows:

Example 11.11

Methane is being released continuously from the top of a vertical stack having an internal diameter of 2 m. The methane temperature and the velocity of release at stack exit are 283 K and 10 m/s, respectively. The ambient air temperature is 298 K, and the wind speed is 4 m/s. Calculate the maximum plume rise.

$$\rho_o = (16/22.4)(273/283) = 0.69 \text{ kg/m}^3$$

$$\rho_a = (29/22.4)(273/298) = 1.19 \text{ kg/m}^3$$

$$b_o = 2 \text{ m},\ w_o = 10 \text{ m/s},\ u = 4 \text{ m/s}$$

Substituting values in Equation (11.32), F_o = 164 m^4/s^3

Substituting the value of F_o in Equation (11.36), x_r = (119) ($164^{0.4}$) = 915 m

Maximum value of x_r from Equation (11.37) = (87) (4) = 348 m, whence from Equation (11.34):

$$\Delta h_B = (1.60)\left(164^{0.333}\right)\left(348^{0.667}\right)/4 = 108 \text{ m}$$

From Equation (11.35), Δh_m = (6) (2) (10)/4 = 30 m. Substituting the values of Δh_B and Δh_m in Equation (11.36), we get

$$\Delta h = \left(108^3 + 30^3\right)^{0.333} \cong 108 \text{ m}.$$

This result shows that the effect of momentum forces, in this case, is negligible.

Such calculations are done following "dispersion modeling" explained in Section 11.3. Several software programs are based on the same methodology for determining plume size considering "passive dispersion" using Gaussian or similar algorithms. Most of the field data on dispersion are based on tests carried out in flat, rural terrains where a surface roughness length is taken as 0.1 m. In urban areas, tall buildings and industrial installations increase local turbulence, thereby facilitating better dispersion; however, these factors are not considered. Though commonly used, the Gaussian model generally gives very conservative results, showing much higher distances than reality.

For realistic dispersion distance calculation, the challenges are as follows:

- Modeling the effects of multicomponent materials with varying physical and chemical properties
- Incorporating the effects of terrain, barriers, slopes, structures, and buildings
- Changes in weather condition during the release
- Near and far-field effects
- When the dispersing gas is significantly heavier than air, e.g., chlorine, the gas cloud will tend to follow surface contours, e.g., ditches, slopes, and streams in a downhill direction (with such materials, it takes a fairly high wind speed to create turbulence).

Some standard, simpler models (such as Gaussian plume models or 2-D slab models used for dispersion modeling) do not account for topography, buildings, and obstructions. Hence, the results from them tend to be overly conservative in the far field and unreliable in the near field. Such predictions could lead to expensive decisions for reducing risks.

11.4 COMPUTATIONAL FLUID DYNAMICS MODELLING

Computational fluid dynamics (CFD) is a branch of fluid mechanics that uses numerical analysis and data structures to analyze and solve complex problems in fluid flow, and simultaneous heat and mass transfer for geometries of arbitrary complexity. CFD modeling techniques are used increasingly as an alternative to the conventional models discussed above. These overcome some of the theoretical challenges described in this chapter, require powerful computers, and provide far more realistic pollutant dispersion estimates. First-principles software is used to perform the calculations required to simulate the free-stream flow of the fluid. Several commercial CFD software packages are available for modeling atmospheric dispersion around the built-up area of a chemical plant: DNV GL – Kameleon FireEx-KFX, PHONIX-GRAD, and FLUIDYN PHOENICS.

REFERENCES

1. AIChE/CCPS: *Guidelines for Use of Vapor Cloud Dispersion Models* (AIChE, New York, 1996).
2. Mannan, S.: *Lees' Loss Prevention in the Process Industries* (4th Ed., Butterworth-Heinemann, Oxford, 2012).

3. Shaw, P. and Briscoe, F.: *Evaporation from Spills of Hazardous Liquids on Land and Water*, Safety and Reliability Directorate (UKAEA) Report No. SRD R 100 (1978).
4. Entec U.K. Ltd., London (Private Communication).
5. AIChE/CCPS: *Guidelines for Chemical Process Quantitative Risk Analysis* (2nd Ed., AIChE, New York, 2000).
6. Britter, R. E. and McQuaid, J.: *Workbook on the Dispersion of Dense Gases*, HSE Contract Research Report No. 17/1988, Sheffield, UK.

12 Hazard Identification

12.1 FRAMEWORK FOR HAZARD MANAGEMENT

Managing safety in a process plant requires a clear understanding of two terms: "hazard" and "risk". The term hazard denotes a significant potential to cause harm. "Process hazard" refers to potential harm arising out of process plant operations. Management of process hazards involves several steps, as shown in Figure 12.1. The first step involves identifying those hazards (chemicals, operating conditions, design systems, etc.) that can harm people, the process plant, or the environment.

"Risk" involves an assessment (qualitative or quantitative) of the probability of the hazard potential being realized. The assessed risk may be acceptable, or measures may be necessary to reduce the risk to an acceptable level.

This chapter deals with various methods for the identification of hazards. Methods of assessing risks and measures to reduce them have been considered in Chapter 13. Chapter 13 also deals with management systems and procedures necessary to ensure operations at an acceptable level of safety and risk.

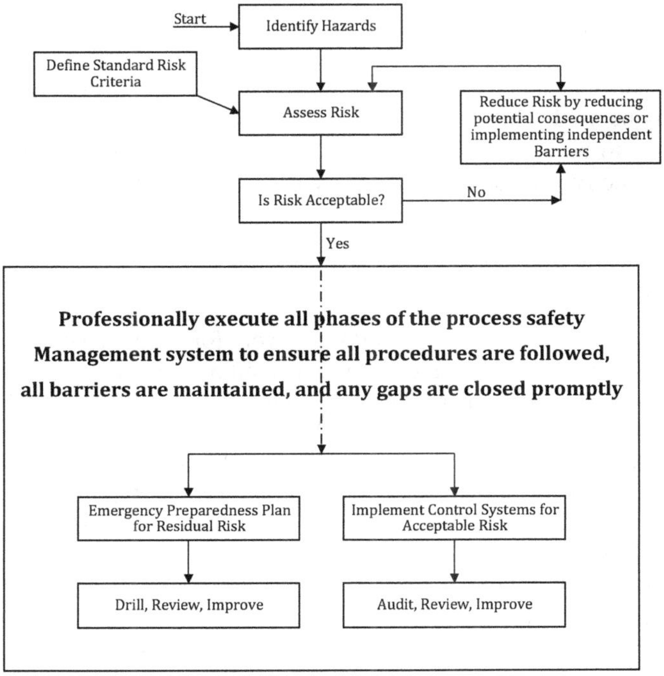

FIGURE 12.1 Framework for management of process plant hazards.

DOI: 10.1201/9781003107873-12

The key word in the definition of hazards is *potential*. The existence of any hazard potential in a process, or system, is not always readily noticeable. Therefore, formal methods are necessary for identifying hazards before an event occurs with harmful consequences. Some forms of hazards that may exist are as follows:

- Material hazards (flammability, reactivity, toxicity, etc.)
- Processing hazards (high pressure, high temperature, high vacuum, etc.)
- Site-related hazards (populated environment, congested layout, etc.)
- Human errors
- Pitfalls in management systems (inadequate training of personnel, deficiencies in safety systems for work, etc.)

These hazard identification techniques are covered in greater detail below.

12.2 HAZARD IDENTIFICATION METHODS

A variety of methods are available for hazard identification; a typical list follows:

- Safety audit
- What-if/checklist
- Hazard and operability (HAZOP) study
- Failure modes and effects analysis (FMEA)
- Fault tree analysis
- Event tree analysis

Each of the above methods has its strengths and limitations. Therefore, one or more of these may be used. The choice depends on the type of facility and potential hazardous events to be considered. Also, the regulatory authority having jurisdiction (AHJ) may specify a specific methodology or even mandate more than one of the approaches listed above.

12.2.1 Safety Audit

Safety audits are essential in the overall framework for the management of hazards. The purpose of an audit is to evaluate the expected performance against applicable standards for any operating plant. These standards may be external, such as government regulations. They may also be internal, as may be specified in company policies and guidelines. The audit team members may be company employees, usually the case with routine plant inspections, or external third parties. External audits are often performed by reputed safety and risk consultants, such as DNV-GL, ABS, GEXCON, LLOYDS, etc., who assign highly qualified and experienced safety professionals to the task. Further, statutory audits are carried by AHJs, whose objective is primarily to check compliance with applicable laws and regulations.

For chemical plants, regulatory safety audits and independent external third-party audits usually are conducted every 2–3 years. Of course, internal safety audits are carried out more frequently, say every six months or annually. Internal audits provide a basis for periodic reviews of safety system performance; they are required to ensure better planning and speedy implementation of any safety-related corrective actions or improvements that may be identified. Checklists ensure comprehensive coverage of such issues. For example, guidelines on internal safety audits are issued by various governmental regulatory authorities, company SHE departments, and international standards. An example would be the ISO/TS 29001 *Standard-Quality Management Certification for Petroleum, Petrochemical, and Natural Gas Industries*, containing exhaustive checklists that cover the following essential items (others may also be required, however):

Organization and administration:

- Systems of work (permit systems, disaster management, occupational health and safety management, management of accidents, near-misses, etc.)
- Employee training
- Fire protection systems
- Plant layouts
- Process safety systems, including relief and disposal systems
- Storage and material movement
- Loading/unloading
- Testing and maintenance of cross-country pipelines

12.2.2 What-If Checklist

In this approach, the audit team comprises analysts who can identify potential accident scenarios based on their experience, expertise, and knowledge. The study team would typically consist of: the plant manager, a hazard analyst, and a specialist highly familiar with the plant/process.

A what-if checklist is often an excellent method for hazard analysis at the conceptual stage when sufficient details of the process or the equipment are still not available. Therefore, the objective is to focus only on critical issues.

The flowchart shown in Figure 12.2 below explains the main steps involved:

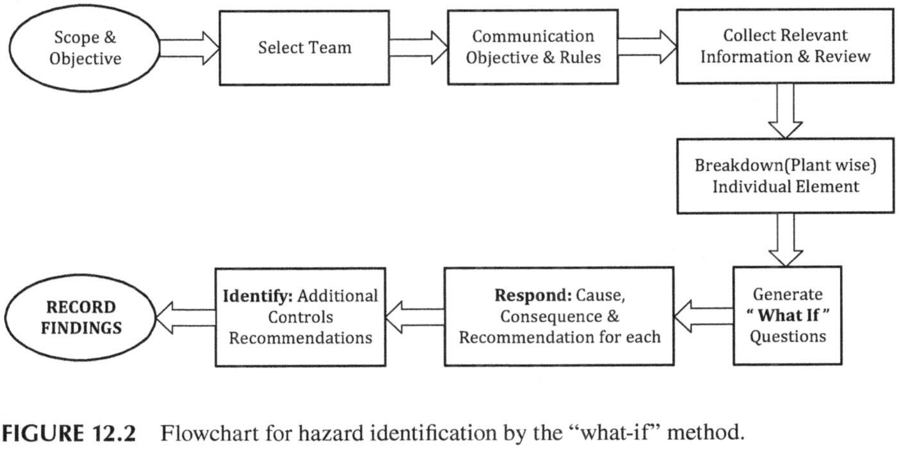

FIGURE 12.2 Flowchart for hazard identification by the "what-if" method.

TABLE 12.1
Simple Format for "What-If" Analysis

"What-If" Hazard Analysis

Co.	By:	
Plant Section	Operation Considered	Date

What If? Scenarios	Answer	Likelihood	Consequences	Recommendations

The procedure is based on examining the process by asking questions about the operational procedure or process design. Typical examples of questions that are asked are as follows:

- What if the pump stops?
- What if the coolant flow stops or if the temperature sensor fails?

Although the starting phrase in the above questions is "what-if", we are free to use other scenarios for inquiries. An example of a what-if hazard analysis format is shown in Table 12.1.

The "what-if" method usually relies on one or more checklists to facilitate free and exhaustive questioning and investigation. These checklists should cover a wide range of questions so that nothing that would be considered obvious is overlooked. A typical checklist would include the following areas:

1. Loss of utilities
    - Steam at various pressures
    - Electric power
    - Cooling water
    - Process water
    - Boiler feedwater
    - Instrument air
    - Natural gas/other fuel gas
    - Effluent water
2. Pressure relief
    - Relief valves
    - Bursting disks
    - Flare header
    - Flare
3. Instruments and controls
    - Local instruments
    - Board-mounted instruments

- Distributed control systems (DCSs)
- Control loops
- Emergency shutdown loops

4. Emergency systems
   - Firewater
   - Fire-fighting equipment
   - External fire
   - Runaway reactions
5. Human factors
   - Operating procedures
   - Training
6. Site location
   - Effluent disposal
   - Location of public water streams
   - Distance from populated areas

A typical example of a structured "what-if" checklist is shown in Table 12.2.

A description of this method has been included by Sutton[1] in his book on process safety.

12.2.3 HAZOP Study

The HAZOP study is a formal, critical examination technique explicitly developed to study hazards in chemical and allied plants. ICI (U.K.) originated this technique in the 1960s for studying the safety and operability of new plant designs.[2,3]

This method has gone through numerous improvements over the decades and is now widely accepted as one of the most useful hazard identification methods. Within ICI, for example, the hazard study became a mandatory requirement for capital projects. The scope was broadened to implement the system in six stages, starting from project definition through commercial production. Table 12.3 shows the objective, timing, and input requirements for these six stages. Stage 3 of the hazard study is the original version of the HAZOP study.

12.2.3.1 Basic Concepts of the Study

The HAZOP study is done by a team consisting of members from multiple disciplines. A typical composition of the team is as follows:

Study leader
Process engineer
Project manager
Mechanical/maintenance engineer
Instrument engineer
Computer systems (DCS) engineer (where the plant involves complex distributed computer controls)
Safety officer
R&D chemist (where specialized process inputs may be necessary)

TABLE 12.2
Structured "What-If" Worksheet

Structured What-if Technique Analysis (Partial for Example Only)

Facility/Operation/Process: CNG Plant /Starting CNG Compressor (XYZ)/Natural Gas Compression

Date --/--/----	AAA – Leader, BBB – Facilitator, CCC – Safety Specialist, EEE – Supervisor and Process Engineer/FFF Mechanical Maintenance Engineer/GGG (Elec. & Instrument Engineer)				
Activity	**A – Gas Compressor Purging and Prestart-Up**				
No.	**What-If**	**Causes**	**Consequences**	**Controls**	**Recommendation**
A1	If gassing up is incomplete, i.e., air/O_2 not entirely removed before start-up	Inadequate amount of purge gas used so as to bring $O_2 < 5\%$	Fire and explosion in compressor	Operator training on gassing up	Must check O_2 level by portable O_2 analyzer after each time gassing up
A2	PLC faulty	Sudden fault in PLC	1. Compressor will not start 2. Compressor damage	Standard operating procedures (SOPs) to include "check PLC before each green start"	Each green start, to be preceded by switching the power supply and PLC check
A3	Crank case oil level low	Oil not topped	High wear of crank shaft main and big end bearings	SOPs to include "check level indicator as well oil level warning in PLC"	SOPs to have checklist to include ticking "oil level ok" by the operator
A4	Compressor discharge valve closed	SOPs not followed, lack of operator training	Lifting of fourth stage/other stage relief valves and possible damage to fourth stage cylinder	Compressor to have "start" interlock with discharge valve	Compressor purchase specification to include this interlock feature under instrumentation and control
A5	Gas and fire detector sensors inside compressor canopy dysfunctional	Poor maintenance and absence of prestart-up check	No Immediate action including FM 200 initiation inside compressor canopy	SOPs to include in prestart-up checklist	Compressor panel to indicate that detectors are functional

The basic assumption in the HAZOP methodology is that the proposed design will work under assumed conditions of operation. HAZOP problems arise when there are DEVIATIONS from design specifications. This method (1) provides a list of GUIDEWORDS used as aids in the questioning process to discover possible

TABLE 12.3
The Six Stages of ICI's Hazard Study System[2,4]

	Objective	Timing	Input Requirements
Hazard study 1	Review safety, health, and environmental criteria to be met according to statutory requirements and company policy. Decide on process parameters, site location, design codes, and standards to meet the criteria	Before budgetary approval	Process block diagrams, raw material, and service requirements, MSDS
Hazard study 2	Identify significant hazards, eliminate them by redesign, or incorporate protective measures to meet SHE criteria agreed in HS1	Before project sanction	Process flow diagrams, basic design data, control systems, risk assessment
Hazard study 3 (HAZOP)	Finalize detailed P&IDs	Detailed design stage	P&IDs, draft operating procedures
Hazard study 4	Pre-commissioning checks for layout; SOPs; completion of actions under HS 1, 2, and 3; and availability of statutory approvals	Before commissioning	Checklists
Hazard study 5	Final checking before commissioning (introducing chemicals into the unit)	Before commissioning	Checklists
Hazard study 6	Compare performance against design	After establishing commercial production	Plant dossier, e.g., P&IDs and SOPs, with all changes incorporated

deviations from design specifications, and (2) examines the causes and consequences of these deviations on safety and workability.

Table 12.4 shows a typical set of guidewords and deviations for continuous and batch processes. Team members may find it necessary to add new guidewords and deviations to this list, depending on the nature of the project. Some of the guidewords for batch processes may need to be used for a continuous process to examine the hazards and operability of the plant during commissioning and regular start-up or shutdown. Table 12.5 gives a checklist of common causes of deviations.

When HAZOP study procedures were developed in the 1960s, chemical plants generally deployed pneumatic control systems, and the guidewords provided were adequate for studying such plants. Modern plants are controlled by programmable electronic systems (PESs), DCSs, and programmable logic controllers (PLCs), all with powerful, special-purpose computers programmed to suit a wide variety of

TABLE 12.4
HAZOP Study Guidewords and Deviations, Continuous Processes

Guideword	Examples of Deviation
No or None	No flow
High	High flow, high temperature, high pressure, high level, high concentration
Low	Low flow, low temperature, low pressure, low level, low concentration
Reverse	Reverse flow
Other than	Presence of contaminants (solids/liquids/gases), static build-up
Start-up/shutdown	Testing/commissioning/maintenance
Utility failure	Power, steam, cooling water, instrument air, etc.
	Additional Guidewords and Deviations for Batch Processes
No or None	Activity not carried out, e.g., reactant not charged, batch not cooled, purging not done, a vessel not cleaned
More of	Increased batch size, increased quantity of an ingredient, increased heating/cooling, etc.
Less of	Less of the above
As well as	Presence of impurities, extra operation carried out
Part of	Operation partly completed, incomplete cleaning between batches
Reverse	Backflow or backpressure, heat rather than cool
Sooner/later than	The activity carried out at the wrong time

TABLE 12.5
Checklist of Common Causes of Deviations

Deviation	Causes
No flow	Pump failure, suction vessel empty, delivery vessel pressurized, valve not opened, diversion of flow at an upstream point, valve/filter blocked
Low flow	Pump failure, poor suction, scaling of delivery line, valve jammed or not fully opened, faulty automatic control
High flow	Delivery pressure lost, suction pressurized, additional pump wrongly put online by the operator
Reverse flow	Pump failure, pump reversed, imperfect isolation, back-siphoning
Low level	Loss of automatic control, equipment failure, operator error
High level	Loss of automatic control, equipment failure, operator error
Low pressure	Equipment failure (pump, compressor, safety valve), utility failure, loss of containment
High pressure	Utility failure, hammer/surge by the rapid change of flow, external fire
Low temperature	Utility failure, flashing
High temperature	Loss of automatic control, exothermic reaction, operator error
High/low concentration	Changes in the proportion of mixture, operator error
High/low mixing	Agitation failure
Static build-up	Earthing failure, high pumping velocity, friction between the human body and clothing
No or partial activity completion	Operator error
Emergency	Failure of safety equipment, coordination/communication system

needs in a highly flexible and reliable manner. Obviously, for such plants, additional criteria need to be considered over and above the standard guidewords and deviations used in a conventional HAZOP study. Some of the essential points that need to be considered for such systems are shown in Table 12.6.

TABLE 12.6
Additional HAZOP Study Points for PESs

Effect of	Points to be Considered
Missing or incorrect measurement	Implications of lack of action?
	False readings or measurements?
	How are faults detected?
	Are alarms run from measurement?
	Are other control devices affected?
Loss of power supply (consider both short- and long durations)	How is loss detected?
	Connections to other systems?
	How are alarms and alarm indications affected?
	Effect on memory?
	Need for uninterruptible power supply (UPS)?
	Need for emergency generators?
	Impact on critical drives?
Resumption of the power supply	Is the system start-up automatic?
	Connections to other systems?
	How is the last state before power loss determined?
	Is there a start-up routine to run through?
	Default parameters downloaded from the computer system on start-up?
External influences	Short/open circuit of input/output signals?
	Power supply failure (e.g., fuses)?
	Common mode failure?
	Others (e.g., environment)?
Loss of communication link	Does the computer system have a relay?
	How does the link fail?
	Status of the computer?
	An indication that the computer link has been lost?
	An indication that controller control is in use?
	Alarms?
Spurious data on the communication link	Error checking?
	Can the device be made read-only?
Application program	Quality assurance?
	Effect on other systems?
Maintenance problems	Integrity/check of replacement components?
	Plant has to be shut down to change the device, or can it be done online?
	Error checking?
Security	Virus protection?
	Back-up systems?
	Password/operator access?
Site emergencies	What is the role of the PES, DCS, or PLC system(s)?
	Impact on total failure?

12.2.3.2 Study Procedure

The study procedure essentially requires systematic questioning of every part of the process. Guidewords help discover how deviations or failures can occur and whether these deviations can give rise to hazards or operability problems. Corrective actions needed to avoid identified hazards or operability problems are documented.

The procedure is explained in Table 12.7 for continuous processes, starting with a vessel in the piping and instrumentation diagram (P&ID). A process engineer intimately familiar with the process is required to explain to the HAZOP study team members the design specifications of all interconnected vessels, piping, and instrumentation.

For a particular vessel, the next step is to select a connected pipeline for study and start the questioning process by selecting a guideword (for example, "No or None") and sequentially examining all possible deviations under that guideword. Against each deviation, possible causes and consequences are considered. Depending on the significance of the consequence, recommendations are agreed upon and documented for action. These recommendations could potentially show the need for changes in plant design or operating procedures.

The procedure is repeated for the selected pipeline using all the guidewords one after another. This phase completes the examination of the selected pipeline, which is marked on the P&ID as having been completed.

The above procedure is repeated for all the lines and auxiliaries connected to the vessel.

TABLE 12.7
HAZOP Study Method for Continuous Processes

1	Select a vessel
2	Explain the general purpose of providing the vessel and its lines
3	Select a line
4	Explain the purpose of the line
5	Apply the first guideword
6	Develop a meaningful deviation
7	Examine possible causes
8	Examine consequences
9	Detect hazards
10	Make a suitable record for action
11	Repeat steps 6–10 for all deviations from the first guideword
12	Repeat steps 5–11 for all the guidewords
13	Mark the line as having been examined
14	Repeat steps 3–13 for each line
15	Repeat steps 3–14 for all auxiliaries (e.g., heating system, relief valve, vent)
16	Explain the purpose of the vessel
17	Repeat steps 5–12 for the vessel
18	Mark the vessel as completed
19	Repeat steps 1–18 for all vessels on the P&ID

The next step is to take up the vessel itself and repeat the procedure using all guidewords and deviations.

The entire process is repeated for all the vessels in the P&ID.

The methodology to be followed for a batch process is shown in Table 12.8. The procedure is broadly similar to the one for continuous processes. However, each vessel and all lines connected to it are examined for several sequential steps. These follow the required steps closely in the manufacturing process (for example, charge materials into a vessel, heat to a specified temperature, add a catalyst, hold for a specified duration, etc.). Also, each pipeline and vessel needs to be examined in both "active" and "inactive" states (active states referring to periods when the equipment is in use). This examination is essential because sections of the plant may contain chemicals in an inactive state that potentially could contribute to hazards, such as corrosion or degradation.

For illustrating this procedure, a real-life operation with considerable hazard potential is considered. This example is the *Operating Instructions for LPG Truck Loading* (refer to Table 12.9). Detailed records of the study are recorded in Table 12.10. Under the team leader's guidance, the officially designated secretary of the team is responsible for screening the discussions, recording what is pertinent, and excluding discussions on trivial issues. The secretary also has the responsibility to ensure that all follow-up actions are taken to complete the agreed list of actions and then documented.

TABLE 12.8
HAZOP Study Method for Batch Processes

1. Select a vessel
2. Select the first step/operation of the process
3. Relate to the appropriate line on the line diagram
4. Explain the purpose of the line when the line is "active"
5. Apply the first guideword
6. Develop a meaningful deviation
7. Examine possible causes
8. Examine consequences
9. Identify hazards
10. Make a suitable record for action
11. Repeat steps 6–10 for all the guidewords
12. Apply the guidewords to the line when the line is in "inactive" state
13. Mark the line as having been examined
14. Repeat steps 3–13 for each step/operation of the process (associated with the selected vessel)
15. Select the auxiliary line (e.g., relief valve, vent, heating system)
16. Explain the purpose of the auxiliary line
17. Repeat steps 3–12 for the auxiliary line
18. Mark the auxiliary line as having been examined
19. Repeat steps 15–18 for all auxiliary lines
20. Mark the vessel as having been examined
21. Consider if any equipment is critical and needs to be registered (e.g., pressure vessels, lifting gear, etc.)
22. Select the next step in the process using the same vessel and continue until all steps have been covered
23. Repeat with all vessels in the P&ID

TABLE 12.9
Operating Instructions for LPG Truck Loading

Activities before actual truck loading:

1. Check that the feed vessel V-101 has been filled with the required quantity of LPG, the liquid level in the storage tank is shown on LIA-01 in the control room, high-level alarm operating at 85% and low-level alarm at 12%
2. Place the road tanker in the filling position. Place suitable barriers at the front and the rear of the vehicle to prevent movement during the loading operation. Check oxygen content in the vapor space of the tanker by taking a sample from the vapor nozzle; this should be less than 1 vol%
3. Connect the liquid and vapor return arms of the loading arm to the respective nozzles of the tanker
4. Connect an earthing clamp to the road tanker to prevent any static discharge during filling of the tanker
5. Purge the liquid and vapor return lines between the tanker and the loading arm with nitrogen. Nitrogen at 7 bar is available in the header

Actual truck filling operation:

6. Open valve ROV-01 at the outlet of the storage tank
7. Open valve ROV-02 on the liquid recirculation line and set FICA-01 at about the middle of the operating range (40–100 m^3/h). The minimum flow rate necessary for running the pump has been specified as 35 m^3/h. If the flow rate drops to 20 m^3/h, FICA low flow alarm will operate in the control room to warn the operator to take corrective action
8. Start the pump (P-01) and check that it is operating on recirculation mode at the set flow rate (note: in case of low suction head, LSLL-03 will prevent the pump motor from starting)
9. Open manual valves (V1) on the vapor return line from the tanker and (V2) on the vapor return arm of the loading arm
10. Open valve ROV-03 on the vapor return line
11. Open manual valve (V3) on the liquid filling line from the tanker
12. Open valve ROV-04 on the liquid filling line
13. Open the manual ball valve (V4) on the liquid arm slowly and set FICA-01 at the desired rate. Continue filling until the desired quantity has been loaded into the tanker, as indicated by the level in the tanker
14. Close the manual ball valve (V4) on the liquid arm slowly
15. Switch off the pump

Postfilling activities:

16. Close valve ROV-04 on the liquid line
17. Close valve ROV-03 on the vapor return line
18. Close the manual valve (V2) on the vapor return arm
19. Close manual valves (V1 and V3) on the tanker
20. Depressurize the liquid line between V3 and V4, and also the vapor return line between V1 and V2, by connecting to the vent stack
21. Purge the liquid line between V3 and V4, and also the vapor return line between V1 and V2, with nitrogen
22. Disconnect the road tanker

12.2.4 Failure Modes and Effects Analysis (FMEA)

The FMEA is a technique for determining how equipment items can fail and the effects of each failure mode. The technique is thus oriented toward equipment failures only, in contrast to a HAZOP study that is open to considering equipment failures, process deviations, and human errors without restriction.

TABLE 12.10
HAZOP Study Proceedings on LPG Truck Loading

Guideword	Deviation	Caused By	Consequence	Preventive/Corrective Measures Provided	Action Required
			Sheet 1 Operating Instructions Steps 1, 2, 4 and 5		
No/none	No liquid in the storage tank	Level not checked (human error)	Damage to pump Loss of operating time	a. LSL-01 to prevent the pump from starting	1. Follow operating instructions strictly
	Barriers not placed before/behind trucks	Human error	Loading arm rupture and LPG leakage possible		Same as 1
	Earthing clamp not fitted	Human error	Possible static discharge and fire		Same as 1
	Nitrogen purging not done	Human error No nitrogen in the header	Formation of a flammable mixture of air and LPG		Same as 1 2. Install PI on nitrogen header in the truck loading area
	No flow through the pump	ROV-01 or ROV02 not opened or failed shut	Pump/motor damage caused by dry-running or churning of blocked liquid in the pump	b. FICA-01 to operate alarm in the control room (when flow < 20 m^3/h) for the Operator to stop the pump through a remotely-operable hand switch	3. Provide open/closed indication in the control room for ROV/SDV valves 4. Provide TSH (setpoint > 50°C) to activate an alarm in the control room and also to stop the pump motor

(Continued)

TABLE 12.10 (*Continued*)
HAZOP Study Proceedings on LPG Truck Loading

Guideword	Deviation	Caused By	Consequence	Preventive/Corrective Measures Provided	Action Required
			Sheet 2 Operating Instructions Steps 6–15		
High	High flow through the pump (>120 m^3/h)	Rupture in liquid recirculation line or tanker loading line	Major LPG leakage/fire	c. Depending on pump characteristic, the pump motor might trip on overload d. Manual hand switch in the control room for the operator to stop the pump e. LPG (40% LEL) detector in the operating area activates an alarm in the control room	5. Provide an interlock system for FICA-01 to trip the pump at flow rate > 120 m^3/h
	High temperature	Not likely unless the LPG storage tank is subjected to external fire			6. Operate a disaster control plan for the site in the event of any major fire or LPG leakage
	High pressure		No hazard is likely as the maximum discharge pressure of the pump is 7 barg, and the pump and the pipeline are designed for 20 barg		
	Overfilling of the tanker	Human error in reading the level gauge of the tanker	The overflow of LPG liquid to vapor return line		7. Provide a load cell or other weight indication system for the truck platform

(*Continued*)

TABLE 12.10 (*Continued*)
HAZOP Study Proceedings on LPG Truck Loading

Guideword	Deviation	Caused By	Consequence	Preventive/Corrective Measures Provided	Action Required
			Sheet 3 Operating Instructions Steps 6–15		
Low	Low flow through the pump	Partial blockage of ROV-01 (not likely with clean LPG liquid)	Pump damage if the flow rate is less than the minimum specified	Same as (b)	
	Low discharge head of the pump	A mechanical problem in the pump	Operability problem during truck loading	f. Regular preventive maintenance	
	Low temperature	Loss of containment in the storage tank	No hazard from low temp as the pump and pipeline are designed for −45°C		Same as 6
Reverse	Reverse flow through the pump	Stoppage of the pump as part of normal operation	Pump damage	g. Nonreturn valve installed at pump delivery	
As Well As	Contaminants	From tanker through vapor return line during tanker loading	Contamination of LPG product in the storage tank		8. Check for the presence of contaminants in the tanker before connecting the loading arm

(*Continued*)

TABLE 12.10 (*Continued*)
HAZOP Study Proceedings on LPG Truck Loading

Guideword	Deviation	Caused By	Consequence	Preventive/Corrective Measures Provided	Action Required
Leakage		Leakage from loading arm swivel joint	LPG leakage	Included in preventive maintenance	
Sheet 4 Operating Instructions Steps 6–15					
Utilities	Power failure		Loss of operating time	All control valves and alarm systems are connected to an uninterruptible power supply (UPS)	
	Instrument air failure	Compressor failure Leakage in compressed air line			9. All ROV/SDV valves to close on air failure (safe mode)
Sheet 5 Operating Instructions Steps 16–22					
No/none	Depressuring (Step 20) not done	Human error	The possibility of LPG entering LPG header		Same as 1
	Purging (Step 21) not done	Human error	Formation of LPG/air flammable mixture Pollution of the workplace environment with LPG		Same as 1

This method starts with defining (1) specific vessels, equipment, and instrumentation that are to be included and (2) the boundary conditions under which they are to be analyzed. In particular, defining these boundary conditions requires documenting the expected performance of the system, process, or equipment and the failure definitions of the equipment items, the process, or the system. British Standard BS 5760[5] and the U.S. Department of Energy Handbook[6] give a good description of this method and provide illustrative examples. A typical list of failure modes of various equipment has also been given by Sutton.[1]

The British Standard considers failure modes at two levels: generic and specific. As an example of generic failure modes, the standard enumerates a list of failures such as the following:

- Failure during operation
- Failure to operate at a prescribed time
- Failure to cease operation at a prescribed time and
- Premature operation.

The standard also provides examples of specific failure modes, such as cracked/fractured, distorted, undersized, etc. While listing various failure modes, it is essential to remember that equipment items could fail in several different ways and that failure may be total or partial. All such situations must be analyzed for a particular item before moving on to the next piece of equipment.

If the equipment is used for different modes of operation, e.g., a batch reactor on a heating or cooling cycle, the equipment analysis should be done for each operational mode separately.

The FMEA worksheet usually includes documentation of failure detection modes, as well as compensating provisions. These can reduce the likelihood of a particular failure or mitigate the consequences and, therefore, are particularly important for ensuring that the evaluation work is done realistically.

The FMEA worksheet requires that the severity class of worst case to smaller consequences be specified in terms of four categories, as follows,[6] see Table 12.11 below.

The method is illustrated below. From Table 12.12 (for Example 12.1), it should be evident that visualizing the possible range of failures and their effects presupposes

TABLE 12.11
Categories of Consequences for FMEA Worksheet

Category I	Catastrophic	May cause death or loss of system or process
Category II	Critical	May cause severe injury, extensive property, or system damage
Category III	Marginal	May cause minor injury, minor property damage, or minor system damage
Category IV	Minor	Is not severe enough to cause injury, property damage, or system damage, but may result in unscheduled maintenance or repair

TABLE 12.12
FMEA Worksheet for LPG Feed Vessel

Sheet 1
Date:
Study Team:

Drawing Reference: Figure 12.3
Component: Valve FCV-02

Operational Mode	Failure Mode	Cause of Failure	Failure Effects	Failure Detection Method	Compensating Provision	Severity Class	Remarks/Action
Vessel filling	Cannot be opened	Jammed/signal failure	No filling	LIA-01 shows no increase in the level	None, except operator action	Negligible (loss of time)	Low probability if scheduled maintenance is done
	Opens partially	Mechanical malfunctioning	Longer filling time	LIA-01 shows a lower filling rate	None, except longer filling time	IV	Unscheduled repair needed
	Does not close tight after the end of filling	Mechanical malfunctioning	Liquid level continuing to rise after apparent closure	LIA-01 shows an increase in the level	LSHH-02 to shut down SDV-02	IV, if the operator acts promptly or SDV-02 operates	Unscheduled repair needed
Truck loading	Valve passes	Mechanical malfunctioning	Liquid level accounting is disturbed		SDV-02 in a shut condition	Negligible if SDV-02 is tightly shut	Unscheduled repair needed

(*Continued*)

Sheet 2
Date: Drawing Reference: Figure 12.3
Study Team: Component: Valve SDV-02

Operation Mode	Failure Mode	Cause of Failure	Failure Effects	Failure Detection Method	Compensating Provision	Severity Class	Remarks/Action
Vessel filling	Cannot be opened	Jammed/signal failure	No filling	LIA-01 shows no increase in the level	None, except operator action	Negligible (loss of time)	Low probability if scheduled maintenance is done
	Opens partially	Mechanical malfunctioning	Longer filling time	LIA-01 shows a lower filling rate	None, except longer filling time	IV	Unscheduled repair needed
	Does not close tight after the end of filling	Mechanical malfunctioning	Liquid level continuing to rise	LIA-01 shows an increase in the level	LSHH-02 to shut down SDV-02	IV, if the operator acts promptly or SDV-02 operates	Unscheduled repair needed
Truck loading	Valve passes	Mechanical malfunctioning	Liquid level accounting is disturbed		FCV-02 in a shut condition	Negligible if FCV-02 is tightly shut	Unscheduled repair needed

Sheet 3
Date: Drawing Reference: Figure 12.3
Study Team: Component: LIA/LT-01

Operation Mode	Failure Mode	Cause of Failure	Failure Effects	Failure Detection Method	Compensating Provision	Severity Class	Remarks/Action
Vessel filling	Float stuck	Rust formation	Level change not shown	LIA-01 shows no increase in the level	LSHH-02 to shut down SDV-02 once 85% level is reached	IV, if LSHH-02 and SDV-02 are operational	The probability of float getting stuck in an environment of clean liquid such as LPG is low
	Transmitter failure		Failure of level indication and alarm function		LSHH-02 to shut down SDV-02 once 85% level is reached	IV, if LSHH-02 and SDV-02 are operational	Modern level systems allow transmitter replacement from outside without shut down
Truck loading	Float stuck or transmitter failure			LI and WI for truck	LSLL-03 against low level	IV, if LSLL-03 is operational	

(*Continued*)

FMEA Worksheet for LPG Storage/Truck Loading Vessel (Sheet 4)

Date: Drawing Reference: Figure 12.3

Study Team: Component: Pressure Safety Valve PSV-01/02

Operational Mode	Failure Mode	Cause of Failure	Failure Effects	Failure Detection Method	Compensating Provision	Severity Class	Remarks / Action
All normal activities, e.g., vessel filling, truck loading, etc.	Jammed shut or opens partially on demand	Mechanical malfunctioning	No protection in the event of uncontrolled pressure rise, e.g., external fire	PI-01 in the control room shows increased pressure	The second PSV is normally on-line except during testing	I (vessel failure and major LPG release)	With scheduled testing in place and considering low demand frequency, hazard rate should be very low

Sheet 5

Date: Drawing Reference: Figure 12.3

Study Team: Component: Vessel V-101

Operational Mode	Failure Mode	Cause of Failure	Failure Effects	Failure Detection Method	Compensating Provision	Severity Class	Remarks/Action
Water draining	The valve cannot be closed on completion of draining	Delayed operator action resulting in flash cooling of LPG	Major LPG leakage	Gas detectors in the plant area	None	I	Two valves provided on the line, but alert operator attention essential
All normal activities, e.g., vessel filling, truck loading, etc.	Over-pressure	External fire	Major LPG leakage	None	Water sprinklers for external cooling	I	
Start-up	No nitrogen purging	Human error (unlikely with team activity)	Internal fire on the introduction of LPG	None	None	I	

a comprehensive technical understanding of the design and operation of the project or proposal under study. The FMEA, therefore, should be done by an authoritative, multidisciplinary team, as with HAZOP studies.

Commercial grade LPG containing approximately 30% propane and 70% butane by volume is stored at atmospheric pressure, refrigerated storage at a temperature of –5°C to 0°C. The product from this storage is pumped through a heater (where its temperature is raised to about 10°to a feed vessel (V-101), from which it is filled into trucks. The entire operation is regulated from a centralized control room.

In pressure vessel V-101, the LPG temperature varies between 10°C and 30°C depending on ambient temperature and storage duration. The truck loading operation involves filling several trucks simultaneously. However, for simplicity, the filling of only one tanker has been considered in this example. A simplified flow diagram for the operation involving filling the feed vessel and loading a truck is shown in Figure 12.3.

Figure 12.3 shows the critical instrumentation and control systems for the feed vessel and the pumping system. All shutdown valves and remotely operable valves shown in the diagram are on-off valves, operated from the control room and other designated locations. Details of control systems involving measuring probes, transmitters, relays, solenoids, etc. are usually part of the detailed design and have not been shown.

V-101 is filled at a rate set in the control room by FIC-02. LIA-01 indicates the liquid level in the vessel. At a level setpoint of 85%, the high-level alarm is triggered for the operator to stop filling. In the event filling continues beyond a level of 85%, the high-high level switch LSHH-02 is provided to automatically close the remotely operable, shutdown valve SDV-02 at the 90% level. No filling of LPG into V-101 is done when the vessel is being used for loading trucks.

Truck loading is a manual batch operation using a loading arm with a liquid arm and a vapor return arm. Pump P-01 is a canned motor centrifugal pump, i.e., a completely closed, hermetically sealed, electromagnetically driven pump/motor unit

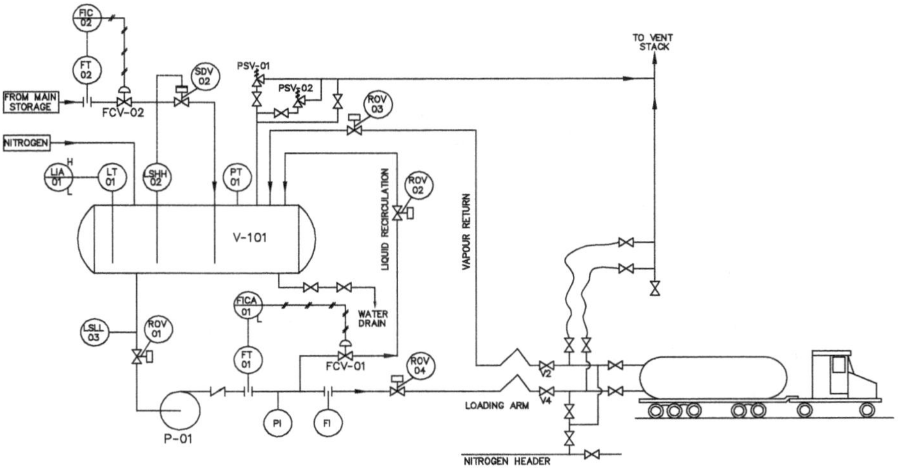

FIGURE 12.3 Simplified flow diagram of LPG feed vessel for LPG loading.

without any shaft sealing. The filling time for a tanker of about 18 metric ton capacity is 30 to 40 minutes, excluding connecting, disconnecting, and check-up activities.

As truck loading continues, the liquid level in V-101 is indicated on LIA-01 in the control room. At a level setpoint of 12%, the low-level alarm operates in the control room for the operator to stop the tanker loading operation. If no action is taken on the alarm, the low-low level switch LSLL-03 is provided to automatically stop the transfer pump P-01 at a 5% level in the vessel.

This straightforward example has been used to illustrate the HAZOP study and FMEA methods. The system chosen for HAZOP study consists of the entire system of pipelines from the bottom of the feed vessel to the road tanker and including the pump, liquid recirculation line, and the vapor return line. The system chosen for illustration of the FEMA method consists of the feed vessel V101, complete with the LPG supply line and including the FIC-02 flow control loop. The scope of FMEA also includes the pressure-relieving system (PSV-01/PSV-02) and the water draining line.

Operating instructions for the LPG truck loading operation are given in Table 12.9. HAZOP study proceedings are given in Table 12.10 and the FMEA worksheet in Table 12.11.

12.2.5 Fault Tree and Event Tree Analysis

Fault tree analysis is a hazard identification method that forms the initial step in a quantitative risk assessment. It is a method used to develop the causes of an event of interest, such as a hazardous event or equipment failure. The event tree analysis considers the development of various consequences or outcomes of a fault. Both techniques have been covered in Chapter 13 as a requirement for determining event frequency.

12.3 COMMENTS ON CHOICE OF THE METHOD

1. The HAZOP study is a comprehensive method of hazard identification covering examination of the effect of equipment failures, human errors, and deviations in operating conditions. The study scope and duration vary depending on the project stage: project definition, detailed design, or plant commissioning. The study requires a strong commitment from several senior engineering professionals, highly conversant with process design, operation, control, and maintenance requirements, to devote considerable time to this effort. These requirements often can become limitations; however, it is imperative to ensure their regular attendance and attention.
2. Methods based on safety audits, checklists, etc. take much less time and can be carried out by a small team. The effectiveness of this kind of study is limited to well-established processes. The study should be done under the guidance of a person highly experienced in the operation of similar plants.
3. FMEA is based on the failure of equipment only. A comprehensive examination of failure modes of various equipment items is possible, as long as

team members have good experience and knowledge of the design, operation, and use of such equipment.

4. Depending on the objectives of a specific study, a combination of several methods is likely to give the best results. For example, in very large, grass-roots project, HAZOP studies may be used for the most vital process units and the checklist method for others. Safety audits are usually done for existing plants. Once potential hazards have been identified, a fault tree analysis should be done for those that are considered critical.

REFERENCES

1. Sutton, I.: *Process Safety Management* (Southwestern Books, Houston, TX, 1997).
2. ICI: *Training Course on Hazard Analysis* (Personal Communication).
3. Canadian Industries Association Limited: *A Guide to Hazard and Operability Studies* (1985).
4. ICI Plc: Based on *Notes on Hazard Studies used within ICI group of Companies* (For Training and Also for Project Safety Evaluation) (the Mid-1970s–1990s).
5. BS 5760-0:2014: *Reliability of Systems, Equipment and Components. Guide to Reliability and Maintainability.*
6. U.S. Department of Energy: *Chemical Process Hazards Analysis, DOE Handbook DOE-HDBK-1100-2004* (August, 2004).

13 Risk Assessment and Control

Once the identification of hazards has been completed by one or more methods, as explained in Chapter 12, the next step is to assess the risks. The terms "hazard" and "risk" are of central importance in process safety engineering and, appropriately, were introduced early in Chapter 1. All hazards pose some risk, and risk is defined as the likelihood or probability of an identified hazardous event with specified consequences. Thus, risk depends both on the probability and consequences arising from undesirable events. Methods for estimating the consequences of hazardous events in process plants (such as fire, explosion, and toxic release) have been covered in Chapters 5–10. Methods of assessing hazards have been explained in Chapter 12. Methods for assessing likelihood are considered in this chapter.

The likelihood of an event can be expressed in terms of a frequency or probability of occurrence. The unit of frequency is the number of times an event occurs in a specified time, usually a year. Thus, if an event occurs, on average, once in 1,000 years, the event frequency is 0.001 per year. Probability, on the other hand, is a dimensionless term. If a person is required to act 100 times in a given period and if he acts 97 times correctly and three times wrongly, then the probability or likelihood of human error is 3 in 100, or 0.03. An event that occurs once a year is more likely to occur than the one which occurs, on average, say once in 1,000 years. That is, the rarer the event, the lower is the likelihood of its occurrence at a given time.

Quantitative risk assessment (QRA) is a time-consuming exercise that requires estimating both the frequency and the consequences of an identified event. Therefore, it is common practice to do an initial screening based on judgmental or semi-quantitative methods and a QRA for high hazard areas.

13.1 METHODS OF EXPRESSING RISKS

13.1.1 Fatal Accident Rate[1,2]

ICI coined the fatal accident rate (FAR) in 1970 to express risks to employees. It is defined as the number of deaths in 10^8 (100 million) hours of exposure. The figure of 10^8 was arrived at by assuming that an operator would work for about 2,500 hours per year in a typical chemical manufacturing site in the U.K. About 1,000 operators were employed who would be expected to work potentially for 40 years.

Table 13.1 shows some typical FAR figures quoted by Trevor Kletz for various industries in the U.K. during the period 1974–1978. Though these data may be somewhat outdated, they are handy as order-of-magnitude values. One can also get a sense from these data about the relative variations in risks to an employee from one industry to another.

DOI: 10.1201/9781003107873-13

TABLE 13.1
FAR Values for Some U.K. Industries

	FAR
Offshore oil and gas	82
Deep-sea fishing	44
Coal mining	10
Construction	7.5
Shipbuilding and marine engineering	5.25
Chemical and allied industries	4.25
All premises covered by the Factories Act	~4
All manufacturing industry	1.15
Vehicle manufacture	0.75
Clothing manufacture	0.25

13.1.2 Individual Risk

Individual Risk is usually defined as the frequency of an event causing a specified level of harm at a given location, regardless of whether anyone is present at that location who might suffer that level of harm. This term is often used as a criterion for controlling significant accident hazards in a factory or plant, where the level of harm is considered to be a fatality. The location usually specified is outside the site boundary, where an individual is assumed to be present all the time.

Death at a specified location may be caused by various accidents occurring in a site independently of each other. The procedure for determining individual risk involves working through all such potential accidents having consequences large enough to cause death at the specified location and summing the frequencies of all such accidents.

Individual risks are often presented as a set of risk contours. These contours are drawn on a plan to show the geographical distribution of individual risk around a site. For drawing the contours, individual risk is calculated for a large number of points around the site. All points of a particular level of risk are then joined to obtain the contour for that risk level. Figure 13.1 shows a typical risk contour diagram with three levels of risk, each differing from the next by a factor of 10.

13.1.3 Average Individual Risk

The term Average Individual Risk is sometimes used when a group of people or a population around a facility is exposed to the risk. It is defined as the individual risk divided by the total number of people at risk. Consider, for example, an accident occurring once in 1,000 years that has the potential to kill all the people (say, 100) living around that facility. Here, the overall exposed population risk is 10^{-3} per year, but the average individual risk is 10^{-3} divided by 100, or 10^{-5} per person per year.

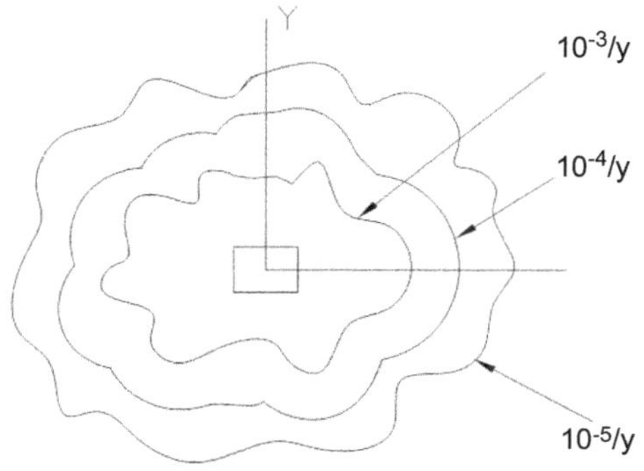

FIGURE 13.1 Typical representation of individual risk contours.

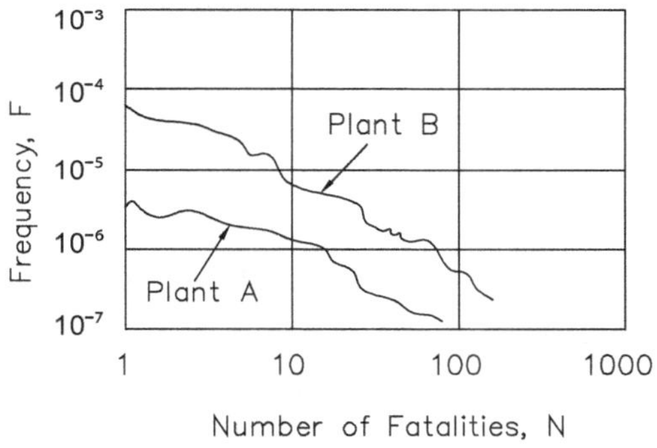

FIGURE 13.2 Typical representation of *F-N* curves.

13.1.4 Societal Risk

Societal risk is used as a measure of risk for multiple casualty events. The societal risk is usually presented in the form of an *F-N* curve, which is a cumulative distribution plot of frequency, *F*, versus the number of fatalities, *N*. In this plot, both *F* and *N* are on a logarithmic scale (see Figure 13.2).

13.2 LAYER OF PROTECTION ANALYSIS[3–6]

Layer of protection analysis (LOPA) is a simplified technique for Risk evaluation and control. It is a semi-quantitative technique used to ensure that process risk is reduced to acceptable levels. Hazard evaluations (HEs) are performed to assess and control risk. It carries out scenario-wise consequence evaluations that recognize and examine the independence of preventive and mitigation measures to determine risk levels. It was originated in the late nineties in the USA.

As a simplified technique of risk assessment compared to QRA, LOPA was adopted quickly by the U.S. chemical industry, thanks to the leadership of the American Institute of Chemical Engineers (AIChE). In 2001, AIChE's Center for Chemical

Process Safety (CCPS) published "*Layer of Protection Analysis – Simplified Process Risk Assessment*", making it convenient for companies in the USA to understand LOPA and practice it systematically. In the U.K., the HSE also introduced the use of LOPA. Of the several methods for performing these risk evaluations, LOPA has, over the recent past, found a much broader application in the chemical industries. Hence, LOPA has been dealt with in detail in this book. Several tables and figures from CCPS publications have been reproduced with the kind permission of CCPS.

LOPA's primary purpose is to examine and ensure that sufficient independent protection layers (IPLs) are provided, as shown in Figure 13.3, against an accident scenario, and to keep risk within limits set by individual companies and the applicable regulatory statutes. In this chapter, we present and discuss the applicability of the LOPA method in determining appropriate safety integrity levels (SILs) for the process industries.

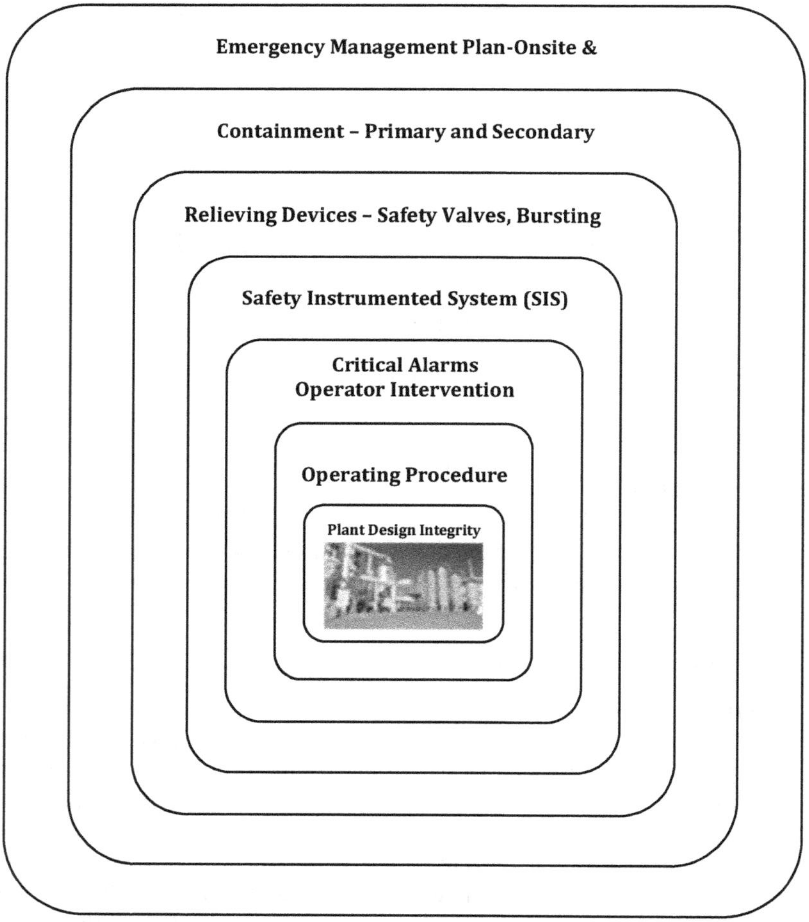

FIGURE 13.3 IPLs.

Because of its simplicity and quick risk assessment approach, LOPA is being adopted and is now used systematically throughout the world's refining and petrochemical sectors. LOPA allows a more detailed consideration of a specific situation and its safeguards than many other methods. Accordingly, LOPA is generally acknowledged as a highly effective tool for reliable and efficient SIL assignments and allocation of risk reduction resources.

In LOPA, individual hazard scenarios defined by cause-consequence pairs are analyzed. Scenario risks are determined by combining scenario frequency and consequence severity. IPLs are analyzed for their effectiveness. The combined effects of the protection layers are then compared to risk tolerance criteria to determine whether additional risk reduction is necessary to reach an acceptable level.

13.2.1 LOPA Process

The LOPA process is shown schematically in Figure 13.4 below.

13.2.2 Select Criteria for Consequence Screening

The first step in LOPA risk evaluation is selecting "consequence criteria" for screening the scenarios envisaged during the HE exercise (preferably, hazard and operability (HAZOP), or similar techniques). The focus would be on the most significant accidents that could cause harm to people and the environment. A LOPA analyst would seek to screen out less consequential accidents. For each scenario, the associated consequences are evaluated using two main approaches:

- To categorize the consequences of a release from "loss of containment", i.e., the quantity of flammable or toxic materials released without explicit consideration of human harm. In this case, companies may categorize consequences based on (1) quantity of toxic and flammable material released from containment, (2) extent of damage of main and auxiliary plants, or (c) financial loss. The categorization is generally company-specific.
- To estimate the consequence qualitatively, but more explicitly, in terms of harm to plant personnel and the public, considering postrelease probabilities of such harm. The consequence is divided into five categories in order of severity, starting with Category 1 (insignificant or negligible risk) to Category 5 (catastrophic risk).

13.2.3 Select Consequence Analysis Scenarios for LOPA

In a LOPA exercise, a scenario comprises a single "initiating event-consequence pair".

The scenarios to be considered are defined before proceeding with the remaining steps of the analysis. The number of scenarios arising from a hazard identification (HAZID) study (say HAZOP) could be moderate to very large.

The bow-tie diagram in Figure 13.5 below shows the relationship between initiating events, layers of protection (LOP) or defense (LOD) to prevent the untoward incidents (i.e., a release capable of doing substantial consequential harm), and layers of protection or defense for mitigation of such consequence.

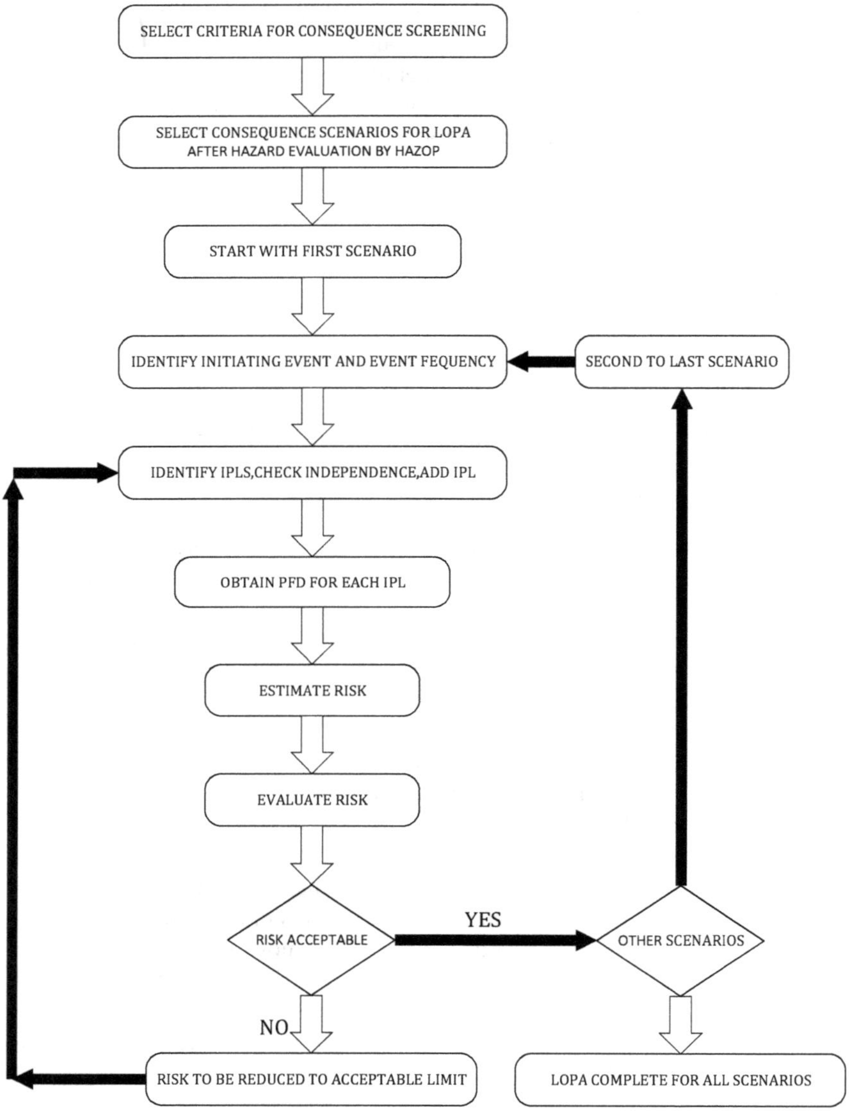

FIGURE 13.4 LOPA process schematic.

13.2.4. Identify Initiating Events and Frequencies

The scenarios to be considered include an initiating event that would lead to the worst-case consequences, as explained in Section 13.2.2 above; complete failure of all of the protective layers would be deemed to have occurred. CCPS[3] mentions three general types of initiating events, several examples of which are listed in Table 13.2 below.

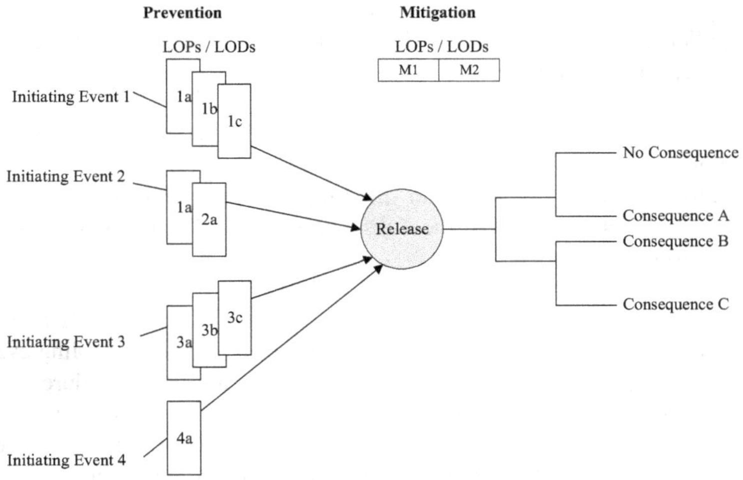

FIGURE 13.5 Initiating events, layers of protection/defense, and consequences.

TABLE 13.2
Types of Initiating Events

Initiating Events	Examples
External events	Fires and explosion in an adjacent plant
	Flooding
	High winds
	Lightening
	Seismic event
	Vehicle impact
	Terrorist activities
	Third-party actions
Equipment failure	Basic process control system component failure
	Software failure
	Utility failure
	Vessel or piping failure (wear, fatigue, or corrosion)
	Vessel or piping failure caused by design, specification, or manufacturing defects
	Vessel or piping failure caused by over- or under-pressure
	Vibration-induced failure
	Failures caused by inadequate maintenance or repair
	Failures resulting from temperature extremes
	Failures resulting from flow surge or hydraulic hammer
	Failures resulting from internal explosions, decompositions, or other uncontrolled reactions
	Failure to execute steps of a task correctly, in the proper sequence, or omitting steps
	Failure to observe or respond appropriately to conditions or other prompts by the system or process
Human Failures	Failure to execute steps of a task correctly, in the proper sequence, or omitting steps
	Failure to observe or respond appropriately to conditions or other prompts by the system or process

CCPS publication on LOPA[3] with permission.

Initiating events often occur because of latent weaknesses in the safety system when a challenge arises, or a demand is made on a system. Latent weaknesses invariably have a root cause:

- "Vehicle impact" is an initiating external event, but the root causes may be "equipment location" and lack of "safety barriers".
- "Vibration-induced failure" is an initiating event (equipment failure), but a root cause is most likely the "absence of dynamic balancing" before commissioning.

The root causes can be useful when assigning frequencies to the initiating event. In specific complex scenarios, it may also be necessary to consider enabling events or conditions. Enabling events or conditions are factors that are neither failures nor protective layers. These do not cause the scenario directly but must be present so that the scenario may proceed. For example, a scenario may involve a delivery hose failure during the transportation of a dangerous substance, when a corrosion-related failure occurred.

For this scenario to occur, the delivery of material must be taking place. Therefore, the initiating event is a combination of such delivery (an enabling condition) and a corrosion failure during unloading. Initiating event frequencies may be obtained from public domain sources, company data, or from simple fault or event trees. The data should be appropriate for the industry or operation under consideration.

Where enabling conditions or factors are present, initiating event frequencies must be modified to take this into account. In general, the initiating event frequency is given by either:

(Enabling condition frequency) (Failure probability)
Or
(Enabling condition probability) (Failure frequency)

When the consequences of the scenario are expressed as a likelihood of a fatality or an expected number of fatalities, then the frequency must be modified to account for factors such as:

- The probability of personnel being present in the affected area
- The probability of fatality given exposure to the material or harmful effect and
- The probability of ignition in the case of flammable releases

This adjustment may be made to either the initiating frequency or the overall scenario frequency calculation.

13.2.5 Identify IPLs

The most critical requirement in the use of LOPA is a clear understanding of IPLs. The CCPS[3] defines IPL as "*a device, system or action which is capable of preventing a scenario from proceeding to its undesired consequence, independent of the initiating event or the action of any other layer of protection associated with the scenario. The effectiveness and independence of an IPL must be auditable.*"

For a device, system, or action to qualify as IPL, it must fulfill the following requirements:

1. It must function and perform as designed to prevent or mitigate a consequence.
2. It must be independent of the initiating event.
3. It must function on its own, entirely independently of any component of any other IPL that may be provided to prevent or mitigate the same consequence.
4. Its effectiveness in consequence prevention must be quantifiable and auditable.
5. Its probability of failure on demand (PFD) must be capable of being validated at required intervals.

All IPLs are thus "safeguards" against consequences; however, all safeguards do not qualify as IPLs: for many safeguards, their effectiveness cannot be quantified or audited owing to an absence of reliable data, uncertainty of independence, and so forth. Table 13.3 below, reproduced from the CCPS Publication,[3] shows examples of safeguards that are not usually considered IPLs.

Active IPLs move from one state to another in response to a change in measurable process parameters (e.g., temperature, pressure, etc.) or a signal from a source (e.g., a push-button, a switch, etc.).

An active IPL comprises the following:

- A sensing device of some type (instrument, mechanical, human)
- A decision-maker (software. logic solver, relay, spring, human, etc.)
- An action (automatic, mechanical, human)

The CCPS publication provides extremely useful guidance on assigning an appropriate PFD for various IPL types, in tabular form with examples. Values are typically quoted as orders of magnitude and are reproduced in Table 13.4 below.

Passive IPLs perform the risk-reducing function but do not have to take action. They usually mitigate the risk and do not prevent the consequence but reduce the "frequency of the high-impact consequence". Table 13.5 below contains examples of IPLs that achieve risk reduction using passive means, with a typical range of PFD values for each type of IPL used in this method. These IPLs achieve the intended function if their process/mechanical design is correct and if they are constructed, installed, and maintained correctly. Examples are tank dikes, blast walls or bunkers, fireproofing, flame or detonation arrestors, etc.

13.2.6 Risk Estimation

In general, the frequency of occurrence of a scenario (initiating event and its consequence) can be obtained from

$$f_i^C = f_i^I \prod_{j=1}^{J} \mathrm{PFD}_{ij} \tag{13.1}$$

TABLE 13.3
Examples of Safeguards Not Normally Considered IPLs[a]

Safeguard	Comments
Training and certification	These factors may be considered in assessing the PFD for operator action but are not – of themselves – IPLs
Procedures	These factors may be considered in assessing the PFD for operator action but are not – of themselves – IPLs
Normal testing and inspection	These activities are assumed to be in place for all HEs and form the basis for judgment to determine PFD. Routine testing and inspection affects the PFD of certain IPLs. Lengthening the testing and inspection intervals may increase the PFD of an IPL
Maintenance	These activities are assumed to be in place for all HEs and form the basis for judgment to determine PFD. Maintenance affects the PFD of certain IPLs
Communications	It is a basic assumption that adequate communications exist in a facility. Poor communication affects the PFD of certain IPLs
Signs	Signs by themselves are not IPLs. Signs may be unclear, obscured, ignored, etc. Signs may affect the PFDs of certain IPLs
Fire protection	Active fire protection is often not considered as an IPL as it is a postevent for most scenarios. Its availability and effectiveness may be affected by the fire/explosion that it is intended to contain. However, if a company can demonstrate that it meets the requirements of an IPL for a given scenario, it may be used (e.g., if an activating system such as plastic piping or frangible switches are used) *Note*: Fire protection is a mitigation IPL as it attempts to prevent a larger consequence after an event that has already occurred. Fireproof insulation can be used as an IPL for some scenarios, provided that it meets the requirements of API and corporate standards
Requirement that information is available and understood	This is a basic requirement

[a] Poor performance in the areas discussed in this table may affect the process safety of the whole plant and, thus, many assumptions in the LOPA process.
CCPS publication on LOPA[3] with permission.

where

f_i^C = frequency of the consequence C associated with the scenario
f_i^I = frequency of the initiating event i leading to consequence C
PFD_{ij} = probability of failure on demand for the jth IPL that protects against consequence C for initiating event i.

$$\left[\prod \text{ indicates a product for PFD}_{ij},\ j = 1 \text{ to } J\right]$$

This equation is valid for low demand situations, that is, where the frequency of the initiating event (f_i^I) is less than twice the test frequency for the first IPL.

TABLE 13.4
Examples of Active IPLs and Associated PFDs[a]

IPL	Comments Assuming an Adequate Design Basis and Inspection/Maintenance Procedures	PFD from Literature and Industry	PFD Used in CCPS Book (For Screening)
Relief valve	Prevents system exceeding specified overpressure. Effectiveness of this device is sensitive to service and experience	10^{-1}–10^{-5}	10^{-2}
Rupture disc	Prevents system exceeding specified overpressure. Effectiveness can be very sensitive to service and experience	10^{-1}–10^{-5}	10^{-2}
Basic process control system	Can be credited as an IPL if not associated with the initiating event being considered (See IEC 61508 (IEC, 1998) and IEC 61511 (IEC, 2001)	10^{-1}–10^{-2} ($>10^{-1}$ allowed by IEC)	10^{-1}
SIFs (interlocks)	See IEC 61508 (IEC, 1998) and IEC 61511 (IEC, 2001) for life cycle requirements and additional discussion		
SIL 1	Typically consists of: Single sensor (redundant for fault tolerance) Single logic processor (redundant for fault tolerance) Single final element (redundant for fault tolerance)	$\geq 10^{-2}$ – $<10^{-1}$	CCPS LOPA publication does not specify a Specific SIL level Examples in the Publication Calculate a required PFD for a SIF
SIL 2	Typically consists of: "Multiple" sensors (for fault tolerance) "Multiple" channel logic processor (for fault tolerance) "Multiple" final elements (for fault tolerance)	$\geq 10^{-2}$ – $<10^{-3}$	
SIL 3	Typically consists of: Multiple sensors Multiple channel logic processor Multiple final elements	$\geq 10^{-3}$ – $<10^{-4}$	

[a] Multiple includes 1 out of 2 (1oo2) and 2 out of 3 (2oo3) voting schemes. "Multiple" also indicates that multiple components may or may not be required, depending upon the architecture of the system, the components selected, and the degree of fault tolerance required to achieve the required overall PFD and to minimize unnecessary trips caused by the failure of individual components; see IEC 61511 (IEC, 2001) for guidance and requirements.

Reproduced from CCPS publication on LOPA[3] with permission.

When the demand exceeds this frequency, the frequency of the consequence or the frequency of demand upon the next IPL in the sequence is given by $2(\text{IPL test frequency, per year})(\text{IPL PFD})$.

The extent to which this calculation needs to be modified depends upon the consequences of interest, as determined.

If the consequences of interest are fatalities, then the value calculated is an individual risk.

TABLE 13.5
Passive IPLs and Associated PFDs

IPL	Comments Assuming an Adequate Design Basis and Inspection/Maintenance Procedures	PFD from Literature and Industry	PFD Used in CCPS Book (For screening)
Dike	Will reduce the frequency of large (widespread) consequences of a tank overfill/ rupture/ spill, etc.	10^{-2}–10^{-3}	10^{-2}
Underground drainage system	Will reduce the frequency of large consequences (widespread spill) of a tank overfill/rupture/spill/etc.	10^{-2}–10^{-3}	10^{-2}
Open vent (no valve)	Will prevent over-pressure	10^{-2}–10^{-3}	10^{-1}
Fireproofing	Will reduce heat input rate and provide additional time for depressurizing/firefighting /etc.	10^{-2}–10^{-3}	10^{-2}
Blast-wall/bunker	Will reduce the frequency of large consequences of an explosion by confining blast and protecting equipment/ buildings/etc.	10^{-2}–10^{-3}	10^{-3}
"Inherently safe" design	If properly implemented, can significantly reduce the frequency of consequences associated with a Scenario. Some companies allow inherently safe design features to eliminate certain scenarios (e.g., vessel design pressure exceeds all possible high-pressure challenges)	10^{-1}–10^{-6}	10^{-2}
Flame/detonation arrestors	If properly designed, installed, and maintained, these should eliminate the potential for flashback through a piping system or into a vessel or tank	10^{-1}–10^{-3}	10^{-2}

Reproduced from CCPS publication on LOPA[3] with permission.

For releases of flammable materials, the calculation becomes

$$IR_{i,\text{flammable}} = f_i^I \left(\prod_{j=1}^{J} \text{PFD}_{ij} \right) p_{\text{ignition}} \, p_{\text{present}} \, p_{\text{fatality}} \tag{13.2}$$

where

$IR_{i,\text{flammable}}$ = individual risk from flammable effect (per year)

f_i^I = frequency of the initiating event i leading to consequence C

PFD_{ij} = probability of failure on demand for the jth IPL that protects against consequence C for initiating event i.

p_{ignition} = probability of ignition of flammable release

p_{present} = probability that an individual is present when an event occurs

p_{fatality} = probability that an individual is killed, given exposure to the event

$\left[\prod \text{ indicates a product for } \text{PFD}_{ij},\ j = 1 \text{ to } J\right]$

In the case of toxic releases, the equation is

$$IR_{i,\text{toxic}} = f_i^I \left(\prod_{j=1}^{J} \text{PFD}_{ij} \right) p_{\text{present}} p_{\text{fatality}} \tag{13.3}$$

where

$IR_{i,\text{toxic}}$ = individual risk from toxic effect (per year)

If the consequence of interest is the number of fatalities, then the value calculated is an expected number of fatalities per year for that scenario. The expected number of fatalities per year is also termed the potential loss of life (PLL). The corresponding equations are

$$\text{PLL}_{i,\text{flammable}} = IR_{i,\text{flammable}} n_{\text{present}} \tag{13.4}$$

and

$$\text{PLL}_{i,\text{toxic}} = IR_{i,\text{toxic}} n_{\text{present}} \tag{13.5}$$

where

$IR_{i,\text{flammable}}$ = individual risk from flammable effect (per year)
$PLL_{i,\text{flammable}}$ = potential loss of life from flammable event (fatalities per year)
$PLL_{i,\text{toxic}}$ = potential loss of life from the toxic event (fatalities per year)
$IR_{i,\text{toxic}}$ = individual risk from toxic effect (per year)
n_{present} = number of persons present and exposed to the event

Note that this method of calculating PLL assumes that exposed individuals are located relatively close together. Where exposed individuals are distributed over a wide area, a different approach to PLL calculations may be required.

The risk contributions from each scenario (that might affect an exposed individual) must be summed to estimate the overall risk at any given location.

As an alternative to performing the calculations described above, the various parameters may be combined within a matrix or decision table. Typically, the table or matrix also embodies the risk criteria for decision making. An example is shown in Table 13.6, extracted from an earlier CCPS reference.[1]

13.2.7 Risk Evaluation

A scenario's risk is compared with a previous, similar scenario that had adopted risk reduction options to lower the consequence frequency. Alternately, the risk can be evaluated by comparing the calculated risk with the company's acceptable risk criteria or as specified by "AHJs (authorities having jurisdiction)".

LOPA specifies four basic categories of criteria that

- place risk characterizations per scenario in matrices, with parameters of frequency and consequence as guides

- specify a maximum allowable risk (e.g., risk of fatality or financial loss) per scenario
- specify a minimum number of IPLs (or IPL credits) for any specific scenario
- specify a maximum cumulative risk for a process or geographical area

Following this evaluation, a judgment is to be made about whether further risk reduction is necessary. The options are

Include the application of additional IPLs,

or

Make a more fundamental change in design to make the process inherently safer (by reducing scenario frequency or consequence),

or

Eliminate the scenario altogether.

13.2.8 LOPA Summary Sheet: An Example

A case considered is the complete rupture of a batch reactor, arising from a runaway reaction. The process is highly exothermic. The maximum number of batches that can be run per year is taken as 100. The required steps are as follows:

Step 1: Categorize the consequence which will fall in "Category 5". For tolerable risk, the frequency for such an event is to be limited to 10^{-5}/year or less, once it is taken as 10^{-5} per year and shown in Table 13.6 below.

Step 2: The initiating event is cooling water failure resulting in the reaction temperature increase rate exceeding a critical value. Referring to the CCPS publication (Table 13.6), the initiating event frequency is taken as 10^{-1} per year.

Step 3: Check for an enabling event or condition that might happen simultaneously with the initiating event, leading to the runaway reaction. In this case, since the runaway reaction can only occur when an acetic acid batch is under preparation, the enabling event probability also needs to be considered. A batch will generally be susceptible to runaway reaction at one stage of the reaction. For 100 batches a year, the enabling condition probability is 100/365 = 0.27.

13.2.9 Advantages of LOPA

LOPA has the following advantages:

- Very systematic approach, widely used throughout the process industries
- All the identified initiating causes in HAZOP are addressed
- Severe consequences are prioritized for risk reduction
- For each initiating cause-consequence pair, IPLs are easily identified
- Facilitates allocation of the minimum required IPLs, thereby helping optimize allocation of risk reduction resources

TABLE 13.6
LOPA Summary Sheet

Scenario No.	Equipment No.	Scenario Title: Cooling Water Failure with Runaway Reaction with Potential for Reactor Overpressure, Leakage, Rupture Injuries, and Fatalities	
Date:........2019	**Description**	**Probability**	**Frequency (per year)**
Consequence Description/category	Rupture of the reactor by runaway reaction		
Risk tolerance criteria (category or frequency)	Maximum Tolerable Risk of a fatality		$<10^{-5}$
Initiating Event (typically a frequency)	Cooling water supply failure		10^{-1}
Loop failure of BPCS LIC. (PFD)			
Enabling event or Condition	Probability of reactor in condition for runaway reaction	0.27	
Conditional modifiers (if applicable)			
	Probability of ignition	N.A	
	Probability of personnel in affected area	N.A	
	Probability of fatal injury	N.A	
	Others	N.A	
Frequency of unmitigated consequence			2.7×10^{-2}
IPLs			
BPCS Alarm and Human Action	BPCS loop high temperature alarm	10^{-1}	
Pressure relief valve	Set at reactor design pressure	10^{-2}	
SIF PFD $=10^{-3}$	Vent valve to open on SIF	10^{-3}	
Safeguards(non-IPLs)	Operator action already credited (BPCS Alarm), no other operator role		
	Emergency cooling water pump (diesel operated) not considered as IPL as too many other common elements (piping, valves, jackets) that could have initiated cooling water failure.		
Total PFD for all IPLs	Note: Including added IPL	10^{-6}	

(Continued)

TABLE 13.6 (*Continued*)
LOPA Summary Sheet

Scenario No.	Equipment No.		Scenario Title: Cooling Water Failure with Runaway Reaction with Potential for Reactor Overpressure, Leakage, Rupture Injuries, and Fatalities
Date:........2019	**Description**	**Probability**	**Frequency (per year)**
Frequency of mitigated consequence			2.7×10^{-8}

Risk tolerance criteria met? (Yes/No): YES, with added SIF

Actions required to meet risk tolerance criteria: Consider adding SIF (see Chapter 8): SIS to be provided on the reactor. Installation of SIF with minimum PFD $=10^{-3}$ for opening vent valve at high temperature. Separate nozzles to be provided for Vent & relief valve. N2 purge added.

Responsible Group/Manager/Date: Technical/DND/.....2019

Notes

1. Operator response to high temperature for IPL to be ensured
2. Reactor Vessel Design, Installation, maintenance to be minimum PFD 10^{-2} level

References (links to originating hazard review, PFD, P&ID, etc.): See designated folder

LOPA analyst (and team members, if applicable):

Reproduced from CCPS publication on LOPA[3] with permission.

- All data used in the analysis are comprehensively summarized and documented
- Improves consistency of SIL assignment
- Offers a rational basis for managing IPLs in an operating plant

13.3 BARRIER ANALYSIS[7]

Barrier analysis defines the hazards, targets, and the pathways through which hazards affect targets and identifies barriers and controls that would block the pathway and maintain the target within the specified range or set of conditions.

Barriers must be in place to ensure the necessary risk reduction through the implementation maintenance of effective barriers. The risk from hazards faced at any given time can be maintained at an acceptable level (as originally planned) by preventing an undesirable incident. This outcome could also be achieved by limiting the consequences should such an incident occur.

Experience shows that catastrophic consequences occur because of multiple barrier failures. Hence, there is a substantial need to ensure that failures of multiple barriers are made near-impossible. If one fails, other barriers will be active in preventing any catastrophic events. That is why one has to ensure that multiple barriers include sufficient redundancy that must remain operational throughout the plant's life cycle. This technique is known as "barrier management".

Norway is one of the countries that introduced, in 2002, the concept of barrier management, through its Petroleum Service Agency (PSA), to control risk in the offshore operation of its oil and gas industry. The paper "*Guidance for barrier management in the petroleum industry*" by Stein Hauge and Knut Øien of SINTEF Technology and Society" is an excellent reference. This document has been extensively referred to, and two tables and a few figures have been reproduced with the authors' kind permission (SINTEF Technology and Society, Norway).

Initially, barrier management focused on single barriers (mostly technical and considered critical) and, to a lesser degree, on operational, organizational, and behavioral conditions. However, several major, catastrophic events in the industry have shown that a focus on single barriers would fail to prevent accidents similar to those characterized by failures of multiple barriers. Barrier management must focus on the entire barrier system, including technical, operational, organizational, and behavioral measures, to avoid potential multiple-barrier failures.

For barrier management, as for the LOPA IPLs, the concept of IPLs is key. The failure of one barrier still leaves in place other independent barrier(s) capable of preventing the incident. "Protective" barriers actually can stop the incident – if they work fast enough, are robust, reliable, and can be proved by testing. Barriers must, therefore, be maintained diligently. The Bhopal example in Chapter 2 is a case where there supposedly were multiple barriers, none of which could, by themselves, prevent the release or completely mitigate it. These barriers were defeated because of a series of disastrous human failures (e.g., corrosion, poor maintenance, no management of change, managerial and supervisory indifference, etc.).

13.3.1 Barrier Failure and Catastrophic Accidents

The UCIL Bhopal catastrophic accident occurred as a result of failures of multiple barriers. Figure 13.6 below demonstrates this pictorially:

13.3.2 Important Definitions Related to Barrier Management

Barrier management: Coordinated activities to decide, select, establish, and maintain barriers to ensure functional performance at all times.

Barrier: Technical, operational, organizational, and behavioral measures intended, individually and collectively, to reduce hazards and the possibility of a specific error (that leads to undesirable consequences) and limit its harmful effects on life and property.

Barrier function: *The task or role of a barrier.* Examples include:

a. prevention of vessel failure,
b. hazardous material spillage,
c. leaks or ignition,
d. reducing fire loads,
e. controlling pressure or temperature fluctuations within limits,
f. preventing and controlling runaway reactions,
g. maintaining operating discipline,
h. strong safety awareness,
i. arresting and suppressing toxic releases,
j. ensuring acceptable personnel evacuation, and
k. preventing/mitigating collateral damage.

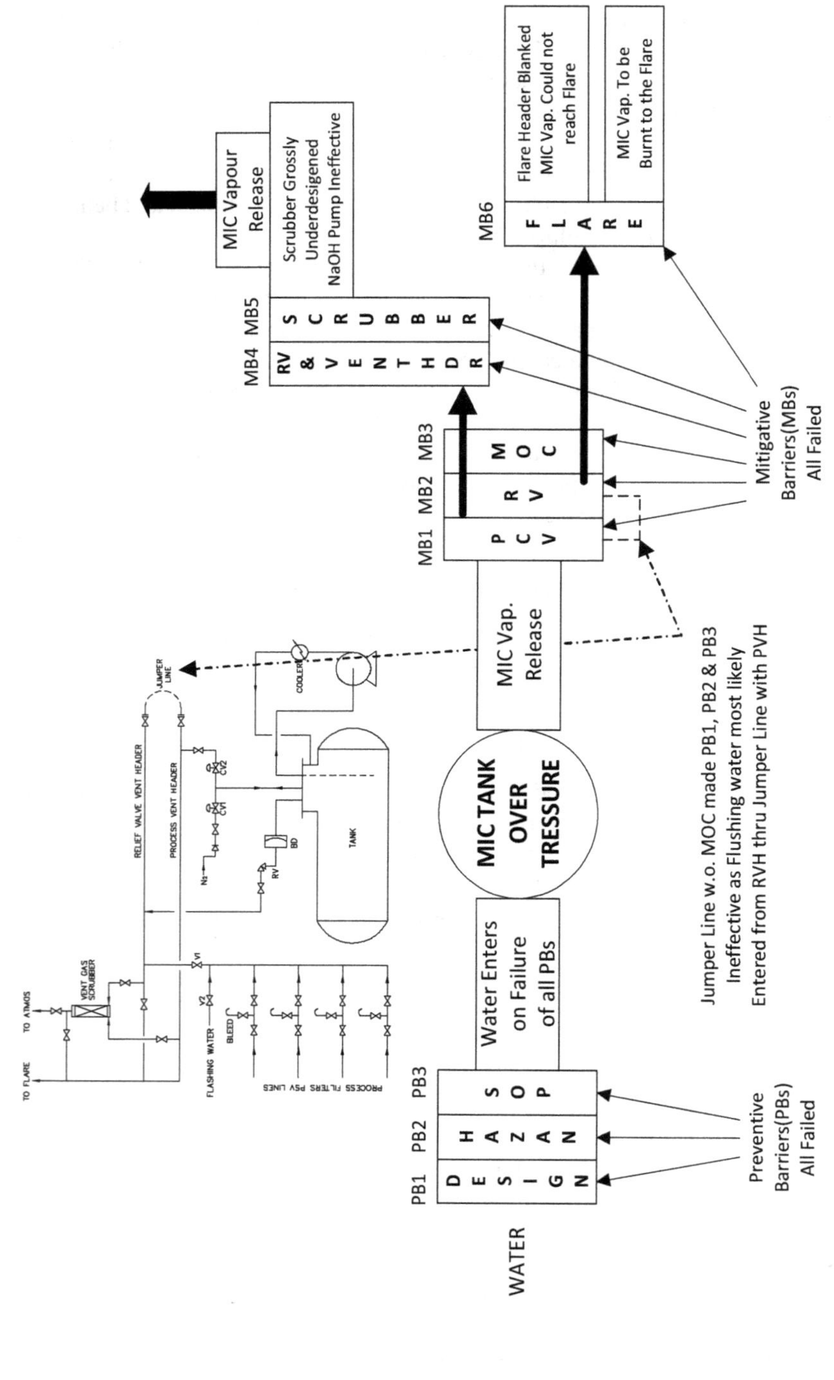

FIGURE 13.6 Multiple barrier failures caused worst-ever chemical accident at Union Carbide, Bhopal.

Examples are "pressure relief" and "high-level trips".

Barrier system: *A system designed and implemented to perform one or more safety functions.* Examples are "pressure control system", "level control system", etc.

Barrier element: Technical/Engineering, operational, organizational, or behavioral measures or solutions that play a role in effectuating a barrier function. There are four types of barrier elements:

1. "Technical": equipment and systems which constitute a part of achieving a barrier function (*What equipment shall be used)*?
2. "Operational": actions and activities personnel have to perform to enable achievement of a barrier function (*What shall be done)*?
3. "Organizational": personnel with defined roles or functions and specific competence who play a role in achieving a barrier function (*Who is doing it)*?
4. "Behavioral": personnel attitudes and actions, under trying circumstances, that enable achievement of a barrier function (*How is one doing it)*?

All assigned personnel must know and understand their roles in the achievement of barrier functions; accordingly, training and drills are quintessential.

Each of the four types of barrier elements listed above prevents escalation of hazards either in a "preventive" or "mitigative" manner. For example, a "pressure safety valve" is a preventive barrier and a "flare stack" is a mitigative barrier".

Barrier strategy: Explains and clarifies the barrier functions and elements to be implemented to reduce risk.

Performance requirements: *Verifiable requirements related to barrier element properties to ensure that the barrier is effective.* These can include capacity, functionality, effectiveness, integrity, reliability, availability, ability to withstand loads, robustness, expertise, and mobilization time.

Performance influencing factors: Conditions that are significant for barrier functions and elements to perform as intended.

Context: External and internal frame conditions that must be considered in barrier management.

Risk management: Coordinated activities to direct and control risk in an organization.

In Figure 13.7, three types of barrier functions are illustrated (note: safety-critical tasks are functions, too):

1. Safety-critical tasks
2. Safety functions, and
3. Safety instrumented functions (SIFs).

It should be noted that barrier sub- and sub-sub-functions often depend on both personnel and technical systems to be achieved in practice. For example, in Figure 13.7, the CCR operator must manually initiate methanol injection; however, success also depends on the proper functioning of both a push button on the operator display and the chemical injection valves in the field.

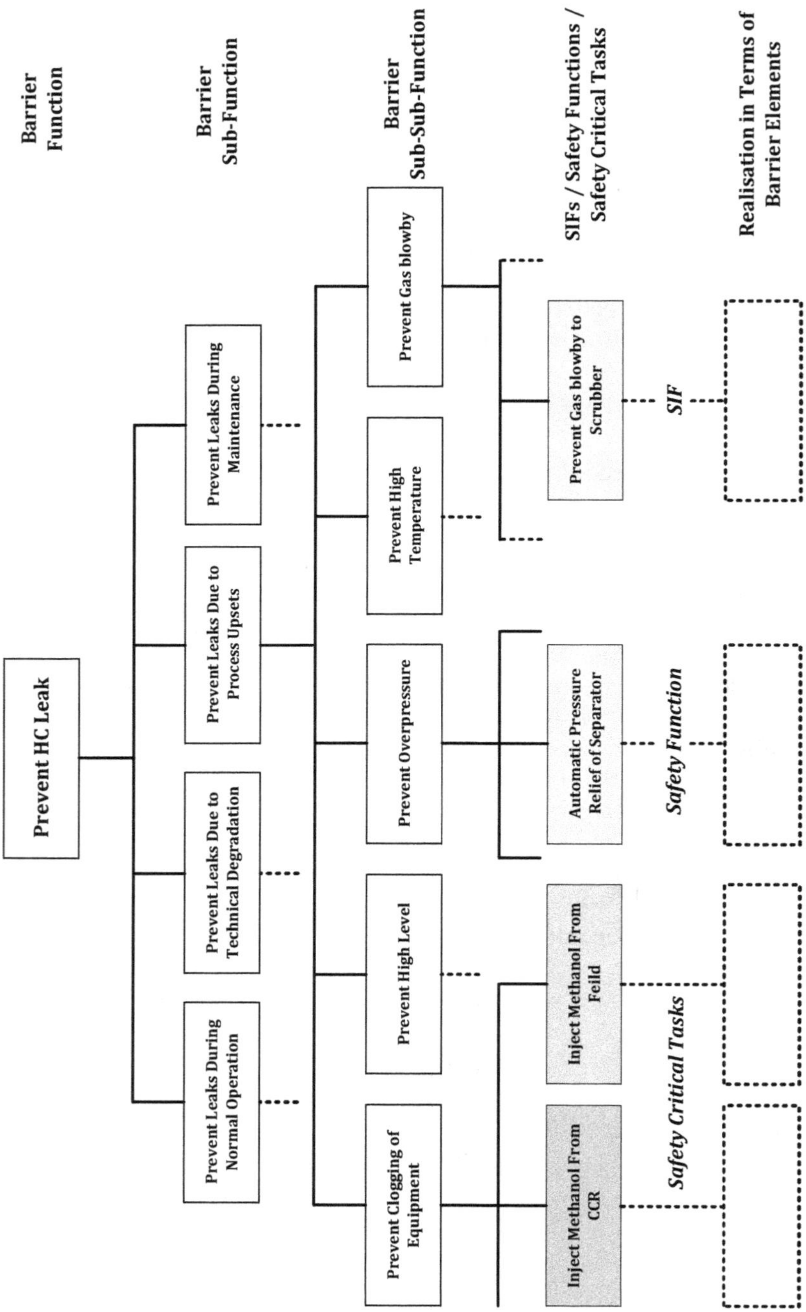

FIGURE 13.7 Types of barrier sub- (or sub-sub-) functions; safety-critical tasks, safety functions, and SIFs.

13.3.3 Independence of Barriers

All barrier elements listed above (Table 13.7) do not qualify as IPLs that may be considered in a LOPA. There are specific functional criteria for the barrier elements (layers of protection) that can be qualified as "independent." A protection layer meets the requirements of being an IPL when designed and managed to achieve the following seven core attributes:

1. Independent
2. Functional
3. Integrity
4. Reliable
5. Validated
6. Maintained and
7. Audited

TABLE 13.7
Summary of the Types and Categories of Barrier

	Technical/Engineering	Operational	Organizational	Behavioral
			Type of Barriers	
Preventive barriers	Following codes	Updating SOPs	Top management to imbibe and maintain a sound safety culture	Safety consciousness of plant personnel
	Following specific guidelines where available	Operator training	Operator refresher training	High moral of the plant personnel
	Safety specific design to avoid "over pressure" and "loss of containment"	Updating of design documents maintaining current update	Institute a sound process safety management training of operating managers to make them follow OSHA/CCPS/EU PSM	
		Updating of engineering documents maintaining current update	Institute sound and fool proof system for MOC	
	Material of construction		Institute fail proof system of noting "near misses", investigation followed by action	
	Corrosion protection		Institute a foolproof accident investigation system and apprising plant personnel	

(*Continued*)

TABLE 13.7 (*Continued*)
Summary of the Types and Categories of Barrier

	Technical/Engineering	Operational	Organizational	Behavioral
	SIS, SIL level SOPS			
	Site safety report with work permit system, hot work permit, HT/LT electric isolation permit etc.			
	Safety specific design to contain releases and leaks within battery limit			
Mitigative barriers	Secondary/tertiary containment	Site periodic internal mock drill	Demonstrate, videos, CSB and other agencies major accident reports portraying "what went wrong" to plant personnel	Plant personnel to be trained emergency responder
	Leak detection systems to energise	Running all mitigation measures, checking and ensuring their full effectiveness		
	ESDs			
	Fire detectors to energise firefighting system			

Access security and management of change: The key criterion for an IPL is independence. For a barrier element to be independent, the barrier's effectiveness must be independent of the following:

- The occurrence or consequences of the initiating event
- The failure of any component of an IPL already credited for the scenario or conditions that caused another IPL to fail; or any other element of the scenario

Further, an IPL is to be auditable and must be designed to enable periodic validation to prevent the consequences of the scenario it protects against effectively.

Some of the preventive barriers listed above as "safe reactor design pressure above run-away reaction pressure", "management of change (MOC)", and "corrosion protection by SCADA", will qualify as IPLs if they are subjected to regular inspection, test, and revalidation. SIL barriers also qualify as IPLs.

Of the mitigative barriers listed in the table as secondary containments, "leak detection system to energize alarm (for operator action)", "ESD (connected to UPS)", and "fire detector to energize deluge valve" will all qualify as IPLs if they are subject to audit and revalidation.

Barriers such as "codes, guidelines, or safety reports" must not be considered as IPLs.

13.3.4 Barrier Management Process

Barrier management activities associated with different life cycle phases are shown in Table 13.8 below.

In the design phases, barriers should be identified and designed to ensure that necessary risk reduction is accomplished during subsequent operations. During the operational phase, diligent follow-up and maintenance of barriers must be emphasized to ensure that they are always available. It is equally vital to implement repairs or compensating measures if barriers are impaired and verify the required performance.

A proper understanding and mastery of the principles and practices for implementing a successful barrier management program is a prerequisite. A dedicated team that possesses the appropriate experience and domain expertise must be appointed. Only then does it become feasible for design, operating, and safety personnel to succeed in ensuring safe and efficient achievement, at the plant level, of the highest corporate safety goals:

- Personnel, process, and equipment safety
- Environmental compliance
- Reliability
- Production
- Product quality and
- Profit

The optimal barrier strategy requires careful consideration of the unique aspects of each process plant or facility. The underlying principles are common to all process

TABLE 13.8
Barrier Management Activities in Specific Life Cycle Phases

Early Design	Detailed Design	Operation
Prepare plan for barrier management	Update plan for barrier management	Prepare plan to assure barrier performance (update if necessary)
Define areas	Verify areas	Review area definition
Perform or review HAZID	Review refined HAZID	Update HAZID (e.g. during modification)
Identify/define major hazards/DSHAs	Revise DSHAs	Review and update risk analyses and DSHAs
Perform barrier analysis	Refine barrier analysis	Update barrier analysis
Establish initial barrier strategy	Refine barrier strategy	Review and update barrier strategy
Establish initial performance standards	Refine performance standards	Review and update performance standards
	Establish system for monitoring of barrier status (e.g. barrier panel)	Monitor barrier status and consider need for compensating measures
	Establish system and process for follow-up of barrier performance	Monitor and verify barrier performance

plants. However, even for the same basic process, there are myriad plant-specific details and regulatory requirements that can vary quite significantly from one location to another. The strategy adopted in each case must reflect a clear understanding of the unique risks to be confronted. This requires safety and reliability studies of the design and engineering details in a professional environment, with in-house and retained outside consultants, as appropriate.

The main steps to establish a barrier strategy and corresponding performance standards include the following:

1. Facility description and area division
2. Area-wise identification of hazards: DSHA (defined situations of accident and hazard)
3. Definition of required barrier functions for the area (based on the risk level)
4. Barrier elements in each area or similar risk level areas
5. Performance requirements for barriers
6. Performance influencing factors
7. Activities for verification of performance requirements

The design phase of barrier management is explained in Figure 13.8 below:

The main criterion for area division in the barrier strategy should be that the risk level is applied uniformly. If the risk level and hence the barrier elements vary within an area, it would be necessary to split the area into several sub-divisions to develop a consistent barrier strategy.

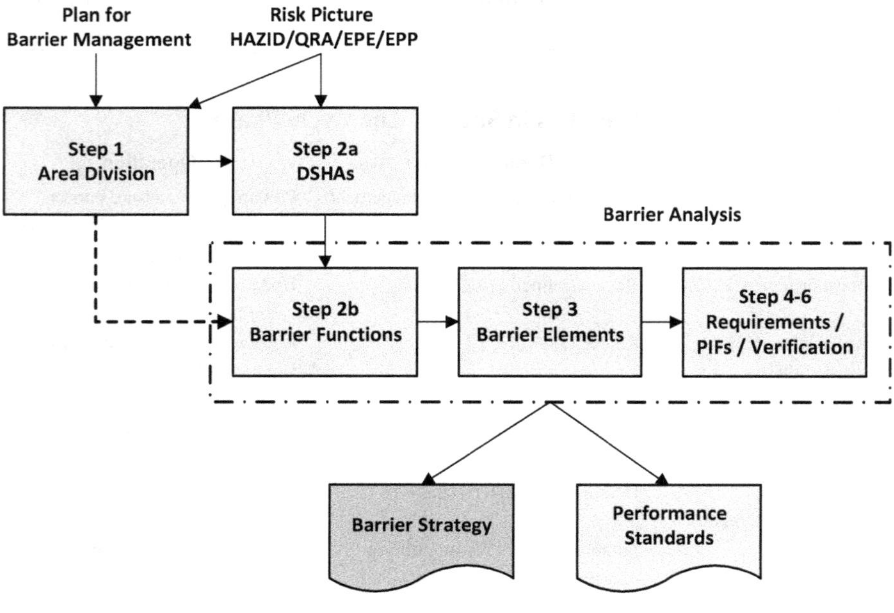

FIGURE 13.8 Barrier management.

For example, the typical "main areas" in an offshore installation (without drilling) are classified as follows:

- Process area
- Riser area (and wellhead area)
- Utility area
- Living quarters
- Shafts
- General functions

Effective barrier management must identify the required barriers, based on a comprehensive understanding of all hazards and associated risk levels related to the project or plant and also all activities under consideration.

In a typical oil and gas development project, hazards are at first determined by HAZID and HAZOP studies. Note that risk levels are defined using both qualitative and quantitative risk assessments. SIL and safety instrumented systems (SIS) analyses, and similar reviews, are also necessary to reduce identified risk levels. In any event, during the detailed engineering phase, the principles of emergency preparedness analysis must be applied.

Comprehensive facility-specific barrier strategies rely on establishing a cause-and-effect relationship between the hazards present in an area and the barriers needed to protect against those hazards. Hence, the barrier strategy document describes a logical relationship between the barrier functions and barrier elements to control and mitigate the hazards identified in the HAZID and HAZOP studies at the design stage. This aspect requires a highly structured approach to ensure that all relevant barriers are covered and that all required barrier elements are identified and described explicitly. This activity typically starts with a HAZID study, where the relevant primary accident hazards are identified.

The next step is to identify and define the barrier functions that need to be in place to prevent and mitigate the hazards. Examples of typical significant accident hazards and associated barrier functions for a floating offshore installation (without drilling) are shown in Table 13.9 below.

As illustrated in Figure 13.9 below, the *organizational barrier element* of a barrier function refers to the personnel directly involved in realizing the function. The *operational barrier element* refers to the procedures that are required to perform the required action immediately.

When identifying barrier elements, the following question must be asked and answered: "what are the necessary technical, operational, and organizational elements to realize this sub-function?".

The primary purpose of the functional breakdown is to clarify and communicate the role of the barrier elements required for realizing the barrier function. Examples of *breakdown structures* for the barrier functions such as "prevent HC leaks from process equipment", "prevent collision with visiting vessel," and "prevent fatalities during evacuation" are illustrated in Figures 13.10–13.12, respectively.

TABLE 13.9
Examples of Typical Major Accident Hazards and Associated Barrier Functions

Major Accident Hazards DSHAs	Associated Barrier Functions
Hydrocarbon (HC) leakage	Prevent HC leak from process equipment
	Prevent HC leak from risers and pipelines
	Prevent HC leak from cargo/slop tank
	Prevent HC leak during offloading operation
	Limit size of HC leak from process equipment
	Limit size of HC leak from risers and pipelines
	Limit size of HC Leak from cargo/slop tank
	Limit size of HC leak from offloading hose
Fire and explosion	Prevent ignition
	Prevent explosive or dangerous atmosphere in cargo/slop tank
	Prevent HC in ballast tank
	Prevent escalation to other equipment
	Prevent escalation to other area
	Prevent fatalities during escape/mustering
	Prevent fatalities during evacuation
Acute pollution	Prevent spill to sea
	Limit consequences of spill to sea
Dropped object	Prevent dropped objects from crane operations
	Limit consequences in case of dropped objects
Ship on collision courses/drifting object	Prevent collision with passing or visiting vessel
Ship collision	Prepare evacuation due to vessel or drifting object on collision course
Loss of buoyancy/stability	Prevent loss of buoyancy/stability
	regain buoyancy/stability
Loss of position	Prevent loss of position
Helicopter accident at installation	Prevent helicopter accident
	Prevent escalation from helicopter accident
Extreme weather	Ensure contingency in case of extreme weather forecast
Structural failure	Prevent loss of structural integrity
	Prevent fatalities upon loss of structural integrity
Non-HC fire	Prevent non-HC fire
	Limit consequences from non-HC fire

An example of a *detailed functional breakdown* of the barrier function "prevent HC leak from process equipment" is shown in Figure 13.13. This barrier function is broken down into sub-functions and, ultimately, safety-critical tasks, safety functions, and SIFs (cf. Figure 13.7). These tasks and functions are then realized for the technical, operational, and organizational behavioral barrier elements.

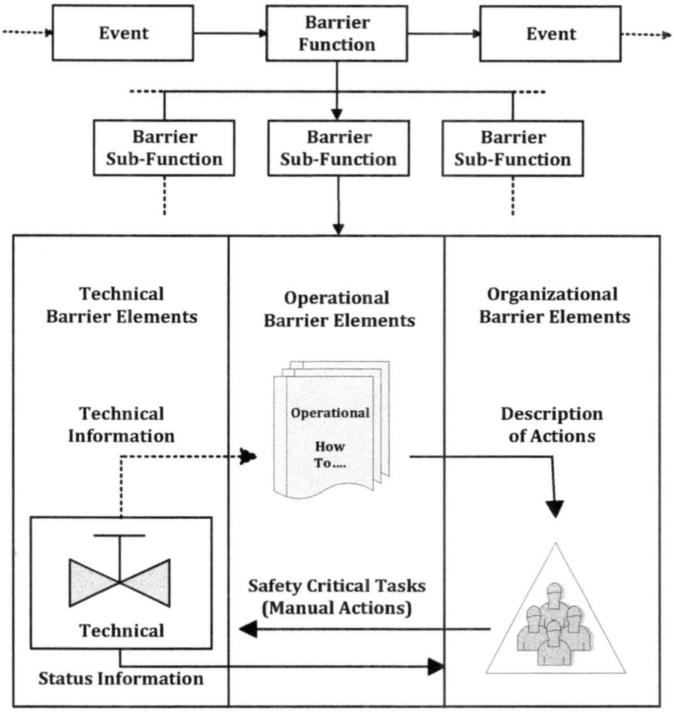

FIGURE 13.9 Barrier functions implemented through barrier elements.

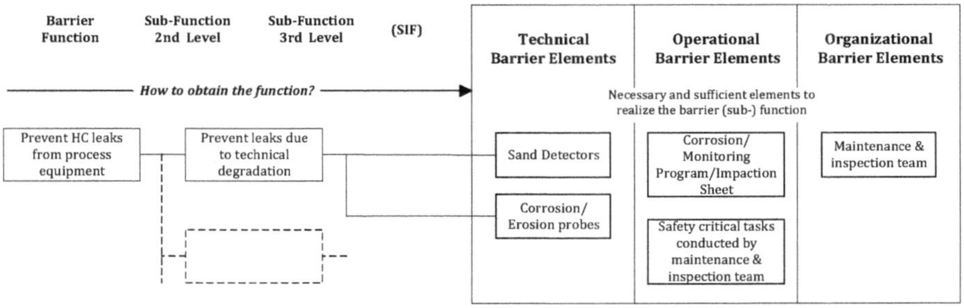

FIGURE 13.10 Breakdown structure for the barrier function "prevent HC leaks" (example).

13.4 QRA

The goal of QRA is to calculate numeric values associated with each component after completion of risk evaluation. It uses empirical or failure history data for the failure of preventive and mitigative protections or barriers. The purpose is to determine the probability of adverse consequences arising from a hazardous initiating event while also considering the effects of enabling events. The adverse consequence

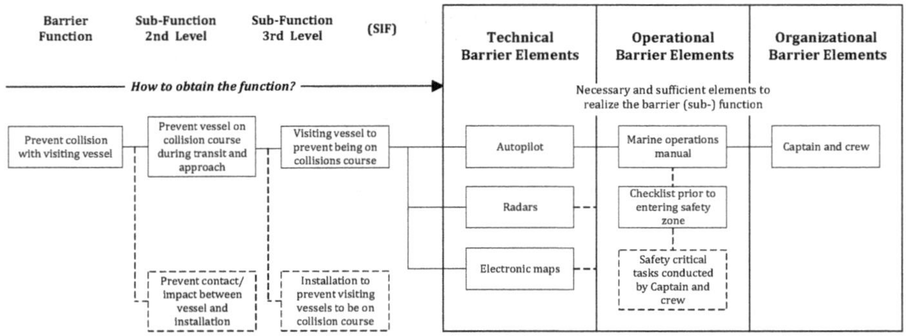

FIGURE 13.11 Breakdown structure for the barrier function "prevent collision with visiting vessel" (example).

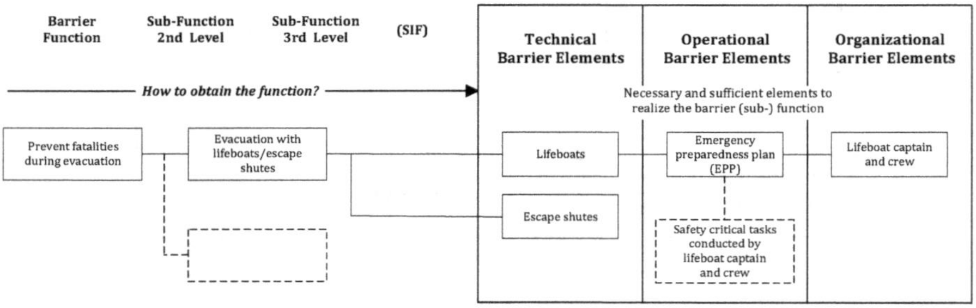

FIGURE 13.12 Breakdown Structure for Barrier Function "prevent fatalities during evacuation" (example).

effects (e.g., fire, explosion, or toxic release) are simultaneously quantified for human injuries, fatalities, and financial losses.

The probabilities of occurrence need to be sufficiently below the prescribed limit of allowable risk level, and also below "as low as reasonably practicable" (ALARP) limits.

13.4.1 Estimation of Frequency of a Hazardous Event

This section covers the methodology for estimating the frequency of an identified hazardous event. We cover two categories of events:

1. The first category covers failure of equipment (major failure of vessels, pipeline rupture, etc.), resulting in the release of hazardous materials
2. The second covers various consequences or outcomes following a release.

The frequency of equipment failure is estimated by analyzing the sequence of causes leading to the event, using a methodology called fault tree analysis (FTA). For the consequences of a release (e.g., vapor cloud explosion, BLEVE, or exposure of a population to a toxic gas cloud), the methodology followed is known as the event tree methodology (ETM).

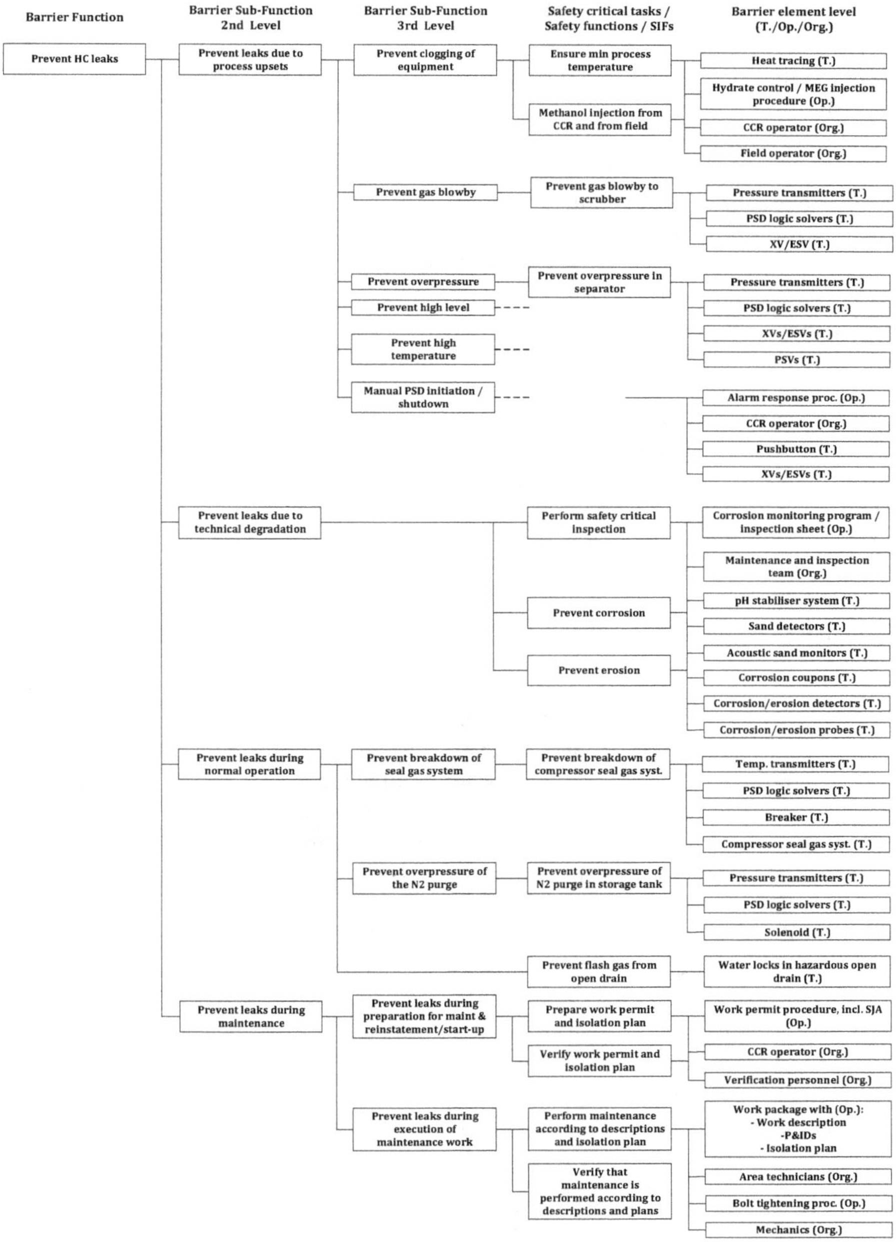

Barrier Function
Barrier Sub-Function 2nd Level
Barrier Sub-Function 3rd Level
Safety critical tasks / Safety functions / SIFs
Barrier element level (T./Op./Org.)
Prevent HC leaks
Prevent leaks due to process upsets
Prevent clogging of equipment
Ensure min process temperature
Methanol injection from CCR and from field
Heat tracing (T.)
Hydrate control / MEG injection procedure (Op.)
CCR operator (Org.)
Field operator (Org.)
Prevent gas blowby
Prevent gas blowby to scrubber
Pressure transmitters (T.)
PSD logic solvers (T.)
XV/ESV (T.)
Prevent overpressure
Prevent high level
Prevent high temperature
Manual PSD initiation / shutdown
Prevent overpressure in separator
Pressure transmitters (T.)
PSD logic solvers (T.)
XVs/ESVs (T.)
PSVs (T.)
Alarm response proc. (Op.)
CCR operator (Org.)
Pushbutton (T.)
XVs/ESVs (T.)
Prevent leaks due to technical degradation
Perform safety critical inspection
Prevent corrosion
Prevent erosion
Corrosion monitoring program / inspection sheet (Op.)
Maintenance and inspection team (Org.)
pH stabiliser system (T.)
Sand detectors (T.)
Acoustic sand monitors (T.)
Corrosion coupons (T.)
Corrosion/erosion detectors (T.)
Corrosion/erosion probes (T.)
Prevent leaks during normal operation
Prevent breakdown of seal gas system
Prevent breakdown of compressor seal gas syst.
Temp. transmitters (T.)
PSD logic solvers (T.)
Breaker (T.)
Compressor seal gas syst. (T.)
Prevent overpressure of the N2 purge
Prevent overpressure of N2 purge in storage tank
Pressure transmitters (T.)
PSD logic solvers (T.)
Solenoid (T.)
Prevent flash gas from open drain
Water locks in hazardous open drain (T.)
Prevent leaks during maintenance
Prevent leaks during preparation for maint & reinstatement/start-up
Prepare work permit and isolation plan
Verify work permit and isolation plan
Work permit procedure, incl. SJA (Op.)
CCR operator (Org.)
Verification personnel (Org.)
Prevent leaks during execution of maintenance work
Perform maintenance according to descriptions and isolation plan
Verify that maintenance is performed according to descriptions and plans
Work package with (Op.):
- Work description
-P&IDs
- Isolation plan
Area technicians (Org.)
Bolt tightening proc. (Op.)
Mechanics (Org.)

FIGURE 13.13 Detailed functional breakdown of barrier function "prevent HC Leak from process equipment" (example).

13.4.1.1 Fault Tree Methodology

A fault tree starts with the event being considered at the top of the diagram; this is called the top event. Typical examples of top events are "fire on top of a tank", "explosion in a reactor", or "release of a toxic gas from a vessel". The tree is then developed downwards, at each stage asking questions such as "how" or "why" to explore possible causes of the event. Thus, various pathways for initiation of the top event are identified, each ending at the bottom in a primary or basic event. A basic event is one whose probability, or frequency of occurrence, is known with sufficient accuracy (typically, from historical data). Alternatively, it can be estimated without requiring any further downward development of the tree. The intermediate events in each pathway are also known as sub-events.

Fault tree analysis is both a qualitative and a quantitative technique. Qualitatively, it is used to identify the individual pathways to the top event. Quantitatively, it is used to estimate the frequency or probability of the top event. Every basic event at the bottom of the tree needs to be assigned a frequency or a probability of occurrence when the tree is used for quantitative analysis. QRA techniques use FTA quantitatively for probability assessment of credible consequence scenarios of severe-to-catastrophic magnitude.

Before developing a fault tree, or trying to understand one that has already been developed, it is essential to become familiar with the meanings of some of the terms used in this methodology:

- *Event*: An event is an unplanned occurrence, e.g., a major uncontrolled emission, fire, or explosion. Such an event may be a cause or a contributing cause of another event, e.g., "cooling water failure" causing "rupture of a reactor". Alternatively, it may be a response to an initiating event, e.g., a fire on the top of a hydrocarbon storage tank following a lightning strike. An explosion or lightning strike are instantaneous events, while fire or emission from a ruptured pipe is an event of finite duration.
- *AND and OR gate*: An FTA, as shown in Figure 13.14, is developed using the concepts of "AND" and "OR" gates. The "OR" gate indicates that a failure event can occur if *any single one* of the measures provided to prevent the event fails, none functioning independently of the other. The probability of a failure event in an "OR" gate is the sum of the probabilities of failure of all the preventive measures.
- An "AND" gate indicates that a failure can occur only if *all* the measures that have been provided to prevent the event fail. The probability of the failure event is arrived at by multiplying the probability of failure of all the preventive measures, each qualifying as an "independent barrier" or "IPL". IPLs should be identified and given due credit in hazard reduction and risk control.
- *State*: The state of a component in a system is the prevailing condition at a given time. For example, equipment in operation may be in a "good" operating state with performance conforming to specifications, or it may be in a failed or "dead" state and no longer capable of performing its specified function.

- *Protective system*: A protective system is provided to prevent an untoward event from occurring when a hazardous situation arises. A typical *instrumented* protective system comprises a sensor, a transmitter, a trip module, and a shutdown valve (SDV). A *manual* protective system would typically include a sensor, an alarm, an operator, and a SDV.
- *Demand*: The dictionary meaning of the word is a request or an order. In control systems terminology, this refers to an undesirable situation that requires a protective system to activate to prevent the mishap or unwanted situation from escalating to a hazardous event.
- *Failure*: This is an event that results in the loss of ability to perform the required function under specified conditions. Examples:
  a. The breakdown of a pump in continuous running duty is a failure in operation;
  b. An alarm failing to act in response to a high-temperature setting is a failure to operate on demand;
  c. A pressure safety valve (PSV) opening prematurely below the operating limit of set pressure is also a failure, but this is a "spurious" or fail-safe failure.
- *Fault*: A fault is the state of any component when it cannot perform its stated function under specified conditions. While constructing a fault tree, it must be noted that a component can only be in one of the two states, namely the "working state" or the "failed state". A failed state can be as follows:
  - "Revealed", as in the case of a stuck float in a level indicating system, or
  - "Unrevealed", as in the case of a high-temperature switch in which a fault can be detected only during testing, or if a demand occurs and the switch fails to act.

  Further, faults can be of two categories: "fail-safe" faults and "fail danger" faults.
- *Frequency and rate*: The statistical frequency of an event is the expected number of events over a given period. Thus, if road accident data in a city show 40 accidents over 10 years, then the city's average road accident frequency is 4 per year. If a machine operates continuously for 2 years and is under maintenance for 6 months, then its failure *frequency*, on average, is 1 in 2.5 years or 0.4 per year, while the failure *rate* of the machine is 1 in 2 years or 0.5 per year.
- *Probability*: By definition, probability is the chance of an event occurring. It is a dimensionless number ranging from 0 to 1, with 0 representing impossibility and 1 representing 100% certainty. In risk assessment terminology, probability refers to the likelihood of success or failure of a protective system to act on demand. However, the frequency of occurrence of an event is also a measure of the expected likelihood of occurrence. For example, if an event occurs on an average once in 100 years, it has a higher probability of occurrence at any time than an event that occurs on an average once in 10,000 years.
- *Fractional dead time (FDT)*: For an instrumented protective system, the probability of being in a failed state is expressed as a FDT. Whenever a potential fault is unrevealed, proof testing must be done routinely to reveal

failures and restore the system to an operational state. Assuming that failures occur, on average, half-way through the proof test interval, the FDT of a simple protection system is given approximately by

$$\mathrm{FDT} = \left(\tfrac{1}{2}\right)(\theta)(T) \tag{13.6}$$

where

FDT = fractional dead time,
θ = failure rate per year
T = proof test interval, years.
Note: $(\theta)(T) \ll 1$

- *Hazard rate*: The hazard rate is the annual frequency of occurrence of a hazardous event. It is obtained by multiplying the demand frequency by the probability of the protective system being in a failed state when that demand occurs.

The symbols commonly used for typical elements of a fault tree are shown in Figure 13.14 below:

The methodology for developing a fault tree is illustrated below through rudimentary Example 13.1. The flow diagram for the process is shown in Figure 13.15. A 20% caustic soda solution is pumped continuously from a storage tank to a level-controlled feed vessel that feeds the processing unit via gravity. The feed pump needs to operate at 50%–60% of its rated capacity to meet its requirements. If the feed vessel's level control system fails, the caustic soda solution will overflow through

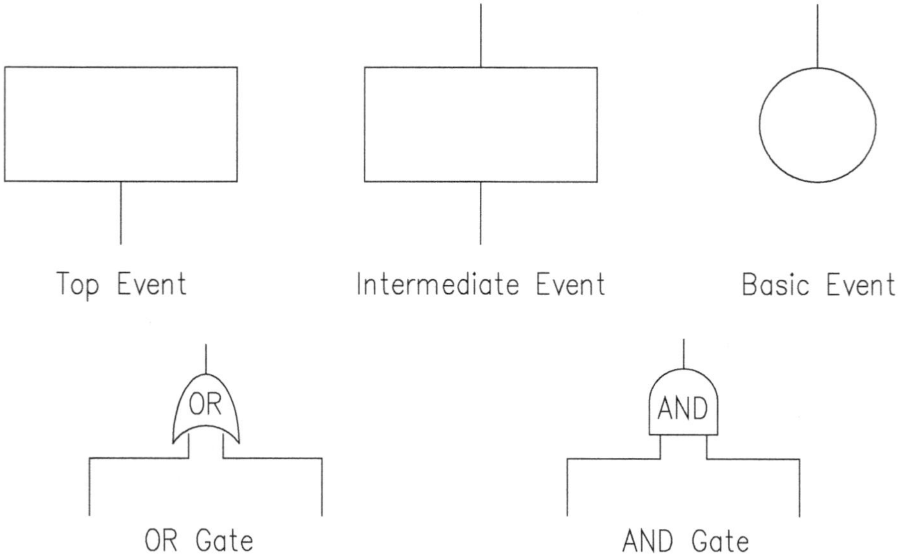

FIGURE 13.14 Symbols used in a fault tree.

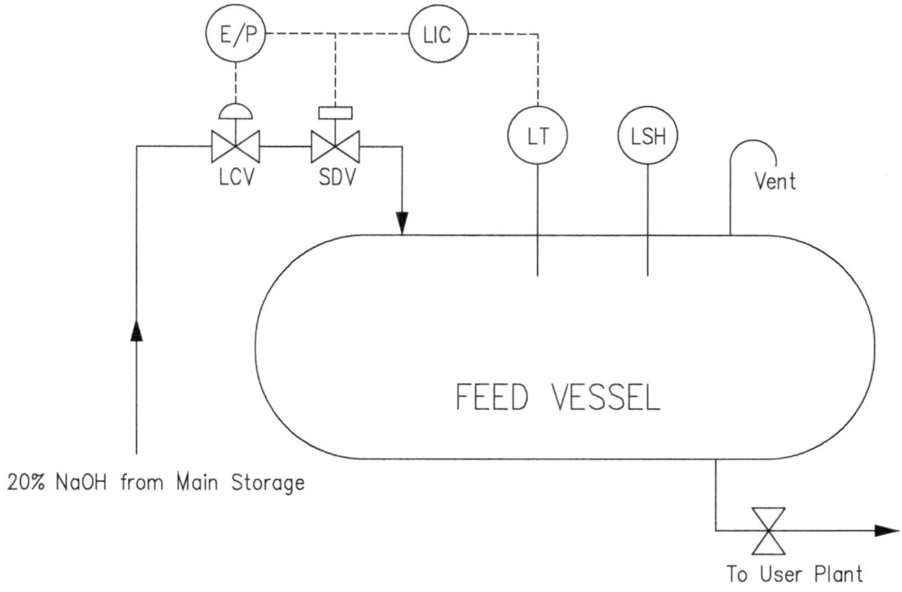

FIGURE 13.15 Caustic soda feeding system for Example 13.1.

the vent, causing a safety/environmental pollution problem. "Overflow of Caustic Through Vent" is, therefore, identified as a hazardous event, and we wish to estimate the frequency of this event.

The construction of a fault tree starts with the identification of the hazardous event (called the top event), identification of demands, and identification of protection systems that are provided for each demand. Starting with the undesired top event, the tree is developed downwards until all the basic events are identified, as shown in Figure 13.16a below.

This part of the tree is called the demand logic, each basic event at the bottom of this diagram being a demand. It should be noted that while drawing the demand logic diagram, no account has been taken of protective devices such as alarms, trips, or operator intervention.

The next step in fault tree development is to examine the probability (FDT) of the protective system being in a failed state when the demand occurs. The protective system provided in this example is simple: a level switch that will stop caustic feeding to the feed vessel by closing the SDV once the preset high level has been reached. In this case, the protective system provided is common for all the demands and should act whenever any component in the level control loop fails. The logic diagram for the protection system is shown in Figure 13.16b below:

The system's fault tree is then drawn, as shown in Figure 13.16c below, by combining the demand logic and the logic diagrams for the protective system.

In Figure 13.16c, all four demands combine through an OR gate, and their frequencies are added up to yield the total demand frequency. Similarly, the fault

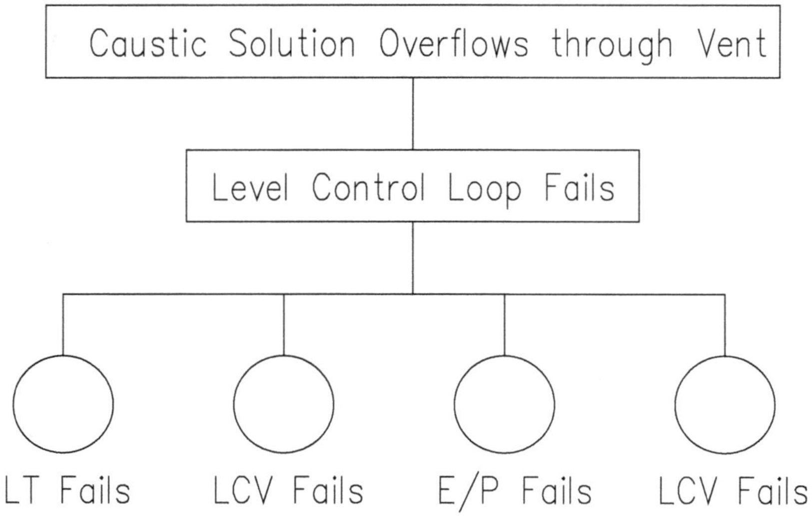

FIGURE 13.16a Demand logic diagram for control system in Figure 13.15.

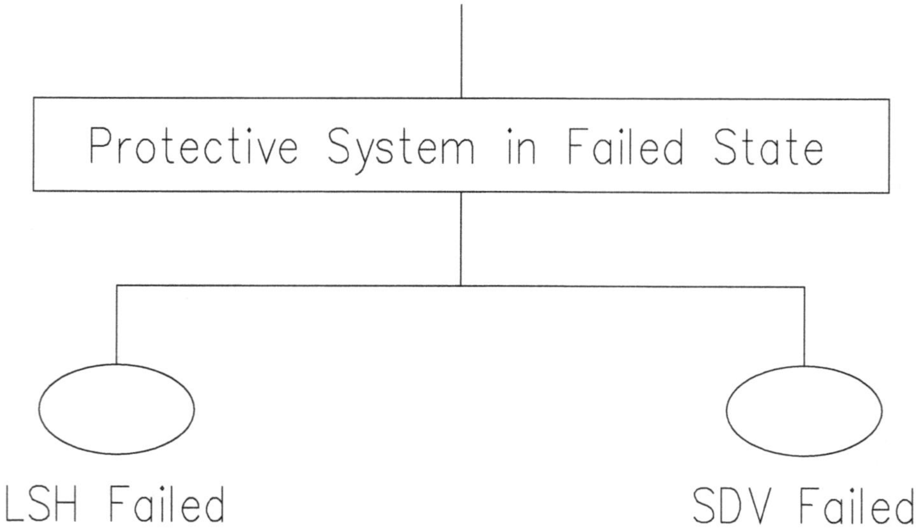

FIGURE 13.16b Logic diagram for the protective system in Figure 13.15.

rates of LSH and SDV pass through an OR gate and, therefore, sum up to yield the total fault rate of the protective system.

The frequency of failure of the control loop is calculated by adding the fail-danger fault rates of the components comprising the loop (Table 13.10a):

Hence, the demand rate (*D*) in this case is 0.7 per year.

For calculating the FDT of the protection system, it is assumed that proof testing is done at intervals of 3 months (Table 13.10b):

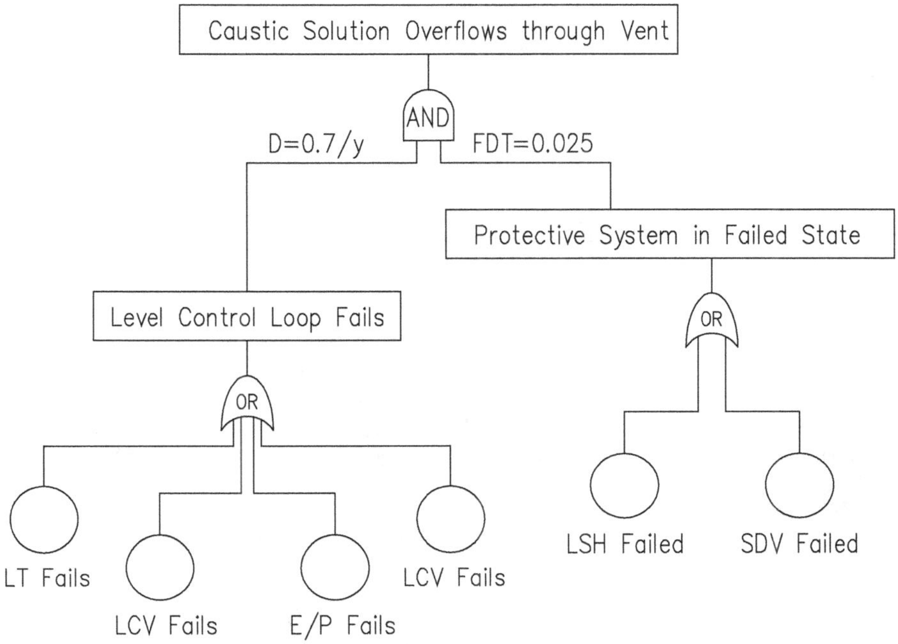

FIGURE 13.16c Fault tree for the level control system in Figure 13.15.

TABLE 13.10a
Fail Danger Fault Level of Components of Control Loop

Component	Description	Fail-Danger Fault Rate (per year)
LT	Electronic level transmitter	0.2
LIC	Level Indicator controller	0.2
E/P	Electronic-to-pneumatic converter	0.1
LCV	The pneumatic level control valve	0.2
	Total failure rate	0.7

TABLE 13.10b
Fail Danger Fault Level of Components of High-Level Trip

Symbol	Description	Fail-Danger Fault Rate (per year)
LSH	Electronic level switch	0.1
SDV	Electronic SDV	0.1
	Total fault rate (*f*)	0.2

Hence, FDT = (½) (0.2) (3/12) = 0.025.

The hazard rate (i.e., the Top Event frequency) is obtained by multiplying the two inputs to the AND gate in Figure 13.16(c). Hence,

$$\text{Hazard rate} = (0.7)(0.025) = 0.0175 \text{ per year (on average, once in 57 years).}$$

The limitations on use of this methodology are as follows:

1. $(f)(T) \ll 1$, and
2. $(D)(T) \ll 1$

In the example above, $(f)(T) = 0.5$ and $(D)(T) = 0.18$.

It should be emphasized that, in deriving the frequency of the top event, some basic rules must be borne in mind while working with both AND and OR gates:

- Probabilities are ratios and hence dimensionless. Frequency is the number of occurrences per year.
- The inputs to an OR gate are added to obtain the output. All inputs can either be probabilities or frequencies but not a combination.
- If all inputs to an OR gate are probabilities, the output is a probability. If all inputs are frequencies, the output is a frequency.
- Inputs to an AND gate are multiplied to yield an output. These inputs can be all probabilities or a combination of frequency and probability. The output of a frequency multiplied by a probability is a frequency. However, the output of a probability multiplied by a probability is also a probability.
- A frequency cannot logically be multiplied by another frequency. Therefore, where inputs to an AND gate are a combination of frequency and probability, the number of frequency inputs is limited to one, and the other inputs must be probabilities. This requirement must carefully be borne in mind while preparing a fault tree.

The simple instrumented protection system considered above is a single channel system. The reliability of such protection systems can be improved by introducing REDUNDANCY; this requires providing identical, additional channels in parallel. A system with two channels is known as a "1-out-of-2 (1oo2) redundant system," and one with three channels is known as a "1out-of-3 (1oo3) redundant system". Expressions for FDT of redundant systems where the channels are completely independent and identical, and proof testing is done simultaneously, are provided in Table 13.11 below:

Although redundancy improves the reliability of a protective system, there is a practical limit to increasing the number of channels: with an increasing number of channels, the number of fail-safe or spurious failures increases, affecting productivity adversely. VOTING logic is introduced to address this issue. A 2-out-of-3 voting system will initiate an output signal only when a signal from one of the channels is confirmed by one of the other two channels.

TABLE 13.11
FDT of Redundant Systems[1,8]

	FDT
1-out-of-1 (1oo1)	$\frac{1}{2} f T$
1-out-of-2 (1oo2)	$\frac{1}{3} f^2 T^2$
1-out-of-3 (1oo3)	$\frac{1}{4} f^3 T^3$
2-out-of-3 (2oo3)	$f^2 T^2$

TABLE 13.12
Estimated Frequencies for Incident Outcomes in Example 13.3

FC	IOC	Incident Outcome	The frequency of IOC (per year)
I	IA	Jet fire	2×10^{-5}
	IB	VCE	6.4×10^{-5}
	IC1	Flash fire toward north-east (*D*/5)	0.4×10^{-5}
	IC2	Flash fire toward north-east (*F*/2)	0.4×10^{-5}
	ID1	Flash fire toward north-west (*D*/5)	0.4×10^{-5}
	ID2	Flash fire toward north-west (*F*/2)	0.4×10^{-5}
	IE	BLEVE	3×10^{-7}

Note: while voting logic reduces spurious shutdowns, it does not provide increased safety. From Table 13.12, an FDT of a 1oo2 system is one-third of the FDT of a 2oo3 system, and hence a 1oo2 system is three times safer than a 2oo3 system.

While providing more than one channel to achieve redundancy, we must guard against potential common mode failure (also called common cause failure), affecting more than one channel simultaneously. For example, in a 1oo2 redundant system, if both channels are measuring pressure in a dirty fluid, there is a possibility that blockage could occur on the impulse lines of both the channels at the same time.

DIVERSITY is a measure introduced against the common-mode problem. It attempts to ensure that different system channels remain independent from one another as far as possible. For example, for pressure measurement, one channel could be based on a pressure sensor, while another channel could be based on a temperature sensor (whose temperature is related to pressure). Similarly, for introducing diversity in shutdowns, one channel could use a SDV, while another channel could use a control valve or a system to trip a pump.

Figure 13.16a–c is a simplified example to introduce the method for developing and analyzing a fault tree. In the risk analysis of an actual chemical plant, the methodology becomes quite elaborate and complex. However, it is possible to identify certain everyday hazardous events across the process industries. It may be possible to use historical failure rate data directly to determine the event frequency without the need to develop a fault tree. This approach is illustrated in Example 13.2 for filling liquefied petroleum gas (LPG) into trucks (Figure 13.17 below).

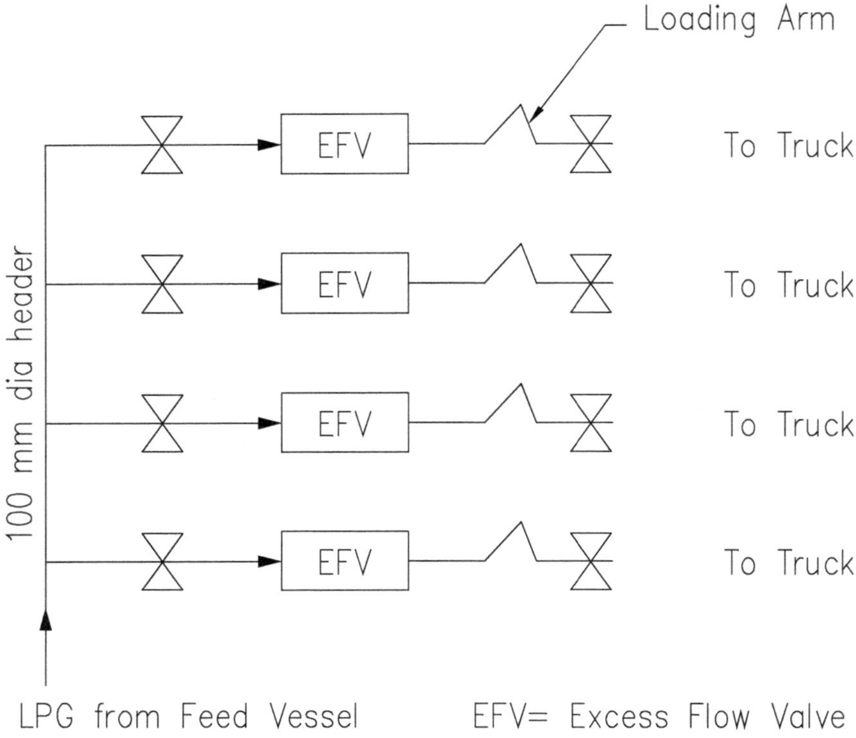

FIGURE 13.17 Schematic for Example 13.2 for loading LPG to trucks.

Example 13.2 considers a scenario where LPG is filled into four trucks parallel to a level-controlled feed vessel at 15°C–20°C. LPG from a refrigerated storage vessel is pumped via a heater to the feed vessel. The feed vessel is schematically similar to the one in Example 1, except that it is a pressure vessel with no vent to the atmosphere. It is protected against overpressure by a PSV that relieves to the atmosphere through a stack of adequate height.

The LPG supply header from the bottom of the feed vessel to the truck loading area is 100 mm in diameter and 100 m long. Near the truck loading area, this header divides into four lines of 50 mm diameter for connecting to four trucks. Each 50 mm diameter line is provided with an excess flow valve and a truck-loading arm, as shown in Figure 13.17.

The total time for filling four trucks is 1 hour, including connecting and disconnecting the loading arms. The actual filling time for each truck is 40 minutes. On average, 16 trucks are filled per day through four loading arms, and filling is done for 300 days in a year.

It is desired to calculate the frequency of a major LPG leakage caused by the failure of any of the loading arms. The frequency of any major leakage from the header and connecting pipelines may be considered negligible.

Data:

i. The failure rate of loading arms is 3×10^{-6} per hour for leakage and 3×10^{-8} per hour for catastrophic rupture.
ii. The failure rate of an excess flow valve to close is 1.3×10^{-2} per demand if tested every year.

The total operating time for each loading arm = (4) (40/60) (300) = 800 h/year.

Therefore, frequency of significant leakage per loading arm = (800) $(3 \times 10^{-6}) = 2.4 \times 10^{-3}$ per year, and the frequency of major leakage per loading arm through catastrophic rupture is (800) $(3 \times 10^{-8}) = 2.4 \times 10^{-5}$ per year. These are *demand rates* per loading arm.

An excess flow valve for protection against high demand is intended to close when the flow exceeds a set value. It is expected that it will operate only in the case of loading arm rupture and not in the case of a significant leakage resulting from the partial failure of the loading arm.

Hence, with four loading arms, estimated *hazard rates* are as follows:

Significant leakage: (4) $(2.4 \times 10^{-3}) = 9.6 \times 10^{-3}$ per year (once in 100 years)
Catastrophic rupture: (4) (2.4×10^{-5}) $(1.3 \times 10^{-2}) = 1.25 \times 10^{-6}$ per year (once in 800,000 years).

Data on failure rates of basic events and components are essential for making quantitative analyses of fault trees. Generic data can be obtained from the published literature. Section 13.5 summarizes data that are often required. The main sources of data are the U.K. AEA,[9] the US Atomic Energy Commission, the Rijnmond Report,[10] the Canvey Island Study,[11] AIChE's CCPS, and other publications in the process industries. Lees[8] has quoted some of these data, together with references. These failure rate data cover pressure vessels, refrigerated storage tanks, process/cross-country pipelines, valves and fittings, flanges, instrument components, human errors, etc.

It is important to note that generic data collections rarely give the conditions under which they have been collected. Presumably, these have been collected over a wide range of conditions representing highly expensive, precise equipment such as nuclear installations on the one hand, and process plants of undocumented environmental conditions and maintenance standards on the other.

These data, therefore, should be used with this limitation in mind. Wherever in-house data or data from similar industries are available, these should be used in preference to published data. Consulting in-house maintenance engineers, who plan the spare parts inventories based on repair and replacement needs under their control, can provide valuable information. Consulting colleagues in similar industries can also be helpful. Many major corporations maintain elaborate failure rate databases, based on accurate maintenance records over many decades of operation.

Finally, it is the analyst's experience and judgment that is most important in the choice of all input data.

Frequently, however, the final result is only marginally affected by modest inaccuracies in the failure rates of some basic events. Ensuring that all relevant factors are taken into account and that no branch or pathway is missed while developing the fault tree is paramount. With such a mistake, the result could be grossly in error. However, if the analysis is carried out by a team of experienced specialists representing the process, maintenance, and instrumentation functions, and under the leadership of an experienced analyst (rather than by a single individual), the likelihood of such errors would be drastically reduced.

13.4.1.2 Event Tree Methodology

An event tree displays possible outcomes of an initial event, such as an accidental release. A release can result in various scenarios depending on the nature of the release and the environmental factors.

Construction of an event tree starts with the initial event placed usually on the left, and branches drawn to the right, each branch representing a different sequence of events and terminating in an outcome.

The methodology for developing an event tree is explained in Example 13.3 (Figure 13.18 below) for accidental release from a spherical storage vessel that contains liquid propane under pressure at ambient temperature. The release is caused by human error during the water draining operation through a bottom outlet from the tank. The frequency of release has been estimated to be 10^{-4} per year, taking into account both the frequency of water draining and the probability of human error. The following consequences (or outcomes) following the release are possible:

- Immediate ignition, resulting in a jet fire
- Escalation of the jet fire into a BLEVE, in case the tank cooling fails
- Delayed ignition, resulting in a flash fire or vapor cloud explosion
- No ignition and safe dispersion

The consequence outcomes are shown in the form of an event tree in Figure 13.18 below:

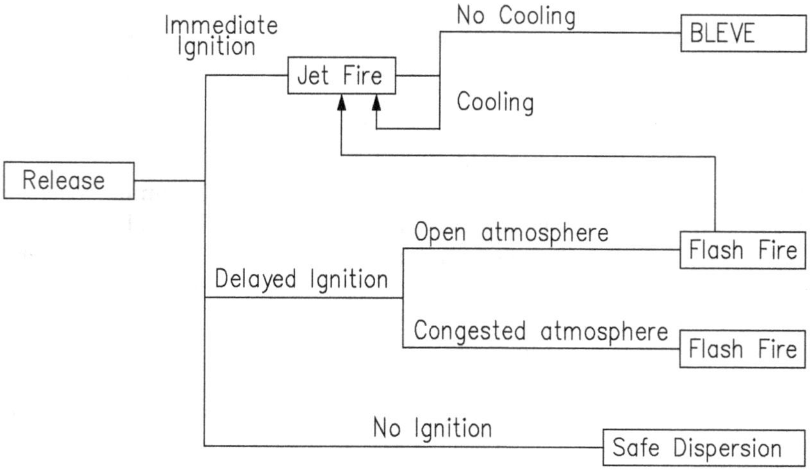

FIGURE 13.18 Event tree for LPG release from a storage tank in Example 13.3

The frequencies of various consequences would depend on the probabilities of occurrence for various environmental factors. For example, in a refinery or a petrochemical plant, reliable safety systems are expected to be in place; the probability of an immediate ignition should be no more than about 20%. Hence, the estimated frequency of a jet fire is about 0.2×10^{-4} to 2×10^{-5} per year. The frequency of development into a massive cloud is, therefore, 0.8×10^{-4} per year. It is assumed conservatively that the probability of no ignition is negligible. Hence, the frequency of delayed ignition of the cloud is 0.8×10^{-4} per year.

Assume also that there would be a fair degree of congestion in the plant layout, and therefore the probability of a vapor cloud explosion should be high, say 80%.

Hence, the estimated frequency of a vapor cloud explosion is

$$(0.8)\left(0.8 \times 10^{-4}\right), \text{ or } 6.4 \times 10^{-5} \text{ per year.}$$

The frequency of a flash fire is

$$(1-0.8)\left(0.8 \times 10^{-4}\right), \text{ or } 1.6 \times 10^{-5} \text{ per year.}$$

Assuming further that the frequency of wind direction is 50% toward the north-east and 50% toward the north-west, the frequency of flash fire toward the north-east is 0.8×10^{-5} per year and that toward the north-west is 0.8×10^{-5} per year.

It is possible that, following a vapor cloud explosion, there could be damage to piping and valves in adjoining spheres as well, resulting in further immediate ignition and jet fire (the domino effect). These are not shown in Figure 13.7, however.

The probability of a jet fire developing into a BLEVE, following an immediate ignition, would depend on whether or not the jet fire impingement heats the affected sphere's vapor space. Assume that (1) there is a 30% probability that the jet flame impinges on the sphere, and (2) in about 5% of the cases, Operations fails to mobilize enough firewater to cool the spheres effectively within 20–30 minutes.

Hence, the estimated frequency of a BLEVE is

$$\left(2 \times 10^{-5}\right)(0.3)\,(0.05), \text{ or } 3 \times 10^{-7} \text{ per year.}$$

Since the flash fire plume size depends on the atmospheric stability class and the wind speed, the incident outcome cases (IOCs) IC and ID should be further sub-divided. Thus, the IOC IC can be split into IC1 and IC2, each having a probability of 0.4×10^{-5} per year, assuming a 50% probability for the atmospheric condition to be a Pasquill stability category *D* at 5 m/s wind speed, and 50% probability for stability category *F* at 2 m/s wind speed. Similarly, IOC ID can be split into ID1 and ID2, each having a frequency of 0.4×10^{-5} per year, assuming a 50% probability for *D*/5 and a 50% probability for *F*/2.

Estimated frequencies for incident outcomes are summarized in Table 13.12.

It should be clear from the above example that quantitative evaluation of frequencies using the event tree methodology requires the use of probability data based on the experience and judgment of the analyst. Since risk assessments are order-of-magnitude estimates only, this is not considered a significant limitation.

13.4.2 Estimation of Risk

Estimation of risk in terms of measures described in Section 13.1 will involve the following steps:

- Identify and list the failure cases (FCs)
- Identify and list the IOCs for each FC
- Estimate the frequency of each FC
- Estimate the frequency of each IOC
- Estimate the number of fatalities for each IOC

In the event tree example summarized in Table 13.13, the number of FCs is 1 (LPG release), and the number of IOCs is 7 (jet fire, vapor cloud explosion, flash fire, and BLEVE). This example also gives the calculated frequencies for the seven IOCs. Calculation of the number of fatalities against each IOC will require data on effect zone figures and the population density in each affected zone. The numbers of fatalities shown in Table 13.13 are arbitrarily assumed figures. In an actual risk analysis, these would need to be assessed based on estimates of consequences.

The risk data shown in Table 13.13 are for one FC, namely the failure of an LPG storage system. There may be other installations within the site that have the potential for hazardous accidents. These may include toxic release cases, as well, for which hazard distances are usually much longer. Therefore, for the site as a whole, Table 13.14 would need to be extended to cover all such identified scenarios. Once this has been done, the next step is to calculate the site risk in terms of measures described in Section 13.1, e.g., individual risk, societal risk, FAR, etc. Obviously, this would be a prolonged exercise.

The methodology for calculating various risk measures for a site is explained below for a simple Example 13.4. Figure 13.19 shows a simplified site layout of a factory for the manufacture of commercial blasting explosives. The explosives mixing and cartridging plant is at location A, an ammonium nitrate manufacturing plant is at location B, liquefied anhydrous ammonia storage is at location C, and the administration building for the site is at location D. There are a few small

TABLE 13.13
Summary of Estimated Risk for the LPG Storage System (in Figure 13.17)

IOC	Frequency (per year)	No. of Fatalities
IA	2×10^{-5}	Negligible
IB	6.4×10^{-5}	2
IC1	0.4×10^{-5}	2
IC2	0.4×10^{-5}	5
ID1	0.4×10^{-5}	1
ID2	0.4×10^{-5}	3
IE	3×10^{-7}	6

TABLE 13.14
Incident Details for Example 13.4

Incident Case No.	Type of Incident	IOC	Incident Frequency (per year)	Incident Outcome Frequency (per year)	Number of Fatalities
I	Explosion	I	10^{-2}	10^{-2}	6 (all plant employees)
II	Explosion	II	10^{-4}	10^{-4}	8 (all plant employees)
III	Toxic release	IIIA: *D*/5 to NE	3×10^{-5}	7.5×10^{-6}	5 (2 from plant)
		IIIB: *F*/2 to NE		7.5×10^{-6}	10 (2 from plant)
		IIIC: *D*/5 to SW		7.5×10^{-6}	2 (all plant employees)
		IIID: *F*/2 to SW		7.5×10^{-6}	3 (all plant employees)

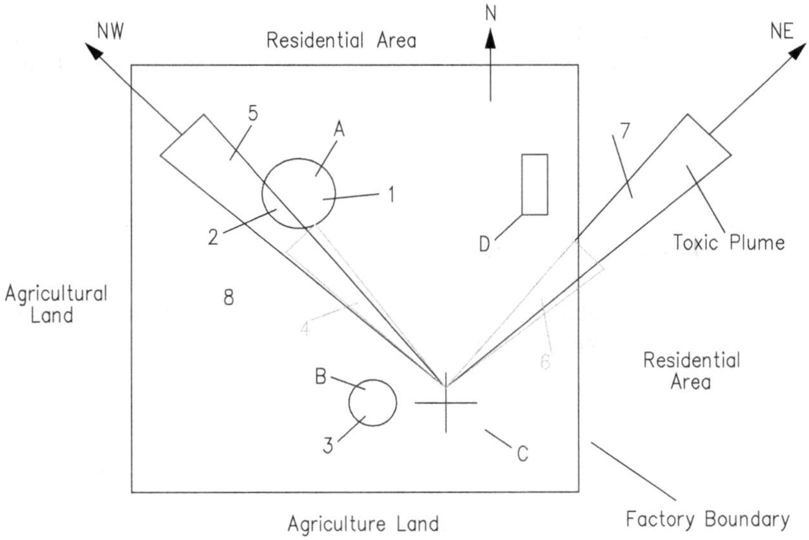

FIGURE 13.19 Simplified layout of an explosive factory.

explosives storage buildings (called magazines) at safe locations within the site that are not shown.

The following undesired accidents have been identified for the site:

- *An explosion in an explosives manufacturing plant*: The estimated frequency is 0.01 per year, the effect zone radius is 200 m, and the estimated number of fatalities is 6 (all plant employees).

- *An explosion in ammonium nitrate plant*: The estimated frequency is 10–^{4} per year, the effect zone radius is 100 m, and the estimated number of fatalities is 8 (all plant employees).
- *A release of anhydrous ammonia as gas from storage*: The estimated frequency of release is 3×10^{-5} per year. There is no ignition, and people in the downwind direction are exposed to ammonia's toxic hazards. Wind direction probability is 50% toward the north-east and 50% toward the south-west. The probability of an atmospheric condition in each wind direction is 50% *D*/5 (Pasquill category *D* at 5 m/s wind speed) and 50% *F*/2 (Pasquill stability category *F* at 2 m/s wind speed). Estimated fatalities are as follows:

Wind Direction	Atmospheric Condition	Hazard Area (Fatality)	No. of Fatalities
North-east	*D*/5	300 m long, 40 m wide	5 (2 plant employees)
North-east	*F*/2	800 m long, 40 m wide	10 (2 plant employees)
South-west	*D*/5	300 m long, 40 m wide	2 (all plant employees)
South-west	*F*/2	800 m long, 40 m wide	3 (all plant employees)

It is desired to estimate the site risk in terms of individual risk, FAR, and the *F*-*N* curve. Values of frequencies, the extent of hazard zones, and the number of fatalities in each case have arbitrarily been assumed to illustrate the methodology for computation of risks.

13.4.2.1 Individual Risk

Everywhere within an area exposed to a single accident outcome, the individual risk is equal to the frequency of that IOC, irrespective of the number of fatalities. At any location exposed to more than one IOC, the individual risk is the sum of the frequencies of each IOC applicable to that location. Accordingly, the calculated individual risks of fatality for various locations within and outside the factory boundary are as follows:

13.4.2.1.1 FAR

Referring to Table 13.6, average individual risk for plant employees is as follows:

$$\frac{\begin{array}{l}\left\{(6)\left(10^{-2}\right)+(8)\left(10^{-4}\right)+(2)\left(7.5\times10^{-6}\right)+(2)\left(7.5\times10^{-6}\right)\right.\\ \left.+(2)\left(7.5\times10^{-6}\right)+(3)\left(7.5\times10^{-6}\right)\right\}\end{array}}{(6+8+2+2+2+3)} = 2.65\times10^{-3}\text{ per year.}$$

Therefore, $\text{FAR} = \dfrac{\frac{\left(2.65\times10^{-3}\right)\left(10^{-8}\right)}{365}}{24} = 30$

Compared to Table 13.1, this FAR is high because the frequency of Incident Case I (explosion in an explosive mixing/cartridging plant) is very high. As a sensitivity study, if this frequency can be reduced to 10^{-3} per year, the FAR comes down to 3.4, and if the frequency can be further reduced to 10^{-4} per year, the FAR is reduced to 0.73. The time, expense, and effort to achieve such improvements would be easy to justify.

13.4.2.2 Societal Risk (*F-N* Curve)

The *F-N* curve is drawn on a log-log paper, the abscissa is the number of fatalities *N*, and the ordinate is the cumulative value of the frequencies of events causing fatalities $\geq N$. Based on the incident outcome frequency and the corresponding fatality data given in Table 13.15, the values of cumulative frequency against *N* have been calculated for this plot. Calculations have been done for two cases:

1. Case 1 (the Base Case), where the frequency of Incident Outcome Case 1 is 10^{-2} per year, and
2. Case 2, where the frequency of Incident Outcome Case 1 is 10^{-2} per year.

The results are summarized below (Table 13.16).

TABLE 13.15
Individual Risk of Fatality at Various Locations (in Figure 13.19)

Location No.	Applicable IOC	Individual Risk (per year)
1	I	10^{-2}
2	I, IIID	10^{-2}
3	II	10^{-4}
4	IIIC, IIID	1.5×10^{-5}
5	IIID	7.5×10^{-6}
6	IIIA, IIIB	1.5×10^{-5}
7	IIIB	7.5×10^{-6}
8	None	0

TABLE 13.16
Calculated Values of Cumulative Frequency vs. *N* for Construction of *F-N* Curve

		Cumulative Frequency (per year)	
Fatalities, *N*	IOCs Having Fatalities ≥ *N*	Base Case	Case 2
2	I, II, IIIA, IIIB, IIIC, IIID	10^{-2}	2.3×10^{-4}
3	I, II, IIIA, IIIB, IIID	10^{-2}	2.23×10^{-4}
5	I, II, IIIA, IIIB	10^{-2}	2.15×10^{-4}
6	I, II, IIIB	10^{-2}	2.08×10^{-4}
8	II, IIIB	1.075×10^{-4}	1.08×10^{-4}
10	IIIB	7.5×10^{-6}	7.5×10^{-6}

13.4.3 Risk Determination

Risk Assessment involves determining the risk posed by hazards in processing activities using various qualitative, semi-quantitative, or quantitative techniques. The important ones (dealt with in detail in this chapter) are utilized, depending on the project stage and purpose for which risk is assessed. In "major" or "highly hazardous" industries, for siting decisions for large greenfield projects and expansion of brownfield projects, the probability of untoward incidents /major accidents with offsite and on-site consequences must be quantified. Such analyses are required for submission to and approval from Federal, State, and Local authorities, and the public Company shareholders and employees also require reassurances on these matters. For these reasons, a "QRA is carried out for the twin purposes of (1) determining the risks of major accidents and (2) demonstrating that the highest risks are at an "acceptable" or "tolerable" level.

Similar analyses are also required in smaller capacity, highly hazardous greenfield projects when considering:

1. Siting
2. Capacity expansion
3. Introduction of new processes
4. New products, and
5. New systems

Small and large projects alike require that the determination of risk must be at acceptable levels. These risk determinations are often carried out using simpler techniques, the most popular one being LOPA.

13.4.4 Risk Acceptability[12]

There are two categories of acceptable risk:

a. *Negligible risk*: Broadly accepted by most people as they go about their daily lives and do not need any control measures. Such individual risks would include the risk of being struck by lightning, brake failure in a car, etc.
b. *Tolerable risk*: Higher than negligible risk level (increased individual risk posed by a hazardous industrial activity); considered reasonably low because, if adequate control measures are employed, their cost would justify the risk reduction. This concept of tolerable risk, in varying forms, is adopted by regulatory authorities in almost all industrialized countries. A risk of between 10^{-4} (one in ten thousand years) to 10^{-6} (one in a million years) is considered tolerable, depending on site location near existing facilities.
c. *Unacceptable risk*: Here, the risk level remains high (even with consideration of control measures). The potential economic and societal benefits are not commensurate with the costs of further risk reduction.

13.4.4.1 Individual Risk – Acceptability Criteria

Individual risk: For a member of the public living in the vicinity of a factory site, the values of the acceptable excess risk threshold from an industrial activity are (1) 10^{-5} per annum as per the HSE, UK, and (2) 10^{-6} per annum for the Netherlands (VROM). The acceptable individual risk for plant personnel is one notch lower (10^{-4} for UK and 10^{-5} for the Netherlands).

In Canada, the risk acceptability criteria are provided in terms of allowable land uses for specified individual risk levels, as shown in Figure 13.20 below.[13]

13.4.4.2 Societal Risk – Acceptability Criteria

These relate to the chances of more than one individual being harmed in an incident.

In the U.K., societal risk is expressed as the relationship between the numbers of fatalities and the frequency with which they are predicted to occur. The HSE proposes that for any single major industrial activity affecting the existing population, "*the risk of an accident causing the death of 50 or more people in a single event should be regarded as intolerable if the frequency is estimated to be more than one in five thousand per annum*[14]". Societal risk criteria are based on the ALARP concept. The *FN* curves show the cumulative frequency *F* (the sum of the frequencies) of all the individual events that could lead to *N* or more fatalities.

13.4.4.2.1 Societal Risk Criteria in the U.K. and the Netherlands

The societal risk criteria in the U.K. and the Netherlands are shown below in Figure 13.21. These lines show, on a log-log plot, the cumulative frequency, *F*, of *N* or more fatalities (per year) vs. the number of fatalities (*N*). Three curves are shown:

1. U.K. Maximum Tolerable
2. Dutch Maximum Tolerable, and
3. U.K. Broadly Acceptable.

Equation constants for these curves are as follows:

$$\log 10\,(F) = A + B * \left(\log 10\,(N)\right):$$

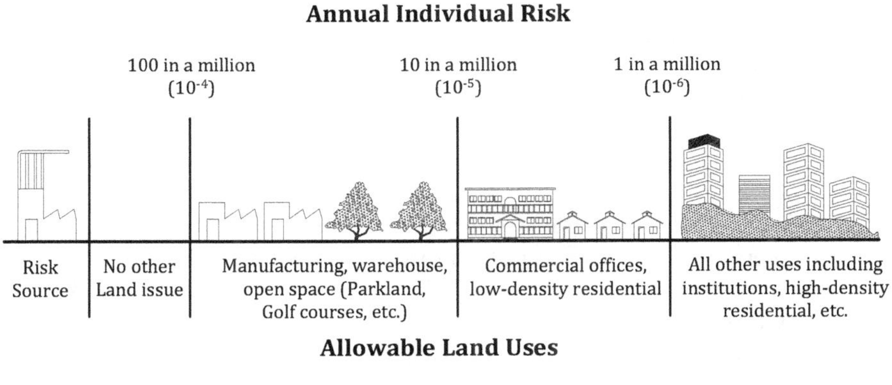

FIGURE 13.20 Individual risk criteria for land use in Canada

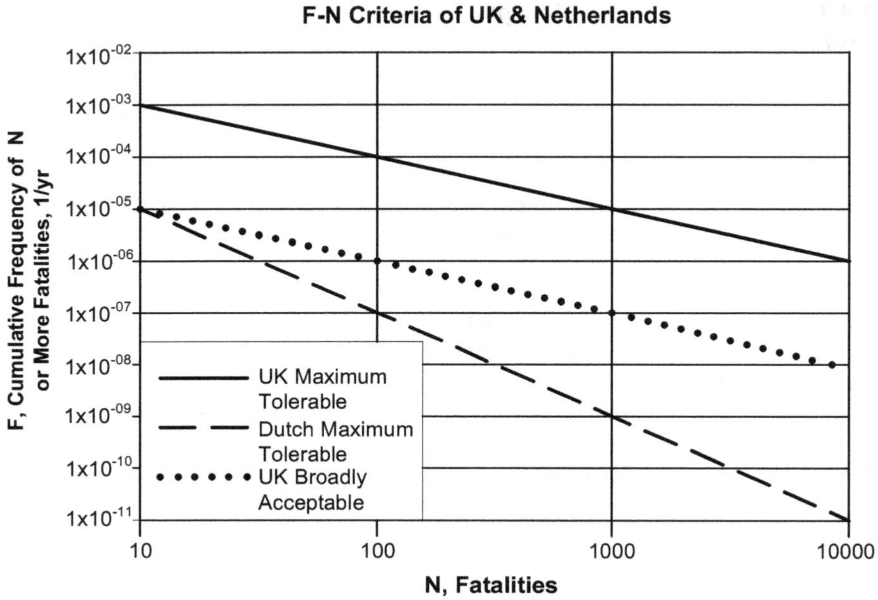

FIGURE 13.21 Societal risk criteria in the U.K. and the Netherlands.

1. U.K. Maximum Tolerable: $A = -2$, $B = 1$
2. Dutch Maximum Tolerable: $A = -4$, $B = 1$
3. U.K. Broadly Acceptable: $A = -3$, $B = -2$

13.4.5 Risk Reduction and ALARP[14]

Whenever the individual risk is at an unacceptable level, control measures must be deployed to lower it to below the highest tolerable limit, following the As Low As Reasonable Practicable (ALARP) concept, Figure 13.22:

Various measures could be considered for reducing risk to a tolerable level:

- Actions at the design stage
- Operational systems and procedures
- Emergency planning and preparedness.

Steps that should be considered at the design stage are as follows:

i. choice of process route that is inherently safer (i.e., involving minimum inventory of hazardous material in process and storage),
ii. providing for adequate safety distances while selecting sites and making plant layouts,
iii. providing multiple levels of controls in plant design, and
iv. minimizing exposure of operators to potential hazards through remote operation from adequately protected control rooms, etc.

FIGURE 13.22 ALARP diagram.

Operational systems and procedures should include the following:

i. written operational and maintenance manuals that are readily available at the shop floor,
ii. training of personnel,
iii. permit to work systems,
iv. authorization procedures for plant and process modifications,
v. safety audits, etc.

Emergency planning should be based on a comprehensive identification of residual hazards in the plant. Emergency actions must be taken promptly to

i. minimize injuries and fatalities,
ii. communication systems and training of operators and public (where off-site effects are envisaged), and
iii. drills to be conducted at regular intervals to ensure the effectiveness of the plan.

The success of risk reduction and risk control measures depends entirely on the sound functioning of the various safety systems deployed, both hardware and software. Failure of these systems to function when needed has resulted in many catastrophic accidents, as documented in Chapter 2. Ensuring the proper and reliable functioning of today's increasingly sophisticated control systems is mandatory and is extremely important in process safety management.

13.5 FUNCTIONAL SAFETY

Functional safety is a vital part of the overall process safety engineering discipline for ensuring that the system and equipment operate correctly in response to all inputs. Functional safety is achieved when every specified safety function is carried out, and the level of performance required of each safety function is met.

In response to the increasing severity and number of industrial accidents, international standards have been adopted that define functional safety requirements. For example, in the Norwegian Oil and Gas industries, the OLF Guideline-070 "*Application of IEC 61508 and IEC 61511 in the Norwegian Petroleum Industry (Recommended SIL Requirements)*" is prominent. The overall goal is to ensure that process plants and equipment can be operated safely. These standards have also forced the process industries to seek new and innovative instrumental solutions designed to improve industrial process safety.

Functional safety relies on *active systems.* An example of functional safety would be the activation of a level switch in a tank containing a flammable liquid when a potentially dangerous level has been reached. This function might cause a valve to be closed to prevent further liquid feed from entering the tank, thereby preventing tank overflow.

Safety achieved by measures that rely on *passive systems* is not functional safety at all. A fire-resistant door or insulation to withstand high temperatures are passive measures that can protect against the same hazards controlled by instrumentation and control systems.

13.5.1 SIS

SIS are an indispensable tool for risk reduction in large process industries such as

i. oil and gas processing,
ii. petrochemicals,
iii. heavy chemicals,
iv. downstream specialty chemicals,
v. metallurgical (steel and nonmetal), and
vi. the utility (power) sector.

These high hazard processes pose substantial risks that must be reduced to a tolerable or ALARP level. An SIS provides the reliability of active protection layers or barriers.

An SIS monitors the safety-related values and parameters in a plant to ensure that they are within acceptable safety limits. As soon as untoward conditions occur, an SIS triggers alarms and automatically initiates direct actions to move the plant to a safe condition. If required, an SIS can result in a plant shutdown.

The SIS systems are, therefore, responsible for ensuring operating safety and executing emergency stops as necessary. The main objective is to avoid onsite and offsite accidents such as the following:

i. loss of containment,
ii. fires,

iii. explosions,
iv. equipment damage,
v. protection of production and property and, above all,
vi. prevent personnel injuries, fatalities, and financial or property losses, and environmental degradation.

In case of any major malfunction or operational problem and SIS should enable achieving a safe condition predictably.

The long history of major process industry accidents, coupled with the evolution of complex processes, has made it mandatory to improve the reliability of active protection layers and barriers. Hence, additional SIS protection layers are required in addition to risk reduction through conventional control systems (now referred to as non-SIS protection layers). These serve the dual function of ensuring functional safety by (1) vastly improving the reliability of a protection layer and (2) by raising the "SIL" and reducing the risk quantitatively to the desired level. This capability brings risk down to ALARP or tolerable levels.

For selecting an SIS, the "residual risk" must first be quantified after a risk assessment, considering only the non-SIS protection layers. Then, if the residual risk is higher than the tolerable risk, or if the tolerable risk is insufficient (perhaps because of the nature of the process or the site location), raising the SIL level by adding SIS protections should be considered.

For process plants, SIS-related global standards apply.

IEC/EN 61508: Functional safety of electrical, electronic, and programmable electronic safety-related systems.[15] Programmable electronic safety systems include the following:

i. PLCs (microprocessor-based systems),
ii. distributed control systems,
iii. sensors, and
iv. intelligent actuators.

IEC 61508 defines requirements for system operation and integrity. Integrity is based on reliability, which is defined as the SIL.

IEC/EN61511: Functional Safety – Safety Instrumented Systems for the Process Industry Sector.[15]

13.5.2 SRS – Safety Requirement Specification

The SRS objective is to specify the following requirements for the SIFs.

- To define the safe state of the process for each identified SIF
- The assumed sources of demand and demand rate (required PFD) on the SIF
- The requirement for proof-test intervals
- The response time requirement for the SIS to bring the process to a safe state
- The SIL and mode of operation(demand/continuous) for each SIF
- A description of SIS process measurements and their trip points

- Requirements relating to energize or de-energize to trip
- A requirement for resetting the SIS after a shutdown
- Maximum allowable spurious trip rate; as per IEC 61511

13.5.3 SIL

SIL determination is done for the following purposes:

- Allocate safety functions to protection layers
- Determine the required SIFs
- Determine, for each SIF, the associated SIL

SIL has three attributes:

1. Applicable to the overall safety function
2. The higher the SIL level, the stricter are the requirements
3. Applicable to technical and nontechnical requirements

There are four discrete SIL levels:

SIL1: risk reduction $\geq$10 and $\leq$100
SIL2: risk reduction $\geq$100 and $\leq$1,000
SIL3: risk reduction $\geq$1,000 and $\leq$10,000
SIL4: risk reduction $\geq$10,000 and $\leq$100,000

13.5.3.1 SIL Verification

The main steps in the SIF conceptual design process are as follows: After the preparation of SRS, based on the SIL assessment exercise, the SIF subsystem is defined. The SIF design is verified to ensure that it meets all functional and integrity requirements.

13.5.3.2 SIL Validation

This stage aims to validate, through inspection and testing, that the installed and commissioned SIS and its SIF will achieve the specified safety requirements.

IEC 61511, as one of the SIL determination methodologies, is effective in illustrating relationships between multiple causes and preventive controls, such as interlocks, alarms, and relief systems for an unwanted event.

To effectively manage critical controls, it is essential to establish a performance requirement detailing the expected performance of all such controls. The performance requirement should include the following information for the life cycle of a critical control element:

- *Design*: Design basis of the critical control
- *Operation*: How the critical control should be operated
- *Maintenance*: Maintenance activities required to maintain the effectiveness of critical controls

- *Change*: Changes that can impact the criteria listed above should go through the MOC process; the critical control performance requirements should then be updated accordingly.

The guidelines for selecting the sensors and final control elements required to implement SIF should consider the following factors:

1. The sensor and final element-to-process interfaces should be included when determining the failure rates and failure modes of the subsystem
2. The sensor and final element subsystem redundancy required to implement the various SIFs should be determined by calculating the average PFD (PFD_{avg}) for each subsystem
3. The sensor and final element hardware "common causes" should be included in the calculation of PFD_{avg}
4. The hardware fault tolerance requirements in Clause 11.4 of IEC 61511-1 must be followed when selecting the sensor and final element redundancy

13.6 DATABASE FOR FAILURE FREQUENCIES AND PROBABILITIES

These values are based on literature citations and must be used with caution since several items are expressed as ranges. Also, a low degree of precision (often just one significant figure) is typical, as these are order-of-magnitude values only.

13.6.1 Failure Frequencies for Tanks and Vessels[8,10]

Type of Vessel	Type of Failure	Failure Frequency, per year
Pressure vessels	Serious leakage	1×10^{-5}
	Catastrophic rupture	1×10^{-6}
Atmospheric tanks	Serious leakage	1×10^{-4}
	Catastrophic rupture	6×10^{-6}
Refrigerated tanks (double-wall, high integrity)	Serious leakage from the inner tank	2×10^{-5}
	Catastrophic rupture of both tanks	1×10^{-6}

13.6.2 Failure Frequencies of Process Pipework[16]

	Frequency/year/m of Pipe Length	
Pipe Diameter, mm	10% of CSA	Full bore rupture
50	1×10^{-5}	1×10^{-6}
100	6×10^{-6}	3×10^{-7}
300	3×10^{-6}	1×10^{-7}

CSA, cross-sectional area.

13.6.3 Failure Frequencies of Cross-Country Pipelines[8]

	Frequency/year/km
Onshore crude oil pipelines in Western Europe (CONCAWE)	0.39×10^{-3}
Onshore petroleum product pipelines in Western Europe (CONCAWE)	1.11×10^{-3}
Oil pipelines in Western Europe (SRD)	2.1×10^{-4}–1.2×10^{-3}
Natural Gas transmission lines in the USA (Federal Power Commission)	1.1×10^{-4}–5×10^{-4}
Gas transmission network (British Gas)	0.65×10^{-3}
Gas transmission network (Gas de France)	3.3×10^{-3}
Chlorine pipelines	3.1×10^{-3}

13.6.4 Failure Rates of Loading Arms[8,16]

	Frequency/hour
Jetty loading arm significant leakage	3×10^{-6}
Jetty loading catastrophic failure	3×10^{-8}
Lorry tankers loading arm, significant leakage	1×10^{-7}
Lorry tankers loading arm, catastrophic failure	1×10^{-8}

13.6.5 Failure Frequencies for Valves[16]

Hole Size	Leak Frequency/year
Full bore	1×10^{-5}
++10% of CSA	1×10^{-4}

13.6.6 Failure Probabilities for Protective Equipment[10]

In this section, failure data are provided for protective equipment either as failure rate per demand or frequency per year. Failure rate per demand is a probability. Where the failure rate has been given as frequency per year, the probability needs to be calculated as FDT, which is defined as follows: FDT = $\frac{1}{2} f T$, where f is the failure frequency/year, and T is the proof-test interval in a year.

		Failure rate
	Valves	
Pressure relief valve	Blocked	0.001/y
	Lifts heavy	0.004/y
	Lifts light/leakage	0.06/y
Vacuum relief valve	Fails to operate	0.005/y
Solenoid valve	Fails to operate	0.3/y
	Blockage	0.3/y
Pneumatic control valve	Fails open	0.3/y
	Fails closed	0.3/y
Manual valve	Blockage	0.1/y
	Seized	0.1/y
Motor operated valves	Fails to operate	1×10^{-3}/D
	Blockage	1×10^{-4}/D
	Measuring Devices	
Level sensor (DP type)	Fails to sense level	0.43/y
Level sensor (Float type)	Fails to sense level	0.41/y
Thermocouple	Fails to sense temp	0.17/y
Resistance thermometer	Fails to sense temp	0.14/y
Pressure sensor (general)	Fails to sense press	0.47/y
Pressure gauge	Fails to sense press	0.09/y
	Trips, Alarms, and Control Systems	
Trip system (general)		0.5/y
Pressure switch	Fails to operate	1×10^{-4}/D
Push-button switch	Fails to operate	4.4×10^{-3}/D
Relays (electrical)	Fails to operate	0.044/y
Relays (pneumatic)	Fails to operate	0.17/y
Impulse lines	Blocked	0.03/y
	Leaking	0.06/y
Audible alarm/siren	Fails to sound	2×10^{-5}/D
Fire alarm system (complete)	Fails to activate	1×10^{-3}/D
Emergency diesel system (complete)	Fails to start	3×10^{-2}/D

Notation/y, per year; /D, per demand.

13.6.7 Probabilities of Human Error[8,17]

	Probability
Error in simple, routine operations	0.001
Correct decision but wrong switches selected when appearances are different	0.001
The error of omission of action embedded in a procedure	0.003
The general error of commission, e.g., misreading the label and hence selecting the wrong switch	0.003
The general error of omission, with no feedback display, e.g., not closing valve after maintenance	0.01
Error in routine operation where some care is required	0.01
Simple arithmetic error with self-checking	0.03
Error in nonroutine operation when other duties present	0.1
Personnel on different shift omit check of plant item	0.1
Error in a nonroutine complicated operation	0.25
General error under very high stress levels, e.g., where dangerous activities are occurring rapidly	0.2–0.3

13.6.8 Ignition Probability of Flammable Liquid Releases[10]

Leak size	Ignition Probability
Minor leak (<1 kg/s)	0.01
Major leak (1–50 kg/s)	0.03
Massive leak (>50 kg/s)	0.08

13.6.9 Ignition of Gas Clouds[11]

Ignition on Release Site		Ignition over Population	
Source of Ignition	Ignition Probability	Ignition Location	Ignition Probability
None readily identifiable	0.1	Edge of un-ignited cloud just reaches a populated area	0.7
Very few sources	0.2	Un-ignited cloud right over the population when ignition occurs	0.2
Few sources	0.5		
Many sources	0.9		

13.7 APPLICATION OF LOPA, BARRIER ANALYSIS, AND QRA

LOPA is a semi-quantitative risk analysis technique that can be applied directly after HAZOP. LOPA originated in the USA as a result of a major initiative undertaken by the AIChE. After the Bhopal tragedy, the CCPS was launched by AIChE in response to the widespread perception that an industry-wide effort was required to investigate deficiencies in industrial process safety practices and devise reliable procedures to improve them. LOPA was developed as a comprehensive risk management tool[5] in 2001. It is most commonly used by the U.S. Chemical industry and has been applied in many cases by the U.K. HSE. The European Process Safety Centre made a detailed evaluation of LOPA and considered certain modifications. LOPA has a distinct advantage in that it integrates the HAZOP analyses into LOPA studies. If a qualified group is convened, study time is reduced, common databases are used that can link causes, consequences, and safeguards, and thus help realize significant time and cost savings.

Barrier analysis is used by the European offshore oil and gas industry. It is a root cause analysis method that considers the pathways through which a hazard can affect a target to characterize the performance of actual or potential barriers and controls interposed to protect the target. The Norwegian Petroleum Safety Authority played a major role in developing and refining the concept of *barrier management* and management of offshore and onshore petroleum industry.

The QRA or CPQRA methodology has evolved since the early 1980s from its roots in the nuclear, aerospace, and electronics industries. To date, its most extensive use has been in the nuclear industry. A QRA quantifies the risk exposure for every possible hazardous event and determines the cost and schedule contingencies needed. QRA is performed for the most severe consequences of events initiated in an industrial unit and is most often used for siting decisions, safe plant layout, assessing public impacts, and, most notably, to compare alternative ways to reduce risks.

REFERENCES

1. Kletz, T.: *What Went Wrong* (3rd Ed., Gulf Publishing Co, Houston, TX, 1995).
2. Gill, D. W.: *Quantitative Risk Assessment* (Presentation at Seminar on LPG Safety and Risk Analysis, Mumbai, 1998); also *Hazard Analysis* (Refresher Training Course in ICI India, July, 1990).
3. *Layers of Protection Analysis: Simplified Process Risk Assessment* (Center for Chemical Process Safety (CCPS), American Institute of Chemical Engineers, New York, 2001).
4. *Lines of Defence/Layers of Protection Analysis in the COMAH Context* – Prepared by Amey VECTRA Limited for the Health and Safety Executive.
5. *Application of Safety Instrumented Systems for the Process Industries* (ANSI/ISA-ISA 84.01-1996, ISA, Research Triangle Park, NC, 1996).
6. *Functional Safety: Safety Instrumented Systems for the Process Sector* (International Electrotechnical Commission (IEC), IEC 61511 Geneva, Switzerland, 2003).
7. *Guidance for Barrier Management in the Petroleum Industry* (Stein Hauge and Knut Øien-SINTEF Technology and Society).
8. Mannan, S.: *Lees' Loss Prevention in the Process Industries* (4th Ed., Butterworth-Heinemann, Oxford, 2012).
9. Bourne, A. J. et al: *Defences Against Common Mode Failures in Redundancy Systems* (SRD R196, Safety and Reliability Directorate, U.K. AEA, January 1981).
10. Rijnmond Public Authority: *Risk Analysis of Six Potentially Hazardous Industrial Objects in the Rijnmond Area, a Pilot Study* (1982).
11. Health and Safety Executive (UK): *Canvey: A Second Report: A Review of Potential Hazards from Operations in the Canvey Island/Thurrock Area Three Years after Publication of the Canvey Report* (1981).

12. INERIS: *Accidental Risks Division: About the Relevance of the Concept of Risk Acceptability in the Risk Analysis and Risk Management Process.*
13. Canadian Society of Chemical Engineering: *Risk Assessment – Recommended Practice for Municipalities and Industries.*
14. HS: *Reducing Risks Protecting People – HSE's Decision Making Process 17. Functional Safety of Electrical/Electronic/Programmable.*
15. IEC/EN61511: *Functional Safety – Safety Instrumented Systems for Process Industry Sector.*
16. Entec U.K. Ltd, London (Private Communication).
17. ICI: *Training Course on Hazard Analysis* (Personal Communication).

14 Human Factors in Process Safety

This chapter is concerned with a vital aspect of process safety management: understanding and eliminating safety hazards attributable to errors in human behavior and actions that violate established standards and training guidelines.

As discussed in Chapter 2, many catastrophic accidents in the process industries have been caused when operators or maintenance personnel behaved in an erratic, undisciplined, or dismissive manner when confronted by potentially dangerous situations. Investigators have often found that, despite significant efforts devoted to safety training, sub-standard attitudes and habits of mind have gone unchecked over lengthy periods. However, such human failures must be distinguished from systematic deficiencies in process and equipment design, insufficient or poorly maintained instrumentation, errors in control system logic, inadequate tuning of automatic controllers, improper or insufficiently detailed operating instructions, and inadequate equipment inspection and maintenance practices, among many others.

Those responsible for plant operations agree on the importance of minimizing, if not eliminating, such attitudes and behaviors that can cause great human suffering and economic losses. This challenge is ubiquitous and remains at the forefront of management concerns. Many innovative solutions to problems caused by operator fatigue and information overload have been devised over the decades. Modern process plants are generally equipped with advanced automation and emergency shutdown systems that mitigate the impacts of significant equipment or even human failures. Nevertheless, the complexity of many process technologies in the petroleum refining, petrochemical, and downstream chemical processes is often daunting. Process plants remain susceptible to accidents caused by human actions, inadvertent or otherwise.

There are many aspects of the human personality that affect performance in stressful situations. We find a worrisome variation in the attitudes and personality traits of plant personnel in most operating plants. Therefore, the challenge faced by plant management is to ensure that rigorous training in the fundamentals of every major required discipline is imparted to all practitioners to maximize operational effectiveness and safety. In many countries, there are extensive programs for secondary education of technicians in diverse areas of manufacturing technology, such as the following:

- Electronics
- Power systems
- Piping and hydraulics
- Maintenance of machinery
- Chemistry and process technology

DOI: 10.1201/9781003107873-14

- Pilot plant operation
- Instrumentation and controls
- Civil and structural work and many others

In some countries, we find less emphasis on secondary education and increased reliance on on-the-job training. In general, however, a combination of formal and on-the-job training or apprenticeships provides industry with the best cadres of trained and motivated technicians and operations specialists. For this reason, we believe that it is incumbent on the industry to support, financially and otherwise, governmental and private educational and industrial training programs that are devoted to advanced training and subject matter education for the technicians who are required for supporting increasingly complex process operations in the chemical industry.

In the discussion that follows, we focus on risk mitigation in all process plants, emphasizing human factors (HFs), attitudes, and behavior.

14.1 ACCIDENTS AND HUMAN FAILURES

The practice of process safety is concerned with identifying and controlling risks posed by hazardous substances and processes. Hazards must be identified, and the risks they pose must be brought down to at least an "acceptable level". Process hazards may be classified as follows:

i. Process parameters (temperature, pressure, level, flow, composition, etc.)
ii. Inherent properties of hazardous chemicals (raw materials, intermediates, and finished products)
iii. Runaway chemical reactions and
iv. Failures of layers of protection and mitigation (both active and passive).

Accidents occur when processing operations are carried out beyond the safe limits that were envisaged during qualitative or/and quantitative risk assessments. Many, if not most, such failures occur when plant personnel act in an unsafe manner or fail to follow established guidelines and procedures. Unexpected human failures may either be direct (maloperation, faulty operating procedure, disregard of management of change (MOC) or process safety directives) or indirect (systematic failures in design, installation, or maintenance of equipment, instrumentation, process controls, and safety shutdown systems).

For example, the U.S. Chemical Safety and Hazard Investigation Board Investigation Report[1] on the accident at the BP Texas City refinery in 2005 found that inadequate emphasis on "safety culture" and "HFs" were two key issues.

In particular, the failures attributable to HFs were listed as follows:

- Procedures were not followed
- Ineffective communication during start-up
- Instrumentation gave misleading information
- Operator fatigue
- Understaffing and lack of supervision

- Ineffective training
- Poorly designed control board displays
- Poor safety culture.

14.2 HUMAN ROLE IN HAZARD CONTROL

Humans are solely responsible for governing, controlling, and accomplishing all of the activities necessary to control and mitigate processing hazards. All critical operating parameters must be maintained consistently within the limits of acceptable risk to prevent accidents. Well-trained and skilled personnel are vital for sustaining safe and profitable process operations, and their knowledge and experience are crucial requirements for long-term success.

Highly skilled professionals are responsible for process and equipment design, digital automation systems and process controls, unit operations and optimization, maintenance, process safety, and environmental stewardship. No aspect of the process life cycle is without some human involvement. The possibility of human error is omnipresent in all domains of plant operations. Some of these have been highlighted in the catastrophic accidents discussed in Chapter 2, including UCC Bhopal, Piper Alpha North-Sea, NASA Columbia, BP-Texas City, BP-Macondo, and others.

14.3 TYPES OF HUMAN ERRORS

Human errors encompass mistakes made when interfacing directly with the process or influencing the behavior of the process. They may be classified as follows:

Unintentional errors: A mandatory, essential step is ignored or misapplied.

Intentional errors: An incorrect step is performed deliberately in contravention of established procedures that are unambiguous. Such intentional errors are often caused by poor judgment and are hardly ever malicious. They are caused mostly when the worker exhibits a lack of awareness of potential hazards that might be created when the erroneous activity is carried out. In some cases, the operator may sometimes be well aware of the risk but (1) is convinced that he or she either knows a better way to accomplish the task or (2) may be deluded into thinking that, since numerous layers of protection exist already, bypassing one such layer will not be harmful.

There are innumerable examples of unintentional errors. A few simple examples are given below for illustration:

- An operator fails to perform a double check to ensure that a critical, manually operated valve is fully open or closed.
- An instrument technician miscalibrates a crucial instrument or analyzer.
- An operator changes the set-point of a controller too rapidly or to an unsafe value when making a significant change in unit throughput.

There also are many kinds of intentional errors; we provide a few examples below:

- An operator thinks that positive isolation has been achieved by closing a valve and fails to install a blind as required by the maintenance manual.
- An operator deliberately ignores a critical start-up or shutdown step or safety check.
- A pipefitter fails to tighten a bolt, using a torque wrench, to the proper tension in a large, high-pressure flange joint.
- A plumber selects a gasket consisting of the wrong material for a flange because the proper gasket was out of stock.

14.4 HUMAN FACTORS IN SAFETY (HFs)

There are many inadequacies in existing facilities that can be held responsible, directly or indirectly, for mistakes by plant personnel that lead to safety incidents. These can be organizational or related to working conditions or even how automation systems are implemented. The process industries have compiled an unenviable record of such deficiencies, and we list some of them below:

Organization:

Safety climate and culture in facilities
Command structure and hierarchy
Communication
MOC culture

Job Conditions:

Working environment
Work schedule
Workload
Operating procedures
Supervision
Punitive or hostile working environment

Process Control Systems:

Instrumentation
Control system design and tuning
Alarm management
Emergency shutdown systems

Individuals:

Competence level
Training

Attitude
Risk perception
Job satisfaction

Numerous studies have shown that HFs are directly or indirectly responsible for the vast majority of process industry accidents.

The principles of process safety management (PSM) systems are based on a recognition of the following truths:

- All accidents are traceable to human errors, including premature failure of equipment.
- Management systems must be designed to minimize the impact of human errors.
- Impacts on safety, health, the environment, and quality/production must be prioritized.

"Near misses" are caused invariably by faulty PSM systems and must never be ignored. They must immediately be recognized and analyzed thoroughly. A systematic process to identify current inadequacies must result in necessary and sufficient corrective actions; these must be implemented expeditiously. Failures to promptly implement such corrective actions have repeatedly led to disastrous accidents throughout the process industries. In Chapter 2, we have identified many such examples of disasters caused by bureaucratic neglect and incompetence.

14.5 HUMAN ERROR IDENTIFICATION

In many instances, finding the cause of an untoward event or accident is a highly nontrivial task, especially if the facilities in question have suffered extensive damage. The complexity of modern process plants is such that it requires trained experts to understand and investigate the myriad disciplines that comingle in the design, fabrication, rigging, construction, and commissioning of the manufacturing complex. Identifying and controlling hazards requires a deep background in the underlying engineering sciences behind each discipline. Ensuring a sound, comprehensive, and integrated approach to diagnosing errors caused by humans, many of whom work under stressful conditions, requires empathy, deep experience in process engineering, and analytical skills. The following list is a useful summary but is by no means complete:

1. Process selection
2. Process and equipment design
3. Process instrumentation and basic process control systems
4. Emergency shutdown systems
5. Piping, plant engineering, and relief systems
6. Construction and installation
7. Equipment inspection
8. Commissioning

9. Operator training
10. Unit-wide Operational Safety Reviews ("What-If")
11. Hazard Identification
12. Layer of protection analysis (LOPA) for risk reduction
13. PSM systems
14. MOC procedures
15. Absence of worker fatigue management
16. Improper, incorrect, or insufficient communications

These types of errors are the root causes of many of the infamous accidents that have been described in Chapter 2. For these reasons, serious attention needs to be devoted to understanding and remediating the kinds of human errors that result in injuries, loss of life, and damage to residential, commercial, and industrial facilities and severe business interruptions.

14.6 HFs – A CORE ELEMENT

In the USA, the Occupational Safety and Health Administration originated a National Emphasis Program for Refineries that includes Human Factors (HFs) as one of the 12 core elements quintessential for oil refinery safety.

It behooves corporate management to ensure that PSM programs consider the importance of HFs in developing and implementing site-specific programs that seek to maximize operating efficiency and profit, while also emphasizing worker safety and the overall safety of the manufacturing complex at all times. The systematic and coordinated functioning of all PSM elements in a well-designed PSM program would reduce human errors and their impacts.

Understanding the frequency and severity of human errors in the past major industrial accidents and their root causes helps guide the development of sustainable process safety programs. Such understanding also helps identify areas where technological innovations would help reduce the risk of injuries and accidents. The study of human reliability analysis (HRA) has evolved based on recognizing that human errors are most often found at the root causes of operational failures and accidents throughout the process industries.

14.7 HUMAN RELIABILITY ANALYSIS (HRA)

HRA begins with recognizing that human behavior is error-prone and that all manufacturing systems (which, after all, are designed by human beings) cannot be assumed to be error-free. Improving reliability requires a deep understanding of how and why human errors occur. Only then can we identify potential safety problems and develop improved risk mitigation strategies. The goal of HRA is to quantify the likelihood of human error in performing a given task. Systematic HRA programs are valuable in identifying vulnerabilities attendant to any task. These programs provide valuable guidance on improving reliability for that task, for example, by deploying the best methods and procedures, tools, instrumentation, or automation.

The ultimate objective of HRA is to evaluate an "operator's contribution to system reliability". In other words, we seek to understand and to predict human fallibility and the rate of occurrence of consequential errors, enabling us to evaluate the degradation in reliability of human-machine systems likely to be caused by human errors for the following:

- Equipment functioning
- Operating procedures and practices
- Other related systems

Human frailties and limitations often have a profound influence on the overall performance of complex systems and processes. Several HRA guidelines and techniques have been developed for use in various industries, many of which are freely available, for example, from the U.K.'s Health and Safety Executive (HSE).[2]

Generally, such HRA tools help calculate the probability of errors for a particular task, based on extensive historical data, while considering the influence of those factors in shaping human performance. Quantitative techniques and computerized software systems have been developed that use extensive databases of the error rates associated with human tasks to estimate the likely probability of an error for a particular task or situation. Qualitative techniques are also available to help experts organize structured discussions designed to estimate the probabilities of failure, given specific information and assumptions about tasks and conditions.

14.8 HRA ADOPTION

The U.S. Nuclear Regulatory Commission has identified the need for improved, traceable, and easy-to-use HRA methods for use with the analytical models associated with their accident sequence precursor (ASP) program. This report documents the most recent update of the Standardized Plant Analysis Risk (SPAR) HRA (SPAR-H) method,[3] which is freely available.

Effective PSM programs must incorporate the methods and procedures that have been perfected in HRA to ensure that the effects of human errors on operational safety and efficiency are minimized at all times. The effectiveness of HRA programs in the nuclear power industry has been proven repeatedly. Despite a few notable and high-profile accidents in the past, nuclear power generating plants maintain exemplary safety records in personnel and equipment safety. Limiting human errors by the conscious implementation of HRA throughout the nuclear industry has served as a good example, worthy of emulation by others.

Developing a credible HRA program as part of a holistic PSM system in any industry requires two necessary steps to be implemented:

- Identify a probability of human error for any operation.
- Evaluate, through appropriate multipliers, the impact of environmental and behavioral factors on this probability.

HRA has also been used to support the development of plant-specific probabilistic risk analysis (PRA) models. Human reliability analysis is an evolving field that addresses the need to account for human errors when:

- Performing PRA safety studies
- Helping to incorporate risk analysis in inspection processes
- Reviewing site-specific or special issues
- Helping to advise regulatory authorities about risks in operating facilities

14.9 HUMAN DEVELOPMENT

Comprehensive process safety management calls for a particular emphasis on human development. Currently, many process safety management programs in the industry have inherent weaknesses in developing those human skills that are crucial for work in today's highly complex and automated process industries.

In our experience, many corporations pay lip service to the HF issue – supervisory exhortations and circulars, emails, and memos from corporate management are entirely insufficient when considering the development of a comprehensive PSM system. What is needed is a far greater emphasis on formal operator training as the most important human development activity. For example, using dynamic (time-dependent) training simulators, coupled with periodic re-certification of board/control room operators, has been demonstrated to be extremely valuable. Using a trainer-trainee environment, where the trainer creates a variety of emergencies and observes the operator's response, can be extremely beneficial in improving process/equipment comprehension and eliminating poor practices or procedures.

There is also a great need for apprenticeship programs led by seasoned and well-trained operators to train recruits. The loss of institutional memory when older operators retire is a recurring and severe problem these days. Many companies seem to be oblivious of the value of the cumulative experience acquired over the years, and at great cost, by their operational staff. Unfortunately, senior management levels in the chemical industry are routinely occupied today by individuals who do not have a scintilla of technical expertise. Notably, they often exhibit a profound lack of knowledge and foresight to understand the value of human development in the process industries, focusing instead on trivial cost containment.

14.10 INDUSTRY RESPONSE

The American Institute of Chemical Engineers (AIChE) is perhaps the most proactive body concerned with the chemical industry's development and well-being globally. A major AIChE article titled *Integrating Human Factors (H.F.) into a Process Safety Management System (PSMS)*[4] is devoted to enumerating the HFs often found to be inadequately considered, or even missing entirely, in many existing process plants. Here, AIChE advocates the development and use of an "integrated process safety management system" (IPSMS) model, based on a thorough review and screening of all major and current PSM frameworks, while integrating the Human Factors Analysis and Classification System. This model describes an implementation strategy emphasizing the following steps: PLAN, DO, CHECK, and ACT.

REFERENCES

1. Final Baker Commission Investigation Report on B.P. Texas City Refinery Explosion: *B.P. Explosion* (December 9, 2005).
2. U.K. HSE Research Report RR 679: *Review of Human Reliability Assessment Methods* (2005).
3. U.S. Nuclear Regulatory Commission: Idaho National Laboratory's Report: *The SPAR-H Human Reliability Analysis Method* (August, 2005).
4. Theophilus, S. C., Nwankwo, C. D., Acquah-Andoh, E., Bassey, E., and Umroen, U.: *Integrating Human Factors (H.F.) into Process Management Systems* (July 30, 2017).

15 Process Safety and Manufacturing Excellence

Broadly defined, there are two aspects of safety in process plants: process safety and personnel (occupational) safety, and it is essential to distinguish between them. A review of the many accidents that have occurred over the years in the process industries shows that, while personnel safety often is a primary concern, the attention devoted to process safety was often cursory or insufficient. The loss of lives, equipment, and property resulting from accidents in the process industry because of inadequate attention to process safety from the top, continues to be a source of great public concern and regulatory focus.

15.1 PROCESS SAFETY LEADERSHIP

All process industries face unique challenges to inculcate a pervasive orientation related to process safety throughout their enterprise. It is worth recapitulating the following cogent observations in the recent review[1] of a recent book co-authored by Berger[2] published under the auspices of the AIChE's Center for Chemical Process Safety (CCPS):

> Releases of chemical and other hazardous materials pose significant – potentially catastrophic – threats. An alarming number of such events, all of which are preventable, occur too often, and reducing the frequency of such incidents is a fundamental responsibility of leadership at all levels. Leadership is a key component to achieving sustained safe operation. Effective and informed leaders provide direction, reinforce commitment, and drive responsibility.
>
> Executives, plant leaders, functional managers, front-line supervisors, and other personnel can use this guide to create a viable culture of safety at their organization, implement and maintain disciplined management systems, and address the risks of process safety deficiencies.

The CCPS guide[2] examines many issues relevant to such challenges: strengthening management system accountability, driving operation within constraints, ensuring corporate memory, verifying execution, etc. A study of these discussions and recommendations is highly recommended for top management.

Process safety should always be at the forefront of all strategies adopted for achieving excellence in manufacturing and process operations. Several well-documented and reliable methodologies for achieving manufacturing excellence are sometimes either not in place or do not command adequate attention on an ongoing basis. Based on our experience in the process industries, we feel that some explanation of these methods and procedures would help draw attention to the many details and inherent complexities.

DOI: 10.1201/9781003107873-15

Accordingly, our book has attempted to isolate several germane issues in the practical implementation of process safety-related initiatives and programs. The approach we have adopted is based on the quantification of some of the scientific, engineering, and technological principles that must be understood by practitioners responsible for implementing process safety management (PSM) initiatives driven by corporate management. While there are many extremely detailed treatments[3,4] of some of these areas, we have chosen to address – in an accessible and understandable way – those aspects that require, in our opinion, the highest level of attention.

15.2 PROCESS SAFETY LAWS AND REGULATIONS

Aspects of personnel safety are regulated by governmental agencies, such as the Occupational Safety and Health Administration (OSHA) in the USA, the Health and Safety Executive (HSE) in the U.K., the European Agency for Safety and Health at work in the E.U., Workplace Safety Law in the People's Republic of China, and the Directorate of Safety and Health (DISH) in India. There are similar regulatory bodies in many other countries. Such regulations can be voluminous and require careful study and interpretation to ensure compliance with both their letter and spirit.

Governmental statutes specify minimum mandatory basic requirements for "Hazard Identification", "Consequence Determination", as well as "Determination and Control of Risk". In the USA, this is done formally under OSHA's PSM (Rule) 1992: *Process Safety Management of Highly Hazardous Chemicals* (Federal Register 29CFR1910.110) and EPA's Risk Management Program (RMP USA) 1995. The primary Process Safety Management (PSM) Regulation in the E.U. is the *Seveso Directive* covering all E.U. member states. The U.K. has now departed from the European Union (Brexit); it has unique process safety regulations and agencies, including the Control of Major Accident Hazards (COMAH) regulation and the HSE. The Chinese PSM regulations are under the State Administration of Work Safety (SAWS) created by the ILO, 2005, and the SAWS PSM regulation, passed in 2010. India has also implemented similar PSM regulations, including the Factories Act (administered by the DISH), and also *Manufacture, Storage and Import of Hazardous Chemicals Rules 1989* (MSIHC 1989), *The Chemical Accidents (Emergency Planning, Preparedness, and Response) Rules 1996*, with subsequent amendments, administered under the Environment Protection Act (EPA) by the Central Government's Ministry of Environment and Forests.

15.3 PROCESS SAFETY VIS-À-VIS PERSONNEL SAFETY

In this book, while we recognize the extreme importance of devices used to provide physical safety for plant personnel, such as personal protective equipment, gas masks, etc., we do not discuss such measures in any detail. We believe that these are already well documented in numerous other books and internet websites. Personnel safety refers to physical measures taken to prevent or minimize plant personnel exposure to physical, chemical, biological, bacterial, and radiological hazards. Such hazards are generally well documented and understood throughout the chemical process industries. There are many types of protective measures and equipment in use, and many sophisticated devices are available internationally from a large number of manufacturers and suppliers.

On the other hand, process safety is concerned with a wide variety of hazards created by processing from high to cryogenic temperatures, extreme pressures, and the presence of toxic or otherwise hazardous chemicals. Further, when exothermic chemical reactions occur, large amounts of energy are released. Many chemicals are toxic, carcinogenic, mutagenic, or teratogenic, and some can be extremely hazardous, even in small quantities.

15.4 THE ROLE OF PROCESS AND EQUIPMENT DESIGN IN ENSURING PROCESS SAFETY

Meticulous design and operational procedures are required to ensure safe and reliable process and equipment designs. Thanks to the diverse nature of the process industries, it would be unreasonable to expect governmental agencies to prescribe specific procedures or methodologies for addressing such issues or potential problems.

It remains the domain of process development and licensing experts carefully to consider such potential hazards. They assume the responsibility of identifying the necessary steps to eliminate such hazards and, as part of their licensed or other proprietary technologies, to specify the installation of safety devices, instrumentation, automated shutdown systems, and other strategies. Collectively, these serve to prevent and mitigate the impact of process safety-related incidents that conceivably could occur. This book has presented several technical methods and procedures for process safety engineering that are often not sufficiently well understood or emphasized, and therefore deserve special attention.

15.5 STRATEGIES FOR IMPLEMENTATION OF PROCESS SAFETY PROGRAMS

A review of the notorious industrial disasters that have been described in Chapter 2 would show clearly that all of these were caused by inadequate attention to process safety, as opposed to inadequate personnel safety measures at the earliest process and equipment design stages. Even today, it is not uncommon to find instances where considerable additional emphasis needs to be placed on enhancing process safety, including safer plant layouts and improved precautionary measures.

To this end, we have chosen to discuss, in some detail, the following items that (among others) have been shown to contribute significantly to a successful process safety program in every major manufacturing complex in the process industries:

Strategy	Objective
1. Sensor validation	Validate measurements, estimate critical parameters
2. Sampling time recording	Build reliable inferential calculations
3. Control system hardware and configuration	Improve disturbance rejection
4. Control valves	Improve final control element performance
5. Regulatory control tuning	Stabilize process operation
6. Online calculations/equipment health monitoring	Improve operator/engineer comprehension, MPC
7. Smart sensors/inferential calculations	Estimate instantaneous online properties
8. Multivariable, optimal predictive control (MPC)	Safely push the unit to most profitable constraints
9. Closed-loop, real-time optimization (CLRTO)	Economic optimization of nonlinear processes
10. Rigorous chemical reactor modeling	Prerequisite for real-time optimization
11. Planning and scheduling optimization	Complex-wide economic optimization
12. Intelligent alarm management	Improve operator performance, reduce accidents
13. Emergency shutdown (ESD) systems	Protect life and property during emergencies
14. Location of process control rooms	Avoid unnecessary personnel exposure to hazards

Each of these strategies is discussed below.

15.5.1 Sensor Validation

In any Distributed Control System (DCS), there are many hundreds, if not thousands, of digital measurements captured from field instrumentation. Primary measurements directly impact the safety, environmental compliance, or productivity of the unit or process. Secondary measurements are essential for process monitoring and control and are often embedded in regulatory control schemes comprising one or more regulatory control loops. Cascaded control loops are generally designed to achieve good regulation of measurements taken at a relatively low frequency, such as product composition, by moving lower-level controllers such as flows.

It is vital to ensure that errors in reported measurements are diagnosed as soon as possible so that process operations are maintained in the desired region. Reliable error detection requires *sensor validation*, a collection of methods and procedures to perform automated analysis and validation of measurements. The objectives are to trap the following kinds of instrumentation errors:

Problem	Symptoms
Stuck values	The value does not change within some tolerance for a specified number of cycles
Rapid change	Values changing at an impermissible rate that exceeds the specified limit
Out of range	Value is outside permissible bounds
Excessive variability	The standard deviation of measurements exceeds the maximum limit

For the validation of each sensor, a decision is first made about the number of sequential measurements that must be analyzed. Simple and highly efficient computations are then performed to diagnose such errors.

Sensor validation is extremely valuable for reliable and safe plant operations and for ensuring the viability of higher-level advanced process control (APC), model-predictive control (MPC), and closed-loop, real-time optimization (CLRTO) applications. These strategies are a primary focus for maximizing plant profitability while simultaneously observing safety limits for equipment and process safety and environmental compliance.

As soon as a sensor is flagged for a validity violation, notification is made to the appropriate instrumentation specialists. The validity flag is not reset until repairs are complete and all dependent control strategies are ready to be commissioned. This practice ensures that reliance on faulty instrumentation is minimized. Several major industrial accidents have occurred as a result of improper reliance on invalid measurements.

15.5.2 Sample Time Recording

Product quality or other laboratory measurements are vital for ensuring safe and reliable plant operations. These are often stored in databases such as the laboratory information management system (LIMS), which could also be connected to the DCS used to run the plant. It is not feasible to develop smart sensors (Section 15.6 below) unless sampling times are recorded reliably.

The best method of ensuring that the sample time is captured is to install a switch near each sample point. When the switch is triggered, the sample time is recorded automatically. The laboratory's subsequent analytical results are then synchronized with the sampling time before being stored in the LIMS/DCS. However, even today, it is common to find instances where the recording of the actual sampling time is not done at all. In many instances, the sampling frequency itself is quite inadequate for ensuring reliable product quality control.

When sample times are recorded accurately and synchronized with the sample results, applications using these to control plant operations run a lot smoother, and product quality is maintained far more reliably. This is especially true if inferential calculations are used for estimating online product quality, as discussed further below.

15.5.3 Control System Hardware and Configuration

In every process plant, it is vital to ensure that the regulatory control system works flawlessly, with meticulous attention to:

a. The selection, sizing, and maintenance of control valves
b. The control logic, configuration, and tuning of the regulatory and cascade control loops throughout the plant.

15.5.4 Control Valves

Excellence in process control requires flawless design, sizing, and operation of control valves. There are many types of control valves, and proper selection and sizing are imperative. However, control valves have many components, and their performance and maintenance is often a source of concern. The issues most often encountered are:

a. Valve type and sizing
b. Actuator type and sizing
c. Hysteresis
d. Valve sticking
e. Loose mechanical linkages.

Many of these issues are addressed in the Control Valve Handbook by Emerson-Fisher,[5] which states as follows: *Properly selected and maintained control valves increase efficiency, safety, profitability, and ecology.* Our experience has shown that this is indeed a vital aspect affecting all plant operations. We know many examples where the benefit from otherwise competent control strategies was diminished dramatically by inadequate attention to control valve performance.

It is common to see situations where a control valve is of a smaller size than the piping in which it is installed. This practice can limit line capacity unnecessarily and should generally be discouraged. It is often observed that the bypass line around some control valves is left partially cracked open by operators, a clear indication of control valve under-sizing. When control valves malfunction, their bypass lines are often used to maintain operation while the valve undergoes maintenance. By the same logic, the bypass line and valve should be of the same size as the main flow line. As a safety precaution, control valves in high-pressure service should have double block valves on both sides to ensure that they can be removed safely, without leaks, for maintenance.

For control valves in hydrogen service, it is well known that gas leakage is quite common. Therefore, double block valves are recommended on both sides of the control valve to ensure positive shutoff whenever the valve is sent for maintenance. Most leaks of hydrogen result in a fire, even when it appears there is no source of ignition. Unfortunately, a hydrogen flame – which burns at an extremely high temperature – is also nearly impossible to see during the day. Nevertheless, we see many instances where double block valves are not used, and a significant leakage hazard is created when such a control valve is taken out for maintenance.

Since control valves move frequently, they are susceptible to erosion over time. Therefore, the materials of construction selected should resist wear and tear and be highly corrosion resistant. The valve sealing system especially deserves careful attention, as leaks are often the cause of fires that lead to valve failure and, ultimately, significant damage to process equipment. Unfortunately, in our experience, it is common to see valve seals that have been overtightened to minimize the possibility of leaks. This practice often results in the sticking of the valve stem. Thanks to *reset windup*, the proportional-integral-derivative (PID) controller moving the valve then begins to exhibit "bang-bang" control: the valve moves from a fully open to a fully closed position cyclically.

Proper control valve sizing is crucial to ensure that regulatory control action is smooth and predictable. Selecting an unduly large size can lead to excessive hunting, control valve chattering, wear and tear, and eventual valve failure. Under-sized control valves must stay close to fully open, often in a region where their flow regulation capability is relatively poor. Inevitably, control performance is degraded to an unacceptable level.

The size and type of control valve actuator must also be examined most carefully. The range of operating pressure and control valve size are both significant. Actuator designs can have quite different operating characteristics, and this aspect requires close attention when selecting any control valve.

It is becoming standard practice in process control to require that control valves be outfitted with valve positioners so that the controller output corresponds closely to the actual valve position in the field. Lack of valve positioners can have a significant deleterious effect on control valve performance.

15.5.5 Control System Configuration

A modern DCS incorporates facilities for enabling excellent regulatory control of major process plants.[6] The power of microprocessors has risen exponentially for over 40 years. There have been concomitant impressive advances in the design and robustness of DCS. Several DCS vendors have made available tools and methodologies for ensuring robust and safe process operations that were scarcely imaginable just a few decades ago.

Nevertheless, the design and configuration of regulatory control functions in the process industries remain as much an art as a science. In the hands of a skilled practitioner, working with an expert in DCS system configuration and integration, we routinely see outstanding results when controlling large-scale, highly interactive, nonlinear processes.

However, many examples can still be found, even in large-scale process plants, where the "logic" of process control loops is far from ideal. In many instances, inadequate control strategies from decades ago are re-implemented in modern DCS. The quality of regulatory process control is, as might be expected, no better than in the past.

For these reasons, it is imperative that staff responsible for control loop configuration have the proper background in chemical engineering principles, a good understanding of process dynamics, and have adequate training and experience in this most vital area. Even today, there are many major plants where such expertise is lacking, and as a result, control system performance is substantially below par. This problem has unexpected but potentially severe process safety implications.

Ultimately, control valves are moved by DCS-resident regulatory controls – typically PID controllers. The process must be capable of stable operation when using just the regulatory control system. (We note that open-loop unstable processes can sometimes be stabilized by the proper selection and implementation of advanced, model-predictive controls.) Advanced control strategies are designed to improve plant profitability while ensuring process and equipment safety and environmental compliance. These applications (which may or may not be resident in the DCS) take over the task of changing the set-points of many, if not most, of the regulatory control loops that are part of their domain.

For this reason, the regulatory control system must be capable of achieving the desired set-points in a speedy yet stable and predictable manner. The basic control elements (control valves), DCS logic, and control loop tuning must function flawlessly. Control valves should be equipped with valve positioners to ensure that the controller output and physical valve position are identical. Any failure to meet these requirements at the regulatory control level generally renders ineffective any higher-level advanced control applications and may even require them to be turned off.

An essential criterion for judging overall control system performance is the percentage of DCS regulatory control loops that are not in "AUTO" or "CASCADE" mode. This assessment needs to be made daily. PID controllers that remain in "MANUAL" mode for more than some reasonable maximum period (say, 2 days) must be flagged for immediate attention and review. Often, the cause is found to be improper control loop configuration or other valve or instrumentation malfunctions. There is no way that loop tuning can compensate for flawed control configuration logic.

A thorough evaluation and repair, as may be needed, should be performed of all sensors that are included in the control loop. The control valve (or valves in case dual-acting controls that are in place) should be tested thoroughly to ensure proper valve type and sizing. Also, valve sticking, hysteresis, or other valve positioner problems must not hamper control loop performance. All such diagnoses require considerable training, skill, and judgment. Plants should be staffed with adequate numbers of such experts that can be relied upon to ensure excellent functioning of the entire control system. Lack of attention to these requirements has often been cited as a proximate cause of accidents in process plants.

15.5.6 Regulatory Control Tuning

Proper tuning of the regulatory control loops for any process is essential to achieve safe and stable unit operations. Historically, loop tuning was performed manually by adjusting the PID controller constants.

The typical procedure was to arrive at a stable controller gain (proportional constant). Next, the integral action would be adjusted while also leaving open the possibility of tweaking the controller gain downward to achieve the steady state in a reliable, speedy, yet stable manner. A compromise in tuning constants would generally be required to ensure efficient response to set-point changes and reject external disturbances. Finally, if the control response were slower than desired, some derivative action would be added. This step would often require turning down the aggressiveness of the proportional and integral actions because derivative action could otherwise potentially destabilize controller performance and cause unacceptable oscillations in the face of significant or sudden disturbances. In most instances, the control specialist would spend a good part of the day for the complex control loops in a single higher-level cascade.

The situation has changed quite dramatically in the last few years. DCS-resident software is now available that can provide at least preliminary regulatory control loop tuning in an automated fashion, with little or no assistance from a control specialist.

However, loop tuning must necessarily provide a *proper balance between speedy response to set-point changes and disturbance rejection*. This balance is generally hard to achieve in a completely automated fashion. Therefore, control specialists must monitor loop performance over time and make tuning adjustments to achieve the right balance. As mentioned previously, attention to the background and training required for performing such work is vitally important. Well-trained control engineers help achieve process safety goals most effectively.

Single-loop PID controllers adjust one manipulated (i.e., independent) variable to control a given variable. However, moving that manipulated variable (MV) could result in changes to other controlled variables (CVs). For example, changing the reflux flow rate to achieve the desired overheads purity in a distillation column also changes the column's material balance. Over time, the distillate and bottoms flow rates and compositions are both affected when the reboiler duty remains unchanged. Accordingly, a change in the reflux flow requires concurrent compensating changes, over time, in the bottoms reboiler duty. Analogously, changing the reboiler duty results in changes to both the bottoms and overheads flows and compositions, requiring coordinated changes to the overhead reflux. This example illustrates the multivariable nature of distillation control. Therefore, it is crucial to recognize that PID controller tuning must account for the multivariable interactions inherent in most chemical engineering processes.

Some vendors now offer single-loop, self-tuning regulatory controllers that do not execute within the DCS and use proprietary sensors and control algorithms. We recommend that a careful review is first performed to ensure that the control task at hand cannot be solved using standard DCS-resident capabilities. One example of such software would be a multiple-input, single-output (MISO) controller to enable more precise regulation of a given CV when that variable is affected by several independent variables.

Another example of such a specialized application is compressor surge control, where the control cycles need to execute at millisecond frequency to ensure a speedy response to the surge phenomenon. This issue is discussed further in Section 15.7.5. Other examples can be found for highly nonlinear control problems, such as pH control in reactors.

15.6 HIGHER-LEVEL MULTIVARIABLE CONTROL AND OPTIMIZATION APPLICATIONS

In general, it is extremely difficult to determine the most profitable operating condition (i.e., the optimal regulatory control set-points) for large plants, taken as a whole, based on current feedstock/utility costs and product values. Economic optimization is, therefore, a highly nontrivial task and generally requires linear or nonlinear control and optimization techniques based on "first-principles" chemical engineering models.

Since the 1970s, there has been remarkable progress in MPC technologies. Chief among these are Aspen Technology's DMC® and Honeywell's RMPCT®. Advanced control strategies use fundamental engineering knowledge about the time-dependent dynamics and the inherently interactive nature of chemical processes. They seek first to determine optimal steady-state targets for the process and then move the regulatory control system set-points in a safe, reliable, and coordinated manner towards those targets. Several variables must be moved simultaneously to enable advanced controls to prevent violations of crucial process safety-related constraints. A compelling test of the design and configuration of such strategies is their ability to reject multiple simultaneous external disturbances (both measured and unmeasured) effectively. We discuss this aspect further under Sections 15.9 and 15.10.

In our experience, developing and maintaining MPC applications is a task that demands the highest engineering skills in the areas of chemical process modeling, process dynamics, and process control. The effort required for a complete application is generally several months in duration and requires close coordination between the engineering staff who execute the project and plant-level operations staff. Also, the dynamic models used for predictive controllers do not remain inviolate over time: changes in feedstock properties, reactor catalysts and operating conditions, equipment changes, changes in process configuration, and alterations in the regulatory control system configuration or tuning can all require that some or all of the MPC dynamic models be rebuilt. Methods that use first-principles dynamic simulation to develop the input–output dynamic models used in MPCs have been developed.[7] These expedite and simplify building or rebuilding MPC models, whenever required, without resorting to invasive and costly plant step tests.

In many instances, plant supervisory and operations teams are understandably reluctant to allow repeated step tests to recalibrate previously developed MPC controllers whose performance has deteriorated over time. There are many reasons why this happens, including changes in feedstocks, reactor operation, changing equipment efficiencies, and so forth. In such instances, these applications invariably get "turned off", and all of their economic benefits are lost, including those that enhance adherence to process safety, environmental, and equipment limits.

All modern DCS manufacturers enable connecting high-performance workstations or personal computers to the regulatory control network. These networked computers provide enormous computing power, unimaginable a few years ago, to supplement the built-in capabilities of the DCS itself. These are configured to access virtually every variable available to the operator or control systems engineer. Many higher-level control applications reside in networked computers and generally run at a lower frequency than the DCS cycle time. Examples include:

- Data historian software
- Multivariable, MPC controllers
- First-principles, CLRTO packages.

Further, these networked computers can be connected bidirectionally to external, secure sources of information. These include corporate planning and scheduling software that optimizes the highest-level economic targets of the company. In this way, the control hierarchy maintains continuity between the various echelons of corporate management and operations within the control room. A preeminent objective of such strategies is to ensure that plant operations adhere stringently to the highest-level process and equipment safety requirements.

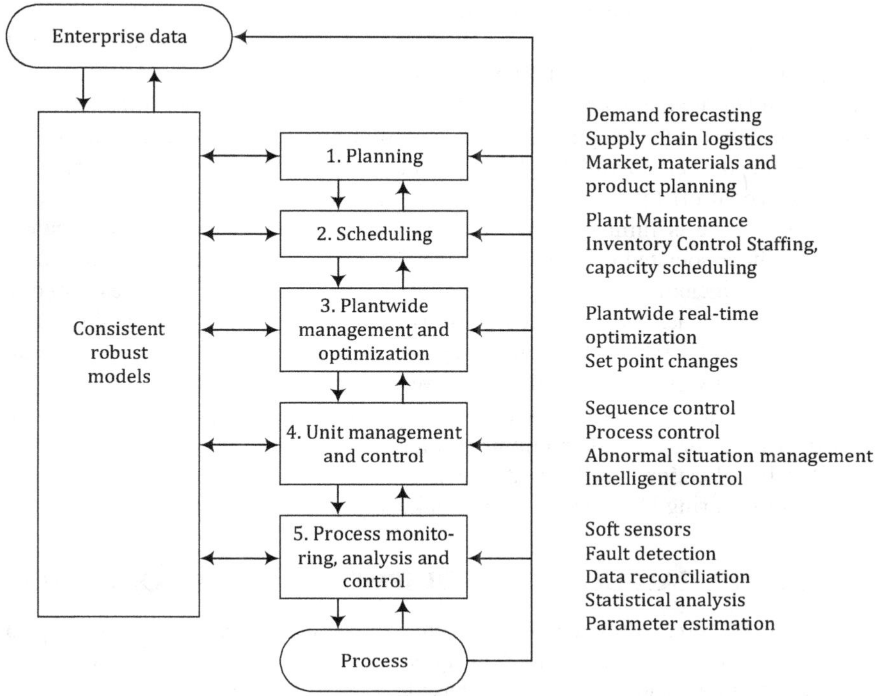

FIGURE 15.1 Hierarchy of planning, scheduling, optimization, and multivariable/regulatory control.[8] (With permission, Edgar, T. F., et al.: *Optimization of Chemical Processes* (2nd Ed., McGraw-Hill, New York, 2001).)

Figure 15.1[8] illustrates this hierarchy between:

- Corporate planning and scheduling
- Unit-wide CLRTO
- Model-predictive, multivariable control
- DCS-resident controls and calculations
- Regulatory control hardware.

In Figure 15.1, the execution frequency goes up as we go down the hierarchy of applications. However, the complexity of the required financial information and the derived economic value generally increase as we go up in the hierarchy. At all levels, reliable mathematical models are required that must be consistent, computationally efficient, and robust in the sense that they are fault-tolerant in case invalid data are encountered.

The planning and scheduling application layers are generally found in corporate computer systems, which can communicate bidirectionally with the plant-wide optimization applications one level below. This information transfer is vital for ensuring that the steady-state optimal targets obtained by the CLRTO applications are consistent. Such information includes the economic value of feeds, utilities, intermediate streams, and products for various CLRTO applications. In turn, the CLRTO applications provide current values for equipment and process constraints at the plant level back to higher-level applications. These include corporate planning and scheduling and linear programming (LP) optimization applications.

The CLRTO applications reside at the plant-wide real-time optimization layer. They generally provide the upper and lower control limits for the multivariable, model-predictive constrained controllers (MPC) at the next lower level (process monitoring, analysis, and control). Often, local mathematical optimizers are found within the MPC layer whose scope is limited to the independent, MVs, the measured disturbances (DVs), and the associated CVs. We note that the effects of unmeasured disturbances on the process, including weather-related or caused by feed quality changes, can only be handled in feedback mode. The data historian software generally also resides at the same level as the CLRTOs. The regulatory control system, sensors, and final control elements (control valves) are found at the lowest level (process) in this hierarchy.

Flawless communications between the various levels in the control hierarchy are vital for ensuring process and equipment safety. This is an important issue when the higher-level applications are located in corporate networks that can read plant-level data but, for ensuring DCS integrity, not make any changes in operations.

15.7 ONLINE CALCULATIONS/EQUIPMENT HEALTH MONITORING

Many process control strategies depend on the availability of online estimates of process parameters, such as product purity, when these are not measured by online instrumentation. All such quantities are based on DCS-level measurements of underlying flows, temperatures, and pressures, coupled with relatively infrequent laboratory measurements. While there is a great variety of such calculations that can and should be implemented within the DCS, we provide below a sampling of a few such examples that have been demonstrated in numerous successful online applications and for equipment monitoring:

- Fired heater radiant section duty
- Heat exchanger duty
- Distillation column pressure-compensated temperature
- Distillation column approach to flooding
- Compressor, turbine, and pump efficiency.

These examples are discussed further below. It should be noted that innumerable other online calculations can be made for tracking equipment performance in real time. The purpose in every case is to:

a. Provide an estimate of how well the equipment is performing
b. Enable examination and evaluation of trends in operational efficiency
c. Provide timely warnings about unacceptable operating conditions or safety issues.

Such routine monitoring and diagnosis make an essential contribution to the overall safety of processes and equipment.

15.7.1 Fired Heater Radiant Section Duty

This application is used for controlling the fuel flow rate to the heater to maintain steady heat duty. It is calculated by multiplying the fuel flow rate, the net (or lower) fuel calorific value, and the radiant section efficiency:

$$Q = W\Delta H_c\eta \tag{15.1}$$

where

Q = radiant section duty, kcal/h
W = fuel flow rate, kg/h
ΔH_c = fuel lower calorific value or net heating value, kcal/kg
η = fractional radiant section efficiency.

Especially with fuel gas firing in petroleum refineries, where gas from various units is routed to the fuel gas system, the calorific value can change substantially over short periods. If fuel gas flow alone is used to control the fired heater, fluctuations in the composition can disrupt the fired duty and, therefore, the heat absorbed in the radiant section.

The process control system reacts to such disturbances, using feedback from affected CVs such as heater outlet or distillation tray temperature. If the fuel gas calorific value is tracked online (e.g., by a chromatograph or Wobbe meter), this can be used to compute the fired duty. The fired duty can then be used as part of a regulatory system cascade loop to proactively adjust the fuel flow. In this way, disturbances caused by fluctuations in the fuel's calorific value can be rejected before disrupting heater performance.

15.7.2 Heat Exchanger Duty

When heat exchangers are used for heating or cooling for process streams, it is often desirable to manipulate the utility flow rate to control overall exchanger duty. A cascaded controller then uses an online calculation of heat duty to manipulate utility flow to maintain the desired duty target.

The heat duty calculation would account for situations where there is no phase change (using specific heat) or when the utility condenses or boils off completely (using the latent heat).

15.7.2.1 No Phase Change

The exchanger duty when there is no phase change is:

$$Q = WC_p\Delta T \tag{15.2}$$

where

Q = heat exchanger duty, kcal/h
W = fuel flow rate, kg/h
C_p = fluid heat capacity or specific heat, kcal/(kg C)
ΔT = change in fluid temperature, °C.

15.7.2.2 Condensing or Boiling

For condensing or boiling applications, the duty is found using:

$$Q = W\Delta H_v \tag{15.3}$$

where

Q = heat exchanger duty, kcal/h
W = fuel flow rate, kg/h
ΔH_v = heat of vaporization for utility fluid, kcal/kg.

While these examples use quite simple underlying calculations, they can be highly effective when used in control strategies for preventing significant and common disturbances from affecting heat exchanger duty and consequently upsetting plant operations.

15.7.3 Distillation Column Pressure-Compensated Temperature

Many distillation columns incorporate controls for tray temperature that manipulate the reflux flow or reboiler duty. However, column pressure is sometimes different from some base value (e.g., because of increased summertime cooling water temperature at the condenser). Pressure may otherwise be disturbed for other reasons, and the tray temperatures would change accordingly. For proper process control, we must compensate for such changes in column pressure on the tray temperature. Failure to provide such compensation can result in unacceptably large disturbances being created by the control scheme itself.

Pressure compensation for the tray temperature is used for handling this issue properly. Generally, for a given tray composition, tray temperatures can be related to the pressure using the Antoine equation[9]:

$$\ln(P) = A - B/(T + C) \tag{15.4}$$

where

P = tray absolute pressure, bar
T = tray temperature, °C
(A, B, and C are all positive constants).

At a base column pressure of P_b, if the measured pressure is P and the temperature is T, then the pressure-compensated temperature T_{pc} is found using the following equation:

$$T_{pc} = -B/\left[\ln(P_b/P) - B/(T + C)\right] - C \tag{15.5}$$

where

T_{pc} = pressure-compensated temperature, °C
P_b = base column pressure, bar
P = online pressure, bar
T = online temperature
(A, B, and C are all positive constants).

Note: When Equation (15.4) is applied to both the base and online conditions, A is eliminated by subtraction to obtain the expression above for T_{pc}.

In the tray temperature control application, the control error would be $(T_{pc} - T_{SP})$, where T_{SP} (the set-point at a base pressure of P_b) is an operator-defined value for the desired tray composition. The pressure-compensated temperature T_{pc} is used instead of the actual temperature T in the control application. In this way, column operation is stabilized by eliminating the effect of varying column pressure.

A simple example of the benefits of this method is useful:

Example 15.1

CALCULATION OF PRESSURE-COMPENSATED TEMPERATURE

For a mixture containing 51.8 mole% propane and 48.2 mole% butane mixture on tray 17 of a distillation column, the tray temperature is expressed as a function of pressure using the Antoine equation:

$$P = e^{\lfloor A - B/(T + C) \rfloor}$$

where

P = vapor pressure, bar,

A = 14.306, B = 3622.38, C = 56.65.

(These were evaluated, on tray 17, at a base tower operating pressure, P_b, of 17.75 bar and a base temperature, T_{base}, of 357 K).

This tray temperature is maintained by a temperature controller that is cascaded to the overhead reflux flow controller. If, during a tower upset, tray 17 pressure changed to 19.1 bar and the temperature to 362.6 K, what would be the pressure-compensated temperature for this stream?

Applying Equation (15.5) for the pressure-compensated temperature, we have:

$$T_{pc} = -3{,}622.38 / \lfloor \ln(17.75/19.1) - 3{,}622.38/(362.6 + 56.65) \rfloor - 56.65, \text{ or}$$

$$T_{pc} = 359.1 \text{ K, which is } 359.1 - 357 = 2.1 \text{ K higher than the base value.}$$

Note: If the temperature had not been compensated, the controller error would have been $T - T_{base}$ = 362.6 − 357 = 5.6 K. This seemingly small uncompensated control error is over 166% larger than that for the pressure-compensated value. Therefore, the tray temperature controller would falsely "detect" a much larger change in tray temperature and change the column reflux much more aggressively than necessary. Potentially, this could cause a highly significant (and wholly unnecessary) upset in tower operation. Such an upset could last for many hours before settling down to a new steady state where both product compositions would be off-specification. Experience shows that installing a pressure compensation for tray temperature invariably delivers lasting benefits for column purity control.

15.7.4 Distillation Column Approach to Flooding

In many distillation columns that are operated close to their capacity limit, the onset of column flooding results in a sudden, dramatic rise in column pressure drop and prolonged loss of fractionation efficiency. Avoiding such severe disruptions is a crucial aspect of process safety. It is useful to develop and use an online calculation for the "flood factor", a measure of the approach to column flooding.

The flood factor is based on the well-known Souder-Brown correlation[10] for distillation column capacity:

$$C_{sb} = U_v\left[\rho_v/(\rho_L - \rho_v)\right]^{0.5} \tag{15.6}$$

where

C_{sb} = Souders-Brown capacity parameter, m/s
U_v = superficial vapor velocity at tray, m/s
ρ_v = vapor mass density at tray, kg/m^3
ρ_L = liquid mass density at tray, kg/m^3.

Often, the liquid density is much larger than the vapor density, and the $(\rho_L - \rho_v)$ term is taken as constant and ignored. Equation (15.6) is thus modified to the following form[11]:

$$F_s = U_v\left[\rho_v\right]^{0.5} \tag{15.7}$$

where

F_s = flood factor, (m/s)(kg/m^3)$^{0.5}$
U_v = superficial vapor velocity at tray, m/s.

In Equations (15.6–15.7) above, the superficial vapor velocity is found using:

$$U_v = W_v/(\rho_v A_n) \tag{15.8}$$

where

W_v = mass flow rate of vapor at tray, kg/s
ρ_v = vapor mass density at tray, kg/m^3
A_n = net column cross-sectional area, excluding downcomers, m^2.

Example 15.2

CALCULATION OF COLUMN FLOOD FACTOR

25% of the cross section of a column of 4.35 m diameter is reserved for its downcomers. A column tray has a vapor load of 1,0918 kgmole/h. The vapor molecular weight is 42.09 kg/kgmole, and the vapor density is 41.2 kg/m^3. Calculate the flood factor F_s. and compare it against the design value of 1.85.

Column cross-sectional area is $A = (\pi/4)\,(D^2) = 3.14159/4/4.35^2 = 14.86\ \text{m}^2$
Net area available for vapor flow is $(A)\,(1 - 25/100) = (14.86)\,(0.75) = 11.15\ \text{m}^2$
Vapor density ρ_v is 41.2 kg/m^3 or 41.2/42.09 = 0.9789 kgmole/m^3.
Vapor flow is 10,918/3600 = 3.033 kgmole/s,
Volumetric flow is 3.033/0.9789 = 3.10 m^3/s
Vapor velocity is 3.1/11.15 = 0.278 m/s.

From Equation (15.7), the flood factor is $F_s = (0.278)\,(41.2^{0.5}) = 1.78$. Therefore, the tray vapor loading is below the design value of 1.85, and the column will not flood at this tray.

It is useful for process control applications to compute the flood factor from Equation (15.7) and store it in the data historian. This parameter should be calculated for all critical trays. It can then be studied over time to see how the tower vapor loading compares with the maximum limits up and down the column. It can also send an alarm to the operator in situations when the tower may begin to flood unless vapor loading is reduced, perhaps by cutting the feed flow rate or changing the reboiler or pumparound duty.

Generally, it is straightforward to estimate the vapor mass flow rate, W_v, at any location in the column using mass and energy balances within an envelope from the tray of interest to the column overheads or bottoms, as described by Watkins[12] and Friedman.[13] Some precautions must be taken to avoid errors caused by the subtraction of large quantities from each other when there exists uncertainty in their true values when the process is not at a steady state.

15.7.5 Pump/Compressor/Turbine Efficiency and Vibration

The most common prime movers in a process plant are rotating machines such as pumps, gas or steam turbines, and centrifugal or reciprocating compressors. Devices that consume or generate many tens of thousands of horsepower are not uncommon in refineries and petrochemical plants. Over time, all such sophisticated devices are subject to a gradual decline in their efficiency. They also can be subject to overheating and vibration problems caused by shaft misalignment or other mechanical issues.

Centrifugal compressors can also experience severe vibration and damage when inlet gas flows go below a safe limit. This phenomenon is called "surge", and the minimum suction flow required to stay out of surge increases with the rotational speed. Compressor surge occurs at a very high frequency (millisecond pulses), causes extreme vibration of the entire machine, and – if not detected and corrected extremely rapidly – can result in catastrophic damage and even utter destruction of the machine and its housing. Such disastrous failures have happened in numerous past accidents.

The best method to keep a compressor out of surge is to recirculate gas around the compressor using a sequence of fast-acting valves that can be opened very quickly, typically in 5–50 ms. The piping length in the recirculation loop needs to be minimized to ensure rapid flow response when the anti-surge valve or valves are opened. Compressor anti-surge control systems are highly specialized in nature. They require high-frequency flow monitoring and control of special-purpose, large, fast-acting recycle gas control valves (milliseconds between open and shut). Generally, this rapid response is not achievable within a typical DCS environment and requires specialized control hardware platforms for fast flow monitoring, anti-surge calculations, and controller action.

Despite the many instances of an outright compressor failure (caused by surge phenomena) that have been documented over the years, it is disheartening to note that compressor installations can still be found where the anti-surge control system is either nonexistent or poorly designed or configured. In some instances, anti-surge controls have been installed in DCS with an execution frequency that is wholly inadequate to handle the extremely rapid surge phenomenon. This deplorable practice results in a false sense of complacency and jeopardizes the safety of the entire plant.

Generally, rotating machinery is expensive and intense efforts are required to maintain safe and efficient operations. Routine mechanical maintenance activities, as prescribed by the manufacturers, must be followed meticulously. It is also beneficial to install online calculations for monitoring their efficiency, especially when fouling or mechanical damage is suspected.

15.7.6 Compressor Efficiency

Suction and discharge conditions for a compressor are related by the following equation[10]:

$$T_2/T_1 = \left(P_2/P_1\right)^{[(n-1)/n]} \tag{15.9}$$

where
- T_2 = discharge temperature, K
- T_1 = suction temperature, K
- P_2 = discharge pressure, bar
- P_1 = suction pressure, bar
- n = polytropic exponent.

The polytropic exponent (n) is defined by the following equation[10]:

$$(n-1)/n = (\gamma-1)/\gamma/\eta_p \tag{15.10}$$

where
- n = polytropic exponent
- γ = average ratio of C_p/C_v for the gas mixture
- η_p = average (fractional) polytropic efficiency.

Also,
- C_p = heat capacity of the gas at constant pressure, J/(kg.K)
- C_v = heat capacity of the gas at constant volume, J/(kg.K).

For ideal gases, these are related by a simple expression[10]:

$$C_p - C_v = R \tag{15.11}$$

where
- C_p = heat capacity of the gas at constant pressure, J/(kg.K)
- C_v = heat capacity of the gas at constant volume, J/(kg.K)
- R = universal gas constant, J/(kg.K).

However, for real (nonideal) conditions, which generally correspond to gases at high pressure or low temperature, Equation (15.11) is not accurate. In such cases, the C_p/C_v (or γ) ratio can only be estimated by solving the appropriate equations of state (EOS) for the mixture, such as the Peng-Robinson (PR), Soave-Redlich-Kwong (SRK),

Lee-Kesler (LK), or other similar EOS that define the P-V-T behavior of the gas mixture reliably. The value of γ is determined uniquely once the gas composition, temperature, and pressure are fixed. In Equation (15.10), γ is an average value from the inlet to outlet conditions, and similarly, the polytropic efficiency (η_p) is also an average for the compressor.

Unfortunately, in several standard references, the value of γ is calculated using the ideal gas assumption, Equation (15.11), even for highly nonideal conditions. These can differ from the true value in some cases by 10%–30%, or even more, depending on the extent of nonideality. In a typical compressor system, the γ value does not change appreciably provided gas composition and inlet/outlet conditions do not change significantly.

Since the inlet and outlet T and P are measured, the polytropic exponent value (n) is found using Equation (15.9). The value of γ is determined using an off-line calculation using the current gas composition and the inlet and outlet temperatures and pressures. The value of the compressor polytropic efficiency (η_p) is then found using Equation (15.10).

It is advantageous to track the value of η_p over time. A steady downward trend generally is a strong indication that the blades on the compressor wheel are fouled or mechanically damaged in some way; the machine then needs to be scheduled for a maintenance outage. In this way, tracking this online parameter enables proactive maintenance on a high-speed machine, an essential aspect of process safety.

15.7.7 Turbine Efficiency

Turbine efficiency is generally defined as the ratio of the actual work done divided by the theoretical maximum (isentropic) work. Efficiency calculations for steam turbines are used to estimate the shaft work delivered by the machine. Similarly, gas turbine efficiency calculations require combustion calculations and a thermodynamic cycle analysis using P-V, T-S, and H-S diagrams. A simplified example of such an analysis is shown by Couper et al.[14] (p. 59).

Alternatively, process simulators are available that can perform such computations rigorously. These applications generally reside in workstations connected to the DCS and are triggered automatically and on-demand. Such details are beyond the scope of the present discussion, however.

15.7.8 Pump Efficiency

The power consumed by a pump is defined as follows:

$$P = WH_p/\eta \tag{15.12}$$

where

P = pump power, W
W = flow rate, kg/s
H_p = polytropic head (the work done per kg of fluid), J/kg
η = polytropic efficiency.

Calculating the head requires measuring the fluid density at flowing conditions and differential pressure across the pump:

$$H_p = \Delta P/\rho \tag{15.13}$$

where

H_p = polytropic head, J/kg
ΔP = differential pressure, Pa
ρ = fluid density, kg/m^3.

The value of H_p is then substituted in the following expression to find the pump efficiency:

$$\eta = W \cdot H_p/P \tag{15.14}$$

where

η = polytropic efficiency
W = flow rate, kg/s
H_p = polytropic head (the work done per kg of fluid), J/kg
P = pump power, W.

These head and efficiency values should be historized. They should be compared routinely against the manufacturer's curves to ensure that the pump runs within acceptable performance limits for the current flow rate. If process conditions have changed since the pump was selected, this kind of computation helps determine if the pump specifications (or pump type) are still satisfactory.

15.8 SMART SENSORS/INFERENTIAL CALCULATIONS

Critical process control and safety-related applications depend on inputs from reliable sensors. When these sensors fail, the applications are turned off, and the level of process safety or reliability is diminished, sometimes unacceptably. Wherever feasible, smart sensors should be designed to provide realistic surrogate values to enable higher-level control or safety strategies to remain online. Sometimes, such applications are also referred to as "inferential calculations".

It is well known that many measurements in a process move consistently with other values that are also measured and available in the DCS. Therefore, it would make sense to examine the feasibility of developing an empirical, statistical correlation that would enable making an estimate of the likely online value for a sensor might have failed. In many instances, such correlations are indeed feasible and of sufficient accuracy to enable control functions to continue to execute until the failed sensor in question is repaired. Developing such correlations requires collecting data at a frequency of, say, once-per-minute – over a sufficient time interval – for several time slices representing different operation modes. Over time, the predicted values must be tracked to ensure that they are reliable and do not suffer from a bias (consistent positive or negative prediction error).

Another issue for managing product quality control arises with laboratory measurements taken irregularly and at a relatively low frequency. These measurements are generally made only once a shift, and sometimes even less frequently. The operator is left to control the process based on a laboratory result for a sample taken many hours earlier. Delayed analyses can create control problems in a situation where the process itself may be subject to disturbances that occur more often than the laboratory sampling frequency.

As a result, operators tend to run the unit conservatively. These keep the unit at a safe distance from conditions where product quality violations may occur. For example, the reflux flow and reboiler duty in a distillation column may be set quite a bit higher than necessary to avoid product quality violations. This practice can waste vast amounts of energy over time while also tying up column capacity. Suppose that either the overhead or bottoms product is in danger of violating a specification. In that case, the operator often cuts the column's feed rate, perhaps by lowering the front-end feed. This practice causes persistent economic losses.

In such cases, an online estimate of the current product quality is hugely beneficial, especially if unpredictable external disturbances are frequent and large in magnitude. Many industrial examples have shown that correlations can be developed to provide reliable instantaneous online estimates for stream compositions that typically are measured only in a laboratory. These enable the unit to be run more safely and profitably, using the inferred product quality estimates between laboratory sample results.

When a laboratory value becomes available, it is compared to the inferred value when the sample was taken. A correction to the online estimate is then made, often using digital filters that guard against a sampling or laboratory error. This correction is often a fraction of the model-predicted error.

Excessive prediction errors are flagged for potential flaws either in laboratory measurements or in the online values used for making the inferential calculations. Also, in cases where the prediction error is of the same sign consistently, it becomes evident that a systematic error – or bias – exists in the underlying correlation, which then needs to be revised.

Many methodologies have been proposed for inferential calculations to estimate product quality in complex fractionators of the type encountered in petroleum refineries (crude and vacuum unit, FCC, hydroprocessing, gas plants, etc.) and petrochemical plants. These include neural networks, statistical correlation and regression, steady-state and dynamic simulation,[15] and online heat and material balances.[13] It is beyond the scope of this book to discuss these methodologies. However, it suffices to say that, once commissioned, none of these correlations should be expected to remain accurate indefinitely. Over time, some correlation parameter adjustments become inevitable as feed, catalyst properties, or other operational or equipment performance-related changes occur. Correlations based on "first-principles" engineering generally require less frequent updating than simplistic polynomial-type expressions.

Properly implemented inferential calculations enable the operator to make changes in set-points far more gradually and reliably than would otherwise be the case. Excessive and rapid set-point changes made by operators are often the cause of unit instability. They are known even to have resulted in unit shutdowns caused by a triggering of the emergency shutdown (ESD) system.

15.9 MULTIVARIABLE, OPTIMAL PREDICTIVE CONTROL (MPC)

Based on over 40 years of experience, multivariable MPC[16,17] has proven to be a technology that improves unit operations and profit dramatically while also keeping plant operations in a safe and sustainable condition. Traditionally, operators chose to run their units far away from all safety constraints, even though improving production or profit invariably requires operation closer to these limits. Since it manages all the significant independent variables (or set points in the control system), while also ensuring that CVs stay within prescribed limits, MPC technology has enabled operation closer to process, equipment, or safety limits. MPC dramatically reduces the likelihood and risk of safety violations, ESDs, and so forth. In this way, these applications are a vital contributor to the goal of achieving manufacturing excellence consistently.

All model-predictive controllers require a dynamic (i.e., time-dependent) input–output model for each pair of independent variables – MVs or DV – and dependent variables (CV). The result is a "matrix" of curves that show the time-dependent response of all the CVs to a unit MV or DV move (e.g., a move of 1°F in a temperature controller). A *dynamic matrix*[16] is just such a collection of individual input–output models from the instant a single MV or DV move was made until each CV reaches its steady state. These curves assume that no other moves in any MV or DV occur at the same time. Such models have traditionally been developed using plant tests that require each independent variable to be moved up and down many times (8–15, typically) to obtain the process response required for developing good input–output models.

Operators resist making large moves in critical plant variables owing to the risk of equipment and process safety limits being violated or products going off-specification. Plant step tests are, therefore, widely seen as highly invasive and disruptive. However, developing good models requires making the MV moves as large feasible without incurring undue risks. Step tests with small-sized moves are a primary reason why derived dynamic models often do not match plant dynamics with a high degree of fidelity. Another issue lies with the control software itself: the dynamic models assume that the process response can be approximated using a system of linear differential equations ("linear models"). These assumptions imply that:

- The magnitude of the process response is proportional to the change in each input variable (proportionality).
- Moving one independent variable is independent of the effect of movement in other variables (superposition).

Real-world chemical engineering principles are known to be highly interactive. They can be quite nonlinear[16,18] in nature (e.g., distillate purity as a function of reflux flow in high-purity columns). As a result, all linear control models predict future behavior with some errors arising from the assumptions listed above. Unmeasured disturbances are another major source of prediction errors (e.g., feed quality, rainstorms). These can affect the process in ways not accounted for within the control models.

Finally, errors in measurements caused by instrument drift or other related phenomena can result in prediction errors being magnified. This is the main reason why sensor validation applications (Section 15.1) should be provided for all critical control variables.

MPC controllers account for prediction errors using a digitally filtered feedback correction (e.g., a Kalman filter[19]) at every control cycle. By keeping track of these model-prediction errors, the control engineer can determine whether they are excessive in magnitude or have a persistently incorrect sign. In either case, some portion of the multivariable dynamic matrix would need to be revised.

15.9.1 Using Dynamic Simulation for Developing MPC Models

An alternative approach, using first-principles steady-state and dynamic simulation to develop these control models,[7] avoids disruptive and costly step testing in the plant. Also, the uncertainties in process-model identification caused by noisy plant data are eliminated, provided it is first ensured that the independent and dependent variables of the plant and these first-principles models are the same.[7]

Model-predictive control has been practiced commercially for well over 25 years. Many papers have been published on the theory and practice of MPC since the technology was first described[16] and then perfected over many years.[17] We summarize below what has been learned in the art of implementing MPC on large-scale fractionation plants, using first-principles steady-state and dynamic simulation methods to enhance the quality of the MPC models. Wherever feasible, we recommend using analytical dynamic simulation methods that avoid the need for the often expensive and challenging step testing alternative.

Traditionally, implementing MPCs in the process industries has required a fixed, linear, dynamic model that relates changes in each input to those in each output. Almost all MPC projects described in the literature have been executed using extensive step tests to develop such linearized control models, using "process-model identification" techniques.[16]

Because such deliberate step tests can be quite costly, disruptive, invasive, and lengthy in duration (often lasting many weeks or months in a large unit), a significant incentive exists to minimize step tests, if not eliminate them altogether.

However, the importance of first ensuring a reliable, stable, and predictable regulatory control scheme cannot be overemphasized. In numerous instances, an inadequate regulatory control configuration in a distillation column has jeopardized any MPC scheme's viability.

When using a step testing approach to build dynamic models for complex (especially, high-purity) distillation systems, it is still possible to obtain good MPC models. Steady-state gains determined using a calibrated steady-state simulation model can sometimes be inserted into existing dynamic models. The shape of the dynamic portion of the curve is then adjusted appropriately.

One would expect that the process-model identification software provided with the common MPC packages would allow users to fix the steady-state gain for one or more MV/CV or DV/CV pairs and identify only the curves' dynamic shapes. Unfortunately, several MPC software packages do not allow this to be done. Fixing gains that have been obtained reliably and independently, as described here, could dramatically shorten the duration of the step tests required for determining the dynamic shape of the MPC models for most projects.

15.9.2 Closing Remarks on Model-Predictive Control (MPC)

Many multivariable control solutions have been reported in the literature that use instantaneous product quality estimates derived from inferential calculations[7,20] It is essential to use periodic laboratory data to ensure the reliability of such calculations, especially when they are connected to higher-level, real-time process control applications running in a closed-loop environment. When appropriately designed and calibrated reliably, such strategies can improve both productivity and safety performance markedly.

While acknowledging that there is no "magic bullet" for improving large-scale process plant performance, model-predictive control technology had established a solid track record. Plant management should review past efforts in this area. Some MPC applications may have floundered for lack of attention or inadequate ongoing efforts to maintain them. These should be targeted for repair and reinstallation.

It has also become clear that superior ongoing maintenance of MPC applications is best achieved when control engineers, who are company employees, are trained thoroughly and held responsible for their maintenance and upkeep. Understandably, management can be reluctant to re-appoint contractors (who developed and commissioned the MPC applications initially) to continue to perform additional work indefinitely to ensure the effectiveness of the MPCs. Rather than turning these applications off, the proper resolution should be to develop the required capabilities in-house and insist that MPC applications continue to perform at their highest level consistently. Those applications that show a percentage online factor below some minimum acceptable value should automatically be reported to plant management. Similarly, if certain MVs or CVs are shown as consistently being turned off by the operator, an advisory message to the control engineers and plant management should be generated.

In summary, the application of multivariable, model-predictive controls is highly recommended to maximize unit-level performance and profit while also respecting the most important production and safety constraints in major process units.

15.10 CLOSED-LOOP, REAL-TIME, OPTIMIZATION (CLRTO)

Historically, chemical plants have presented numerous challenging economic optimization problems. Today, the steady-state and dynamic behavior of chemical processes can be described adequately, in most cases, by first-principles nonlinear models. It has been proven that seemingly insignificant changes in process operations can improve safety, profitability, product quality, environmental impacts, and online efficiency. Formal optimization using nonlinear programming (NLP) is an essential tool for chemical process optimization. In our experience, such tools have proven most successful in maximizing the profitability of process plants in a safe, consistent, predictable, and reliable manner.[20] Economic optimization based on nonlinear models is also found increasingly in corporate planning and scheduling applications.

As mentioned by Biegler,[21] addressing the modeling and solution of large-scale process optimization problems requires:

- Selecting the NLP methods that are best suited for specific applications
- Understanding how to formulate large-scale chemical engineering optimization problems and which aspects should be addressed prominently
- Exploiting the specific structure of large-scale optimization software[22] to extend current modeling capabilities, for example, by using "open-equation" modeling techniques.

Open-equation modeling (also known as *equation-based modeling*) requires formulating the equations in a way that moves all the variables and constants to one side of an equation, with the right side being equal to zero.

15.10.1 Open-Equation Modeling for a Counter-Flow Heat Exchanger

We illustrate the principles of equation-based modeling for a counter-flow heat exchanger.

The heat duty is expressed as a function of selected process variables:

$$Q = UAF\Delta T_{lm} \tag{15.15}$$

where

Q = Exchanger duty, W
U = Overall heat transfer coefficient, W/(m^2.K)
F = Factor to correct for deviations from true counter-flow
ΔT_{lm} = Logarithmic mean temperature difference (LMTD), K.

The ΔT_{lm} is defined as follows:

$$\Delta T_{lm} = (\Delta T_1 - \Delta T_2)/\ln(\Delta T_1/\Delta T_2) \tag{15.16}$$

$$\Delta T_1 = T_{hi} - T_{co} \tag{15.17}$$

$$\Delta T_2 = T_{ho} - T_{ci} \tag{15.18}$$

where

T_{hi} = temperature of the hot fluid inlet, K
T_{ho} = temperature of the hot fluid outlet, K
T_{ci} = temperature of the cold fluid inlet, K
T_{co} = temperature of the cold fluid outlet, K
ΔT_1 = stream temperature difference on the hot side of the exchanger, K
ΔT_2 = stream temperature difference on the cold side of the exchanger, K.

The F factor is 1 for true counter-flow exchangers but is always lower than 1. Its value depends on the geometrical configuration for exchangers that deviate from the counter-flow ideal (e.g., 1 shell pass and 2 tube passes). F can be a complex function of the terminal temperatures and the ratio $W_h\ C_{ph}/(W_c\ C_{pc})$. In this discussion, we assume $F = 1$ for simplifying the problem since our purpose here is to focus on the principles of equation-based modeling rather than heat exchanger design.

We also have the heat balance for each stream:

$$Q = W_h C_{ph} (T_{hi} - T_{ho}) \qquad (15.19), \text{ and}$$

$$Q = W_c C_{pc} (T_{co} - T_{ci}) \qquad (15.20)$$

In an open-equation modeling environment, Equations (15.15) to (15.20) would be re-written as follows:

$$Q - UA\,\Delta T_{lm} = 0 \qquad (15.21)$$

$$\Delta T_{lm} - (\Delta T_1 - \Delta T_2)/\ln(\Delta T_1/\Delta T_2) = 0 \qquad (15.22)$$

$$\Delta T_1 - (T_{hi} - T_{co}) = 0 \qquad (15.23)$$

$$\Delta T_2 - (T_{ho} - T_{ci}) = 0 \qquad (15.24)$$

$$Q - W_h C_{ph} (T_{hi} - T_{ho}) = 0 \qquad (15.25)$$

$$Q - W_c C_{pc} (T_{co} - T_{ci}) = 0 \qquad (15.26)$$

In effect, all terms have been moved to the left-hand side of each equation. These equations are written in *residual form.* This means that the residual error on each equation's right-hand side must be driven close to zero at the final solution.

These six equations could be solved by any software package that solves nonlinear equations simultaneously. This explains why this method of defining such reformulated equations is also called *equation-based* modeling.

However, in large-scale chemical processes, it can readily be seen that many hundreds of thousands of simultaneous nonlinear equations would be required to model the whole system. One interesting fact would then emerge: each equation carries only a few variables. Solving such problems efficiently requires large-scale optimizers that exploit the *sparse nature of the equation set.*[20]

The heat exchanger problem above, however, is not sparse. Equations (15.21) to (15.26) have 4, 3, 3, 3, 5, and 5 variables in them, respectively.

This problem has a total of 14 variables in 6 equations:

$$Q,\ U,\ A,\ \Delta T_{lm},\ \Delta T_1,\ \Delta T_2,\ T_{hi},\ T_{co},\ T_{ho},\ T_{ci},\ W_h,\ C_{ph},\ W_c,\ \text{and}\ C_{pc}$$

Therefore, if any eight of them were fixed, we would be left with six equations in six unknown variables.

If W_h, C_{ph}, W_c, C_{pc}, T_{hi}, T_{ci}, U, and A were specified (typical exchanger rating problem), the solution of the six residual equation above would provide the solution for the remaining six unknowns Q, ΔT_{lm}, ΔT_1, ΔT_2, T_{co}, and T_{ho}.

A different problem formulation could fix Q, W_h, C_{ph}, C_{pc}, T_{hi}, T_{ci}, T_{co}, and U. Here, we would be asking the question: how much heat exchanger surface is required, and how much cold fluid could be heated from a given inlet temperature to a specified outlet temperature, if the hot fluid flow rate and inlet temperature are fixed (design problem)?

In this case, too, the same six equations would be solved for a different set of six unknowns, with a different set of eight variables being specified. Thus, we see that the *exchanger model* remains the same; what changes from case to case are the definitions of fixed and unknown variables. When the number of equations and the number of variables are the same, the system is called *square*.

15.10.2 Building Successful Plant-Wide CLRTO Applications

In a running plant, there invariably are more variables than the total number of model equations and, therefore, the system is "non-square". In this situation, the value of some of the extra variables must somehow be determined differently. The economic optimization problem is an example of a non-square system. Here, we formulate:

- An objective function to be maximized (e.g., total profit)
- Several equality and inequality constraint functions define the upper or lower limits for several variables, such as safety-related or environmental limits.

We would then invoke a mathematical optimization package to solve this problem.

There are other significant engineering benefits to using equation-based methods for modeling process plants. In many cases, even when the plant is running close to steady state, the overall material and heat balances do not "close". This *data reconciliation problem* can also be solved, with a change of variables, using equation-based methods. Further, the performance-related parameters for various equipment pieces, such as heat transfer coefficients, fouling factors, compressor efficiencies, etc. (as discussed above), can be updated using *parameter estimation* in an open-equation system based on reconciled data.

Therefore, using the same model, an open-equation formulation is a highly effective way to solve four classes of industrially significant problems by changing only what is fixed and what is calculated:

1. Simulation
2. Data reconciliation
3. Parameter estimation
4. Economic optimization.

Equation-based optimization has been carried out successfully, over many decades, for major petroleum refinery units, such as crude and vacuum, fluid catalytic cracking, catalytic reforming, alkylation, and delayed coking. Plant-wide optimizations have also been reported for several major chemicals (e.g., ammonia, methanol, styrene, bisphenol-A) and petrochemical plants (olefins, polyvinyl chloride).

This economic benefits of this approach are enormous as such projects generally have a payback of under 1 year, often better. The economically optimized solution is implemented automatically online by downloading the upper and lower limits of the various multivariable controllers located one level below in the control hierarchy. No operator intervention or approval is required, making the process truly *closed-loop* and far more reliable and error-free.

A major side-benefit of CLRTO is that it helps resolve technical disagreements or disputes among plant personnel. There are often competing ideas about how best to run a plant. Resolving these can be extremely difficult, particularly when opinions are held too firmly or when senior management is misinformed about technical issues. We have seen several instances where the plant-level disagreements were resolved satisfactorily after a CLRTO was implemented and commissioned successfully.

All economic optimizations using CLRTO obey process safety constraints, a matter of paramount importance to corporate management.[21] Unfortunately, while CLRTO technology has been developed over many decades in refining and petrochemicals, it has not found widespread adoption in many other quite similar continuous (as opposed to batch) process industries. Generally, numerous prior CLRTO applications have been successful from an engineering, economic, and process safety perspective. However, it must be appreciated that maintaining CLRTO applications successfully requires a high degree of chemical engineering expertise, both for initial implementation and for ongoing maintenance.

15.10.3 Challenges in Rigorous Chemical Reactor Modeling

The real-time optimization approaches described above all require creating a calibrated, plant-wide steady-state simulation model as a prerequisite. Using modern equation-based optimization software, this can readily be done for distillation columns and most other process equipment. The exception is for catalytic chemical reactors with complex kinetics where the reactor effluent composition is far from thermodynamic equilibrium.[23] Equilibrium reactor models generally can be formulated far more conveniently than nonequilibrium kinetic models that describe catalytic phenomena. The latter typically cannot be used to estimate the relevant gains. Fortunately, most major catalytic reactors have relatively quick settling times (i.e., the time for control to reach steady state). Identifying the relevant MPC models using simple step tests is not unduly arduous.

The complexity of first-principles chemical reactor modeling is well documented in the enormous literature on this subject. At present, this is one of the primary technical challenges in plant-wide CLRTO. Today, some vendors license quite elaborate first-principles, open-equation reactor models (developed by major industrial corporations over many years) for use in such CLRTO projects.[20] To ensure acceptable accuracy, all such models require careful calibration against steady-state plant data representing normal plant operations.[15]

Implementing MPC is a prerequisite for the CLRTO layer. The level of effort required for successful MPC and CLRTO at a major petroleum refinery unit, such as a fluid catalytic cracker, can be 12-18 months, requiring a sustained effort by two teams of 4–5 chemical engineers that are highly skilled in multivariable process control and process modeling/optimization.

The engineering and mathematical skills required to develop CLRTO applications are not much higher than that required for traditional steady-state simulation of refining and petrochemical processes. However, the essential concepts of equation-based modeling for chemical engineering applications are not taught in typical undergraduate programs. Such indoctrination would enable recent graduates to better understand and participate in higher-level corporate strategies to maximize profit safely and reliably.

15.11 PLANNING AND SCHEDULING OPTIMIZATION

Planning and scheduling applications in major enterprises command the highest level of attention from corporate management.[8,21] The smooth and most profitable operation of their chemical process plants is a very high priority, and a myriad of related issues must be addressed:

- Selecting and balancing feedstock and utility purchases
- Seasonal reconciliation of production targets against product demands
- Adjustment of operations during planned maintenance turnarounds or unexpected plant outages
- Product storage and shipment
- Evaluating the benefits and costs of installing additional units or modifying existing ones.

Such problems are solved routinely using planning and scheduling software. Typically, these are LP applications with many thousands of variables and constraint equations. However, the underlying models are not rigorous, first-principles models of the type used in optimization at the CLRTO layer.

In practice, such models are simplified for the variables of interest, including the functional relationships between them. Linearized approximations are used for all such functions. In some instances, piecewise linear segments are used to describe nonlinear behavior.

Note that the objective function and all LP constraint functions could include any variables from the complete set of problem variables. All functions could be linear (linear programming, LP), or one or more could be nonlinear (nonlinear programming, NLP). If even one of the many thousands of functions is nonlinear, the problem becomes an NLP. In general, NLP problems are much more challenging to solve than linear problems, especially for large-scale systems. The use of piecewise linearization can enable the continued use of LP software to solve NLP problems.

However, an additional major complication in this effort arises from the fact that many decision variables are not continuous. Examples include tier-pricing for products or utilities. Another possibility is that some of the variables can only be allowed to have discrete, integer values. An example would be the number of ships required to transport products by sea. Another issue arises with the proper handling of feed and product tankage as part of the optimization effort.

When such issues arise, the problem is handled using mixed-integer, linear programming (MILP) solver technology. In other instances, a few of the functions may still be required to remain nonlinear. These more complex MINLP problems are solved by a different class of optimization algorithms.[8,21] As of this writing, there are severe limitations in the size of MINLP problems that can be solved with commercially available software. Accordingly, the linearized MILP software options remain the preferred ones for planning and scheduling applications.

Problem formulation for planning and scheduling optimization remains a formidable technical challenge and requires the sustained attention of the highest levels of corporate and production management. There is often a secondary planning and scheduling optimization layer at the plant level with fewer variables that devotes a lot more attention to unit performance parameters.[21] These two levels of optimization must be synchronized effectively to ensure successful integration with the CLRTO layer.

Another challenge requiring constant attention is that the corporate-level planning and scheduling software must be updated to reflect current plant operating conditions. Such updating must be done routinely to ensure that the optimization results downloaded to the CLRTO layer are realistic and corporate optimizers do not violate current process safety or environmental constraints at the plant level. The CLRTO develops a solution that is consistent only with the current condition of the plant. Therefore, it is feasible and highly beneficial to enable the CLRTO applications to update the appropriate *LP vectors* periodically to the higher-level planning and scheduling applications.[20] This can be done automatically or on-demand. One example is a set of utilities to bridge CLRTO sensitivity vector data with standard planning LP applications.

Excellent commercial packages are available for planning and scheduling optimization in chemical plants and refineries. The effort required for a complete petroleum refinery can extend over approximately two years and require a team of 3–4 full-time process planning professionals for the duration. Early projects benefit significantly from the participation of the software vendors to train engineering staff.

15.12 INTELLIGENT ALARM MANAGEMENT

A common problem in many plants is the incidence of *alarm flooding.* Alarm floods are usually caused by a critical sensor or control valve's failure at the lowest control hierarchy level. Many higher elements or control strategies in the cascade also have alarms triggered when their functionality is compromised. In this situation, operators must contend with multiple high-level alarms that require acknowledgment and diagnosis. Especially during emergencies, this can be a huge distraction, and the effectiveness of an operator handling an event such as a cooling water failure could quickly deteriorate unacceptably.

Another example might be partial or total electrical power failure in the plant when the operator is scrambling to implement standard shutdown procedures but is afraid to ignore all high-level alarms. This issue raises the question of what could be done automatically to acknowledge and silence many high-level, related alarms.

We have seen one instance at a simple hydrocarbon fractionation plant where the operator had to acknowledge more alarms every day than are expected to be seen in an entire petroleum refinery. This situation was so egregious that the operator would hold one hand an inch or so above the alarm acknowledgment button almost permanently. The problem had persisted for many years without any intervention by plant management. Upon examination of the alarm journal, it was found that a novice had written some code (that contained abysmal logic) for performing analyzer validation. Whenever a situation arose that the programmer didn't understand or know how to ameliorate, he triggered an alarm. Correcting this gross error took just a few hours, and the operator was relieved from the burdens imposed by this tedious and highly distracting situation.

While it is true that this is an extreme example, it remains evident that many plants have not yet recognized the seriousness of the problem and have not purchased the software systems to enable the implementation of this capability. Failure to install any intelligent alarm management capabilities remains a significant cause of confusion and can easily lead to a human operator making faulty diagnoses in emergencies. The implications for severe lapses in process safety are apparent.

Over the last 20 years or so, significant progress has been made in intelligent alarm management.[24] A thorough review of alarm interdependency is first conducted throughout the unit. A logic tree that contains all the information about alarm cascades is constructed. Alarm management software is then configured to incorporate all this information. When lower-level faults occur, all higher-level alarms are identified immediately and acknowledged. A brief message is sent to the operator describing the original fault and the likely cause in every case.

A modern alarm management program would encompass the following steps[24]:

1. Alarm Philosophy Development
2. Data Collection and Benchmarking
3. Bad Actor Alarm Resolution
4. Alarm Documentation and Rationalization
5. Alarm Audit and Enforcement
6. Real-Time Alarm Management
7. Alarm System Control and Maintenance.

Investment in alarm management software is highly recommended, considering the benefits obtained by minimizing the safety risks posed by high-frequency alarm floods. Also, since the effort required to build the logic trees and perform software configuration is modest, we recommended that all process plants implement intelligent alarm management capabilities urgently.

A team, led by a senior process control specialist, should be assembled to carry out the steps outlined above – with the participation of 3–4 representatives from operations, instrumentation, process control, and, especially, the alarm management software vendor. This effort typically takes 3–4 months for a major process unit. For example, several unit-level teams would need to be appointed for an entire petroleum refinery to enable the work to be done simultaneously over approximately 1 year.

15.13 EMERGENCY SHUTDOWN SYSTEMS (ESD)

Throughout the process industries, ensuring safe and orderly shutdowns to protect life and property has always remained the highest priority. The regulatory consequences of chemical releases to the environment are grave, and loss of public confidence and goodwill invariably impacts any company severely. Worst of all, we continue to observe highly publicized instances of improper explanations by corporate spokespeople and failure to accept responsibility for such untoward accidents when they occur.

Unfortunately, it is not uncommon to see disastrous fires, explosions, chemical releases, and other process plant accidents. These unfortunate and lamentable events continue to occur at an utterly unacceptable frequency, despite the advances made over many years in plant monitoring, control, optimization, and emergency management systems. This situation has caused a great deal of controversy and public alarm and has resulted in ever-increasing regulatory scrutiny and punitive actions.

While the corporate and local management of every process plant professes their dedication to process safety and protection of life and property, a study of the systems that have been built into their manufacturing operations often reveals severe inadequacies that require immediate correction. Many of these crucial issues have been addressed in the preceding sections of this chapter. We now turn to the subject of ESD systems that are the last resort to protect against disastrous accidents leading to loss of life and property damage.

ESD systems have been implemented from the early days of process plant operations, even when all process control hardware was pneumatically driven. In modern digital DCS, the ability to monitor and control unit operations has increased exponentially. *However, it has increasingly become clear that ESD systems must use redundant measurement hardware (instrumentation) and final control elements such as control valves that are independent of the standard DCS-resident process control system.* The main reason is that, during an emergency, it is most unwise to depend on a DCS, already burdened with an abnormal load, also to shut down the plant flawlessly and rapidly. There have even been instances when all the instrumentation and cabling (for providing information to the DCS and move control valves) were destroyed during a fire. Consequently, the DCS-based ESD system failed for lack of redundancy.

A modern ESD system requires the use of extremely reliable, high-frequency programmed logic control (PLC) hardware and redundant, dedicated fast-acting control valves specially designed for very rapid action when a shutdown situation is detected. The entire ESD system (PLC, instrumentation, control valves) must be verified to be completely independent and separate from the regular DCS control network. It is entirely unacceptable to use any DCS information for triggering an ESD.

Building an ESD requires following several mandatory steps:

- Nominate a team of process, operations, maintenance, and safety experts to provide the knowledge required for the design and implementation of the entire ESD system.
- Assign a single project manager to be responsible for ensuring the integrity of the project.

- Based on prior maintenance records, enumerate all possible sources of failures for instrumentation, control hardware, or equipment. This is a major and laborious undertaking, requiring the utmost attention to technical detail. For new facilities, invite external experts of long-standing (who have compiled a verifiable, successful record in similar industries) to provide advice for the project duration. Such experts are often available through the companies that offer DCS or PLC systems.
- Develop logic trees reflecting the causal chain of events that would follow any failure at the lowest level. In many instances, several events must coincide, in series or in parallel, to trigger an ESD action. The use of combinations of IF-THEN-ELSE-AND-OR-NOT logic is required to identify many unsafe scenarios and the corrective ESD actions that would be required. This process requires close collaboration between the ESD team members and can require several months to complete in a petroleum refinery or chemical complex.
- Configure the ESD control system to implement all ESD scenarios that were developed. Some ESDs affect only a portion of the plant (e.g., a heater shutdown), while others ensure a plant-wide shutdown.
- Develop a plan for off-line testing of the logic and flawless execution of all ESD scenarios, including simultaneous triggering for several combinations of scenarios (factory acceptance test, or FAT). Ensure that all coterminous events are appropriately accounted for and tested. This test is generally performed at the vendor's facilities.
- Develop a maintenance plan to install the required redundant hardware and instrumentation required for the plant's ESD system at the next available opportunity.
- Install all of the required ESD systems in the plant. Many piping modifications may need to be made, during unit turnarounds, if provisions for such redundant instrumentation or valve connections were not made at the time when the plant was first built.
- Relocate the ESD system to the plant site and make the cabling and other connections to the PLC hardware in the field. For maximum reliability, locate the ESD cabling in cable trays that are also dedicated to ESD purposes (wherever feasible).
- Plan and schedule a series of tests, including all FAT scenarios, for the ESD system as part of the maintenance shutdown (site acceptance test, or SAT). All ESD triggers must also provide backward notification so that plant management would also be alerted.
- Ensure that the proper personnel notification alarms (sonic and visual) are installed and tested individually for all ESD events throughout the plant.
- Perform all ESD tests, including several repetitions. Ensure that all coterminous events are correctly accounted for, tested, and field-verified. Each test result must be certified by the responsible professional staff and the project manager.
- Provide intensive training regarding the ESD system to all board operators, control supervisors, unit operations specialists, and control engineers. The

location, triggering requirements, and protection against accidental triggering for all kill switches must receive special attention.

- Provide up-to-date documentation for all aspects of the ESD system. Develop a maintenance record for all activities related to the ESD system.
- Develop a plan for periodic reviews of the entire ESD system to ensure its ongoing functionality and reliability in the face of process or equipment changes.

Implementing, testing, and commissioning an ESD system for a major process unit require 3–4 qualified operations experts, board operators, process control engineers, and representatives from the hardware and software vendor for the chosen ESD system. For example, the elapsed time would be 6–12 months for a fluid cat cracker in a petroleum refinery.

15.14 LOCATION OF PROCESS CONTROL ROOMS

It is imperative to consider the issue of the location of plant control rooms in hazardous areas. Many process plants that are still operational were designed and built over 50 years ago when process instrumentation and control systems were relatively rudimentary. In those days, it was felt that the plant operators needed to be close to the operating equipment physically (1) to observe plant areas physically, or (2) in case some manual intervention was required. Today, we have readily available, relatively inexpensive DCSs and safety-related process equipment. These include incredibly versatile and robust instrumentation systems that markedly improve our ability to monitor process areas remotely using cameras, sound monitors, flame and heat detectors, vibration monitors, sensors for minute concentrations of numerous chemicals and complex flammable mixtures, deflection or deformation of equipment, and other similar devices.

Additionally, there are wide varieties of remotely actuated valves that can be substituted for manually operated block valves. These ensure rapid and safe opening and closing, an important safety consideration, while also minimizing plant personnel exposure to potential hazards. An example would be valves used for compressor surge control. Another example would be tight shutoff valves (for isolation) that could prevent the proliferation of chemical or hydrocarbon releases in the event of outbreaks of fire, piping or valve leaks, vessel overpressure, or outright vessel failures.

All such systems and related hardware and software enable plant operators and shift engineers to operate the plant very well remotely. Hence, the control rooms of today's highly automated plants must be located at safe distances from hazardous areas and hazard sources. This strategy maximizes process and personnel safety by enabling operators safely to shut down the plant during major upsets or accidents. It also enables operators to evacuate the plant area quickly without panic or confusion.

Safety systems can add significantly to the overall cost of plant construction or upgrading. However, these costs are minuscule compared to the cost (and damage to corporate reputation) caused by industrial accidents, environmental releases, personnel injuries, or worse.

REFERENCES

1. *Book Review*, Chemical Engineering Progress (CEP), p. 56, (AIChE, September 2019).
2. Berger, S.: *Process Safety Leadership from the Boardroom to the Frontline* (Center for Chemical Process Safety (CCPS), AIChE-Wiley, New York, 2019).
3. Crowl, D. A. and Louvar, J. F.: *Chemical Process Safety: Fundamentals with Applications* (Prentice Hall, New York, 1990).
4. Mannan, S.: *Lees' Loss Prevention in the Process Industries* (4th Ed., Butterworth-Heinemann, Oxford, 2012).
5. Emerson Electric: *Fisher Control Valve Handbook* (2017).
6. Nise, N. S.: *Control Systems Engineering* (6th Ed., Wiley, Hoboken, NJ, 2011).
7. Mathur, U. and Conroy, R. J.: *Successful Multivariable Control without Plant Tests – First-Principles* – Dynamic Simulation Models Can Be Used Instead, pp. 55–65, *Hydrocarbon Processing*, (Gulf Publishing Co., Houston, TX, June, 2003).
8. Edgar, T. F., Himmelblau, D. M., and Lasdon, L. S.: *Optimization of Chemical Processes* (2nd Ed., McGraw-Hill, New York, 2001).
9. Smith, J. M., Van Ness, H. C., and Abbott, M. M.: *Introduction to Chemical Engineering Thermodynamics* (7th Ed., McGraw-Hill, New York, 2005).
10. McCabe, W. L., Smith, J. C., and Harriott, P.: *Unit Operations of Chemical Engineering*, (5th Ed., p. 574, McGraw-Hill, New York, 1993).
11. Billet, R.: *Distillation Engineering* (Chemical Publishing Co., New York, 1979).
12. Watkins, R. N.: *Petroleum Refinery Distillation* (2nd Ed., Ch. 2–7, Gulf Publishing Co., Houston, TX, 1979).
13. Friedman, Y. Z.: *Control of Crude Fractionator Product Qualities During Feedstock Changes by Use of a Simplified Heat Balance* (American Control Conference, Boston, MA, 1985).
14. Couper, J. R., Penney, W. R., Fair, J. R., and Walas, S. M.: *Chemical Process Equipment – Selection and Design* (3rd Ed., Elsevier, New York, 2012).
15. Mathur, U., Rounding, R. D., Webb, D. R., and Conroy, R. J.: Use Model-Predictive Control to Improve Distillation Operations, *Chemical Engineering Progress*, 104, pp. 35–41 (2008).
16. Cutler, C. R. and Ramaker, B. L.: *Dynamic Matrix Control – A Computer Control Algorithm* (Proceedings of the Joint Automatic Control Conference, San Francisco, CA, June, 1980).
17. Lahiri, S. K.: *Multivariable Predictive Control: Applications in Industry* (Wiley, Hoboken, NJ, 2017)
18. Henson, M. A. and Seborg, D. E.: *Nonlinear Process Control* (Prentice-Hall, Englewood Cliffs, NJ, 1997).
19. Grewal, M. S. and Andrews, A. P.: *Kalman Filtering: Theory and Practice with MATLAB* (4th Ed., Wiley-IEEE Press, New York, 2014).
20. Schneider Electric: *Rigorous Online Modelling and Equation-Based Optimization (ROMeo)*.
21. Biegler, L. T.: *Nonlinear Programming – Concepts, Algorithms, and Applications to Chemical Processes* (Society for Industrial and Applied Mathematics and the Mathematical Optimization Society, Philadelphia, PA, 2010).
22. *Harwell Scientific Subroutine Library* (Atomic Energy Research Establishment, Oxford, UK).
23. Ancheyta, J.: *Modeling and Simulation of Catalytic Reactors for Petroleum Refining* (Wiley, Hoboken, NJ, 2011).
24. Hollifield, B. and Habibi, E.: *Alarm Management Handbook* (2nd Ed., Plant Automation Services, Houston, TX, 2010).

Index

For Product Safety Concerns and Information please contact our EU
representative GPSR@taylorandfrancis.com
Taylor & Francis Verlag GmbH, Kaufingerstraße 24, 80331 München, Germany

www.ingramcontent.com/pod-product-compliance
Lightning Source LLC
Chambersburg PA
CBHW060423220326
41598CB00021BA/2270
* 9 7 8 0 3 6 7 6 2 0 8 9 9 *